Kurt Gödel

COLLECTED WORKS

Volume V

(JvN) Gödel

Notre-Dame, 20./III. 1939.

Lieber Herr v. Neumann!

Vielen Dank für Ihren Brief vom 28./II. Das Resultat von Kondo ist von grossem Interesse für mich u. wird jedenfalls den Beweis der Widerspruchsfreiheit von 3. und 4. des beiliegenden Separatums erheblich zu vereinfachen gestatten.

Was meine Princetoner Vorlesung betrifft, so bin ich eben dabei, das Manuskript vor der Vervielfältigung noch einer gründlichen Revision zu unterziehen. Damit hoffe ich aber im Laufe dieser Woche fertig zu sein u. werde es Ihnen dann sofort zuschicken.

Mir geht es hier soweit ganz gut, bloss das Klima finde ich unangenehmer als in Princeton.

Herzliche Grüsse u. beste Empfehlungen an Ihre Frau Gemahlin.

Ihr Kurt Gödel

Kurt Gödel
COLLECTED WORKS

Volume V
Correspondence H–Z

EDITED BY
Solomon Feferman
John W. Dawson, Jr.
(Editors-in-chief)
Warren Goldfarb
Charles Parsons
Wilfried Sieg

Prepared under the auspices of the Association for Symbolic Logic

CLARENDON PRESS • OXFORD

OXFORD
UNIVERSITY PRESS

Great Clarendon Street, Oxford OX2 6DP
Oxford University Press is a department of the University of Oxford.
It furthers the University's objective of excellence in research, scholarship,
and education by publishing worldwide.

Oxford is a registered trade mark of Oxford University Press
in the UK and in certain other countries

First published 2003
First published in paperback 2014

Published in the United States of America by Oxford University Press
198 Madison Avenue, New York, NY 10016, United States of America

ISBN 978-0-19-850075-9(Hbk)
ISBN 978-0-19-968962-0(Pbk)

Preface

This is the fifth and final volume of a comprehensive edition of the works of Kurt Gödel. Volumes I and II comprised all of his publications, ranging from 1929 to 1936 and from 1937 to 1974, respectively. Volume III consisted of a selection of unpublished papers and texts for individual lectures found in Gödel's *Nachlaß*, together with a survey of the *Nachlaß*. The present volume and its predecessor, being published simultaneously with this one, are primarily devoted to a selection of Gödel's scientific correspondence and the calendars thereto. In all cases our criterion for inclusion was that letters should either possess intrinsic scientific, philosophical or historical interest or should illuminate Gödel's thoughts or his personal relationships with others. This volume also contains a full inventory of his *Nachlaß*.

There were several sources for the correspondence from which the selection in this and the preceding volume was made. The primary one was, of course, Gödel's *Nachlaß*; we have also solicited and obtained from other archives and from individuals (or their estates) copies of correspondence that filled gaps therein. The section on Permissions, below, contains a full list of all those sources, to whom we are, of course, greatly indebted. The total number of items of personal and scientific correspondence in the Gödel *Nachlaß* alone is around 3500, distributed over 219 folders.

In the main body of these two volumes we have selected correspondence with 50 individuals from the indicated sources. The most prominent correspondents among these are Paul Bernays, William Boone, Rudolf Carnap, Paul Cohen, Burton Dreben, Jacques Herbrand, Arend Heyting, Karl Menger, Ernest Nagel, Emil Post, Abraham Robinson, Alfred Tarski, Stanisław Ulam, John von Neumann, Hao Wang, and Ernst Zermelo. In addition, the reader will find in Appendix A to the present volume several letters written on behalf of Gödel to others by Felix Kaufmann, Dana Scott and Hao Wang. In all cases our criterion for inclusion was that letters should either possess intrinsic scientific, philosophical or historical interst or should illuminate Gödel's thoughts or his personal relationships with others.

There are two major correspondents of Gödel who declined to allow us to publish their side of the exchanges, namely Paul Cohen and Georg Kreisel. In the latter case, except for one item from Gödel which is not included here, the correspondence was entirely a one-way street, but it is revealing of many topics of discussion of mutual interest, and thus we regret that it could not be represented. In the case of Cohen, as the reader will see, the nature of the correspondence could be fully reconstructed from the items found in Gödel's *Nachlaß*.

As in the first three volumes of this edition, all the original material was written in German (sometimes in the Gabelsberger shorthand) or English, sometimes in both; those items originally in German are accompanied by facing translations. Credit for the work on the translations is contained in the Information for the reader below. Also as in the first three volumes, a significant component is played by introductory notes and, to a greater extent than previously, these notes have been written by the editors. We are additionally indebted in this respect to Michael Beeson, Jens Erik Fenstad, Akihiro Kanamori, Øystein Linnebo, Moshe Machover and David Malament. The purpose of the notes themselves is to provide historical context to the correspondence, explain the contents to a greater or lesser extent, and, where relevant, discuss later developments or provide a critical analysis. Because of these requirements, several of the introductory notes turned out to be quite extensive; in those cases the reader is advised to consult them in tandem with the correspondence itself.

Once more our endeavor has been to make the full body of Gödel's work and thought as accessible and useful to as wide an audience as possible, without compromising the requirements of historical and scientific accuracy. As with the preceding volumes, this one is expected to be of interest and value to professionals and students in the areas of logic, mathematics, computer science, philosophy and even physics, as well as to many nonspecialized readers with a broad scientific background. Naturally, even with the assistance of the introductory notes, not all of the material to be found here can be made equally accessible to such a variety of readers; nonetheless, the general reader should be able to gain some appreciation for what is at issue in the various exchanges.

Work on this volume and its predecessor was supported in its entirety by a grant from the Sloan Foundation, whose generosity and flexibility were indispensable to their successful completion. We are also grateful to the Department of Philosophy at Harvard University for its generous assistance with some last-minute expenses. In addition, by helpful arrangement with William Joyce and Donald C. Skemer of the Princeton University library system, and with Marcia Tucker of the library of the Institute for Advanced Study, the Foundation completely underwrote the preservation microfilming of Gödel's *Nachlaß*, in an effort to prevent further deterioration. That lengthy and delicate task was actually carried out by Preservation Resources Company of Bethlehem, Pennsylvania.[a] As with the *Nachlaß* itself, one copy of the microfilms is housed at the Rare Books and Manuscripts Division of Firestone Library at Princeton University; the

[a]In conjunction with the preparation for the microfilming, the papers were reorganized and the original finding aid was revised. An HTML version of that revision may be found online at http://libweb.princeton.edu/libraries/firestone/rbsc/aids/godel/.

inventory of Gödel's *Nachlaß* to be found in this volume includes the finding aid for that preservation microfilm. Abridged copies, excluding correspondence, are distributed by IDC Publishers, Inc., 350 Fifth Avenue, Suite 1801, New York, NY 10118 (web address: http://www.idc.nl).

The editorial board for these volumes consists of the undersigned as editors-in-chief, together with Warren Goldfarb, Charles Parsons and Wilfried Sieg. We are especially thankful to Cheryl Dawson, who performed extraordinarily sustained, careful and thorough service as managing editor over a time period that was far longer than any of us anticipated. Under her supervision, the volumes themselves were set in camera-ready copy using the TEX system, and later AMS-TEX, in a form that had been developed for the previous volumes by Yasuko Kitajima; after significant initial TEX work by Kitajima, most of the remaining TEX work in the present volumes was carried out (promptly and with diligence—often under pressure) by Bruce S. Babcock, with the balance done by Cheryl Dawson. We were also ably assisted by Montgomery Link, who joined the effort late in the process to complete the work of researching reference citations; his work under pressure is much appreciated.

From the outset with volume I of these *Works*, the project to produce the volumes has been sponsored by the Association for Symbolic Logic, and the grants under which they were carried out were ably administered for the Association at the hands of its Secretary-Treasurers, C. Ward Henson and, since the beginning of the new millennium, Charles Steinhorn. Clerical support was provided by the Department of Mathematics of Penn State York, with special assistance, especially for the extensive photocopying required, by Carole V. Wagner. Our editor Elizabeth Johnston at Oxford University Press (in Oxford, England) has been both encouraging and very patient.

We mourn the loss of our dear friend, Stefan Bauer-Mengelberg, who died of a heart attack on 19 October 1996, at the age of 69. Noted for his unusual multifaceted career as mathematician, symphony conductor and lawyer, Stefan had both given us legal advice and provided us with considerable assistance on the translations in the previous volumes. Following the publication of volume III of these *Works* in 1995, he was much looking forward to further work with us to help bring the entire project to completion.

Solomon Feferman and John W. Dawson, Jr.

Information for the reader

Copy texts. A basic tenet of documentary editing is that published texts of letters should represent what recipients actually saw. In particular, recipients' copies of letters should be used as copy texts whenever they are available, and readers should be made aware of authorial errors and emendations. In the correspondence reproduced herein both of those precepts have been followed. In cases where the recipient's copy of a letter was unobtainable but the author's retained copy has been preserved, we have used the latter as our copy text. Details concerning the copy texts and their sources are provided in the calendar for each correspondent. Errors and emendations are indicated by a variety of devices, chosen with the aim of facilitating proofreading and of distracting as little as possible from readability. They are described in detail below.

Arrangement of letters. Letters are grouped alphabetically according to the names of Gödel's correspondents, and within each group by date.

Dating of letters. The date and author's return address, when included as part of a letter, are placed flush right above the salutation and text, even if not so positioned in the original. Undated letters are identified as such and have been placed in sequence on the basis of annotations on the copy texts, postmarks on retained envelopes or internal references to other correspondence or events of known date. Conjectural dates are enclosed in double square brackets (⟦, ⟧).

Editorial apparatus. Original pagination of letters, except for the first page of each, is indicated by small numbers in the outer margins of these pages. The symbol | in the text indicates the location of page breaks. Authorial errors and emendations are indicated as follows:

1. Letters or symbols that should have been deleted are backslashed (like this). Spaces that should be deleted are indicated by a ligature symbol (‿) below the space in question.
2. Letters or symbols that should be replaced by other letters or symbols (including capitalization errors) are backslashed, and the symbols that should replace them are printed in small type in the inner margin of the page on the corresponding line. A "square cup" symbol (⊔) placed in the margin indicates that a symbol is to be replaced by a blank space.
3. Letters or symbols inserted by the editors are enclosed within double square brackets (⟦, ⟧). Authorial insertions are enclosed within single pointed brackets (⟨, ⟩). (We have not distinguished among insertions made above, below, or on the line.) A caret below the line, together with the "square cup" symbol in the margin, indicates a place where a space should be inserted.

4. Material crossed out by the author is printed with a horizontal overstrike (~~thus~~).
5. Errors or emendations not falling in the above categories, including endorsements (annotations by the recipient or a third party), as well as ancillary details, such as the use of letterhead stationery, are noted in textual notes or editorial footnotes.

Translations. Textual errors in German originals, such as misspellings of names, have been corrected silently in the translations. Authorial deletions have not been translated. Authorial insertions have been translated, but are not indicated as such in the translations, in view of the lack of one-to-one correspondence between the structure of German and English. Likewise, German salutations and closings, which are more varied than those customary in English and would seem affected or obsequious if rendered literally, have been translated by conventional English phrases.

Overall responsibility for the preparation and accuracy of translations was shared by John Dawson and Wilfried Sieg. In general, the editor responsible for preparing the introductory note to each body of correspondence was also responsible for the translation thereof. Where that was not the case, credit for the drafting and/or revision of the translations is indicated at the end of the introductory note.

Introductory notes. To a greater extent than in previous volumes of these *Works*, the introductory notes have been written by the editors. The authorship of each note is given in the Contents and at the end of the note itself. As in the earlier volumes, the notes aim (i) to provide historical context, (ii) to explain the contents of the texts to a greater or lesser extent, (iii) to discuss later developments and, in some cases, (iv) to provide a critical analysis, either of the contents of individual letters or of the correspondence as a whole. The notes also provide biographical information about Gödel's correspondents, cross-references to related correspondence, and, where relevant, indications of the contents of letters not selected for inclusion.

Drafts of each note were circulated among the editors and were subsequently revised by their authors in response to criticisms and suggestions. No attempt was made to impose uniformity of style or point of view. The lengths of the notes vary, depending on the extent and significance of each particular body of correspondence and on how familiar the correspondents were thought likely to be to readers.

Introductory notes are distinguished typographically by a running vertical line along the left- or right-hand margin and are boxed off at their end.[a]

[a]A special situation occurs when the note ends in mid-page before facing German and English text. Then the note extends across the top half of the facing pages and is boxed off accordingly.

Footnotes. We use a combination of numbering and lettering, as follows. Authorial footnotes in the letters and their translations are numbered, even in cases where an author (especially Gödel) employed non-numeric (and sometimes idiosyncratic) footnote symbols. Editorial footnotes, on the other hand, used to provide reference citations, to supply ancillary information and to alert the reader to textual issues of various sorts, are lettered and placed below a horizontal line at the bottom of the page. (Where texts are accompanied by facing-page translations, such footnotes are divided evenly across the two pages.) When the number of editorial footnotes extends beyond 26, double letters, ordered lexicographically, are employed.

Editorial annotations and textual notes. Editorial annotations within any of the original texts or their translations are signaled by double square brackets: ⟦, ⟧. Editorial emendations other than those indicated by the editorial apparatus described above are discussed either in editorial footnotes or in the textual notes at the back of the volume. In addition, the following kinds of changes have been made uniformly in the original texts: (i) authorial footnotes have been numbered sequentially within each letter; (ii) spacing, used for emphasis in German texts, as well as underlining, have been rendered by italics; (iii) initial subquotes in German have been raised, e.g., „engeren" becomes "engeren"; (iv) inside addresses to Gödel at the I.A.S. have been omitted.

References. The list of references in volumes IV and V is restricted to items cited therein; however, all citation codes for references listed in earlier volumes remain unchanged. Citation codes consist of the name(s) of the author(s) followed by a date with or without a letter suffix, e.g., "1930", "1930a", "1930b", etc.[b] When the author is clear from the context, that part of the code may be omitted. Where no name is specified or determined by the context, the reference is to Gödel's bibliography, as, e.g., in "Introductory note to *1929*, *1930* and *1930a*". For each reference, the date is that of publication, where there is a published copy, or of presentation, for unpublished items such as a speech. (In the case of works by Gödel published posthumously in volume III of these *Works*, the date is that of presentation or composition; the code for such works is preceded by an asterisk.) A suffix is used when there is more than one publication by an author in the same year; however, the ordering of suffixes does not necessarily correspond to the order of publication within that year. A question mark is used when a date, or some part thereof, is uncertain. For works whose composition or publication extended over a range of dates, the starting and ending dates are both given, separated by a slash.

[b] "200?" is used for articles whose date of publication is not yet known.

Except in the reference list, citation codes are given in italics. They are employed in the introductory notes, editorial footnotes and textual notes. Where citations occur within letter texts and translations, however, they are reproduced as the author gave them. An accompanying editorial footnote then provides the citation code.

References to page numbers in Gödel's publications are to those of the textual source. References to other items in these *Works* are cited by title, volume number and page number within the volume.

To make the reference list as useful as possible for historical purposes, authors' names there are supplied with first and/or middle names as well as initials, except when the information could not be determined. Russian names are given both in transliterated form and in their original Cyrillic spelling. In some cases, common variant transliterations of the same author's name, attached to different publications, are also noted.

Logical symbols. The logical symbols used by authors of letters are here presented intact, even though these symbols may vary from one letter to another. Authors of introductory notes have in some cases followed the notation of the author discussed and in other cases have preferred to make use of other, more current, notation. Also, logical symbols are sometimes used to abbreviate informal expressions as well as formal operations. No attempt has been made to impose uniformity in this respect. The following is a brief glossary of logical symbols that are used in one way or another in these volumes, where 'A', 'B' are letters for propositions or formulas and '$A(x)$' is a propositional function of x or a formula with free variable 'x'.

Conjunction ("A and B"): $A.B$, $A \wedge B$, $A \& B$
Disjunction ("A or B"): $A \vee B$
Negation ("not A"): $\overline{A}$, $\sim A$, $\neg A$
Conditional, or Implication ("if A then B"): $A \supset B$, $A \rightarrow B$
Biconditional ("A if and only if B"): $A \equiv B$, $A \sim B$, $A \leftrightarrow B$
Universal quantification ("for all x, $A(x)$"): $(x)A(x)$, $\Pi x A(x)$, $x\Pi(A(x))$, $(\forall x)A(x)$
Existential quantification ("there exists an x such that $A(x)$"): $(Ex)A(x)$, $\Sigma x A(x)$, $(\exists x)A(x)$
Provability relation ("A is provable in the system S"): $\mathrm{S} \vdash A$

Note: (i) The "horseshoe" symbol is also used for set-inclusion, i.e., for sets X, Y one writes $X \subset Y$ (or $Y \supset X$) to express that X is a subset of Y. (ii) Dots are sometimes used in lieu of parentheses, e.g., $A \supset .B \supset A$ is written for $A \supset (B \supset A)$.

Calendars. Separate calendars of Gödel's correspondence with each of ten major correspondents in this volume are included in this volume. Those calendars list all extant letters known to us, whether or not selected for inclusion herein, as well as the source archives for each letter and details concerning the form of each document (whether typed or handwritten, signed or unsigned, etc.). In addition, each volume contains a general calendar that lists all items of correspondence included in either volume.

Finding aid to the Gödel Nachlaß. In 1998, funds from the Alfred P. Sloan Foundation enabled the preservation microfilming of Gödel's *Nachlaß*, held by the Rare Books and Manuscripts Division of the Firestone Library at Princeton University. At that time a revision was prepared of the finding aid to the *Nachlaß* compiled by John Dawson in 1984. A further revision of that document by Cheryl Dawson, including references to the location of items on the microfilm reels, is included in this volume for the benefit of scholars who may wish to consult the originals or order copies from the microfilm.[c]

Appendices. Rounding out this volume are two appendices to volume V. The first contains a small number of letters written by others on Gödel's behalf. The second provides an alternate version of Remark 3 of item *1972a*, published in volume II of these *Works*, p. 306.

Typesetting. These volumes have been prepared using the TeX computerized mathematical typesetting system devised by Donald E. Knuth of Stanford University, as described in the preface to volume I. (For the first three volumes, camera-ready copy was delivered directly to the publisher.) The computerized system was employed because: (i) much material, including the introductory notes and translations, needed to undergo several revisions; (ii) proofreading was carried on as the project proceeded; (iii) in the case of previously published letters, texts could be prepared in a uniform format, incorporating our editorial apparatus, instead of being photographed from the original sources. Choices of the various typesetting parameters were made by the editors in consultation with the publisher. After significant initial work by Yasuko Kitajima, primary responsibility for the typesetting in these volumes lay with Cheryl A. Dawson and Bruce S. Babcock.

Photographs. Primary responsibility for securing these lay with John Dawson. Their various individual sources are credited in the Permissions section, which follows directly.

[c]A selective subset of the microfilms, *excluding* all correspondence and photographs, is available for purchase from IDC Publishers (Leiden). Further information about that edition is available online at http://www.idc.nl/catalog/index.php?c=375.

[illegible] edition, each volume contains a general [illegible] calendar [illegible] lists an[illegible] of correspondence included in either volume.

[illegible] 1985. Items from the Albert B. [illegible] microfilming of [illegible] held by the Rare Books and Manuscripts Division of the Firestone Library at Princeton University. [illegible] was the result of the finding [illegible] the Veblen [illegible] compiled by [illegible] Dawson in [illegible]. A further revision [illegible] references to the location [illegible] included in this volume for the benefit [illegible] the originals in order [illegible]

[illegible] are two appendices to volume V [illegible] with [illegible] on Gödel's [illegible] version of Remark 3 of [illegible] published [illegible] Collected Works, [illegible]

[illegible] the TeX computer [illegible] Stanford [illegible] the first three volumes [illegible] directly to the publisher [illegible] (i) much material [illegible] needed to undergo several [illegible] the project proceeded. (iii) [illegible] be prepared in a [illegible] a result of using [illegible]. Choice of the various typesetting [illegible] in consultation with the publisher. [illegible] Marko Kitaguchi [illegible] responsible [illegible] and [illegible] A. Brown [illegible]

[illegible] with John [illegible] for the [illegible] which [illegible]

[illegible] photographs, [illegible] about that [illegible]

Copyright permissions

The editors are grateful to the Institute for Advanced Study, Princeton, literary executors for the estate of Kurt Gödel, for permission to reproduce and translate items of correspondence written by him. In addition, we thank the following individuals and institutions for providing us with copies of letters written by Gödel, and for allowing their publication herein:

The Heyting Archief, part of the State Archives, Haarlem, for letters to Arend Heyting;

the Illinois Institute of Technology, Chicago, for letters to Karl Menger;

the Ernest Nagel Papers, Rare Book and Manuscript Library, Columbia University, New York, for letters to Ernest Nagel;

the late Phyllis Post Goodman, for the letter to her father, Emil L. Post;

Constance Reid, for the letters to her;

Yale University Library, New Haven, for letters to Abraham Robinson;

Katherine Salzmann, Curator of Manuscripts, the Library of Living Philosophers Collection, Special Collections, Morris Library, Southern Illinois University, Carbondale, for letters to Paul A. Schilpp;

Dr. Beat Glaus, Dr. Robert Schulman and the ETH-Bibliothek, Zürich, for the letters to Carl Seelig;

an anonymous private source, for the letter to Thoralf Skolem;

Dr. Wolfgang Kerber , director of the Zentralbibliothek für Physik in Wien, for the letter to Hans Thirring;

the American Philosophical Society, Philadelphia, for the letter of 8 November 1957 to Stanisław Ulam;

Ralph L. Elder, Director, the Center for American History, University of Texas at Austin, for letters to Jean van Heijenoort;

the Library of Congress, Washington, DC, for letters to John von Neumann;

Renate Maurer, on behalf of the Universitätsbibliothek, Freiburg im Briesgau, for the letter of 12 October 1931 to Ernst Zermelo.

For permission to reproduce and translate the texts of letters by others included in this volume we thank the following individuals and institutions:

Professor Anne S. Troelstra, on behalf of the Heyting Archief, for the letters of Arend Heyting;

Ralph Hwastecki, for his letter;

Staatsbibliothek zu Berlin–Preußischer Kulturbesitz, from *Nachlaß* 335 (H. Behmann), for the letter from Felix Kaufmann to Heinrich Behmann;

Leon M. Despres, Trustee of the Trust Agreement from Karl Menger, dated 25 May 1967, as amended, for the letters of Professor Menger;

Professor Alexander Nagel, son of Ernest Nagel, for letters of Ernest Nagel;

the Ernest Nagel Papers, Rare Book and Manuscript Library, Columbia University, New York, for Nagel's letter of 22 August 1957 and for quotations from his letters to Allan Angoff;

the late Phyllis Post Goodman, for the letters of her father, Emil L. Post;

Professor Wolfgang Rautenberg, for his letter;

Renée Robinson, widow of Abraham Robinson, and the Yale University Library, New Haven, for Professor Robinson's letters;

Constance Reid and Frederick W. Sawyer III, for their letters;

Katherine Salzmann, Curator of Manuscripts, the Library of Living Philosophers Collection, Special Collections, Morris Library, Southern Illinois University, Carbondale, for the letters of Paul A. Schilpp;

Professor Dana S. Scott, for his letter on Gödel's behalf to Hao Wang and Burton Dreben;

Jan Tarski, for the letters of his father, Alfred Tarski;

Dr. Wolfgang Kerber , director of the Zentralbibliothek für Physik in Wien, and Professor Walter Thirring, son of Hans Thirring, for the letter of Hans Thirring;

Françoise Ulam, widow of Stanisław Ulam, for the letters of Professor Ulam;

Ralph L. Elder, Director, the Center for American History, University of Texas at Austin, for the letters of Jean van Heijenoort;

Professor Marina von Neumann Whitman, daughter of John von Neumann, and the Library of Congress, Washington, D.C., for the letters of John von Neumann;

the late Hao Wang, and his widow, Hanne Tierney, for the letters of Professor Wang;

Renate Maurer, on behalf of the Universitätsbibliothek, Freiburg im Breisgau, for the letters of Ernst Zermelo.

We are grateful to the following sources for supplying portraits of Gödel's correspondents and/or granting permission for their use as illustrations in this volume:

Professor Solomon Feferman, for the photograph of Arend Heyting;

Springer-Verlag/Wien, for the photograph of Karl Menger, used as the frontispiece in the 1998 reprinting of Menger's *Ergebnisse eines mathematischen Kolloquiums*, edited by E. Dierker and K. Sigmund;

University Archives Columbiana Library, Columbia University, for the photograph of Ernest Nagel;

Birkhäuser Boston, for the photograph of Emil L. Post, used as the frontispiece for *Solvability, provability, definability: the collected works of Emil L. Post*, edited by Martin Davis;

George Bergman and Constance Reid, for the portrait of her;

Renée Robinson and the Yale University Library, for the photograph of Abraham Robinson;

Diane Sikora, for the photograph of Paul A. Schilpp. Reprinted by permission of Open Court Publishing Company, a division of Carus Publishing Company, from *Essays in honor of Paul Arthur Schilpp. The abdication of philosophy: philosophy and the public good*, edited by Eugene Freeman, ©1967 by Open Court Publishing Company;

Professor Steven Givant, for the photograph of Alfred Tarski;

Cambridge University Press, for the photograph of Stanisław Ulam, taken from the book *From cardinals to chaos: reflections on the life and legacy of Stanislaw Ulam*, edited by Necia Grant Cooper;

Anita Burdman Feferman, for the photograph of Jean van Heijenoort;

the MacMillan Company, for the photograph of John von Neumann, published in *Adventures of a mathematician*, by Stanisław Ulam;

the Institute for Advanced Study, for the photograph of Gödel and Hao Wang, from Gödel's *Nachlaß*;

Springer-Verlag and the late Mrs. Gertrud Zermelo, for the photograph of Ernst Zermelo, published as the frontispiece in *Zermelo's axiom of choice: its origins, development and influence*, by Gregory H. Moore.

Contents

Volume V

List of illustrations

Kurt Gödel

COLLECTED WORKS

Volume V

Leon Henkin

Writing in 1972 to Leon Henkin, an editor for the then forthcoming *Proceedings of the Tarski Symposium* (*Henkin et alii 1974*),[a] Gödel first expresses regret that he has no paper suitable for publication and then urges the expeditious publication of the proceedings because of the interesting results therein. Gödel expresses particular interest in a result of Robert Solovay's, that one "can derive from Tarski's $\sim C_2$" (here $\sim$ is a symbol for negation) the existence of cardinals α such that $2^\alpha = \alpha^+$. Mostly notably, Gödel writes that "This seems to me the most interesting set⟦-⟧theoretical result about cardinal numbers since Koenig's Theorem." The reference to the classical result emanating from *König 1905* dramatically emphasizes the lack of progress in cardinal arithmetic. Gödel's remark reflects his discussion of the continuum problem in *Gödel 1947*, *1964* both with respect to the lack of progress made and to his interest in strong axioms of infinity (large cardinal axioms).

C_2 is from *Keisler and Tarski 1964* and denotes the class of cardinals that, in modern parlance, are not strongly compact. Strong compactness is a global large cardinal property, first formulated in *Tarski 1962*, that directly generalizes the compactness property of first-order logic to certain infinitary languages. In large cardinal theory strong compactness has been largely supplanted by the enhanced property of supercompactness as the prominent global large cardinal property from which many relative consistency results are derivable. Solovay in his paper *Solovay 1974* for the Tarski Symposium established that if κ is strongly compact, then for every singular strong limit cardinal $\lambda > \kappa$, $2^\lambda = \lambda^+$. Thus, strong compactness implies that the cardinals satisfying instances of the generalized continuum hypothesis (GCH) form a proper class.

Although of intrinsic interest, Solovay's result has remained a somewhat isolated accomplishment. The result was applied in *Magidor 1976* toward establishing the relative consistency of the least strongly compact cardinal being also the least supercompact cardinal; however, this relative consistency has since been established by other means. Perhaps Gödel's phrase "the most interesting set⟦-⟧theoretical result about cardinal numbers since Koenig's Theorem" would better have been applied

[a]Leon Henkin (1920–), at the time Professor of Mathematics at the University of California, Berkeley, will be known to readers of these *Works* for his work in mathematical logic, particularly his early work on the completeness of first- and higher-order logic and his work in algebraic logic.

to Jack Silver's result, established soon after Solovay's: If λ is a singular cardinal of uncountable cofinality and $2^\alpha = \alpha^+$ holds for every cardinal $\alpha < \lambda$ (or just stationarily many such cardinals), then $2^\lambda = \lambda^+$ (*Silver 1975*). This was a surprising result in ZFC about how instances of the GCH holding below a cardinal entail that it holds at the cardinal as well. It inspired a wave of refinements in ZFC as well as new and sophisticated consistency results relative to large cardinals about powers of singular cardinals. However, there seems to be no record of Gödel's awareness, in his late years, of Silver's result. While Solovay's result may have reinforced Gödel's enthusiasm for the possibility of new axioms settling questions in set theory, Silver's result, being a fruitful result of ZFC, may have tempered that enthusiasm.

Akihiro Kanamori

1. Gödel to Henkin

April 18, 1972

Professor Leon Henkin
Department of Mathematics
University of California
Berkeley, California 94720

Dear Professor Henkin:

I am terribly sorry I have no paper ready that would be suitable for publication in the Tarski volume.

In view of the interesting results that, apparently, will be contained in it I hope the volume will come out very soon.

I am especially interested in Solovay's result that there exist cardinals for which $2^\alpha = \alpha_+$,[a] which, he claims, he can derive from Tarski's $\sim C_2$. This seems to me the most interesting set⟦-⟧theoretical result about cardinal numbers since Koenig's Theorem.

Sincerely yours,

Kurt Gödel

[a]What Gödel wrote is α_+; what is surely intended is α^+.

Jacques Herbrand

The correspondence between Kurt Gödel and Jacques Herbrand consists of two remarkable letters that are focused on two fundamental issues, namely, the extent of finitist methods and the effect of Gödel's incompleteness theorems on Hilbert's consistency program. Gödel and Herbrand expressed sharply contrasting views on the latter issue. The correspondence is also intimately linked to a wider discussion of these theorems that involved most directly Johann von Neumann, Paul Bernays, and members of the Vienna Circle. Characterizing the extent of finitist methods is for Herbrand very much a matter of circumscribing the extent of the concept of finitist function.[a] A historically and conceptually fascinating question is related, namely, what effect did Herbrand's discussion of finitist functions have on the definition of general recursive functions as given in *Gödel 1934*? Gödel remarked in note 34 of his *1934* that Herbrand had suggested a central part in private communication. When queried about this remark by Jean van Heijenoort in a letter of 25 March 1963, Gödel responded on 23 April 1963 that the suggestion had been communicated to him in a letter of 1931, and that Herbrand had made it in exactly the form in which *1934* presented it. But Gödel was unable to find the letter among his papers.[b] John Dawson discovered the letter in the Gödel *Nachlaß* in 1986, and it became clear that Gödel had misremembered a crucial feature of Herbrand's discussion.[c]

Herbrand was born in Paris on 12 February 1908. At the age of only 23, he died in a mountaineering accident at La Bérarde (Isère) on 27 July 1931.[d] He defended his doctoral thesis *Recherches sur la théorie de la démonstration* on 11 June 1930, spent the academic year 1930–1931 in Germany on a Rockefeller Scholarship and intended to go for the next

[a]I take it that Herbrand used "intuitionist" as synonymous with "finitist"; cf. also *Herbrand 1931b*. On pp. 116–118 of his *1985*, van Heijenoort, following Gödel's lead, examines very carefully the possibility of giving "intuitionist" in Herbrand's work a broader interpretation than "finitist". The outcome is inconclusive at best. In my view, the examination does not provide any evidence for such a broader interpretation; see section 2.2 of *Sieg 1994*.

[b]The exchange between Gödel and van Heijenoort is also published in this volume of these *Works*.

[c]The background and the content of the Herbrand–Gödel correspondence were first described in *Dawson 1993*. The crucial feature Gödel had misremembered concerns the computability of finitist functions; see the discussion in the last part of this note.

[d]For biographical details, see *Chevalley 1934* and *Chevalley and Lautman 1931*.

academic year to Princeton University. In his report to the Rockefeller Foundation he wrote that his stay in Germany extended from 20 October 1930 to the end of July 1931: until the middle of May 1931 he had been in Berlin, then for a month in Hamburg and for the remainder of the time in Göttingen. In these three cities, he had mainly worked with von Neumann, Artin and Emmy Noether.[e] Concerning his stay in Berlin he continued that in Berlin he had worked with von Neumann on questions in mathematical logic, and that "my research in that subject will be presented in a paper to be published soon in the 'Journal für reine und angewandte Mathematik'."[f] The paper to which he alluded is his *1931*, *Sur la non-contradiction de l'arithmétique*, notably comparing his own results with those of Gödel, as his friend Claude Chevalley put it.[g]

Indeed, Herbrand had learned of the incompleteness theorems from von Neumann shortly after his arrival in Berlin.[h] In a letter of 3 December 1930 he wrote to Chevalley:

> The mathematicians are a very strange bunch; during the last two weeks, whenever I see ⟦von⟧ Neumann, we have been talking about a paper by a certain Gödel, who has produced very curious functions; and all of this destroys some solidly anchored ideas.[i]

This sentence opens the letter. Having sketched Gödel's arguments and reflected on the results, Herbrand concluded the logical part of his letter with: "Excuse this long beginning; but all of this has been pursuing me, and by writing about it I exorcise it a little."[j] When Herbrand wrote to Gödel on 7 April 1931 he had actually read the galleys of *Gödel 1931*; von Neumann had received them at the beginning of January 1931, but it seems that Herbrand had obtained access to them only more "recently"

[e]The remark in the report reads in French: "Dans ces trois villes, j'ai travaillé surtout avec M. von Neumann, Artin et Fr. Noether."

[f]A Berlin, j'ai surtout travaillé avec M. von Neumann, sur des questions de Logique Mathématique, et mes recherches dans cette branche seront exposeés dans un mémoire qui paraitra prochainement dans le 'Journal für reine und angewandte Mathematik'.

[g]*Chevalley 1934*, p. 25.

[h]For von Neumann's role in the early discussion of Gödel's theorems, see the introductory note to his correspondence with Gödel, in this volume.

[i]Les mathématiciens sont une bien bizarre chose; voici une quinzaine de jours que chaque fois que je vois ⟦von⟧ Neumann nous causons d'un travail d'un certain Gödel, qui a fabriqué de bien curieuses fonctions; et tout cela détruit quelques notions solidement ancreés.

[j]Excuse ce long début; mais tout cela me poursuit, et de l'écrire m'en exorcise un peu.

through Bernays, with whom he also had contact during his stay in Berlin.[k]

On the very day he wrote to Gödel, Herbrand sent a note as well to Bernays, enclosed a copy of his letter to Gödel and contrasted his consistency proof with that of Ackermann (which he ascribed mistakenly to Bernays):

> In my arithmetic the axiom of complete induction is restricted, but one may use a variety of other functions than those that are defined by simple recursion: in this direction, it seems to me, my theorem goes a little farther than yours.[l]

The central issue of the letter to Gödel is formulated for Bernays as follows: "I also try to show in this letter how your results can agree with these of Gödle ⟦sic⟧."[m] All of this information puts into sharper focus the remark in Herbrand's *1931c*, which according to Goldfarb's introductory note to that item in *Herbrand 1971* was submitted to Hadamard at the beginning of 1931.

> Recent results (not mine) show that we can hardly go any further: it has been shown that the problem of consistency of a theory containing all of arithmetic (for example, classical analysis) is a problem whose solution is impossible. ⟦Herbrand is here alluding to *Gödel 1931*.⟧ In fact, I am at the present time preparing an article in which I will explain the relationships between these results and mine ⟦this article is *1931*⟧.[n]

It seems quite clear that Herbrand's attempt to come to a thorough understanding of the relationship between Gödel's theorems and ongoing proof-theoretic work, including his own, prompted the specific details in his letter to Gödel as well as in his *1931c*.

[k]During the 1920s Bernays spent the semester breaks mostly in Berlin with his family. Gödel had sent the galleys to Bernays' Berlin address in early January, but Bernays received them only in mid-January in Göttingen; see Bernays' letter to Gödel of 18 January 1931.

[l]Bernays, in his letter to Gödel of 20 April 1931, pointed out that Herbrand had misunderstood him in an earlier discussion: he, Bernays, had not talked about a result of his, but rather about Ackermann's consistency proof. The German text in Herbrand's letter to Bernays reads: "In meiner Arithmetik ist das Axiom der Vollständigen Induktion beschränkt, aber man darf allerlei andere Funktionen benutzen als diejenige die durch einfache Rekursion definiert sind: in dieser Richtung scheint es mir dass mein Theorem etwas weiter geht als das Ihrige."

[m]Cf. previous note, as to the results to which Herbrand is referring. The German text is: "Ich suche auch in diesem Brief zu zeigen wie Ihre Ergebnisse mit diesen von Gödle übereinstimmen können."

[n]*Herbrand 1931c* (*Herbrand 1971*, p. 279). The remarks in double brackets are due to Warren Goldfarb, the editor of *Herbrand 1971*.

At issue is the extent of finitist or, for Herbrand synonymously, intuitionist methods, and thus the reach of Hilbert's consistency program. Herbrand's letter can be understood, as Gödel in his response quite clearly did, to give a sustained argument against Gödel's assertion in his *1931* that the second incompleteness theorem does not contradict Hilbert's "formalist viewpoint":

> For this viewpoint presupposes only the existence of a consistency proof in which nothing but finitary means of proof is used, and it is conceivable that there exist finitary proofs that *cannot* be expressed in the formalism of P (or of M and A).[o]

Herbrand introduces a number of systems for arithmetic, all containing the axioms (I) for predicate logic with identity and the Dedekind–Peano axioms for zero and successor. The systems are distinguished by the strength of the induction principle, whether it is available for all formulas or just quantifier-free ones, and by the class F of finitist functions for which recursion equations are available. The system with full induction and recursion equations for functions in F is denoted by $\mathrm{I} + 2 + 3\,\mathrm{F}$; if induction is restricted to quantifier-free formulas, the resulting system is denoted by $\mathrm{I} + 2' + 3\,\mathrm{F}$. The defining axioms for elements $f_1, f_2, f_3, \ldots$ in F must satisfy, according to Herbrand, the following conditions:

> (I) The defining axioms for f_n contain, besides f_n, only functions of lesser index.
>
> (2) These axioms contain only constants and free variables.
>
> (3) We must be able to show, by means of intuitionistic proofs, that with these axioms it is possible to compute the value of the functions univocally for each specified system of values of their arguments.

As examples for classes F he considers the set E_1 of addition and multiplication, as well as the set E_2 of all primitive recursive functions from Gödel's *1931*. He asserts that the functions definable by his own "general schema" include many other functions, in particular, the Ackermann function (which he calls the Hilbert function). Furthermore, he argues that one can construct by diagonalization a finitist function that is not in E, if E is a set of functions satisfying axioms such that "one can always determine, whether or not certain defining axioms are among these axioms".

The fact of the open-endedness of (a finitist presentation of) the concept of finitist function is crucial for Herbrand's conjecture that one

[o] *Gödel 1931*, p. 197; in these *Works*, vol. I, p. 195. P is the version of the system of *Principia mathematica* in Gödel's 1931 paper, M is the system of set theory introduced by von Neumann and A is classical analysis.

cannot *prove* that all finitist methods are formalizable in *Principia mathematica.* But he claims that every finitist proof, as a matter of fact, can be formalized in a system of the form $\mathrm{I} + 2' + 3\,\mathrm{F}$ with a suitable class F (that depends on the given proof) and, thus, also in *Principia mathematica.* Conversely, he insists that every proof in the quantifier-free part of $\mathrm{I} + 2' + 3\,\mathrm{F}$ is finitist. He summarizes his reflections by saying in the letter (and in almost identical words in *1931* [*Herbrand 1971*, p. 297]):

> It reinforces my conviction that it is impossible to prove that every intuitionistic proof is formalizable in Russell's system, but that a counterexample will never be found. There we shall perhaps be compelled to adopt a kind of logical postulate.

The conjectures and claims are strikingly similar to those von Neumann communicated to Gödel in his letters of 29 November 1930 and of 12 January 1931. We know of Gödel's response to von Neumann's dicta not through a letter from Gödel, but rather through the minutes of the meeting of the Schlick Circle that took place on 15 January 1931. These minutes report what Gödel viewed as questionable, namely, the claim that the totality of all intuitionistically correct proofs is contained in *one* formal system. That, he emphasized, is the weak spot in von Neumann's argumentation.[p]

In response to Herbrand's letter, Gödel makes more explicit his reasons for questioning the formalizability of finitist considerations in a single formal system, say in *Principia mathematica.* He agrees with Herbrand on the indefinability of the concept "finitist proof". However, even if one accepts Herbrand's very schematic presentation of finitist methods and the claim that every finitist proof can be formalized in a system of the form $\mathrm{I} + 2' + 3\,\mathrm{F}$, the question remains "whether the intuitionistic proofs that are required in each case to justify the unicity of the recursion axioms are all formalizable in *Principia mathematica.*" He continues:

> Clearly, I do not claim either that it is certain that some finitist proofs are not formalizable in *Principia Mathematica*, even though intuitively I tend toward this assumption. In any case, a finitist proof not formalizable in *Principia Mathematica* would have to be quite extraordinarily

[p]The minutes are found in the Carnap Archives of the University of Pittsburgh. Part of the German text is quoted in *Sieg 1988*, note 11, and more fully in *Mancosu 1999a*, pp. 36–37. For other accounts of early reactions to Gödel's results, see *Dawson 1985* and *Mancosu 1999a.* Interestingly, *Bernays 1933* uses "von Neumann's conjecture" to infer that the incompleteness theorems impose fundamental limits on proof-theoretic investigations.

complicated, and on this purely practical ground there is very little prospect of finding one; but that, in my opinion, does not alter anything about the possibility in principle.

Gödel had changed his views significantly by late December 1933 when he gave an invited lecture to the Mathematical Association of America in Cambridge, Massachusetts. In the handwritten text for this lecture, Gödel's **1933o*, he sharply distinguishes intuitionist from finitist arguments, the latter constituting the most restrictive form of constructive mathematics. He also insists that the known finitist arguments given by "Hilbert and his disciples" can all be carried out in a certain system *A*.[q] In turn, he asserts, proofs in the system *A* "can be easily expressed in the system of classical analysis and even in the system of classical arithmetic, and there are reasons for believing that this will hold for any proof which one will ever be able to construct".[r] The direct consequence of this observation and the second incompleteness theorem is that classical arithmetic cannot be shown to be consistent by finitist means. Gödel had anticipated that consequence by stating earlier: "But unfortunately the hope of succeeding along these lines ⟦of trying to establish consistency by means that satisfy the restrictive demands of system *A*⟧ has vanished entirely in view of some recently discovered facts."

Nevertheless, Gödel formulates on the next page of his **1933o* a theorem of Herbrand's as the most far-reaching among interesting partial results in the pursuit of Hilbert's consistency program: "If we take a theory which is constructive in the sense that each existence assertion made in the axioms is covered by a construction, and if we add to this theory the non-constructive notion of existence and all the logical rules concerning it, e.g., the law of excluded middle, we shall never get into any contradiction." The result, mentioned in Herbrand's letter as Remark 2 (on p. 3), can be understood in just this way; it foreshadows of course the central result of Herbrand's *1931*. Gödel conjectures that Herbrand's method might be generalized, but emphasizes again (on p. 27) that "for larger systems containing the whole of arithmetic or ana-

[q]The restrictive characteristics of the system *A* are formulated on pp. 23 and 24 of **1933o*: (i) universal quantification is restricted to totalities whose elements can be generated by a "finite procedure"; (ii) negation cannot be applied to universal statements; (iii) notions have to be decidable and functions must be calculable. As to condition (iii), Gödel claims, "such notions and functions can always be defined by complete induction"; cf. note s below and also Gödel's own note 3 of *1934*.

[r]*Gödel *1933o*, p. 26; in these *Works*, vol. III, p. 52. This issue is discussed also in Feferman's introductory note to **1933o*, these *Works*, vol. III, pp. 40–42, and in the correspondence with Bernays, in particular in Gödel's letter of 24 January 1967.

lysis the situation is hopeless if you insist upon giving your proof for freedom from contradiction by means of the system *A*".[s]

There is one prima facie puzzling remark in Herbrand's letter, when he claims in point 3: "In general, if we want to apply your methods to an arithmetic that has the functions of a set *F*, we need a larger set of functions. (This can be proved precisely: it is very easy.)" At the end of point 5 Herbrand refers in a parenthetical remark to this issue; he maintains that it is the function obtained by diagonalization that forces the consideration of larger classes of functions. Gödel finds point 3 "not completely comprehensible"; after all, he adds, a consistency proof forces us to go beyond the system being studied, but the proof of the statement (*) "If the system is consistent, then the proposition given by me is unprovable." can be given in the system. Herbrand, as if anticipating Gödel's rejoinder, claims in *1931* that for the proof of just this statement one needs the function enumerating all elements of *F*. Consequently, Gödel's argument cannot be carried out in the system.

> But, to carry out Gödel's argument, we have to number all objects occurring in proofs; we are thus led to construct the ⟦enumeration⟧ function of two variables $f_y(x)$; this justifies what we were saying above, namely, that it is impossible, in an arithmetic containing the hypotheses C′, to formalize Gödel's argument about this arithmetic.[t]

Herbrand's specific assumption—that a finitist metamathematical description of an arithmetic like his, even when restricted to a definite

[s]This systematic context allows us to calibrate the strength of the system *A* in **1933o* and, thus, Gödel's views about the extent of finitist methods at this time. In Gödel's judgment, Herbrand had given a finitist consistency proof for a theory of arithmetic with quantifier-free induction and a large class *F* of calculable functions that included the Ackermann function; Gödel was thoroughly familiar with that theory, as he used it—with full induction—in his *1933e*. The system *A* is consequently stronger than primitive recursive arithmetic.

From the details of the consistency proof it is clear that the functions in *F* must be available in the finitist theory, and that in particular the Ackermann function is finitist. Gödel was at this time not alone in considering the Ackermann function as a finitist one; Herbrand obviously did, and so did von Neumann as witnessed by his letter of 29 November 1930 to Gödel. Indeed, Mark Ravaglia makes in his doctoral dissertation the case that Hilbert and Bernays view (in their *1934*) extensions of "rekursive Zahlentheorie" by Ackermann-type functions as finitist.

In part III of his Lecture at Zilsel's, **1938a*, Gödel distinguishes three constructive systems that all satisfy the most stringent constructivity requirements, and it is here that he introduces another system, also called *A*, that clearly is primitive recursive arithmetic. He claims on p. 3 that Hilbert "*wanted to carry out the proof* ⟦of consistency⟧ *with this*". (I do not have a conjecture, why Gödel changed his views.)

[t]*Herbrand 1931*, p. 296. The hypotheses C′ are "a definite group of schemata of type C", i.e., a definite group of recursion equations for the functions in *F*—that allows a finitist determination of which recursion equations are involved.

set of recursion equations, uses necessarily an enumeration function—is not correct.[u] However, the implicitly underlying general point is worth emphasizing: the proof of (*), and thus the proof of the second incompleteness theorem, is based delicately on additional assumptions concerning the proof predicate. Those assumptions were formulated as *derivability conditions* in the second volume of Hilbert and Bernays' *Grundlagen der Mathematik.* In his own further reflections on the generality of his theorems, Gödel seems to focus exclusively on the analysis of "mechanical procedures" or "effective calculability", i.e., a general characterization of formal theories; but as we will see below that is not quite correct either.

This issue leads naturally to a discussion of the role this correspondence played for the origins of recursion theory. From the very beginning, Gödel attributed to Herbrand the inspiration for the definition of general recursive function in his 1934 Princeton Lectures. In those lectures Gödel strove, as indicated even by their title *On undecidable propositions of formal mathematical systems*, to make his incompleteness results less dependent on particular formalisms. In the introductory §1 he discussed the notion of "a formal mathematical system" in some generality and required that

> the rules of inference, and the definitions of meaningful formulas and axioms, be constructive; that is, for each rule of inference there shall be a finite procedure for determining whether a given formula B is an immediate consequence (by that rule) of given formulas $A_1, \dots, A_n$, and there shall be a finite procedure for determining whether a given formula A is a meaningful formula or an axiom.[v]

He used, as in his *1931*, primitive recursive functions and relations to present syntax, viewing the primitive recursive definability of formulas and proofs as a "precise [condition which in practice suffices as a substitute for the unprecise] requirement of §1 that the class of axioms and the relation of immediate consequence be constructive".[w] But a notion that would suffice *in principle* was really needed, and Gödel attempted to arrive at a more general notion. He considered the fact that the value of a primitive recursive function can be computed by a finite procedure for each set of arguments as an "important property" and added in footnote 3:

[u] See *Rose 1984* for a contemporary presentation of such theories.

[v] *Gödel 1934*, p. 1; in these *Works*, vol. I, p. 346.

[w] *Gödel 1934*, p. 19; in these *Works*, vol. I, p. 361. The bracketed text was added by Gödel for the publication of the lecture notes in *Davis 1965*.

> The converse seems to be true if, besides recursions according to the scheme (2) ⟦i.e. primitive recursion as given above⟧, recursions of other forms (e.g., with respect to two variables simultaneously) are admitted. This cannot be proved, since the notion of finite computation is not defined, but it can serve as a heuristic principle.[x]

What other recursions might be admitted is discussed in the last section of the notes under the heading "general recursive functions". Gödel described in it the proposal for the definition of a general notion of recursive function that (he thought) had been suggested to him by Herbrand:

> If ϕ denotes an unknown function, and $\psi_1, \ldots, \psi_k$ are known functions, and if the ψ's and ϕ are substituted in one another in the most general fashions and certain pairs of resulting expressions are equated, then, if the resulting set of functional equations has one and only one solution for ϕ, ϕ is a recursive function.[y]

He went on to make two restrictions on this definition. He required, first of all, that the left-hand sides of the equations be in a standard form with ϕ being the outermost symbol and, secondly, that "for each set of natural numbers $k_1, \ldots, k_l$ there shall be exactly one and only one m such that $\phi(k_1, \ldots, k_l) = m$ is a derived equation". The rules that were allowed in giving derivations are simple substitution and replacement rules. This proposal was taken up for systematic development in *Kleene 1936*.

We should distinguish then, as Gödel did, two features: first, the precise specification of *mechanical* rules for deriving equations, i.e., for carrying out computations, and second the formulation of the *regularity condition* requiring calculable functions to be total. That point of view was also expressed by Kleene who wrote in his *1936* with respect to the definition of general recursive function of natural numbers:

[x] *Gödel 1934*, p. 3; in these *Works*, vol. I, p. 348. Gödel added later: "This statement is now outdated; see the Postscriptum, pp. 369–371." He refers to the Postscriptum appended to the lectures for *Davis 1965*.—It should also be emphasized that Gödel did not intend to formulate (a version of) Church's Thesis; cf. *Davis 1982*, p. 8. It is of interest to note, however, that already in **1933o*, p. 24, Gödel asserts that functions that "can be calculated for any particular element" can always be defined by complete induction.

[y] *Gödel 1934*, p. 26; in these *Works*, vol. I, p. 368. *Kalmár 1955* pointed out that the class of functions satisfying such functional equations is strictly greater than the class of general recursive functions; see also the exchange of letters between Gödel and Büchi in these *Works*, vol. IV.

> It consists in specifying the form of the equations and the nature of the steps admissible in the computation of the values, and in requiring that for each given set of arguments the computation yield a unique number as value.[z]

In his letter to van Heijenoort, dated 14 August 1964, Gödel asserted that "it was exactly by *specifying* the rules of computation that a mathematically workable and fruitful concept was obtained". When making this claim Gödel took for granted that Herbrand's suggestion had been "formulated *exactly* as on page 26 of my lecture notes, i.e. without reference to computability".[aa] As was noticed, Gödel had to rely on his recollection which, he said, "is very distinct and was still very fresh in 1934". On the evidence of Herbrand's letter it is clear that Gödel misremembered. This is not to suggest that Gödel was wrong in his broad assessment, but rather to point to the most important step he had taken by disassociating recursive functions from the epistemologically restricted notion of intuitionistic proof in Herbrand's sense.

Gödel later on dropped the regularity condition altogether and emphasized "that the precise notion of mechanical procedures is brought out clearly by Turing machines producing partial rather than general recursive functions".[ab] The very notion of partial recursive function, of course, had been introduced in *Kleene 1938*. At this earlier historical juncture, however, the introduction of an equational calculus with particular computation rules was important for the mathematical development of recursion theory as well as for the underlying conceptual analysis. It brought out clearly what Herbrand, according to Gödel in his letter of 23 April 1963 to van Heijenoort, had failed to see, namely "that the computation for *all* computable functions proceeds by *exactly the same rules*". In addition, the rules needed are of a remarkably elementary character due to the general symbolic character of the computation steps. It seems that Gödel was right, for stronger reasons than he put forward, when he cautioned in the same letter that Herbrand had *foreshadowed*, but not *introduced*, the notion of general recursive function. In a way, the mathematical development of computability theory based on this general analysis provided an important fact for responding

[z] *Kleene 1936*, p. 727.

[aa] See Gödel's letter to van Heijenoort of 23 April 1963, letter 2 in this volume.

[ab] *Wang 1974*, p. 84.

to Herbrand's issue concerning the proof of the second incompleteness theorem that was mentioned above. Gödel formulated in section 6 of *1934* a number of "conditions that a formal system must satisfy in order that the foregoing arguments apply", i.e., the arguments for the incompleteness theorems. The very first condition states:

> Supposing the symbols and formulas to be numbered in a manner similar to that used for the particular system considered above, then the class of axioms and the relation of immediate consequence shall be ⟦primitive⟧ recursive.[ac]

The condition becomes superfluous, Gödel writes in his 1964 Postscriptum, if formal systems are viewed as mechanical procedures for producing formulas, and if Turing's analysis of such procedures is accepted. The antecedent of this conditional provides the basis for a proof "that for any formal system provability is a predicate of the form $(Ex)x\mathfrak{B}y$, where $\mathfrak{B}$ is primitive recursive". Together with the introducibility of all primitive recursive functions in elementary number theory, the latter fact is crucial for the detailed proof of the second incompleteness theorem (more specifically, for the verification of the third derivability condition) by Hilbert and Bernays, which "carries over almost literally to any system containing, among its axioms and rules of inference, the axioms and rules of inference of number theory".[ad]

Wilfried Sieg[ae]

[ac] *Gödel 1934*, p. 19; in these *Works*, vol. I, p. 361.

[ad] Postscript to *Gödel 1934*, these *Works*, vol. I, p. 370.

[ae] I am most grateful to Cathérine Chevalley, who had provided me already in 1991 with Herbrand's remarkable letter to her father (as well as Herbrand's reports on his stay in Germany); the letter was used in my *1994* to understand better the early reception of Gödel's talk in Königsberg. I also want to thank John Dawson, Solomon Feferman and, in particular, Charles Parsons for suggestions that led to real improvements of earlier drafts of this note.

1. Herbrand to Gödel

Berlin, den 7. April 31

Sehr geehrter Herr Gödle! el

Ich schicke Ihnen gleichzeitig Sonderabdrücke einiger meiner Arbeiten von mathematischer Logik. Herr v. Neumann hatte mir von Ihren Arbeiten gesprochen, und vor Kurzem hat ⟨mir⟩ Herr Bernays ~~mir~~ eine Korrektur Ihrer nächsten Abhandlung mitgeteilt. Sie hat mich sehr interessiert, und ich möchte hier einige Bemerkungen machen, die Ihre Ergebnisse mir zum Gedanken bringen. k

Betrachten wir die Arithmetik; ich möchte zuerst ihre Axiome genau schreiben. Wir haben darin nur ein Variablen-Typus, eine Konstante, 0, eine Funktion $x+1$, und einen primitiven Ausdruck $x = y$. Wir haben die üblichen Axiomen: 1

$$\left.\begin{array}{c} \text{alle die logischen Axiome} \\ x = x \qquad x = y \,.\supset .\, y = x \qquad x = y \times y = z \,.\supset .\, x = z \\ x + 1 \neq 0 \\ x + 1 = y + 1 \,.\equiv .\, x = y \end{array}\right\} \quad \text{(I)}$$

und das Axiom der vollständigen Induktion:

$$\Phi\, 0 \times .(x).\Phi\, x \supset \Phi\, x + 1 : \supset .(x)\Phi\, x \qquad (2)$$

Nennen wir dieses Axiome (2′) wenn wir fordern dass in $\Phi\, x$ *keine gebundenen* Variablen (keine Sein- oder All-zeichen) stehen.

Ausserdem, haben wir in Arithmetik andere Funktionen, zum Beispiel 2 ~~Fu~~ durch Rekursion definierten Funktionen, die ich werde | mit folgenden Axiomen definieren. Nehmen wir an[,] dass wir alle die Funktionen $f_n(x_1, x_2, \ldots x_{p_n})$ einer gewissen endlichen oder unendlichen Menge F definieren wollen. Jede $f_n(x_1 \ldots)$ wird gewisse Definitionsaxiome haben, alle diese Axiome werde ich die Axiome (3 F) nennen. Diese Axiome werden folgende Bedingungen genügen: J

I) Die Definitionsaxiome von f_n enthalten, ausserdem f_n, nur Funktionen von kleinerem Index.

2) Diese Axiome enthalten nur freie Variablen, und Konstanten.

3) Man muss, mit intuitionistischen Beweisen, zeigen können, dass es möglich ⟨ist⟩ das ⟨Wert der⟩ Funktionen für jedes bestimmtes Wertsystem ihrer Argumente mit diesen Axiomen eindeutig zu berechnen.

Zum Beispiel, haben wir folgende Beispiele:

a) Die Funktionen $x + y$ und $x.y$, die die Menge E_1 machen, die

1. Herbrand to Gödel

Berlin, 7 April 1931

Dear Mr. Godel,

I am sending you, at this same time, reprints of some of my papers in mathematical logic. Mr. von Neumann had spoken to me of your works, and recently Mr. Bernays showed me a set of proofs of your next paper. It was of great interest to me, and here I would like to make a few remarks that your results bring to my mind.

Let us consider arithmetic; I would first of all like to write down its axioms precisely. In them we have only one type of variables, a constant 0, a function $x+1$, and a primitive expression $x=y$. We have the usual axioms:

$$\left.\begin{array}{c} \text{all the logical axioms} \\ x=x \qquad x=y\,.\supset.\,y=x \qquad x=y \times y=z\,.\supset.\,x=z \\ x+1\neq 0 \\ x+1=y+1\,.\equiv.\,x=y \end{array}\right\} \tag{I}$$

and the axiom of complete induction:

$$\Phi\,0 \times .(x).\Phi\,x \supset \Phi\,x+1 :\supset .\,(x)\Phi\,x \tag{2}$$

Let us call this axiom (2′) when we require that *no bound* variables (no existential or universal quantifiers) occur in $\Phi\,x$.

In arithmetic we have other functions as well, for example functions defined by recursion, which I will | define by means of the following axioms. Let us assume that we want to define all the functions $f_n(x_1, x_2, \ldots x_{p_n})$ of a certain finite or infinite set F. Each $f_n(x_1 \ldots)$ will have certain defining axioms; I will call all these axioms (3 F). These axioms will satisfy the following conditions: 2

I) The defining axioms of f_n contain, besides f_n, only functions of smaller index.

2) These axioms contain only free variables and constants.

3) We must be able to show, by means of intuitionistic proofs, that it is possible to compute with these axioms the value of the functions univocally for each particular system of values of their arguments.

We have, e.g., the following examples:

a) The functions $x+y$ and $x.y$, which constitute the set E_1, that

~~definiert sind~~ mit folgenden Axiomen definiert sind:

$$\left.\begin{array}{rl} x+0=x & x+(y+1)=(x+y)+1 \\ x.0=0 & x.(y+1)=xy+x \end{array}\right\} 3' \text{ oder } 3\,E_1$$

b) Ihre ~~R~~ rekursiven Funktionen, $\phi_i(x_1, x_2, \ldots x_{n_i})$, (ich werde ihre Menge durch E_2 bezeichnen), mit folgenden Axiomen definiert:

$$\left.\begin{array}{r} \phi_i(0, x_2 \ldots x_{n_i}) = \psi(x_2 \ldots x_{n_i}) \\ \phi_i(k+1, x_2 \ldots x_{n_i}) = \mu(k, \phi(k, x_2 \ldots x_{n_i}), x_2, \ldots x_{n_i}) \end{array}\right\} 3'' \text{ oder } 3\,E_2$$

c) Aber in meinem allgemeinen Schema, können viel andere Funktionen definiert sein: nehmen wir zum Beispiel die Hilbert'sche Funktion (in seiner Abhandlung "~~U~~Über das Unendliche"[a]), $\phi(a, a, a)$, die mit folgenden Axiomen definiert ist: Ü

Die Hilbertsche Funktion ist $\phi(a, a, a)$, wo man hat:

$$\left\{\begin{array}{r} \phi(n+1, a, b) = \phi(n, a, \phi(n+1, a, b-1)) \\ \phi(n, a, 0) = a \\ \phi(0, a, b) = a+b \end{array}\right.$$

I) Bemerken wir zuerst⟦,⟧ dass jeder intuition~~n~~istische~~r~~ Beweis kann~~,~~ in einer Arithmetik geführt werden, ~~mit~~ ⟨die⟩ nur die Axiome I, 2′ and 3 F, für eine gewisse Menge F von Funktionen (die von dem Beweis abhängt) besitzt.[b] | Und jeder Beweis in dieser Arithmetik, d~~ie~~r keine gebundene Variablen besitzt, ist intuition~~n~~istisch: diese~~r~~ Tatsache hängt von der Definition unserer Funktionen ab, und man kann es unmittelbar sehen. 3

Eine Theorie mit den Axiomen I, 2′ und 3 F, werde ich durch I + 2′ + 3 F bezeichnen.

2) Es folgt unmittelbar aus meiner Methoden⟦,⟧ dass alle diese Theorien I + 2′ + 3 F *widerspruchslos* sind (wenn wir gebundene Variablen benutzen).

3) Sie haben bewiesen, dass, wenn man Ihre Methoden einer Arithmetik mit den Funktionen E_1 anwenden kann, man die Funktionen E_2 braucht (um Ihre Funktionen zu bauen). Im allgemein, wenn man Ihre Methoden einer Arithmetik mit den Funktionen einer Menge F anwenden will, braucht man eine grössere Menge von Funktionen (man kann es genau beweisen: es ist sehr leicht).

[a] *Hilbert 1926.*

are defined by means of the following axioms:

$$\left.\begin{array}{ll} x+0=x & x+(y+1)=(x+y)+1 \\ x.0=0 & x.(y+1)=xy+x \end{array}\right\} 3' \text{ oder } 3\,E_1$$

b) Your recursive functions, $\phi_i(x_1, x_2, \ldots x_{n_i})$, (I shall denote their set by E_2) defined by means of the following axioms:

$$\left.\begin{array}{r} \phi_i(0, x_2 \ldots x_{n_i}) = \psi(x_2 \ldots x_{n_i}) \\ \phi_i(k+1, x_2 \ldots x_{n_i}) = \mu(k, \phi(k, x_2 \ldots x_{n_i}), x_2, \ldots x_{n_i}) \end{array}\right\} 3'' \text{ oder } 3\,E_2$$

c) In my general scheme, however, many other functions can be defined; let's take, for example, the Hilbert function $\phi(a, a, a)$ (in his paper "On the infinite"[a]), which is defined by means of the following axioms:
The Hilbert function is $\phi(a, a, a)$, where we have:

$$\left\{\begin{array}{c} \phi(n+1, a, b) = \phi(n, a, \phi(n+1, a, b-1)) \\ \phi(n, a, 0) = a \\ \phi(0, a, b) = a + b \end{array}\right.$$

I) We note first of all that every intuitionistic proof can be carried out in an arithmetic that has only the axioms I, 2′, and 3 F for a certain set F of functions (which depends on the proof).[b] | And each proof in this arithmetic, which has no bound variables, is intuitionistic—this fact rests on the definition of our functions and can be seen immediately. 3

I will denote a theory with the axioms I, 2′, and 3 F by I + 2′ + 3 F.

2) It follows immediately from my methods that all these theories I + 2′ + 3 F are *consistent* (if we use bound variables).

3) You have proved that, if your methods can be applied to an arithmetic that has the functions E_1, the functions E_2 are needed (in order to construct your functions). In general, if we want to apply your methods to an arithmetic that has the functions of a set F, we need a larger set of functions. (This can be proved precisely: it is very easy.)

---

[b]Here Gödel had penciled a vertical bar in the left-hand margin from "1) Bemerken" to "besitzt."

4) Sie beweisen, dass I + 2 + 3″ mit I + 2 + 3′ equivalent ist (aber man kann nicht beweisen, dass I + 2′ + 3″ mit I + 2′ + 3′ equivalent ist). Herr Bernays hat mir gesagt⟦,⟧ dass er die Widerspruchlosigkeit von I + 2 + 3′ bewiesen hat;[c] aus Ihren Methoden folgt⟦,⟧ dass seiner Beweis nicht in I + 2 + 3′ formalisierbar ist, und auch nicht in I + 2 + 3″; mit anderen Worten, in seinem Beweis mussen gewisse Funktionen liegen, die nicht der Addition und der Multiplikation (wie Ihre rekursiven Funktionen) reduzierbar sind. In meinem Beweis, im Gegenteil, braucht das nicht der Fall zu sein.

Aber man sieht leicht⟦,⟧ dass viele andere Funktionen als die rekursiven Funktionen der Addition und der Multiplikation reduzibel sind. Es ist mir gar nicht gelungen⟦,⟧ eine solche Funktion zu ersinnen⟦,⟧ die nicht diese Eigenschaft besitzt. Ein Beispiel einer solchen Funktion (der aus dem Bernays'schen Beweis ausgezogen sein muss), würde sehr interessant sein.

4 | 5) Ich bin so der folgenden Bemerkung geführt: ich verstehe gar nicht wie es möglich sei, dass es intuitionnistische Beweise gibt, die nicht im Russel⟦l⟧'schen System formalisierbar sind; mit anderen Worten wie es eine Arithmetik mit den Axiomen I + 2 + 3 F[d] geben kann, die nicht in diesem System übersetzbar ist. Nur ein Beispiel konnte mich davon überzeugen; aber ich glaube nicht⟦,⟧ dass man beweisen kann⟦,⟧ dass jeder intuitionnistischer Beweis in Russel⟦l⟧'schen System formalisierbar ist, und auf folgenden Gründen:

Nehmen wir an⟦,⟧ dass wir eine gewisse Menge E von Funktionen haben, mit gewissen Axiomen, so dass man immer feststellen kann⟦,⟧ ob gewisse Definitionsaxiome unter diesen Axiomen sind, oder nicht. Man kann immer, unter diesen Bedingungen, andere Funktionen definieren, die von allen den vorigen verschieden sind. Man sieht dass mit dem Diagonalverfahren: man kann immer, intuitionnistisch, ein Verfahren beschrieben um alle ⟨die Funktionen⟦,⟧ die nur eine Variable haben und mit den Funktionen E (durch Einsetzung) gebaut werden können⟩⟦,⟧ ~~unsere~~ Funktionen[e] unter dem Form $f_n(x)$ $(n = 1, 2, 3, \ldots)$ zu schreiben; dann ist die Funktion $f_n(n) + 1$ von allen ⟨den anderen und von allen ihren Kombinationen⟩ verschieden. (Um Ihre Methoden einer Arithmetik⟦,⟧ die alle die Funktionen der Menge E enthält⟦,⟧ zu anwenden, muss man immer ~~diese~~ ⟨solche⟩ Funktion⟨en⟩ bauen: daraus kommt was ich in 2)[f] sagte.)

ei

---

[c]Herbrand had misunderstood Bernays, who had reported in their conversation on Ackermann's work; see Bernays' letter of 20 April 1931 to Gödel.

[d]Given that Herbrand claims in 1) that every intuitionistic proof can be carried out in a system of the form I + 2′ + 3 F (for suitable F), it is plausible that he intended to refer to such a system (and not to I + 2 + 3 F).

4) You prove that $\mathrm{I} + 2 + 3''$ is equivalent to $\mathrm{I} + 2 + 3'$ (but one cannot prove that $\mathrm{I} + 2' + 3''$ is equivalent to $\mathrm{I} + 2' + 3'$). Mr. Bernays told me that he has proved the consistency of $\mathrm{I} + 2 + 3''$.[c] By your methods it follows that his proof is not formalizable in $\mathrm{I} + 2 + 3'$, and also not in $\mathrm{I} + 2 + 3''$; in other words, in his proof there must be certain functions that are not reducible to addition and multiplication (as your recursive functions are). In my proof, on the contrary, that need not be the case.

We readily see, however, that many functions other than the recursive functions are reducible to addition and multiplication. I haven't at all succeeded in devising a function that does not have this property. An example of such a function (which must be extracted from Bernays' proof) would be very interesting.

| 4 5) I am thus led to the following remark: I do not at all understand how it is possible that there are intuitionistic proofs that are not formalizable in Russell's system; in other words, how there can be an arithmetic with the axioms I + 2 + 3 F[d] that is not translatable into that system. Only an example could convince me of that; I do not believe, however, that it can be proved that every intuitionistic proof is formalizable in Russell's system, and that for the following reasons:

Let us assume that we have a certain set E of functions, with certain axioms, such that we can always ascertain whether certain defining axioms are among those axioms or not. Under these conditions, we can always define other functions that differ from all the preceding ones. We see that by means of the diagonal procedure: we can always describe, intuitionistically, a procedure for writing all the functions that have only one variable and are built up by means of the functions E[e] (through substitution) in the form $f_n(x)$ $(n = 1, 2, 3, \dots)$; the function $f_n(n) + 1$ then differs from all the others and from all their combinations. (In order to apply your methods to an arithmetic that contains all the functions of the set E, we must always construct such functions—from that follows what I said in 2.)[f]

[e]Herbrand may have intended to cross this word out; it was preceded by a crossed-out "unsere" and the clause preceding it was an insertion at the side of the page.

[f]Herbrand must have intended a reference to remark 3) above.

Mit anderen Worten, ist es unmöglich alle die Verfahren Funktionen intuition~~n~~istisch zu bauen genau zu beschreiben: wenn man solche Verfahren beschreibt, gibt es immer Funktionen die nicht mit diesen Verfahren gebaut sein können: man kann nicht die intuition~~n~~istischen Methoden mit einer endlichen Zahl von Worten beschreiben. Diese~~r~~ Tatsache scheint mir sehr merkwürdig.

Sie verstärkt meine ~~Ü~~berzeugung⟦,⟧ dass es unmöglich ist ~~ein axiomatisches System zu finden~~ zu beweisen⟦,⟧ dass jeder intuition~~n~~istische~~r~~ Beweis im Russel⟦l⟧'schen System formalisierbar ist, aber dass man nie ein Gegenbeispiel finden wird. Man wird vielleicht | gezwungen sein, dort eine Art von logische~~n~~ Postulat anzunehmen.

Ü

5

m

Entschuldigen Sie diese langen Überlegungen, die vielleicht, wegen meiner schlechten Kenntnis der deutschen Sprache, nicht vollkommen klar sind. Aber es sind in diesen Fragen noch viel geheimnisvolle Tatsachen, und diese Frage der Formalisierung der intuition~~n~~istischen Beweise scheint mir sehr wichtig zu sein, für die philosophische Meinung der Metamathematik.

Hochachtungsvoll,

Ihr sehr ergebener

J. Herbrand

Ich bin nur noch ~~ein~~ für ein kurzes Zeit in Berlin; meine gewöhnliche Adresse ist:

~~1~~0 Rue Viollet le Duc,
Paris (9)

1

2. Gödel to Herbrand

Wien, 25./VII. 1931

Sehr geehrter Herr Herbrand!

Besten Dank für Ihr interessantes Schreiben sowie für die freundliche Übersendung der Separata Ihrer Arbeiten. Ihre "Thèses"[a] waren mir schon früher bekannt und die darin entwickelten Methoden für Wider-

[a] *Herbrand 1930*.

In other words, it is impossible to describe exactly all the procedures for constructing functions intuitionistically: When we describe such procedures, there are always functions that cannot be constructed by means of those procedures; the intuitionistic methods cannot be described in a finite number of words. This fact seems very remarkable to me.

It reinforces my conviction that it is impossible to prove that every intuitionistic proof is formalizable in Russell's system, but that a counterexample will never be found. There we shall perhaps | be compelled to adopt a kind of logical postulate. 5

Please excuse these long considerations, which, because of my poor knowledge of the German language, are perhaps not completely clear. But in these questions there are still many mysterious facts, and this question of the formalization of intuitionistic proofs seems to me to be very important for the philosophical significance of metamathematics.

Respectfully yours,

J. Herbrand

I am in Berlin for only a short time yet; my regular address is:

10 Rue Viollet le Duc,
Paris (9)

2. Gödel to Herbrand

Vienna, 25 July 1931

Dear Mr. Herbrand,

Thank you very much for your interesting letter, as well as for kindly sending ⟦me⟧ the reprints of your papers. I was already familiar with your "Thèses"[a] earlier, and the methods developed therein for consis-

spruchsfreiheitsbeweise scheinen mir sehr wichtig und bisher die einzigen zu sein, welche für ausgedehntere Systeme zu positiven Ergebnissen geführt haben.

Ich möchte jetzt auf die Frage der Formalisierbarkeit der intuition⟦ist⟧ischen Beweise in bestimmten formalen Systemen (etwa den Princ. Math.) eingehen, da hier eine Meinungsverschiedenheit zu bestehen scheint. Ich glaube, wofern man dieser Frage überhaupt einen präzisen Sinn zuerkennt (wegen der Undefinierbarkeit des Begriffs | "finiter Beweis" könnte man mit Recht daran zweifeln), kann der einzig korrekte Standpunkt nur der sein, daß man zugibt, darüber nichts zu wissen. Sie sind ja (ebenso wie ich) der Meinung, daß eine erschöpfende Definition für den Begriff "finiter Beweis" unmöglich ist; d. h. aber | Math.[b] (oder auch nur dem der Arithmetik $\mathrm{I} + 2 + 3'$ in Ihrem Brief) hinausführt, aber daraus folgt doch gar nichts über den weiteren Verlauf des Prozesses, den man in seiner ganzen Ausdehnung eben nicht überblickt. Auch Ihre Darstellung der finiten Arithmetik hat mich keineswegs vom Gegenteil überzeugt. Denn selbst wenn man zugibt, daß jeder intuit. Beweis in einem der Systeme $\mathrm{I} + 2' + 3$ F geführt werden kann (was mir durchaus nicht selbstverständlich erscheint[1]), so bleibt noch immer die Frage offen, ob die intuit. Beweise, welche jeweils erforderlich sind, um die Eindeutigkeit der Rekursionsaxiome sicherzustellen, sämtlich in den Princ. Math. formalisierbar sind. Selbstverständlich behaupte ich auch nicht, es sei sicher, daß irgendwelche finite Beweise in den | Princ. Math. nicht formalisierbar sind, wenn ich auch gefühlsmäßig eher zu dieser Annahme neige. Jedenfalls müßte ein in den Princ. Math. nicht formalisierbarer ~~Beweis~~ finiter Beweis ganz außerordentlich kompliziert sein und es besteht aus diesem rein praktischen Grunde sehr wenig Aussicht einen zu finden, aber das ändert nach meiner Meinung nichts an der prinzipiellen Möglichkeit.[b]

Zu Ihrer Note in den Comptes rendus vom 14. Oktober 1929[c] möchte ich noch folgendes bemerken: Ich vermute, daß Sie beim Beweise des Satzes ~~2.~~ ~~a. und~~ b.) unter 2. ~~die Voraussetzung benutzt ha~~ vorausgesetzt haben, daß der Widerspruchsfreiheitsbeweis und das Entscheidungsproblem mit den logischen ~~M~~ Mitteln der Princ. Math. gelöst sind. Nun ist

[1]Die von Brouwer u. Heyting als intuitionistisch einwandfrei zugelassenen Schlußweisen gehen ja sicher über die System⟦e⟧ $\mathrm{I} + 2' + 3$ F hinaus; aber selbst wenn man den Intuitionismus enger einschränk, wäre das möglich.

---

[b]The text of an alternate page 3 follows this letter; see below. Gödel changed his letter at this point significantly; that is indicated not only by the alternate text, but also by the fact that only half of the second page is actually used. It is reasonable to conjecture that Gödel left out part of a sentence in rewriting matters.

tency proofs seem to me to be very important and up to now the only ones which have led to positive results for more extended systems.

I would like now to enter into the question of the formalizability of intuitionistic proofs in certain formal systems (say that of *Principia mathematica*), since here there appears to be a difference of opinion. I think, insofar as this question admits a precise meaning at all (due to the undefinability of the notion | "finitary proof", that could justly be doubted), 2 the only correct standpoint can be that we admit not knowing anything about it. Indeed, you are of the opinion (just as I am) that an exhaustive definition of the notion "finitary proof" is impossible. That is, | ⟦to 3 say, it⟧ goes beyond the bounds ⟦of⟧ mathematics[b] (or also just that of the arithmetic I + 2 + 3′ in your letter); but from that nothing at all follows about the further course of the process, whose overall extent we just do not survey. Even your presentation of finitary arithmetic in no way convinces me otherwise. For even if we admit that every intuitionistic proof can be carried out in one of the systems I + 2′ + 3 F (which seems not at all obvious to me[1]), the question still always remains open whether the intuitionistic proofs that are required in each case to justify the unicity of the recursion axioms are all formalizable in *Principia mathematica*. Clearly I do not claim either that it is certain that some finitary proofs | are not formalizable in *Principia mathematica*, even though in- 4 tuitively I tend toward this assumption. In any case, a finitary proof not formalizable in *Principia mathematica* would have to be quite extraordinarily complicated, and on this purely practical ground there is very little prospect of finding one; but that, in my opinion, doesn't alter anything about the possibility in principle.

Concerning your note in the *Comptes rendus* of 14 October 1929,[c] I would like to add the following remark: I assume that in the proof of proposition b.) under 2. you have presupposed that the consistency proof and the decision problem are settled by the logical means of *Prin-*

[1]The rules of inference allowed as intuitionistically unobjectionable by Brouwer and Heyting certainly go beyond the system I + 2′ + 3 F; but even if intuitionism be more narrowly restricted, that would be possible.

---

[c]*Herbrand 1929.*

aber durch meine Arbeit gezeigt, daß diese Voraussetzung niemals zutreffen kann. Ein | Beweis des Satzes ~~a.)~~ b.) ohne die genannte Voraussetzung d. h. lediglich unter der Annahme, daß das Widerspruch⟦s⟧freiheits- und das Entscheidungsproblem auf intuit. Wege gelöst sind, scheint mir kaum möglich.

Nicht ganz verständlich war mir in Ihrem Brief die Bemerkung, daß zur Anwendung meiner Methoden auf ein arithmetisches System immer Funktionen erforderlich sind, die über dieses System hinausgehen. Dieses Hinausgehen ist doch wohl nur für den Widerspruchsfreiheitsbeweis erforderlich. Dagegen kann man den Satz: "Wenn das System widerspruchsfrei ist, dann ~~sind die~~ ist der von mir angegebene Satz unbeweisbar" innerhalb desselben Systems beweisen.

Bezüglich unseres Briefwechsels möchte ich Ihnen vorschlagen, daß jeder von uns in seiner Muttersprache | schreibt, um unnötige Mühe zu ersparen.[d]

Ihr ergebener Kurt Gödel

P.S. Ich übersende Ihnen heute gleichzeitig Separata meiner Arbeiten.

⟦The following is the text of the alternate page 3 mentioned in note b. Presumably this was replaced by page 3 as printed above.⟧

Math. ~~hi~~ (oder auch nur dem der Arithmetik) hinausführt und daß man auch heute keinen Weg sieht, auf dem er hinausführen könnte, aber daraus darf man doch nicht schließen, daß ein solcher Weg nicht existiert. Auch Ihre Darstellung der finiten Arithmetik hat mich keineswegs vom Gegenteil überzeugt. Denn selbst wenn man zugibt, daß jeder finite Beweis in einem der Systeme I, 2′, 3 F geführt werden kann (was mir durchaus nicht selbstverständlich erscheint), so bleibt noch immer die Frage offen, ob die intuit. Beweise, welche jeweils erforderlich sind, um die Eindeutigkeit der Rekursionsaxiome sicherzustellen, sämtlich in den Princ. Math. formalisierbar sind.

---

[d] The correspondence was not to continue. The only response received by Gödel was a letter from Herbrand's father dated 13 September 1931, the text of which reads:

> Le 26 Juillet vous avez écrit une longue lettre á mon fils, Jacques Herbrand.
>
> Le 27, mon fils est tombé au cours d'une excursion dans les Alpes et s'est tué.
>
> Ainsi j'explique son silence. Mon fils aurait été heureux, sans aucun doute, d'entretenir une correspondance avec vous sur le sujet qu'il ai incite.
>
> Veuillez croire, Monsieur, à mes sentiments respectueux et vivement attristis. ⟦Translation opposite⟧

cipia mathematica. But now, through my work it is shown that this presupposition can never apply. A | proof of proposition b.) without the 5 aforementioned presupposition, that is, solely under the assumption that the consistency problem and the decision problem are settled by intuitionistic means, hardly seems possible to me.

I did not completely understand the remark in your letter that for the application of my methods to an arithmetic system, functions going beyond that system are always required. Rather, this going beyond is only required for the consistency proof. By contrast, the proposition "If the system is consistent, then the proposition stated by me is unprovable" can be proved within that same system.

With regard to our correspondence, I would like to suggest that each of us write in his mother tongue, | in order to spare unnecessary trouble.[d] 6

I thank you again for your kind letter, the belated response to which I ask you to excuse.

With highest regards,

Yours sincerely, Kurt Gödel

P.S. I am sending you today, at the same time, reprints of my works.

⟦The following is the text of the alternate page 3 mentioned in note b. Presumably this was replaced by page 3 as printed above.⟧

math (or also just that of arithmetic), and that even today we see no way by which we can go beyond it; but from that we may not at all conclude that such a way does not exist. Even your presentation of finitary arithmetic in no way convinces me otherwise. For even if we admit that every intuitionistic proof can be carried out in one of the systems I, 2′, 3 F (which seems not at all obvious to me), the question still always remains open whether the intuitionistic proofs that are required in each case justify the unicity of the recursion axioms are all formalizable in *Principia mathematica.*

The 26th of July you wrote a long letter to my son, Jacques Herbrand.

The 27th my son fell during an excursion in the Alps and was killed.

Thus I explain his silence. My son would have been delighted, without any doubt, to keep up a correspondence with you on the subject that he had begun.

Please believe, Sir, my respectful and deeply sorrowful sentiments.

⟦Signature illegible⟧

Arend Heyting

Arend Heyting

Arend Heyting (1898–1980) is known as the first to produce formal systems for intuitionistic logic and mathematics.[a] As the successor of L. E. J. Brouwer in Amsterdam, he became the leader of intuitionism in the early post-war period. However, at the time of his more extended correspondence with Gödel in 1931–1933, Heyting was teaching secondary school in Enschede in the eastern Netherlands.

The correspondence between Heyting and Gödel consists of a number of letters between August 1931 and September 1933, some correspondence in 1957 (not included here) concerning an invitation to lecture at the IAS, and an exchange of single letters in 1969. The origin of the first correspondence lies in the interest of the editors of the review journal *Zentralblatt für Mathematik und ihre Grenzgebiete* in including in their new monograph series *Ergebnisse der Mathematik und ihrer Grenzgebiete* a survey of recent developments in the foundations of mathematics. Otto Neugebauer, then Managing Editor of the *Zentralblatt*, wrote to Heyting on 25 June 1931 inviting him to write this monograph.[b]

He invited Heyting to confine himself to intuitionism but said he would prefer for him to sketch "the whole of foundational research". Heyting was in principle ready to accept the invitation for a comprehensive survey; however, he wrote

[a] *Heyting 1930*, *1930a* and *1930c*, and for questions of the interpretation of the connectives, *1930b*, *1931* and *1934*. *Troelstra 1981* gives an account of Heyting's life and work, accompanied by the bibliography *Niekus, van Riemsdijk and Troelstra 1981*.

[b] Otto Neugebauer (1899–1990) began as a mathematician in Germany but already in the 1920s turned to the study of ancient, especially Babylonian, mathematics. He was the first managing editor of the *Zentralblatt* and continued in that role after being fired from his academic position in Göttingen for refusing to take an oath of loyalty to the Nazi regime; in the course of his correspondence with Gödel and Heyting he moved to Copenhagen. From 1939 on he was at Brown University, where he was a founding editor of *Mathematical reviews* and was a leader in the study of ancient mathematics and science. (See *Pyenson 1999*, on which the above is largely based. On his exit from the *Zentralblatt* and role in *Mathematical reviews*, see *Pitcher 1988*, pp. 71–75.)

It may be surprising that Heyting was asked to write this survey, since he had received his doctorate only in 1925 and did not have a university position (although his career path was not unusual for Dutch mathematicians at the time). It might be reasonable to conjecture that his talk at the Königsberg conference (*Heyting 1931*) made a good impression. Possibly he was suggested by B. L. van der Waerden, who was an editor of the *Zentralblatt*.

> Unfortunately I don't feel sufficiently at home in Russell's direction and related tendencies such as Wittgenstein and Ramsey, so that I would have to transfer that part to a collaborator.[c]

Details were to be discussed when Heyting visited Göttingen for a lecture on 21 July; possibly it was then that it was proposed to invite Gödel as the collaborator Heyting suggested. Neugebauer wrote that day to invite Gödel. On 5 August he wrote to Heyting that Gödel had accepted.

The correspondence from Heyting's letter of 22 August (letter 1) through the year 1932 consists of a matter-of-fact discussion of the organization of the essay, the division of labor between the two authors, and various specific issues. The reader notices, however, that although Heyting sent Gödel virtually the whole of his own contribution, Gödel at no point sent any of his to Heyting. And after November 1932 Gödel fell silent. Although he eventually replied in May 1933, the rest of what he wrote consists of excuses and proposals of ever more distant dates when his contribution was to be ready. Early in 1934 Heyting, Neugebauer and the publisher[d] gave up on Gödel, at least as a co-author with Heyting. They agreed on a separate publication, which appeared as the well-known *Heyting 1934*. The title, *Mathematische Grundlagenforschung. Intuitionismus. Beweistheorie*, shows that it was not the comprehensive survey that was originally planned.

The story has some interesting details. In his opening letter Heyting proposed an overall structure for the monograph. The elements of it that were to be his responsibility proved remarkably durable: nos. 4 (Intuitionism), 5 (Formalism), 6 (Other standpoints) and 8 (Mathematics and natural science) survive as chapter headings in *Heyting 1934*. At the end of the letter Heyting thanked Gödel for reprints that must have included *Gödel 1931*. His comment may indicate that he was less surprised by the suggestion that intuitionistic mathematics might contain methods of proof that are not available in Hilbert's formal system than by the fact that this was backed up by such a rigorous proof.

Heyting asked Gödel to undertake the first three chapters: a brief historical introduction, a chapter on the paradoxes and one on the "logical calculus" and logicism.[e] That Gödel should handle logicism was

[c] Letter to Neugebauer, 28 June 1931, my translation. The correspondence of Neugebauer and Heyting that is cited here is in Heyting's papers. That of Neugebauer and Gödel cited here is in Gödel's papers.

[d] The Verlag von Julius Springer, predecessor of the present Springer-Verlag.

[e] Throughout I have rendered *Abschnitt* as "chapter", although it would be more literally rendered as "section". The *Abschnitte* of *Heyting 1934* would in all probability have been called chapters had the work been written in English. More generally, an *Abschnitt* in a German work is very often a larger division than what is usually called a section in English, which is often closer to the German *Paragraph*.

evidently part of the initial arrangement. But he resisted the idea of an initial chapter on the paradoxes, proposing instead that they should be treated in the chapter on logicism "in whose development they play a decisive role" (letter 3). He may have been already inclined to play down the importance of the paradoxes for the foundations of mathematics in general. He proposed a separate chapter on metamathematical results, many of which he clearly thought did not fit neatly into the three tendencies of logicism, intuitionism and formalism. That chapter would include his own work. Heyting agreed with Gödel's proposals, dissenting only on a minor point concerning the discussion of Poincaré's views (letter 4). The result was basically the division of the monograph that was envisaged for the rest of their collaboration (see letter 9). There is no further correspondence until mid-1932.

The proposed chapter on metamathematics posed an organizational problem, where its boundary with that on "formalism" should fall, since the consistency proofs the Hilbert school aimed at were metamathematical results, and Gödel's own incompleteness theorems bore on the Hilbert program. This issue led to some discussion in the summer of 1932 (letters 5–9) and seems to have been resolved by a compromise: Heyting would present the consistency proofs more or less directly coming from the Hilbert school (*Ackermann 1924* and *von Neumann 1927*), while Gödel would take up that of *Herbrand 1931*, the comparison of different proofs, and the relevance of the incompleteness theorems.

In three installments in August and October 1932, Heyting sent Gödel the chapters on Formalism and Intuitionism and what he thought he should write on other tendencies and on mathematics and science. However, the discussion of "other tendencies" was minimal; he says he has discussed only two views held by mathematicians, one of them being Gerrit Mannoury.[f] On 15 November Gödel responded with extensive comments (letter 14). He also says he has written about half of his part, the chapter on logicism and a little of the chapter on metamathematics. But he apologizes for not being able to send any of it to Heyting because it exists only in shorthand.[g] The most significant of Gödel's comments concern consistency proofs. He remarks that Ackermann's proof has "nowhere ⟦been⟧ presented in corrected form" and that it "cannot be adequate for ⟦full⟧ number theory". He interprets what he has heard about what must have been Bernays's lecture at the International Congress of Mathematicians in Zürich in September as

[f]The other must have been the geometer Moritz Pasch; see *Heyting 1934*, chapter 3, §3. It is likely that this chapter was essentially unchanged in *Heyting 1934*.

[g]The editors of volume III of these *Works* were not able to construct from what was found in Gödel's *Nachlaß* a text that would be organized enough to be considered for publication. Cf. *Dawson 1997*, p. 85.

admitting that "the consistency of number theory has up to now been proved only with a restriction on complete induction given by Herbrand", that is, that only quantifier-free instances of induction be admitted.[h] Commenting on a place in the discussion of intuitionism, Gödel remarks that he has only proved that "for every system, for which one has a finitary consistency proof, there are unprovable propositions that are intuitionistically provable." A system without such a consistency proof might contain the whole of intuitionistic mathematics. One might ask why Gödel did not substitute "intuitionistic consistency proof". It would then follow that if a formal system does contain all of intuitionistic mathematics, then its consistency cannot be proved intuitionistically.

In the letter Gödel says he will not be able to supply the brief discussion of logicism and the application of mathematics to science that Heyting had asked for, questions the necessity of the historical introduction that had been planned originally, and also says it seems to him that attempts to solve the paradoxes outside the three main tendencies are "scarcely of interest", although he says he has discussed Behmann's in the chapter on logicism.[i]

These remarks suggest an attempt to reduce his commitments. From a later letter (letter 18, 16 May 1933) we learn that Gödel had done little on the project since the beginning of the winter semester in Vienna the previous October. Letter 18 was, as noted above, the first word Heyting had from Gödel since the November letter. He says he will try to hasten to finish his work, since he is to go to Princeton for a year. In an apparently lost letter from September (see Heyting's letter 20, 30 September) he may have broached the possibility that Heyting might publish his part separately,[j] but he seems also to have said that he could have his part ready by the beginning of 1934, which Heyting found acceptable.

Heyting's reply was the end of their known correspondence on the matter. The rest of the story can be learned from correspondence of the two with Neugebauer. On 26 September 1933 he wrote to Heyting that he too had received Gödel's promise "to spare no effort to get his part finished by the end of the year". But on 3 January 1934 he wrote that after some dunning Gödel had written that he would not be able to finish the work in America but, after his return in May, would be able

[h]*See Herbrand 1931*, p. 5 or *van Heijenoort 1967*, p. 624. In his published abstract Bernays states just that about the validity of Ackermann's and von Neumann's proofs (*1933*, p. 342).

[i]Cf. *Behmann 1931* and the introductory note to the correspondence with Behmann in volume IV of these *Works*.

[j]Neugebauer wrote Gödel on 26 September opposing this suggestion.

to deliver the manuscript in July. This prompted Neugebauer to sound out Heyting on the question of publishing his part separately. Heyting seems to have been favorable, but the publisher resisted.[k] Before this was resolved, Neugebauer wrote on 10 March that he would prefer to boot Gödel out completely (*ganz herausschmeißen*). On 14 March he confirmed that the publisher had accepted a separate publication. Heyting may well have done by then what was necessary to make his manuscript sufficiently self-contained, since by September *Heyting 1934* had appeared.[l]

Although the collaboration of Gödel and Heyting ended early in 1934, the story does not quite end there. Someone at Springer wrote to Gödel a few times in 1934 reminding him of his commitment to what was now to be a separate publication. On 11 December 1934 Neugebauer wrote to him to the same effect, asking for a progress report. It seems that Gödel did not reply to any of these letters. After a few more letters from Springer, Neugebauer wrote on 24 May 1935 asking Gödel for a definitive decision. Gödel replied quickly that he could not supply what he had agreed to in the foreseeable future. He could in a relatively short time produce about twenty pages on consistency proofs and suggested asking Tarski and Carnap to write about metamathematics and logicism, respectively. This suggestion was apparently not followed up.

That *Heyting 1934*, widely read in its time, was a torso was not so far as I know much noticed. The introduction does end with the following statement:

> *Not* treated are *first*, the logicist construction of mathematics, *second*, the problems of pure logic such as the decision problem, *third*, those general metamathematical investigations that take quite arbitrary calculi as their object. A special review of these matters is planned for this series (p. 2, my translation).

Later, the relevance of Gödel's theorem to the project of consistency proofs is briefly discussed, and Heyting says that "a more detailed discussion of his methods must be reserved for the special report on logicism".[m] Other traces of the original project can no doubt be found. In spite of Gödel's comments, the statements on the scope of the existing consistency proofs are not very clear. Heyting does remark that "a complete proof of the consistency of number theory has not been

[k]Neugebauer to Heyting, 12 January 1934.

[l]The Harvard library copy bears a stamp, "Harvard College Library, Sep 18 1934." That is very likely the date of receipt.

[m]P. 47. The word rendered "logicism" is *Logistik*, which is often better translated simply as "symbolic logic". Heyting may well have been exploiting this ambiguity, given what Gödel's part of the monograph was intended to cover.

published up to now" (p. 47), but his remarks on von Neumann's proof two pages later leave the impression that it does cover all of first-order arithmetic. Heyting adopts Gödel's formulation of what his second incompleteness theorem implies about intuitionistic provability (p. 56), but the context is not that of the possibility of a formalism encompassing all of intuitionistic mathematics. He did not follow Gödel's advice to give examples of classical arguments that would not be intuitionistically acceptable.

Someone familiar with Gödel's biography will be inclined to compare the unsuccessful collaboration with Heyting with other episodes in which Gödel either after a long time failed to deliver promised work (such as the paper on Carnap for *Schilpp 1963*[n]), delivered it late after agonizing about details (as in the case of *Gödel 1944*[o]), engaged in long negotiation, or got into actual conflict about the publication of his own writing, as in the case of the failed project to publish *1931* and *1934* with *Nagel and Newman 1958*.[p] We should note, however, some significant differences between the present situation and these others. The correspondence up until November 1932 is straightforward enough. If Gödel was held back either by exceptional meticulousness or by the hesitation about some philosophical issues that he displayed later, he did not reveal it. It is true that the state of the manuscript found in his *Nachlaß* (see note g) shows that he had accomplished less than the reader of his letter of 15 November would think. But one explanation for his lack of progress would be that he was simply busy with other tasks. In the years between 1931 and 1934 he published a remarkable number of papers given his restraint about publication in general, and he had a number of other tasks including working to become a *Privatdozent*. Given the immediate pressures of his environment, a general expository work probably simply did not assume high priority. It does seem clear that, for whatever reason, he did not wish to work on the project during his time at the Institute for Advanced Study, when the pressure of the Vienna environment could be expected to be much less. Evidently he did not work on it in the fall, and on writing to Neugebauer at the beginning of 1934 he promised only to return to the task after the end of his stay in America. In the fall he did, however, write another important expository work, the lecture **1933o*. That text may suggest a way in which the project with Heyting could have been uncongenial:

[n]See, in the correspondence with Paul Arthur Schilpp in this volume, the portion from 1953 to 1959.

[o]See the portion of the same correspondence in 1942 and 1943.

[p]See the correspondence with Ernest Nagel and its introductory note in this volume, as well as the related correspondence with Allan Angoff and Wilson Follett in volume IV of these *Works*.

In **1933o* he gives to the problem of the foundations of mathematics an organization that is his own, in which the framing of mathematics in set theory and Hilbert's program are the main organizing conceptions. In the project with Heyting he had to adapt to a structure imposed by existing work in foundations and a then standard classification into logicist, intuitionist, and formalist tendencies, as well as to Heyting's intention to deal with "formalism" himself. The foundations of set theory were reserved for another monograph in the *Ergebnisse* series. Furthermore, that collaborative writing was not congenial to Gödel is suggested by the fact that he published only one jointly authored paper in his lifetime, and that a short note.[q] Considerations such as these may well be sufficient to explain Gödel's failure to carry through the enterprise with Heyting.

On 2 January 1969, after Heyting's retirement from Amsterdam, he wrote to Gödel (letter 21) on behalf of the editors of the series Studies in Logic and the Foundations of Mathematics, published by North-Holland. He asks if it is true, as they have been told, that Gödel is interested in publication of his collected works, and urges him to "open negotiations on publication in our series". Gödel's reply denies that he has been considering such an edition, and expresses coolness about the idea because "practically all my papers (and, at any rate, all of my important papers) are readily available", at least since the appearance of *Benacerraf and Putnam 1964*, *Davis 1965* and *van Heijenoort 1967*. It is hard to know to what extent this reply reflected Gödel's real reasons for being cool to the proposal. In any event, no further attempt seems to have been made to persuade him.[r]

Charles Parsons

A complete calendar of the correspondence with Heyting occurs on pp. 447–448 of this volume. The editors wish to thank Delia Graff for preparing a typescript of the correspondence and Mark van Atten for checking our readings of Gödel's letters against the originals in the Rijksarchief Noord-Holland. The translation is by John W. Dawson, Jr., revised using suggestions of Charles Parsons and Thomas Teufel.

[q] *Gödel 1933h*, with Karl Menger and Abraham Wald. Dawson notes the relevance of this fact in his account of the episode (*1997*, p. 86).

[r] I am indebted to John Dawson for helpful comments and to Dawson and Ti-Grace Atkinson for assistance.

1. Heyting to Gödel

ENSCHEDE, den 22. August 1931

Lieber Herr Gödel,

Mit Freude erfuhr ich durch Herrn Neugebauer, dass Sie bereit sind, mit mir zusammen den Bericht über Grundlagenforschung zu schreiben. Wie denken Sie über die Einteilung und Abgrenzung des Stoffes, wie sie die folgenden Stichwörter andeuten?

1. Kurze historische Einleitung. (Poincaré).
2. Die Paradoxien; Klärungsversuche ausserhalb der drei Hauptrichtungen.
3. Der Logikkalkül; seine Weiterführung (Amerikaner); der Logizismus.
4. Der Intuitionismus.
5. Der Formalismus.
6. Andere Standpunkte.
7. Beziehungen zwischen den verschiedenen Richtungen.
8. Mathematik und Naturwissenschaft.

Ich möchte Sie bitten, die ersten drei Abschnitte zu übernehmen; natürlich ist die Einteilung als sehr vorläufig gemeint und es ist durchaus nicht meine Absicht, Ihre Tätigkeit auf die genannten Abschnitte zu beschränken, wie auch ich mir vorbehalten möchte, die Grenze gelegentlich zu überschreiten. Leider kann ich der Versammlung in Bad Elster nicht beiwohnen, so dass eine mündliche Besprechung wohl nicht möglich sein wird.

Eingeschlossen sende ich Ihnen einen Vertragsentwurf, den ich vom Verlag erhalten habe. Die Zeit bis zum 1. April ist mir viel zu kurz; Herr Neugebauer sprach zuerst von einem Jahr und ich möchte es nicht kürzer nehmen. Wenn Sie noch Bemerkungen über diesen Vertrag haben, bitte ich sie mir mitzuteilen; wollen Sie ihn sonst dem Verlag (Berlin W9, Linkstrasse 23/24) zurückschicken?

2 | Ich bedenke, dass ich Ihnen noch für die Zusendung der Separate Ihrer Arbeiten danken muss; ich brauche nicht zu sagen, dass sie mich sehr interessiert haben. Wenig konnte ich in Königsberg vermuten, dass Ihre Bemerkung, es könne in der intuitionistischen Mathematik Beweismittel geben, die in Hilberts System nicht vorkommen, einen so tiefen Hintergrund verbarg.

Mit besten Grüssen,

Ihr sehr ergebener

A. Heyting

1. Heyting to Gödel

ENSCHEDE, 22 August 1931

Dear Mr. Gödel,

I was delighted to learn from Mr. Neugebauer that you are prepared to write the report on foundational research together with me. What do you think of the division and delimitation of the subject matter suggested by the following key words?

1. Short historical introduction. (Poincaré).
2. The paradoxes; attempts at resolution apart from the three principal directions.
3. The calculus of logic; its further development (Americans); logicism.
4. Intuitionism.
5. Formalism.
6. Other standpoints.
7. Relations between the different directions.
8. Mathematics and natural science.

I would like to ask you to undertake the first three chapters; of course, the division is intended as very provisional and it is not at all my intention to restrict your activity to the aforementioned chapters, as I would also like to reserve for myself ⟦the right⟧ occasionally to overstep the boundaries. Unfortunately I cannot attend the meeting in Bad Elster, so that an oral consultation will probably not be possible.

I am sending you, enclosed, a draft contract that I received from the publisher. The time between now and 1 April is much too short for me; Neugebauer initially spoke of a year, and I would not like to make it any shorter. If you have further remarks about this contract, I ask that you communicate them to me; otherwise, would you send it back to the publisher (Berlin W9, Linkstrasse 23/24)?

I remember that I still must thank you for sending the offprints of your works; I need not say that they have interested me very much. Little could I imagine in Königsberg that your remark that there could be means of proof in intuitionistic mathematics that do not occur in Hilbert's system concealed such a deep background.

With best wishes,

Yours very sincerely,

A. Heyting

2. Gödel to Heyting. Draft.[a]

Wien /VIII. 1931

Sehr geehrter Herr Heyting!

Wie Ihnen Herr Neugebauer bereits mitgeteilt haben dürfte, habe ich mich bereit erklärt, den Bericht über die Grundlagenforschung der Mathematik für das Zentralblatt in Zusammenarbeit mit Ihnen zu schreiben, und ich möchte nun die Aufteilung des Stoffes mit Ihnen besprechen.

Da es ja feststeht, daß der Intuitionismus von Ihnen u. der Logizismus von mir behandelt wird, so handelt es sich hauptsächlich dann, von wem die formalistischen (metamathematischen) Arbeiten der Göttinger u. der Warschauer Schule besprochen werden sollen. Ich ersuche Sie mir Ihre diesbezüglichen Absichten mitzuteilen; ich selbst bin der Meinung, daß die Metamathematik | (als Theorie der Sprachformen) sich von der Logistik nur schwer trennen läßt u. schlage daher vor, daß ich sie zur Bearbeitung übernehme. Eine Ausnahme würde selbstverständlich die Formalisierung des Intuitionismus bilden, welche von Ihnen zu behandeln wäre.

2

Was die halbintuitionistishe Richtung (Borel, Weyl "Das Kontinuum"[b] etc.) betrifft, so möchte ich gleichfalls vorschlagen, daß sie von Ihnen behandelt wird.

---

[a]It appears that this draft is of a letter that was never sent. Nothing that would be a finished version was preserved in Heyting's papers. No reference is made to it in the subsequent correspondence. The previous letter shows that by 22 August Heyting had already been informed by Neugebauer of Gödel's agreement to the collaboration; he

3. Gödel to Heyting

Wien 3./IX. 1931

Lieber Herr Heyting!

Besten Dank für Ihr Schreiben von 22./VIII. Mit der geplanten Disposition unseres Artikels bin ich im wesentlichen einverstanden, doch möchte ich folgendes bemerken: Ich glaube, die Paradoxien sollte man nicht am Anfang u. in einem eigenen Abschnitt behandeln, sondern in

2. Gödel to Heyting[a]

Vienna, August 1931

Dear Mr. Heyting,

As Mr. Neugebauer may have already informed you, I have declared myself ready to collaborate with you in writing the report on research in the foundations of mathematics for the Zentralblatt, and now I would like to discuss the apportionment of the subject matter with you.

Since it's understood that intuitionism will be handled by you and logicism by me, it is principally the question, by whom should the formalist (metamathematical) works of the Göttingen and Warsaw schools be discussed. I request that you inform me of your intentions in that regard; I myself am of the opinion that metamathematics (as the theory of linguistic forms) may only with difficulty be separated from logistic, and I therefore propose that I undertake to work on it. The formalization of intuitionism would of course constitute an exception, which would be treated by you.

As for the semi-intuitionistic direction (Borel, Weyl's "Das Kontinuum",[b] etc.), I would likewise propose that it be treated by you.

---

makes no reference to having heard about the matter directly from Gödel. Most likely Gödel received Heyting's letter before sending the present one and then thought his draft outdated.

[b] *Weyl 1918.*

3. Gödel to Heyting

Vienna, 3 September 1931

Dear Mr. Heyting,

Thank you for your letter of 22 August. I am essentially in agreement with the planned disposition of our article, yet I would like to note the following: I think the paradoxes should not be treated at the beginning and in a separate chapter, but rather in the chapter on logicism,

dem Abschnitt über Logizismus, in dessen Entwicklung sie eine entscheidende Rolle spielen. Die Lösungsversuche außerhalb der 3 Richtungen wären dann in Abschn. 6 zu behandeln, wo sie sich gut einfügen würden. Auch die Poincaré-sche Kritik (imprädikat. Funktionen etc.) möchte ich lieber in dem Abschnitt über Logizismus behandeln. Im ersten Abschn. möchte ich statt (oder neben) einem kurzen historischen | Überblick einiges über die Ziele u. Probleme der Grundlagenforschung im allgemeinen sagen.

2

Nicht ganz im klaren bin ich mir, wo die gerade in den letzten Jahren an Bedeutung zunehmenden "metamathematischen" Arbeiten einzureihen wären, welche ~~nicht~~ die Theorie des Kalküls als Selbstzweck betreiben u. sowohl mit dem Logizismus als mit dem Formalismus Berührungspunkte haben. Ich denke dabei z. B. an die Arbeiten von Bernays u. Post über den Aussagenkalkül, ferner an Tarski, Łukasiewicz, Le~~s~~niewski, an Skolem zum Teil auch Herbrand u. a.

Ł ś

Da auch meine Arbeiten in dieses Gebiet fallen, möchte ich es gerne zur Behandlung übernehmen und zwar ~~entweder~~ in einem eigenen Abschnitt ~~oder in ein~~ anschließend an den Logizismus. In dem Abschnitt über Formalismus wären dann die prinzipiellen Fragen des formalistischen Stand|punktes, besonders die Widerspruchsfreiheitsbeweise zu besprechen. Ich möchte Sie bitten, mir Ihre Ansicht über diesen Punkt mitzuteilen.

3

Mit dem Vertragsentwurf bin ich einverstanden u. sende ihn an Herrn Neugebauer weiter.

Ich sehe mit Interesse Ihrer Antwort entgegen und bin mit den besten Grüßen

Ihr ergebener Kurt Gödel

P.S. Vielen Dank für die seinerzeit übersandten Separata.

4. Heyting to Gödel

Enschede, 24. IX. 1931.

Lieber Herr Gödel,

Besten Dank für Ihren wertvollen Brief vom 3. d. M.; ich bitte zu entschuldigen, dass durch Arbeit anderer Art meine Antwort etwas verzögert ist. Sie sind natürlich frei in der Einteilung der Abschnitte, die Sie

in whose development they play a decisive role. The attempts at resolution apart from the three ⟦principal⟧ directions would on that account be treated in chapter 6, where they would fit in well. I would also rather treat Poincaré's criticism (impredicative functions, etc.) in the chapter on logicism. In the first chapter I would like to say a bit about the aims and problems of foundational research instead of (or in addition to) a short historical overview.

I am not quite clear where to subsume those "metamathematical" works, of increasing significance just in the last few years, that pursue the theory of the ⟦logical⟧ calculus as an aim in itself and that have points of contact with logicism as well as with formalism. I am thinking thereby, for example, of the works of Bernays and Post on the propositional calculus, and furthermore of ⟦the work⟧ of Tarski, Łukasiewicz, Leśniewski, part of ⟦the work of⟧ Skolem, and also Herbrand and others.

Since my own work also falls in this area, I would be glad to undertake to treat it, and in fact in a chapter of its own adjoined to the one on logicism. In the chapter on formalism the fundamental questions of the formalistic standpoint, especially the consistency proof, would then be discussed. I would like to ask you to inform me of your view about that point.

I am in agreement with the draft contract and am sending it on to Mr. Neugebauer.

I look forward with interest to your reply and am, with best wishes,

Sincerely yours,

Kurt Gödel

P.S. Many thanks for the offprints transmitted previously.

4. Heyting to Gödel

Enschede, 24 September 1931

Dear Mr. Gödel,

Thanks very much for your valuable letter of the 3rd; I apologize for the fact that my reply is somewhat delayed by work of another sort. You are of course unconstrained in the division of the chapters that you treat;

behandeln; meine Einteilung war nur zum Zweck der Abgrenzung entworfen. Schreiben Sie bitte auch den Abschnitt über die "Metamathematiker"; wir können später entscheiden, ob dieser sich gleich nach dem Logizismus oder besser nach dem Formalismus einreihen lässt. Auch mit der Behandlung der Paradoxien, wie Sie sich diese denken, bin ich einverstanden.

Poincaré hat durch seine Kritik nicht nur den Logizismus, sondern auch durch die Forderung der Widerspruchslosigkeit den Formalismus und durch die Betonung der vollständigen Induktion als intuitives Moment in mathematischen Denken den Intuitionismus stark beeinflusst. Ich glaube ~~nicht~~, dass man seiner Bedeutung nur gerecht werden kann 2 durch eine einheitliche Darstellung | seines Standpunktes; daneben braucht eine eingehendere Behandlung seiner Gedanken, wo diese die verschiedenen Richtungen beeinflusst haben, in den betreffenden Abschnitten, keine Wiederholungen zu ergeben. Ich muss aber gestehen, dass es mir durch Mangel an historischem Sinn schwer fallen würde, diesen Abschnitt über Poincaré zu verfassen; ich hoffe darum, dass Sie ihn noch in ihre Einleitung einfügen wollen.

Mit Besten Grüssen,

Ihr sehr ergebener

A. Heyting

5. Heyting to Gödel

Enschede, 11 Juni 1932.
Parkstraat 25.

Lieber Herr Gödel,

Ich möchte mich noch über einige Einzelheiten unseres Artikels mit Ihnen besprechen. Erstens möchte ich wissen, ob Sie damit einverstanden sind, die Grundgedanken jeder Richtung ziemlich ausführlich zu beschreiben und anschliessend möglichst zusammenhängend aber kurz über die einschlägigen Arbeiten zu berichten.

Zweitens kann ich einige Arbeiten nicht mit Sicherheit einteilen. Das gilt für diejenigen Arbeiten, die ohne bestimmte philosophische Voraussetzungen die formale Logik behandeln, wie z. B. die Amerikanischen Untersuchungen über Axiomatik der "Boolean algebra", die Arbeiten von

my division was only drawn up with the aim of delimiting boundaries. Please do also write the chapter on "metamathematicians"; we can decide later whether that is to be placed directly after logicism or ⟦would be⟧ better after formalism. I am also in agreement with the treatment of the paradoxes as you conceive of it.

Through his criticism Poincaré not only strongly influenced logicism, but also formalism, through the demand for consistency, and intuitionism, through the emphasis on complete induction as the intuitive factor in mathematical thought. I think one can do justice to his significance only through a unified presentation of his viewpoint; moreover, a detailed treatment of his ideas, where they have influenced the various directions, need not result in repetition in the corresponding chapters. But I must confess that for want of historical sensibility, I would find it difficult to write that chapter on Poincaré. I hope, therefore, that you are still willing to fit it in to your preface.

With best wishes,

Yours very sincerely,

A. Heyting

5. Heyting to Gödel

Enschede, 11 June 1932
Parkstraat 25

Dear Mr. Gödel,

I would still like to discuss with you some details of our article. First, I would like to know whether you agree with the proposal of reporting the basic ideas of each direction rather thoroughly and then ⟦reporting⟧ on the relevant works briefly, linking them as coherently as possible.

Second, I cannot classify a few works with certainty. That is the case for those works that treat formal logic without definite philosophical presuppositions, as for example the American investigations about axiomatics of "Boolean algebra", the works of Bernays and Schönfinkel

Bernays u. Schönfinkel und Herbrand[a] über das Entscheidungsproblem, usw. Ich habe den Eindruck, dass diese sich am besten im Anschluss an den Logizismus behandeln lassen. Auch Hilbert–Ackermann, "Theoretische Logik" gehört zu dieser Gruppe.

Rechnen Sie Chwistek zu den Logizisten oder wollen Sie ihn unter 2 "Andere Standpunkte" bringen? In diesem | Fall möchte ich Sie bitten, den betr. Paragrafen zu schreiben; sein Standpunkt und seine Symbolik sind mir schwer verständlich.

Dankbar wäre ich Ihnen für einige Seiten über das Verhältnis des Logizismus zur Naturwissenschaft, die ich in das Schlusskapitel einschalten könnte.

Herr Neugebauer wünscht am Schluss ein Literaturverzeichnis "mit einer eineindeutigen Beziehung der Arbeiten auf die ganzen Zahlen" (ich vermute: positiv und unterhalb einer endlichen Schranke). Ich schlage vor, die Arbeiten nach Abschnitten einzuteilen, die Nummer 1–20 für allgemeine Werke zu reservieren, 21–200 für Ihre Abschnitte; wenn Sie mehr Raum brauchen, bitte ich um Nachricht, sonst fange ich mit 201 an.

Ich habe die Absicht in September nach Zürich zu kommen und hoffe dort Ihnen wieder zu begegnen.[b]

Mit freundlichen Grüssen,

Ihr sehr ergebener

A. Heyting

---

[a]Evidently *Bernays and Schönfinkel 1928* and *Herbrand 1931a.*

6. Gödel to Heyting

Wien 1./VII. 1932

Lieber Herr Heyting!

Besten Dank für Ihren Brief vom 11./VI. Die von Ihnen erwähnten und andere Arbeiten, welche keine bestimmte philosophische Einstellung voraussetzen, möchte ich in einem eigenen Abschnitt behandeln, der am besten nach den beiden Abschn. über Logizismus u. Formalismus einzuschalten wäre. In diesem Abschnitt möchte ich auch die Widerspruchfrei-

and Herbrand[a] on the decision problem, and so on. I have the impression that these are best treated following logicism. Hilbert–Ackermann's "Theoretische Logik" also belongs to this group.

Do you reckon Chwistek among the logicists, or do you want to place him under "other viewpoints"? In that case I would like to ask you to write the corresponding sections; his viewpoint and his symbolism are hard for me to understand.

I would be grateful to you for a few pages on the relation of logicism to natural science, which I could insert into the final chapter.

Neugebauer desires a bibliography at the end "with a one-to-one relation of the works to the whole numbers" (I suppose: positive and below a finite bound). I propose to subdivide the works according to chapters, reserving the numbers 1–20 for general works and 21–200 for your chapters; if you need more space, let me know, otherwise I will begin with 201.

I have the intention of coming to Zürich in September and hope to meet you again there.[b]

With cordial greetings,

Yours very sincerely,

A. Heyting

---

[b]Heyting evidently refers to the International Congress of Mathematicians which took place in Zürich on 4–11 September 1932. Gödel did not attend, as his remark at the end of the following letter indicates.

6. Gödel to Heyting

Vienna, 1 July 1932

Dear Mr. Heyting,

Thank you very much for your letter of 11 June. I would like to treat the works you mentioned, and other works that presuppose no definite philosophical focus, in a chapter of their own, which would best be inserted after the two chapters on logicism and formalism. In that chapter I would also like to discuss consistency proofs (insofar as such exist),

heitsbeweise (soweit solche existieren) bzw. die Unmöglichkeit der Widerspruchfreiheitsbeweise besprechen. Denn diese Fragen sind methodisch so nahe mit dem Entscheidungs- u. Vollständigkeitsproblem verwandt (vgl. z. B. Herbrand, Thèses[a]), daß mir eine Trennung sehr künstlich vorkommt. In dem von Ihnen zu schreibenden Abschn. über Formalismus wären dann die Grundgedanken u. die philosophische Einstellung
2 | dieser Richtung samt den darauf bezüglichen Arbeiten zu besprechen. Ebenso würde ich in dem Abschn. über Logizismus nur die prinzipiell wichtigen Fragen (Antinomien, Typentheorie, Reduzibilitätsaxiom u. Versuche zu seiner Ausschaltung etc.) besprechen u. die axiomatischen Untersuchungen über den Aussagen, Funktionenkalkül u.s.w. in den neuen Abschn. aufnehmen. In diesem wäre dann alles zusammengestellt, was vom rein mathematischen Standpunkt an Logizismus u. Formalismus interessant ist. Chwistek muß man wohl zu den Logizisten rechnen (er dürfte übrigens für jeden Menschen schwer verständlich sein).—Zwischen sachlicher Besprechung u. Referaten über die Arbeiten, möchte ich keine scharfe Trennung vornehmen, doch will ich Sie natürlich in keiner Weise hindern es in Ihren Abschnitten zu tun. Die Nummern von 21–200 genügen mir vollkommen.

3 | Die Besprechung des Verhältnisses von Logizismus u. Naturwissenschaften kann ich wohl übernnehmen; ich glaube aber es würde über das Thema "Grundlagen der Mathematik" hinausgehen, näher auf die einzelnen Probleme (deren vollständige Klärung übrigens zum Teil noch aussteht) einzugehen u. möchte mich daher auf eine Darstellung der Hauptgesichtspunkte auf etwa 3–4 Maschinschreibseiten beschränken.

Nach Zürich werde ich aller Wahrscheinlichkeit nach leider nicht kommen können.

Es grüßt Sie bestens

Ihr sehr ergebener Kurt Gödel

---

[a] *Herbrand 1930*.

or respectively, the impossibility of consistency proofs. For methodologically these questions are so closely related to the decision problem and the completeness problem (see, e.g., Herbrand's Thesis[a]) that a separation seems to me very artificial. In the chapter on formalism, to be written by you, the fundamental ideas and the philosophical position of that direction would then be discussed, together with the works relating to it. Similarly, in the chapter on logicism I would discuss only the important questions of principle (antinomies, type theory, axiom of reducibility and attempts toward its elimination, etc.) and take up the axiomatic investigations about the propositional and functional calculi, and so on, in the new chapter. In this ⟦chapter⟧ everything which, from a purely mathematical standpoint, is interesting in logicism and formalism would then be placed together. Chwistek ought probably to be reckoned among the logicists (he might, by the way, be hard for everyone to understand).—I would like to make no sharp separation between substantive discussion and reports on the works, but of course I do not want to hinder you in any way from doing so in your chapters. The numbers from 21–200 are entirely sufficient for me.

I can probably undertake the discussion of the relation of logicism and natural science; but I believe it would go beyond the theme "foundations of mathematics" to go in more detail into the particular problems (whose complete clarification, after all, is in part still outstanding) and I would therefore like to restrict myself to a presentation of the principal points of view in perhaps 3–4 typewritten pages.

In all probability I will unfortunately not be able to come to Zürich.

Best wishes to you,

Yours very sincerely,

Kurt Gödel

7. Heyting to Gödel

Enschede, 17 Juli 1932
Parkstraat 25.

Lieber Herr Gödel,

Besten Dank für Ihren Brief. Ihr Vorschlag, die Widerspruchfreiheitsbeweise in dem Abschnitt über Metamathematik zu behandeln, hat vieles Anziehende, aber auch Nachteile. Wenn ich die Beweise selbst streichen soll, ist es schwierig zu verhelfen, dass meine Betrachtungen über ihren Sinn und Tragweite, über die Notwendigkeit intuitionistischer Strenge usw. in der Luft schweben. Aus zwei Gründen stimme ich dennoch Ihrem Vorschlag zu. Erstens ist es begreiflich, dass der Gegenstand Sie anzieht; zweitens befriedigt mich meine Darstellung nicht. Es ist mir nicht gelungen, in genügender Vollständigkeit zu entscheiden, an welchen Punkten je einer dieser Beweise weiter führt als die andern. Auch ohne die gegenseitige Abhängigkeit der als widerspruchslos erwiesenen Axiomensysteme vollständig zu untersuchen, hätte ich doch die Teile der klassischen Mathematik die sie aufzubauen gestatten, genauer umgrenzen mögen. Ich darf wohl hoffen, dass Sie darin | weiter kommen als ich. Wenn auch Sie die Beweise nur so nebeneinander stellen könnten, würde ich die Vorteile, sie in einem Abschnitt zu vereinigen, nicht einsehen.

Der Paragraph über das Verhältnis von Logizismus und Naturwissenschaft, wie Sie sich ihn denken, entspricht genau meinem Wunsch.

Mit besten Grüssen,

Ihr sehr ergebener

A. Heyting

Vom 23. Juli bis zum 3. August:
Banstraat 15, Amsterdam Z.

8. Gödel to Heyting

Wien 20./VII. $\overline{32}$.

Lieber Herr Heyting!

Aus Ihrem letzten Briefe glaube ich zu entnehmen, daß Sie den Abschnitt über Formalismus schon fertiggestellt haben, und ich will Sie dann natürlich keinesfalls veranlassen, eine Umarbeitung vorzunehmen,

7. Heyting to Gödel

Enschede, 17 July 1932
Parkstraat 25

Dear Mr. Gödel,

Thanks very much for your letter. Your proposal to treat the consistency proofs in the chapter on metamathematics has many attractions, but also drawbacks. If I should delete the proofs themselves, it would be hard to prevent my considerations about their meaning and significance, about the necessity of intuitionistic rigor, etc. from floating in the air. Nevertheless, I consent to your proposal for two reasons. First, it is understandable that the subject attracts you; second, my own presentation doesn't please me. I have not succeeded in determining with sufficient completeness at which points any one of these proofs goes further than the others. Also, without completely investigating the mutual dependence of the axiom systems shown to be consistent, I would still have liked to delimit more precisely the parts of classical mathematics that they allow to be built up. I may perhaps hope that you will advance further in the matter than I. If you too were able only to so juxtapose the proofs, I would not see the advantage of uniting them in one chapter.

The section on the relation of logicism and natural science, as you conceive it, corresponds exactly to my wish.

With best wishes,

Yours very sincerely,

A. Heyting

From 23 July until 3 August:
Banstraat 15, Amsterdam Z.

8. Gödel to Heyting

Vienna, 20 July 1932

Dear Mr. Heyting,

From your last letter I gather that you have already finished the chapter on formalism, and of course in no case do I then want to cause you to

um so weniger als wir ja seinerzeit vereinbart haben, daß Sie die W-Beweise behandeln werden. Übrigens fällt ja auch jeder Grund für eine Änderung fort, wenn Sie gegen eine Aufteilung der W-Beweise in zwei verschiedene Kapitel nichts einzuwenden haben, denn das Argument aus meinem letzten Brief bezog sich ja nur auf den Herbrandschen Beweis.

Sehr dankbar wäre ich Ihnen, falls Sie mir (wenn möglich noch vor Ihrer Abreise) eine Kopie Ihres Abschnittes über Formalismus zuschicken
2 könnten, | damit ich daran anknüpfen kann und um Wi⟦e⟧derholungen in dem Kapitel über Metamathematik zu vermeiden.

Schließlich möchte ich Ihnen noch mitteilen, daß es mir kaum möglich sein wird, den Termin 1. Sept. einzuhalten, da ich in den letzten beiden Semestern sehr beschäftigt war u. daher die Arbeit nur langsame Fortschritte machte.

Mit den besten Grüßen

Ihr ergebener Kurt Gödel

9. Heyting to Gödel

Amsterdam, 26 Juli 1932
Banstraat 15.

Lieber Herr Gödel,

Es war mir leider unmöglich, Ihren Brief, den ich rechtzeitig erhalten habe, früher zu beantworten. Über den Formalismus habe ich eine Kladde fertig; ich bin gern bereit, Ihnen sofort nach der Ausarbeitung einen Durchschlag zuzuschicken; ich werde dasselbe mit den andern Abschnitten tun. Nach erneuter Überlegung scheint es mir im Interesse des Artikels am besten, dass ich die Grundgedanken der Ackermannschen und v. Neumannschen Widerspruchfreiheitsbeweise stehen lasse; alle Einzelheiten sowie den Herbrandschen Beweis und die Vergleichung der Beweise untereinander hinsichtlich ihrer Tragweite überlasse ich Ihnen.[1] Um Ihnen schon jetzt einigen Anhalt zu bieten, gebe ich hierunter durch einige Schlagwörter den Inhalt meiner diesbezüglichen Ausführungen an.

Einteilung der Zeichen in Veränderliche und Konstante; in Aussagenzei-
2 chen und Dingzeichen. Operationsregeln. | Für Aussagen- und Funktionenkalkül verweise ich auf Sie. Das Hilbertsche ϵ; seine Bedeutung für die

[1] Für ein Lehrbuch würde die Sache anders stehen und wäre überhaupt eine scharfe Trennung der Richtungen wohl nicht durchführbar.

undertake a reworking, all the less since we agreed before that you will treat the consistency proofs. Moreover, every reason for a change also becomes void if you have no objection to a splitting of the consistency proofs into two different chapters, for the argument of my last letter referred only to Herbrand's proof.

I would be very grateful to you if you could send me (if possible, even before your departure) a copy of your chapter on formalism, so that I can refer to it and avoid repetitions in the chapter on metamathematics.

Lastly, I would like to inform you in addition that it will hardly be possible for me to keep to the deadline of 1 September, since I was very busy during the last two semesters and thus the work progressed only slowly.

With best wishes,

Yours sincerely,

Kurt Gödel

9. Heyting to Gödel

Amsterdam, 26 July 1932
Banstraat 15

Dear Mr. Gödel,

It was unfortunately impossible for me to reply sooner to your letter, which I received in good time. On formalism I have a notebook ready; I am happy to send you a carbon copy ⟦of a typescript⟧ immediately after working it out. I will do the same with the other chapters. After fresh consideration it seems to me in the best interest of the article that I let the basic ideas of Ackermann's and von Neumann's consistency proofs stand as is; all details, as well as the proof of Herbrand and the comparison of the proofs among each other with respect to their scope I leave to you.[1] In order to offer you some clues already, I state below by means of some key words the content of my exposition concerning them.

Division of the symbols into variables and constants; into symbols for propositions and symbols for objects. Rules of operation. I refer to you for the propositional and functional calculi. Hilbert's ϵ; its significance

[1] The matter would be different for a textbook, and a sharp separation of the directions would in general probably not be feasible.

Logik und Mengenlehre. Das Axiomensystem für die Analysis. Der Ackermannsche Beweis (es wird ein Verfahren angegeben, das alle Formeln ⟨eines Beweises⟩ in numerische verwandelt und das bestimmten Bedingungen genügt). Der v. Neumannsche Beweis (andere Auffassung von "Axiom"; Begriff der "Teilwertung").

Ich benutze diese Gelegenheit, Ihnen einige Vorschläge über Einzelheiten zu machen; diese Punkte sind an sich nicht wichtig und ich gebe das Folgende für Besseres gern auf; es ist aber notwendig, dass wir hierin einheitlich vorgehen. I

Einteilung des Artikels: in Abschnitte (I. Einleitung; ~~2~~ II. Logizismus; ~~3~~ III. Formalismus; ~~4~~ IV. Metamathematik; ~~5~~ V. Intuitionismus; ~~6~~ VI. Andere Standpunkte; ~~7~~ VII. Beziehungen zwischen den verschiedenen Richtungen; ~~8~~ VIII. Mathematik und Naturwissenschaft. Jeder Abschnitt zerfällt in grössere Paragraphen. Einzelne Absätze oder Gruppen von solchen werden durch Schlagwörter bez. Schlagsätze gekennzeichnet; ich möchte es der Druckerei überlassen, diese in margine oder in den Text drucken zu lassen.

Literaturverweise durch eingeklammerte Zahlen, wenn nötig mit Seitenangabe, z. B. (54, S.115; 55, S. 28). Verweisungen auf andere Abschnitte wie folgt: (II §3, S. 25); auf denselber Abschnitt (§4, S. 110).

Logische Zeichen: $\supset, \vee, \cdot, \sim$.[2] In der intuitionistischen Logik weiche ich hiervon ab.

3 | Auch mir wird es schwer fallen, vor dem 1. September fertig zu sein; wenn Sie nichts dagegegen haben, werde ich den Verlag um einen Aufschub von 2 Monaten bitten.

Mit den besten Grüssen,

Ihr sehr ergebener

A. Heyting

[2] $(x), (Ex), Fx$ = Aussagefunktion F von x.

10. Gödel to Heyting

Wien 4./ VIII. 1932

Lieber Herr Heyting!

Besten Dank für Ihren Brief vom 26 d. M. Mit Ihren Vorschlägen hinsichtlich der äußeren Form des Artikels bin ich im ganzen einverstanden.

for logic and set theory. The axiom system for analysis. Ackermann's proof (a procedure is given that converts all formulas of a proof into numerical form and that satisfies specified conditions). Von Neumann's proof (another conception of "axiom"; notion of "partial valuation").

I take this opportunity to make a few suggestions to you about details; these points are in themselves not important, and I will gladly give up what follows for something better; it is necessary, though, that we proceed uniformly in this.

Division of the article: into chapters (I. Introduction; II. Logicism; III. Formalism; IV. Metamathematics; V. Intuitionism; VI. Other viewpoints; VII. Relations among the different directions; VIII. Mathematics and natural science.) Each chapter is split up into larger sections. Individual paragraphs or groups of such are distinguished by key words or key sentences; I would like to leave it to the printer ⟦whether⟧ to have these printed in the margin or in the text.

Bibliographic references ⟦indicated⟧ by numbers in parentheses, if necessary with page citations, for example (54, p. 115; 55, p. 28). References to other chapters as follows: (II §3, p. 25); to the same chapter (§4, p. 110).

Logical signs: $\supset$, $\vee$, $\cdot$, $\sim$.[2] In intuitionistic logic I deviate from this.

It will also be difficult for me to be ready before 1 September; if you have nothing against it, I will ask the publisher for a delay of two months.

With best wishes,

Yours very sincerely,

A. Heyting

[2] $(x), (Ex), Fx =$ propositional function F of x.

10. Gödel to Heyting

Vienna, 4 August 1932

Dear Mr. Heyting,

Thank you very much for your letter of the 26th of this month. I am on the whole in agreement with your suggestions concerning the outer

Nur möchte ich den Abschnitt IV entspr. seinem Inhalt "Logikkalkül und Metamathematik" nennen; ferner[1] Aussagefunktionen mit kleinen griechischen Buchstaben bezeichnen, um in Übereinstimmung mit Russell zu bleiben.

Was den Ablieferungstermin betrifft, möchte ich Sie bitten, den Verlag um einen Aufschub von 3 Monaten zu ersuchen.

Mit den besten Grüßen

Ihr ergebener

Kurt Gödel

[1]wenn Sie nichts dagegen haben

11. Heyting to Gödel

Enschede, den 15. August 1932.
Parkstraat 25.

Lieber Herr Gödel,

Besten Dank für Ihren Brief vom 4.d.M. Ich habe den Verlag um einen Aufschub von drei Monaten gebeten; die zustimmende Antwort und das Konzept meiner Antwort schliesse ich hierbei ein. Ich kann Ihnen jetzt auch meine Ausführungen über die Beweistheorie senden; es fehlt nur noch die Besprechung der beiden letzten Arbeiten Hilberts (Math. Ann. 104 und Göttinger Nachrichten 1931)[a] und des Lösungsversuches zum Kontinuumproblem (Math. Ann. 95).[b] Ich möchte die betr. Arbeiten noch einmal zu Rate ziehen; sie sind mir aber nicht sofort zugänglich. Das Manuskript, das ich Ihnen heute sende, umfasst die zweite Hälfte des Abschnittes III "Axiomatik und Formalismus"; im ersten Teil werde ich die prinzipiellen Fragen der axiomatischen Methode, auch in ihrer Anwendung auf die Mengenlehre, behandeln. Ueber die "Grundlagen der Mengenlehre" im allgemeinen hat Herr Neugebauer eine gesonderte Darstellung geplant. Was die Bezeichnung der Veränderlichen betrifft, habe ich noch die Hilbertsche Methode gefolgt; die Russellsche ist mir

[a]*Hilbert 1931* and *1931a*.

form of the article. Only I would like to call chapter VI "The calculus of logic and metamathematics", in accordance with its content; furthermore[1] I would like to denote propositional functions with lower case Greek letters in order to remain in agreement with Russell.

As concerns the delivery deadline, I would like to ask you to request a delay of three months from the publisher.

With best wishes,

Yours sincerely,

Kurt Gödel

[1]if you have nothing against it

11. Heyting to Gödel

Enschede, 15 August 1932
Parkstraat 25

Dear Mr. Gödel,

Many thanks for your letter of the 4th of this month. I have asked the publisher for a delay of three months. I am enclosing herewith the concurring answer and the draft of my reply. I can now also send you my remarks on proof theory; it still lacks only the discussion of Hilbert's last two papers (Math. Ann. 104 and Göttinger Nachrichten 1931)[a] and the attempts to resolve the continuum problem (Math. Ann. 95).[b] I would like to consult the works in question once again; but they are not immediately available to me. The manuscript that I am sending you today comprises the second half of chapter III, "Axiomatics and Formalism"; in the first part I will treat the questions of principle for the axiomatic method, also in their application to set theory. Neugebauer planned a separate exposition on the "foundations of set theory" in general. As regards the designation of variables, I have still followed Hilbert's method;

[b]*Hilbert 1926*, pp. 180–190.

nicht geläufig genug, um sie auf die höheren Veränderlichengattungen Hilberts ausdehnen zu können. Wenn Sie mir dazu irgendeinen Leitfaden an die Hand geben können, will ich gern versuchen, Einheit in der Bezeichnung zu erreichen. Ferner bitte ich Sie auch für das Folgende darum, wenn Sie irgendeine Kritik an meiner Arbeit üben können, sei es was die Auswahl und Anordnung des Stoffes, sei es was die Sprache betrifft, mir diese freimütigst mitzuteilen; Sie brauchen dabei für keine Reizbarkeit meinerseits zu fürchten.

Ich füge noch das Literaturverzeichnis zum Abschnitt III bei; es ist wohl am besten, dass Sie es in Ihr Verzeichnis verarbeiten. In welcher Weise Sie diese Titel einordnen wollen, bleibt Ihnen natürlich überlassen; nur bitte ich Sie, mir zu gelegener Zeit mitzuteilen, welche Nummern sie bei Ihnen erhalten haben, damit ich diese im Text anführen kann. Zu diesem Zweck schliesse ich ein Duplikat der Liste ein. Wollen Sie bitte auf S.1 nach Geiger noch d~~ie~~ as folgende ~~n~~einschalten?

2 | Hertz, P. Ueber Axiomensysteme für beliebige Satzsysteme. Math. Ann. 87, S. 246–269 (1922), 89, S. 76–102 (1923), 101, S. 457–514 (1929).[c]

Mit den besten Grüssen,

Ihr Sehr Ergebener

A. Heyting

---

[c] *Hertz 1922, 1923* and *1929*.

12. Heyting to Gödel

Enschede 27 Augustus 1932.
Parkstraat 25.

Lieber Herr Gödel,

Jetzt kann ich Ihnen erstens den Abschnitt über Intuitionismus mit Literaturverzeichnis senden, zweitens einige Ergänzungen zur Beweistheorie. Es schien mir besser, doch den "Gödelschen Satz", dessen Besprechung ich Ihnen überlassen hatte, mit einigen Worten zu erwähnen.

Es ist mir aufgefallen, dass Hilbert und Ackermann Widerspruchsfreiheit schreiben, v. Neumann dagegen Widerspruchfreiheit ohne s. Berich-

I am not fluent enough with Russell's to be able to extend it to Hilbert's higher variable types. If you can provide some clue to me about that, I am quite willing to try to achieve unity in the notation. Furthermore, I ask you also for the following: If you have any criticism of my work, be it concerning the selection and ordering of material or be it concerning the language, communicate that to me candidly; you need fear no sensitiveness on my part for that.

I am enclosing as well the list of references for chapter III; it is probably best that you incorporate it into yours. In what way you want to arrange these titles remains of course up to you; I only ask you to tell me at the appropriate time which numbers they have received in your [scheme], so that I can cite them in the text. Toward that end I am enclosing a duplicate of the list. On p. 1 after Geiger would you please insert the following as well?

Hertz, P. Ueber Axiomensysteme für beliebige Satzsysteme. Math. Ann. 87, pp. 246–269 (1922), 89, pp. 76–102 (1923), 101, pp. 457–514 (1929).[c]

With best wishes,

Yours very sincerely,

A. Heyting

---

12. Heyting to Gödel

Enschede 27 August 1932
Parkstraat 25

Dear Mr. Gödel,

Now I can send you, first of all, the chapter on intuitionism with ⟦its⟧ list of references, ⟦and⟧ secondly, some supplemental material on proof theory. It seemed better to me after all to devote a few words to Gödel's Theorem, whose discussion I had left to you.

It has struck me that Hilbert and Ackermann write 'Widerspruchsfreiheit', ⟦while⟧ von Neumann, on the contrary, ⟦writes⟧ 'Widerspruchfrei-

ten Sie mir bitte gelegentlich, welches Sie für das Richtige halten. Die gleiche Frage erhebt sich für Widerspruch(s)losigkeit.

Mit besten Grüssen,

Ihr sehr ergebener

A. Heyting

13. Heyting to Gödel

Enschede, 29 Oktober 1932.

Lieber Herr Gödel,

Ich sende Ihnen hierbei die noch fehlenden Paragraphen unseres Artikels. Der erste Teil von Abschnitt III (Axiomatik) befriedigt mich nicht; ich sehe aber nicht, wie es besser kann ohne ausführlicher auf die Grundlagen der Mengenlehre einzugehen. Vielleicht wird es möglich sein, im Anschluss an den von Ihnen bearbeiteten Abschnitten, mit denen der Stoff viele Berührungspunkte hat, einiges zu streichen oder zu bessern.

Die Schluss~~betrachtungen~~abschnitte habe ich sehr kurz gehalten. Verschiedene Bemerkungen über die Beziehungen der Richtungen zueinander k~~o~~nnten doch besser über die anderen Abschnitte verteilt werden; ö was noch über das Verhältnis von Intuitionismus und Formalismus zu sagen blieb, habe ich in einem Schlussparagraphen zu Abschnitt V untergebracht, so dass Abschnitt VII wegfallen konnte. Von den "anderen Richtungen" habe ich nur zwei von Mathematikern verteidigte Auffassungen behandelt. Die erste, von Mannoury, versucht eine Art Synthese der wichtigsten Standpunkte; der Inhalt des betr. §2 ist | von Herrn Mannoury selbst geschrieben. 2

In Abschnitt VII habe ich §2 für Sie offen gelassen.

Es fehlt jetzt nur noch §1 von Abschnitt VI, in dem ich die Arbeiten einiger Mathematiker und Philosophen, die ausführlicher über unsern Gegenstand gearbeitet haben, nennen und womöglich durch einige Stichwörter kennzeichnen will. Hier können untergebracht werden:

J. König, Neue Grundlagen der Logik, Arithmetik und Mengenlehre.

O. Hölder, Die mathematische Methode.

~~E~~. Meyerson, du ch~~e~~minement de la pensée.[a] É é

Für Beiträge zu diesem Paragraphen wäre ich Ihnen dankbar.

[a] *König 1914*, *Hölder 1924* and *Meyerson 1931*.

heit', without the 's'. Please tell me, when convenient, which you hold to be correct. The same question arises for 'Widerspruch(s)losigkeit'.

With best wishes,

Yours very sincerely,

A. Heyting

13. Heyting to Gödel

Enschede, 29 October 1932

Dear Mr. Gödel,

I am sending you herewith the sections of our article that were still lacking. The first part of chapter III (Axiomatics) doesn't satisfy me, but I don't see how it can be bettered without going more thoroughly into the foundations of set theory. Perhaps it will be possible, in connection with the chapters dealt with by you, with which the material has many points of contact, to delete some things or to improve them.

I've kept the final chapters very short. Various remarks about the relations of the directions to one another could perhaps better be distributed among the other chapters; what still remained to be said about the relation of intuitionism and formalism I have placed in a final section in chapter V, so that chapter VII could be omitted. Of the "other directions" I have treated only two conceptions defended by mathematicians. The first, by Mannoury, attempts a kind of synthesis of the most important standpoints; the content of the corresponding §2 is written by Mr. Mannoury himself.

In chapter VII I have left §2 open for you.

Still lacking now are only §1 of chapter VI, in which I mention the works of a few mathematicians and philosophers who have worked more thoroughly on our subject matter and, where possible, ⟦I⟧ want to distinguish ⟦them⟧ by means of a few key words. Here can be placed:

J. König, Neue Grundlagen der Logik, Arithmetik und Mengenlehre.

O. Hölder, Die mathematische Methode.

É. Meyerson, Du chéminement de la pensée.[a]

I would be grateful for contributions to this section.

Ist es Ihnen recht, dass Sie Ihr Manuskript an mich senden und ich es weiterschicke? Ich kann dann die beiden Teile zu einem Ganzen vereinigen. Sie werden mich sehr verpflichten, wenn Sie mir schon jetzt einen Abzug von den fertigen Paragraphen senden wollen, damit ich wo notwendig meine Arbeit ihnen anpassen kann. Es wird mir nicht möglich sein, das alles zugleich in der letzten Woche zu machen; es sind in jeder Woche nur wenige Stunden, die ich für wissenschaftliche Arbeit frei machen kann.

Mit besten Grüssen,

Ihr sehr ergebener

A. Heyting

14. Gödel to Heyting

Wien 15./XI. 1932.

Lieber Herr Heyting!

Besten Dank für die freundliche Übersendung der 3 Partien Ihres Manuskriptes und die beiliegenden Briefe. An Springer habe ich selbst geschrieben, daß der von Ihnen übernommene Teil beinahe fertig ist und von meinem etwa die Hälfte (es ist der Abschnitt über Logizismus[1]). Gleichzeitig habe ich den Verlag ersucht, mir eine eventuelle Verspätung von einigen Wochen zu konzedieren.

Ich bitte Sie zu entschuldigen, wenn ich Ihnen noch keine Abzüge der fertigen Partien übersenden kann, da sie erst im Sten~~n~~ogramm existieren; ich glaube übrigens nicht, daß irgendwelche Änderungen in Ihrem Teil nötig sein werden, mit Ausnahme des Abschn. über Axiomatik. Hier bitte ich Sie, die Theorie der endlichen Mengen zu streichen, welche ich anläßlich der logischen Theorie der nat. Zahlen besprochen habe. Ferner möchte ich die Axiomensysteme *als Theorien innerhalb der Mathematik* 2 | in Abschn. IV behandeln (etwa im Sinn von Carnap, Erkenntnis I p 303[a]). Es wäre daher vielleicht besser, wenn Sie Ihre diesbez. Ausfüh-

[1] u. einzelne § von Abschn. IV

[a] *Carnap 1930a.*

Is it all right with you that you send your manuscript to me and I send it on? I can then combine the two parts into a whole. You will put me very much in your debt if you will already send me a copy of the finished paragraphs, so that I can, where necessary, adapt my work to them. It will not be possible for me to do all that at once in the last week; in each week there are only a few hours that I can make free for scientific work.

With best wishes,

Yours very sincerely,

A. Heyting

14. Gödel to Heyting

Vienna, 15 November 1932

Dear Mr. Heyting,

Thanks very much for the friendly forwarding of the three parts of your manuscript and the accompanying letters. I myself wrote Springer that the part undertaken by you is almost finished and that mine is about half done (it is the chapter on logicism[1]). At the same time I asked the publisher to concede to me a possible delay of a few weeks. I ask you to excuse that I can still forward no copies of the finished parts to you, since they exist only in shorthand; anyway, I don't think that any sort of changes in your part will be necessary, with the exception of the chapter on axiomatics. There I ask you to delete the theory of finite sets, which I have discussed in connection with the logical theory of the natural numbers. Furthermore I would like to treat the axiom systems *as theories within mathematics* in chapter IV (approximately in the sense of Carnap, Erkenntnis I p. 303[a]). It would therefore perhaps be better if

[1] and a single § from chapter IV.

rungen (insbes. über Unabhängigkeit u. Vollständigkeit) auf das nötigste beschränken würden (ich erbitte mir darüber Ihre Mitteilung).

Nun noch einige Bemerkungen über den Inhalt Ihres Manuskripts:

III p. 12 Zeile 5 v.o.: Der Ackermannsche Beweis (der übrigens in korrigie⟦r⟧ter Form noch nirgends dargestellt ist), kann auch für die Zahlentheorie nicht ausreichen. Bernays hat in seinem Züricher Vortrag[b] (soviel mir bekannt) auch zugegeben, daß man die Widerspruchsfreiheit d. Zahlentheorie ⟨bisher⟩ nur mit einer von Herbrand gegebenen Einschränkung für die vollständige Induktion beweisen kann.

III. Zusatz zur Vollständigkeitsfrage Zeile 6 v.o.: In Ann. 104[c] behandelt Hilbert die Vollst.-Frage nur für die Zahlentheorie u. hat auch hier (trotz dem neuen Axiom) nur eine kleine Teilfrage gelöst. (Ich habe übrigens in einem Brief an Bernays[d] gezeigt, daß es auch in dem erweiterten System | unentscheidbare Sätze gibt). Ich glaube überhaupt, Sie beurteilen Hilberts letzte Arbeiten etwas zu günstig. z. B. ist doch in Göttinger Nachr. 1931[e] kaum irgend etwas bewiesen.

3

V. p. 6 Zeile 10. v.o. Warum die einzelne reelle Zahl nicht erreicht werden kann, sollte vielleicht durch ein paar Worte erläutert werden.

V. p. 12 Zeile 6. v.u. Dieses Argument scheint mir verfehlt, denn man könnte doch, nachdem man die intuitionistische Mathematik schon teilweise aufgebaut hat, sich fragen, ob alle ihre Sätze aus endlich vielen Axiomen folgen.

~~III §1 p 3~~ V. §6 p. 3 Zeile 17 v.u. Ich habe nur bewiesen, daß es für jedes System, *für das man einen finiten Widerspruchfreiheitsbeweis hat,* un~~entscheidbare~~beweisbare Sätze gibt, die intuitionistisch beweisbar sind. Dagegen könnte in Systemen ohne finiten Wid-Beweis (z. B. dem der Analysis oder Mengenlehre) vielleicht die ganze ~~Intu~~ intuit. Mathematik enthalten sein (d. h. das Gegenteil ist nicht bewiesen).

III §1 p 4 Zeile 7 v.o. Dies scheint mir dasselbe zu besagen wie: "jede Realisierung des einen *ist ~~auch~~ zugleich* eine | des andern⟦"⟧ (wenigstens für den einzig wichtigen Fall "formaler" Axiome~~)~~ im Sinn von Carnap l.c.)

4

[b]See the introductory note.

[c]*Hilbert 1931a.*

you were to *restrict* your remarks pertaining thereto (in particular, about independence and completeness) to what is most necessary (I solicit your advice about that).

Now still a few remarks about the content of your manuscript:

III p. 12 line 5 from the top: Ackermann's proof (which, by the way, has still not been presented anywhere in corrected form) can also not be extended to number theory. In his Zürich lecture[b] (as far as I know), Bernays also granted that up to now the consistency of number theory can only be proved with a restriction on complete induction given by Herbrand.

III. Addendum to the completeness question, line 6 from the top: In Ann. 104[c] Hilbert treats the completeness question only for number theory and even there (despite the new axiom) solves only a small subquestion. (Moreover, in a letter to Bernays[d] I have shown that there are also undecidable sentences in the extended system.) In general I think you judge Hilbert's last papers somewhat too favorably. For example, in Göttinger Nachr. 1931[e] surely hardly anything is proved.

V. p. 6, line 10 from the top. Why the particular real number can't be reached should perhaps be elucidated by a few words.

V. p. 12 line 6 from the bottom. This argument seems to me unsuccessful, for one could after all ask, after intuitionistic mathematics has already partially been built up, whether all its theorems follow from finitely many axioms.

V. §6 p. 3 line 17 from the bottom. I have only proved that for every system *for which one has a finitary consistency proof* there are undecidable sentences that are intuitionistically provable. On the contrary, the whole of intuitionistic mathematics can perhaps be encompassed in systems without finitary consistency proof⟦s⟧ (for example, analysis or set theory)—that is, the contrary has not been proven.

III. §1 p. 4 line 7 from the top. This seems to me to say the same as "every realization of the one *is at the same time* one of the other", at least for the only important case, "formal axioms" in the sense of Carnap (l.c.).

---

[d]See Gödel's letter to Bernays of 2 April 1931 in these *Works*, vol. IV.

[e]*Hilbert 1931a*.

V p 23 Zeile 8 v.o. Dies bedeutet doch, daß man die Regeln, nach denen man Eigenschaften bilden kann, nicht erschöpfend formulieren kann, sondern immer von neuem an die Intuition appellieren muß? Vielleicht sollten Sie das deutlich sagen.

Für sehr instruktiv für den Leser würde ich es halten, wenn Sie einige von den Intuitionisten abgelehnte Schlußweisen an Beispielen erläutern würden; ferner wenn Sie etwas über das Verhältnis von Hilberts Finitismus (wie er ihn in seinen früheren Arbeiten vertreten hat) u. dem Intuitionismus sagen würden. Der erstere dürfte wohl noch weniger Schlußweisen u. Begriffsbildungen anerkennen als der letztere.

Zu Abschn. VI werde ich leider unmöglich Beiträge liefern können, da ich mit dem übrigen vollauf beschäftigt sein werde. Halten Sie eine historische ~~Einleitu~~ Übersicht in der Einleitung für nötig? Ich glaube, sie würde nur wiederholen, was sonst schon gesagt ist. Wie steht es mit den Lösungsversuchen der Paradoxien außerhalb der 3 Hauptrichtungen? ~~s~~ie S scheinen mir kaum von Interesse zu sein (Behmann habe ich natürlich in Abschn. II besprochen)⟦.⟧

5 | Ich bin natürlich einverstanden, wenn Sie das Manuskript an die Monatshefte schicken, zweifle aber, ob dadurch eine frühere Besprechung erfolgen ~~wird~~ kann, weil ja wahrscheinlich erst das Juliheft 1933 dafür in Betracht kommt.[f]

Widerspruchsfreiheit schreibt man m. E. besser mit s.[g]

Anbei noch ein Separatum einer Note, die Sie vielleicht interessieren wird.[h]

Mit besten Grüßen

Ihr ergebener Kurt Gödel

---

[f] This paragraph is puzzling; it appears that Gödel misunderstood Heyting's proposal that he should send the manuscript in for publication (presumably to the *Zentralblatt*, i.e. to Neugebauer) and took him to be proposing to send it to the *Monatshefte für Mathematik und Physik* so that the latter could commission a review in advance of its publication.

V. p 23 line 8 from the top. Doesn't this mean that the rules according to which properties can be formed cannot be formulated exhaustively, but rather must always appeal anew to the intuition? Perhaps that should be clearly stated.

I would regard it as very instructive for the reader if you would elucidate by examples some of the inference rules rejected by the intuitionists; moreover, if you would say something about the relation of Hilbert's finitism (as he advocated it in his earlier works) and intuitionism. The former might perhaps recognize even fewer inference rules and concept formations than the latter.

Unfortunately I will not possibly be able to provide contributions to chap. VI, since I will be completely occupied with the rest. Do you regard the historical overview in the introduction as essential? I think it would only repeat what has already been said elsewhere. What is the status of the attempts to resolve the paradoxes outside of the three principal directions? They appear to me hardly to be of interest (I have of course discussed Behmann in chap. II).

Of course I agree to your sending the manuscript to Monatshefte, but I doubt whether an earlier review can thereby result, because the July 1933 volume is probably the first to come into consideration for it.[f]

'Widerspruchsfreiheit', in my opinion, is preferably written with 's'.[g] Enclosed is another offprint of a note that will perhaps interest you.[h]

With best wishes,

Yours sincerely,

Kurt Gödel

---

[g]Nonetheless, earlier in the letter, in commenting on V §6 of Heyting's manuscript, Gödel used "Widerspruchfreiheit".

[h]In all probability *Gödel 1932*; see the addition dated 26 November to the following letter.

15. Heyting to Gödel

Enschede, 24 Nov. 1932
Parkstraat 25.

Lieber Herr Gödel,

Besten Dank für Ihren Brief, vor allem für die wertvollen Bemerkungen zu meinem Manuskript. Die Theorie der endlichen Mengen kann ich ohne Schwierigkeit streichen; ich bitte Sie aber, noch zu bestätigen, dass Sie auch die *mengentheoretische* Definitionen der natürlichen Zahlen besprochen haben; das ist aus Ihrem Brief noch nicht zweifellos zu lesen. Es ist richtig, dass meine Ausführungen über Unabhängigkeit usw. von Axiomensystemen an vielen Stellen mit Ihren Behandlung der innermathematischer Axiomatik zusammenfallen werden; das Beste scheint es mir, dass ich nach Empfang Ihres Manuskripts eventuelle Wiederholungen durch Hinweisungen ersetze.

Ihre Bemerkungen, dass Ackermann nicht einmal vollständig die Widerspruchsfreiheit der *Zahlentheorie* bewiesen hat und dass Hilbert (Math. Ann. 104[a]) sich auf die Zahlentheorie beschränkt, sind richtig. Nur habe ich in den formalistischen Literatur nirgends eine genaue Definition der 2 Zahlentheorie gefunden. Wäre es richtig, zu sagen: Die Zah|lentheorie ist derjenige Teil der Mathematik, in welchem nur solche Funktionen auftreten, deren Werte natürliche Zahlen sind?

Es war nicht leicht, über die letzten Arbeiten von Hilbert zu schreiben. Sie sind doch immerhin von Hilbert! Dass Hilbert das Vollständigkeitsproblem gelöst habe, war ein Schreibfehler; es wurde sofort durch die darauf folgenden Zeilen widerlegt. Es ist gut, dass Sie es bemerkt haben.

Ihre Bemerkung zu V S. 12 Z. 6 v.u. verstehe ich nicht. Sie betrifft doch nur die Anwendung der axiomatischen Methode innerhalb der intuitionistischen Mathematik. Selbst wenn man beweisen könnte, dass jeder richtige Satz der intuitionistischen Mathematik sich aus gewissen Axiomen ableiten liesse, so könnte dieser Satz doch erst nach der vollständigen Begründung der intuitionistischen Mathematik ausgesprochen und bewiesen werden. Ist es vielleicht deutlicher, wenn ich sage "Eine genaue Aufzählung ... ist schon darum als Begründung der intuitionistischen Mathematik nicht zulässig, weil ..."?[b]

---

[a] *Hilbert 1931*.

[b] Cf. *Heyting 1934*, p. 12: "Eine genaue Aufzählung der in der Mathematik zulässigen Grundbegriffe und Elementarschlüsse ist schon darum als Begründung der intuitionistischen Mathematik unzureichend, weil der Begriff der Aufzählung schon einen wesentlich mathematischen Kern enthält und wir so wieder in einen Zirkel geraten würden." ("A

15. Heyting to Gödel

Enschede, 24 Nov. 1932
Parkstraat 25

Dear Mr. Gödel,

Thanks very much for your letter, above all for the valuable remarks on my manuscript. I can delete the theory of finite sets without difficulty; but I ask you to confirm that you have also discussed the *set theoretic* definitions of the natural numbers; from your letter that is still not to be gathered without doubt. It is correct that my exposition about independence and so on of axiom systems coincides in many places with your treatment of intramathematical axiomatization; it seems best to me that after receipt of your manuscript I replace eventual repetitions by references.

Your remarks that Ackermann has not once completely proved the consistency of *number theory* and that Hilbert (Math. Ann. 104[a]) restricts himself to number theory are correct. Only I have never found a precise definition of number theory in the formalist literature. Would it be correct to say: Number theory is that part of mathematics in which only those functions occur whose values are natural numbers?

It was not easy to write about the last papers of Hilbert. They are, after all, by Hilbert! That Hilbert solved the completeness problem was a slip of the pen; it was contradicted at once by the lines following it. It is good that you noticed it.

I don't understand your remark about V. p. 12 line 6 from the bottom. It really concerns only the application of the axiomatic method within intuitionistic mathematics. Even if one could prove that every correct statement of intuitionistic mathematics may be derived from certain axioms, that theorem could still only be enunciated and proved after the complete grounding of intuitionistic mathematics. Is it perhaps clearer if I say "A precise enumeration ... is for that very reason ⟦schon darum⟧ not admissible as a grounding of intuitionistic mathematics, because ..."?[b]

---

precise enumeration of the basic concepts and elementary inferences admissible in mathematics is already insufficient as a grounding of intuitionistic mathematics for this reason: since the concept of emumeration already contains an essentially mathematical core and we would thus again fall into a circle.")

Ich habe die Erfahrung gemacht, dass es ausserhalb der Hauptrichtungen, wenigstens mathematisch betrachtet, fast nichts bemerkenswertes gibt; das gilt auch für die Lösungsversuche der Paradoxien. Auch glaube ich mit Ihnen, dass eine historische Übersicht in dem Umfang den wir dazu disponibel haben, nur geringen Wert hätte.

3 | 26 November.

Besten Dank auch für die Zusendung Ihrer interessanten Note. Es ist, als ob Sie ein boshaftes Vergnügen daran hätten, die Zwecklosigkeit von Untersuchungen Anderer zu zeigen. Aber nützlich im Sinn der Oekonomie des Denkens ist diese Arbeit gewiss, und dazu kommt noch die besondere Schönheit Ihres kurzen Beweises. Es interessiert mich ausserordentlich, wie Sie beweisen können, dass $\mathfrak{A} \vee \mathfrak{B}$ nur dann beweisbar sein kann, wenn entweder $\mathfrak{A}$ oder $\mathfrak{B}$ beweisbar ist. Ich habe rein formal nicht einmal zeigen können, dass $\neg a \vee \neg\neg a$ unbeweisbar ist.

Sind Sie damit einverstanden, dass Sie Ihr Manuskript an mich einsenden? Ich kann dann an meinem Teil wenn notwendig noch etwas ändern und die beiden Manuskripte zu einem Ganzen vereinigen.

Mit besten Grüssen,

Ihr ergebener

A. Heyting

16. Heyting to Gödel

Enschede, 15 April 1933.

Lieber Herr Gödel,

Da ich auf meine Karte[a] noch keine Antwort erhalten durfte, erlaube ich mir nochmals die Bitte, mir doch mitzuteilen, ob die Vollendung Ihrer Arbeit für das Referat jetzt vor der Tür steht. Das Manuskript in der Lade lässt mir keine Ruhe.

---

[a]The card referred to was evidently sent in mid-March (see the following card) and appears to be lost.

I have had the experience that outside of the principal directions there is almost nothing worthy of note, at least considered mathematically; that holds true also for the attempts at resolving the paradoxes. I also think, with you, that a historical overview of the extent that we have available for it would have only meager value.

26 November

Thanks very much, too, for sending your interesting note. It is as if you had a malicious pleasure in showing the purposelessness of others' investigations. But in the sense of economy of thought this work is certainly useful, and in addition to that comes the particular beauty of your short proof. It interests me exceedingly how you could prove that $\mathfrak{A} \vee \mathfrak{B}$ can only be provable if either $\mathfrak{A}$ or $\mathfrak{B}$ is provable. I have not even been able to show purely formally that $\neg a \vee \neg\neg a$ is unprovable.

Do you agree that you should send your manuscript on to me? I can then, if necessary, still change something in my part and can combine the two manuscripts into a whole.

With best wishes,

Yours sincerely,

A. Heyting

16. Heyting to Gödel

Enschede, 15 April 1933

Dear Mr. Gödel,

Since I have not yet had the privilege of receiving a reply to my card,[a] I take the liberty to request once again that you inform me whether the completion of your work for the report is near at hand. The manuscript in the drawer leaves me no peace.

Vor einiger Zeit sandte mir Herr Gentzen aus Göttingen ein Manuskript, das Sie vielleicht auch interessiert. Er beweist u. A. folgenden Satz : "Wenn die intuitionistische Arithmetik widerspruchsfrei ist, so ist auch die klassische Arithmetik widerspruchsfrei." Unter der "intuitionistischen Arithmetik" ist ein bestimmtes formales System verstanden, dessen richtige Formeln sämtlich durch intuitionistisch richtige Sätze interpretiert werden können. Es ergibt sich also, dass die Begründung der Arithmetik auf den von Hilbert eingeschlagenen Weg, die sich mit 2 den von Hilbert gebrauchten "finiten" Hilfsmit|teln höchstwahrscheinlich nicht durchführen lässt, bei Heranziehung weiterer Teile der intuitionistischen Mathematik möglich wird.

Ich weiss nicht, wann und wo die geplante Publikation des Artikels erfolgen wird.[b]

Mit besten Grüssen,

Ihr sehr ergebener

A. Heyting

---

[b]Gentzen's manuscript was never published; it was withdrawn when he learned of *Gödel 1933e*; see the introductory note to *1933e* in these *Works*, vol. I, p. 284.

17. Heyting to Gödel (postcard)

Enschede, 7 Mai 1933.
Parkstraat 25.

Lieber Herr Gödel,

Ich durfte noch keine Antwort erhalten auf meine Karte von Mitte März und meinen Brief von Mitte April. Entschuldigen Sie bitte diese Karte, die keinen anderen Zweck hat als diesen, zu untersuchen, ob meine Sendungen Sie noch erreichen.

Mit besten Grüssen,

Ihr sehr ergebener A. Heyting

Some time ago Mr. Gentzen from Göttingen sent me a manuscript that perhaps also may interest you. He proves among other things the following theorem: "If intuitionistic arithmetic is consistent, classical arithmetic is also consistent." By "intuitionistic arithmetic" is understood a definite formal system, all of whose correct formulas can be interpreted by means of intuitionistically correct statements. It thus yields that the grounding of arithmetic in the way entered upon by Hilbert, which, it is highly likely, cannot be carried through by means of the "finitary" resources employed by Hilbert, is possible by drawing upon further parts of intuitionistic mathematics.

I don't know when and where the planned publication of the article is to take place.[b]

With best wishes,

Yours very sincerely,

A. Heyting

---

17. Heyting to Gödel (postcard)

Enschede, 7 May 1933
Parkstraat 25

Dear Mr. Gödel,

I have still not had the privilege of receiving an answer to my card of mid-March or my letter of mid-April. Please excuse this card, which has no other aim than that of investigating whether my mailings still reach you.

With best wishes,

Yours very sincerely,

A. Heyting

18. Gödel to Heyting

Wien 16./V. 1933.

Lieber Herr Heyting!

Ich danke Ihnen bestens für Ihren Brief vom 15./IV. und bitte Sie, zu entschuldigen, daß ich erst heute dazukomme, Ihnen betreffs unseres Buches zu schreiben. Im Oktober 1932 hatte ich von meinem Teil schon etwa drei-Viertel abgefaßt, mußte aber dann bei Semesterbeginn die Arbeit daran unterbrechen, weil ich anderweitig viel zu tun hatte (ich habe mich inzwischen in Wien habilitiert); auch war ich einige Zeit krank. So kommt es, daß ich seit letztem Herbst nicht wesentlich weiter gekommen bin. Ich werde mich aber jetzt nach Möglichkeit beeilen, schon deshalb, weil ich im Oktober 193~~2~~3 für ein Jahr nach Princeton fahre, und bis dahin auch schon sämtliche Korrekturen erledigt sein müßten.

Daß man die klassische Zahlentheorie durch die intuitionistische interpretieren kann (wodurch sich selbstverständlich auch ein intuit. Widerspruchfreiheitsbeweis ergibt), ist mir bekannt und dürfte auch in Göttingen | schon seit Juni 1932 bekannt sein. Um diese Zeit habe ich nämlich darüber im Mengerkolloquium referiert und zwar in Anwesenheit O. Veblens, der kurz darauf nach Göttingen fuhr. Meine diesbezügliche Arbeit wird im Laufe der nächsten Wochen im Bericht über das Mengerkolloquium erscheinen u. ich werde Ihnen dann ein Separatum zuschicken.

2

Mit besten Grüßen

Ihr ergebener K. Gödel

19. Heyting to Gödel

Enschede, 24 Augustus 1933.

Lieber Herr Gödel,

Ich danke Ihnen für den Abdruck Ihrer Arbeit,[a] die auch in der Methode weitgehend mit derjenigen von Herrn Gentzen übereinstimmte. Ich muss annehmen, dass der Bericht über Ihren Vortrag doch nicht genügend nach Göttingen durchgedrungen war.

---

[a]Evidently *Gödel 1933e*; see the preceding two letters.

18. Gödel to Heyting

Vienna, 16 May 1933

Dear Mr. Heyting,

I thank you very much for your letter of 15 April and apologize for the fact that only today am I getting around to writing you concerning our book. In October 1932 I had already composed about three quarters of my part, but then at the start of the semester I had to interrupt the work on it, because I had too much else to do. (In the meantime I habilitated in Vienna.) I was also ill some of the time. So it turns out that since last fall I've gotten essentially no further. But I will now hasten as much as possible, not least because I am going to Princeton in October 1933 for a year, and by then all the proof sheets already have to be finished too.

That one can interpret classical number theory by means of the intuitionistic ⟦theory⟧ (whereby of course an intuitionistic consistency proof also results) is known to me and should also already have been known in Göttingen since June 1932. In particular, around that time I spoke about it in the Menger colloquium, and in fact in the presence of O. Veblen, who shortly thereafter went to Göttingen. My paper on the subject will appear in the course of the next few weeks in the report of the Menger colloquium and I will then send you an offprint.

With best wishes,

Yours sincerely,

Kurt Gödel

19. Heyting to Gödel

Enschede, 24 August 1933

Dear Mr. Gödel,

I thank you for the offprint of your paper,[a] which agrees throughout in the method with that of Mr. Gentzen. I must assume that the report about your lecture had not sufficiently got through to Göttingen after all.

Seit Ihrem Brief vom 16. Mai habe ich Ihr Manuskript erwartet; ich hoffe, dass es nicht Krankheit war, die Sie verhinderte zu arbeiten. Leider wird es mir unmöglich sein, auch wenn ich jetzt Ihr Manuskript erhielte, die Korrekturen vor Oktober abzuschliessen. Es wäre wohl Schade, nachdem Sie einen beträchtlichen und schwer ersetzbaren Teil fertiggestellt haben, wenn Ihre Arbeit nicht erscheinen könnte. Ist es nicht möglich, Sie soweit zu führen, dass das fehlende von einem Dritten ergänzt werden kann? Sie könnten dafür einen suchen, der die Zeit dafür finden kann und geben will. Ich werde auch an Herrn Neugebauer in diesem Sinn schreiben.

2 | Mit besten Grüssen,

Ihr ergebener

A. Heyting

20. Heyting to Gödel

Enschede, 30 Sept. 1933.
Parkstraat 25.

Lieber Herr Gödel,

Besten Dank für Ihren Brief.[a] Es ist mir viel wert, wenn unser Bericht als ein Ganzes erscheinen kann. Auch würde es mir einige Zeit kosten, meinen Anteil so abzurunden, dass es getrennt publiziert werden könnte. Wenn Sie also im Anfang 1934 fertig sein können, will ich diesen Termin lieber abwarten. Es scheint mir am einfachsten, dass Sie mir Ihr Manuskript zusenden, damit ich meiner Anteil daran anschliessen kann. Für frühere Zusendung einzelner Abschnitte wäre ich dankbar.

Ich wünsche Ihnen ferner einen recht angenehmen und fruchtbaren Aufenthalt in Amerika zu.

Mit besten Grüssen,

Ihr

A. Heyting

⟦ From here the correspondence continues in English.⟧

[a]This letter appears to be lost.

I have been expecting your manuscript since your letter of 16 May; I hope it was not illness that prevented you from working. Unfortunately, even if I received your manuscript now, it will not be possible for me to finish with the proof sheets before October. It would indeed be a pity, after you have prepared a considerable and hardly replaceable part, if your work could not appear. Isn't it possible for you to lead on so far that what is missing can be supplied by a third party? You could seek someone for that who can find and is willing to give the time for it. I will also write to Neugebauer to that effect.

With best wishes,

Yours sincerely,

A. Heyting

20. Heyting to Gödel

Enschede, 30 Sept. 1933
Parkstraat 25

Dear Mr. Gödel,

Thanks very much for your letter.[a] It is worth a lot to me if our report can appear as a whole. Also, it would cost me some time to polish my portion so that it could be published separately. So if you can be finished at the beginning of 1934, I would prefer to wait until that time. It seems simplest to me that you send me your manuscript, so that I can add my portion to it. I would be grateful for the earlier forwarding of individual chapters.

I wish you in addition a quite pleasant and fruitful sojourn in America.

With best wishes,

Yours,

A. Heyting

21. Heyting to Gödel[a]

Castricum, 2 januari 1969

Dear Professor Gödel,

I am writing in the name of the editorial committee of the series "Studies in Logic". We were told that you consider the publication of your collected works. We are convinced that such a publication will be very useful and we shall be happy to publish the book in our series. We therefore ask you if it is true that you would like to have your collected papers published. In this case we beg you to open negotiations on publication in our Series. I expect hopefully your favourable answer.

I take this opportunity to send you my best wishes for 1969.

Yours sincerely

A. Heyting

P.S. Please note that my address is no longer at the Mathematical Institute, but

Prinses Margrietstraat 1
Castricum, Netherlands

[a]This letter is an Aerogramme (air letter). Above the address to him on the outside Gödel wrote "My Collected Works", underlined. Below this and to the left he wrote "5/I.69", evidently the date of receipt.

22. Gödel to Heyting

March 12, 1969

Professor A. Heyting
Prinses Margrietstraat 1
Castricum, Netherlands

Dear Professor Heyting:

Thank you very much for your letter and New Years' Wishes. I have so far never been considering an edition of my collected works. In fact, I am very doubtful about the usefulness of such a project, since practically all my papers (and, at any rate, all of my important papers) are readily available: especially since the books: "Philosophy of Mathematics", edited by P. Benacerraf and H. Putnam,[a] "The Undecidable", edited by M. Davis,[b] and "From Frege to Gödel", edited by Jean van Heijenoort[c] have been published in 1964, 1965, 1967 respectively. There are only a few notes[1] in the "Ergebnisse eines mathematischen Kolloquiums", edited by K. Menger which, perhaps, are hard to get. But I think they are, at present, more of a historical and biographical, than of a logical interest.

With best wishes and kind regards,

Yours sincerely,

Kurt Gödel

[1]Heft 2 (1929/30), p. 27; Heft 3 (1930/31), p. 20; Heft 4 (1931/32), p. 9,[d] p. 34, p. 39, p. 40;[e] Heft 5 (1932/33), p. 1.[f] Moreover there are some remarks of mine in a discussion in Erkenntnis 2 (1931/32), p. 147–151.[g]

---

[a]*Benacerraf and Putnam 1964*, which contains *Gödel 1964* and the reprint *1964a* of *1944*.

[b]*Davis 1965*, which contained the first publication of *Gödel 1934* and *1946* and translations of *1931, 1933e* and *1936a*.

[c]*van Heijenoort 1967*, containing translations of *Gödel 1930, 1930b, 1931* and *1932b*.

[d]*Gödel 1932a, 1932c* and *1933a*, respectively. Note that Gödel in this note cites the volumes of the *Ergebnisse* by the year of the colloquium rather than by the year of publication.

[e]Apparently *1933d, 1933f* and *1933n*, respectively. *1933e* is also on p. 34 of this volume but is covered by the reference to *Davis 1965* (see note b).

[f]*1933g*.

[g]*1931a*.

Ted Honderich

⟦For an introductory note see the correspondence with Hao Wang, p. 379.⟧

1. Gödel to Honderich[a]

June 27, 1972

Professor Ted Honderich
Department of Philosophy
University College
Gower Street
London W.C.1, England

Dear Professor Honderich:

I am sending you herewith the revised pages of chapter VI of Professor Wang's book.[b]

As far as section 7 of chapter X is concerned, I am sorry I need a little more time for revising it.

Of course I could not agree either to these things being reported as my views without my having revised them carefully. I believe, moreover, that several items in this section should be omitted, because they are too remote from the theme of the book and would have to be explained much more thoroughly if they are published at all. I can send you an abbreviated version of this section within a week or perhaps ten days.

Sincerely yours,

Kurt Gödel

P.S.[c] I am sure Prof. Wang has informed you that I wish to retain the right of having the pages ⟨written or⟩ revised by me (~~⟨which are⟩ men-~~

---

[a]The copy text is a carbon on paper containing the word "COPY" in large letters roughly where the text of the letter falls. From inquiries with Professor Honderich and with Routledge, it is very highly probable that the originals of this letter and of letter 2 are no longer extant.

[b] *Wang 1974*. The chapter is the well-known one on the concept of set. The revisions must have concerned the remarks attributed to Gödel on the axiom of replacement (p. 186) and on the principles for setting up axioms (pp. 189–90). See Wang's letters to Gödel of 10 and 26 April 1972 in this volume.

[c]This P.S. is handwritten and evidently a draft. After "P.S." is an insertion "Zum Brief 27/VI.72" ("to the letter of 27/VI.72"), with "Zum Brief" in shorthand. This was evidently a note to himself and not intended to be part of the postscript.

~~tioned in the preface of the book~~) reprinted if an occasion ~~should arise~~ for ~~exposing~~ ⟨stating⟩ my views on these matters should arise.

2. Gödel to Honderich

Princeton, July ~~12~~ 19, 1972

Dear Professor Honderich,

Thank you very much for your letter of June 30. I am sending you herewith the new version of sect. 7 of chapter X.[a] I have retained only what has some bearing on the theme of the chapter as stated in its title. Therefore, I had to rewrite the whole section and I also changed its title.

Some of the other items Professor Wang reported in this section might fit into another chapter of the book. But unfortunately I don't have the time now to give a satisfactory account of these things. I first discussed them in the Gibbs Lecture of 1951.[b] But I believe the arguments and results given there can, and ought to be, strengthened considerably, especially in view of the fact that these questions are very controversial.

There is one more minor change which I would request, but failed to mention to Professor Wang. Namely on p. 12, line 7 from below, of the Introduction the phrase "deeply felt" should be replaced by "firmly held."

I am sorry it took me ~~4~~ ⟨3⟩ weeks to finish the material I promised. I hope I have not caused any inconvenience thereby.

Yours sincerely,

Kurt Gödel

[a]In *Wang 1974*, this section is entitled "Gödel on minds and machines." See the introductory note and Appendix B.

[b]*Gödel *1951.*

Ralph Hwastecki

Kurt Gödel, who rarely taught and never did so below the college level, would seem an unlikely person from whom to solicit an opinion on the teaching of mathematics in the elementary grades. Nevertheless, on 17 March 1971 Ralph Hwastecki, a special student in education at Elmhurst College in Illinois, wrote to Gödel seeking his reaction to the idea that "the worth and beauty of ... mathematics ... should be brought out to all beginning students" in the hope of stimulating their interest in the field. Hwastecki never received a reply, but Gödel did scribble a very rough longhand draft of a response on two sheets and the back of an envelope, as reconstructed below. His opinions are of interest in the context of the New Math initiative, whose effects were just then beginning to be observed.

John W. Dawson, Jr.

1. Hwastecki to Gödel

219 W. Hickory Road
Lombard, ILL. 60148
March 17, 1971

Dear Dr. Gødel;　　ö

I am a special student in Education at Elmhurst College in Elmhurst, Illinois and as part of the curriculum we are required to do a unit on mathematics for the elementary grade level.[a]

My thinking on this is that mathematics is of value to all of us in all walks of life; therefore, my unit will be written with this thought in mind. I would like to use as an introduction to the unit a short note from a prominent and respected mathematician, such as yourself, explaining this to students at the elementary level. Therefore, could you please find the time to reply to this letter with that thought in mind?

I do realize this is a great demand on your valuable time. However, as a student and future teacher, my feelings are very strong on this subject,

[a]The underlining here is presumably Gödel's.

that is, that there should be an emphasis placed on the worth and beauty of all mathematics and that it should be brought out to all beginning students in mathematics with the hope that it will stimulate and motivate them in this field.

I repeat any help that you can give to this thought will be deeply appreciated.

Sincerely;

Ralph Hwastecki

2. Gödel to Hwastecki[a]

Dear Mr.

What should be pointed out ⟨to beginning students in math (but of course only *after* they have learned *some* math)⟩ ~~then~~ ⟨it seems to me⟩ is the ⟨truly⟩ astonishing number of simple & nontrivial theorems and relationships that ~~exist~~ ⟨prevail in math⟩. (see e.g. the | laws ~~that prevail~~ for the elements ⟨&⟩ subsidiary lines in a triangle[1] or cnf. Peano's formal. des math.) ⟨In my op.⟩ this property ⟨of math⟩ somehow mirrors the order ⟨and regularity⟩ which prevails in the whole world⟦,⟧[b] which

[1]There are ~~much~~ ⟨many⟩ more than ⟨~~is~~⟩ are pointed out to the students. ~~see e. g. Peano Form. des~~ ⟨To name only one example:⟩ it is little known that in a triangle the center of gravity & the intersection of the altitudes & center of the Umkreise lie on a straight line. ~~See Generally see~~ Peano's Form. des math ~~cont.~~⟨⟦?⟧⟩ ~~such ⟨arithmetical⟩ relations of~~ many elem. arithm. rel. of this kind.

[a]Undated draft of a reply to Hwastecki's letter of 17 March 1971. A crossed-out introductory passage reads: It seems to me the worth & beauty of mathematics ~~shou~~ should not be pointed out to *beginning* students at the elementary grade level because one has ⟨first⟩ to know some mathematics ~~at in order to per~~ ⟨before one can⟩ appreciate such a statement. ⟨Perhaps⟩ one might do ~~it~~ so toward the *end* of this stage of teaching. ~~and~~

A note on an accompanying envelope reads: Either the human mind is not a machine or number theory is a science which infinitely transcends the powers of the human mind. The greatest prin value of math in my opinion is that it shows incontrovertibly to every unbiased ~~prej~~ observer ⟦written above: "person"⟧ that a realm ⟦above: "whole world"⟧ of immaterial ⟦above "im-": "non"⟧ objects ~~much larger~~ ⟨with its own fact⟦s⟧ & laws⟩ exists & ~~even is infinitely larger than the realm of all material objects of our world~~ ~~moreover~~ ingeniously combine⟦s⟧ beauty & necessity.

[b]A crossed-out passage reads "and ⟨in the laws of physics, psychol., soc., etc. and ⟦?⟧ etc. & which in my opinion⟩".

turns out to be in~~finitely~~ ⟨comparably⟩ greater than would appear to the superficial ⟨ ~~& even to the present-day scientific~~ ⟩ observer.[c]

⟨Briefly ⟨speaking⟩ one may say: In the world of math everything is well/poised and in perfect order. Shouldn't the same be (expected) ⟨assumed⟩ for the world of reality, contrary to appearances?⟩[d] Incident. it seems to me that generally speaking abstract considerations are started too soon in today's schools (while formerly they were started to⟦o⟧ late or omitted altogether). E.g., I learned with dismay (and in fact refused to believe it) that the v. Neum. integers are ~~used~~ ⟨introduced⟩ already at the elementary grade level.

Sinc. yours ...

[c] Another passage was inserted here and then crossed out. It reads "The appearance of a chaos is also deceptive. The world is very orderly despite it's often chaotic appearance."

[d] Just above the beginning of this paragraph (which is an insertion to the text) there appear the words "Every chaos is merely a wrong appearance."

Karl Menger

Karl Menger

Karl Menger's long association with Kurt Gödel began in the fall of 1927, when Gödel enrolled as a student in Menger's course on dimension theory. Not long afterward Menger was invited to join the group, centered around Moritz Schlick, that became known as the Vienna Circle, and there, too, he encountered Gödel, whose abilities immediately impressed him.

In the succeeding academic year Menger founded a mathematical colloquium that was devoted to the presentation of research results in logic, set theory, topology and geometry. Proceedings of the sessions, entitled *Ergebnisse eines mathematischen Kolloquiums*,[a] were published in eight annual volumes covering the years 1928–1936. Late in 1929 Menger invited Gödel to participate in the colloquium, and from then until its disbanding in 1937 Gödel did so most actively: aside from discussion remarks, he spoke before the colloquium on at least eleven occasions, contributed an equal number of articles to the proceedings, and assisted in the editing of all but one of the volumes.

Menger spent the months between the summer of 1930 and the fall of 1931 on an extended visit to America, and it was during that time that his correspondence with Gödel commenced. From then until 1968 at least 35 letters were exchanged between the two, all but one of which are preserved in the *Nachlässe* of one or the other of the correspondents. (A detailed inventory is given in the accompanying calendar.) Of those, 26 are reproduced here; excluded are cover letters accompanying offprints, acknowledgments of receipt, and a few routine communications (holiday cards, picture postcards, etc.) of no scientific or biographical interest.

In terms of subject matter, the letters fall into four main groups: those dealing with Gödel's incompleteness theorems and his other contributions to the colloquium; those concerning Gödel's results in set theory; those regarding Menger's invitation to Gödel to come to Notre Dame; and those focussing on Menger's work in geometry.

Chronologically, there are several notable gaps in the correspondence. That between 1933 and 1937 spans the period both of Gödel's incapacitation due to depression (1934–1936) and of Menger's emigration to America (1937). The one-year gap in 1940 marks the year of Gödel's own emigration and his resettlement in Princeton. And the cessation of the correspondence after 1968 reflects Gödel's physical and mental debilitation during the final decade of his life.

[a]Reissued in book form as *Menger 1998*.

What is most striking, however, is the long lapse in correspondence after 1943. Subsequent exchanges were sporadic. They were always initiated by Menger, and they are concerned almost exclusively with his own writings. The tone remains cordial, but the content of the letters is more professional than personal in nature.

The abrupt shift in the character of the correspondence suggests that a rift of some sort had developed between the two men. There is no indication of any particular disaffection on Gödel's part—his failure to sustain the correspondence may be attributed in large part simply to his ever-increasing reclusiveness. But in his memoir of Gödel (*Menger 1994*), Menger confesses that, though he was "immensely relieved" to know that Gödel had escaped from the Nazis, he found it hard to rekindle the warmth he had once felt for him—presumably because of Gödel's determination to return to Austria to seek the restoration of his *Dozentur* and his insistent protestations about the Nazis' violations of

1. Menger to Gödel

1931

Lieber Herr Gödel,

Ich beantworte Ihren lieben Brief[a] in einem fahrenden Zuge und deshalb mit Maschine. Schon im Herbst hatte mir Nöbeling von Ihrer großen Entdeckung geschrieben. Ich habe Ihren Artikel mit dem aller größten Interesse gelesen und hier sofort ein Referat über ihn gehalten. Ich rechne Ihre Leistung zu den größten der modernen Logik und beglückwünsche Sie aufs aller herzlichste.

Es freute mich sehr von Nöbeling to hören, daß Sie auch in meiner Abwesenheit zu unseren Abenden kommen. Ich wäre Ihnen sehr verbunden, wenn Sie den Inhalt Ihres interessanten Briefes an mich dort vortragen wollten.

Ich habe Hahn vorgeschlagen, im nächsten Jahre im Seminar neuere Ergebnisse der Logik zu behandeln. ~~Selbs~~ Ich habe mich im letzten Jah-

[a]No longer extant.

his personal rights, in apparent disregard of the much worse plights of others.[b]

Biographically, Gödel's correspondence with Menger is an important source of information that helps to date events during several seminal periods in Gödel's life.

John W. Dawson, Jr.

A complete calendar for the correspondence with Menger appears on pp. 449–450 of this volume.

[b]In a footnote to that memoir, the translator notes that in a conversation shortly before Menger's death in 1985, Menger claimed he had then "only recently learned... that Gödel had ⟦been⟧ married...in 1938". That claim is belied, however, by Menger's letters to Gödel of mid-October and December 1938 (items 18 and 19 below), in which he congratulated Gödel on his marriage and inquired why Adele had remained in Austria.

1. Menger to Gödel

1931

Dear Mr. Gödel,

I am replying to your charming letter[a] in a moving train and therefore by typewriter. Nöbeling had already written me last autumn about your great discovery. I read your article with the utmost interest and immediately delivered a report on it here. I rank your achievement among the greatest of modern logic and send you ⟦my⟧ heartiest congratulations.

I'm very pleased to hear from Nöbeling that you are also coming to our evening sessions in my absence. I would be much obliged to you if you cared to lecture there on the content of your interesting letter to me.

I have suggested to Hahn to take up more recent results of logic in the seminar next year. During the past year here I have familiarized myself

re hier ziemlich in die diesbezügliche Literatur eingearbeitet. Ihre Teilnahme an~~n~~ der Leitung des Seminars wäre natürlich nicht nur mir, sondern sicher auch Hahn sehr wertvoll.

Ich verlasse Texas Ende Mai und reise~~n~~ dann aller Wahrscheinlichkeit westwärts, d. h. über Californien, Hawai, Japan, China, Indien heim und werde in Wien gerade zu Vorlesungsbegin~~n~~ ankommen. Es wird mir eine besondere Freude sein, Sie dann bald zu sehen. Inzwischen bin ich mit den besten Grüßen

Ihr

Karl Menger

Leider habe ich Ihre Adresse verloren & sende Ihnen diesen Brief daher durch Herrn Nöbeling.

2. Gödel to Menger

Wien 30./III. 1931

Sehr geehrter Herr Professor!

Die von Ihnen aufgeworfene Frage über die ~~B~~ Axiomatik des Aussagenkalküls[a] habe ich mir überlegt und bin zu folgendem Resultat gekommen: Aus den Axiomen a), b) 1–4[b] folgt zwar noch nicht die Existenz zweier Klassen $\mathfrak{S}_0$, $\mathfrak{S}_1$ mit den verlangten Eigenschaften (was Sie ja, wie mir Herr Nöbeling erzählte, inzwischen schon selbst festgestellt haben); postuliert man aber außer den 5 Axiomen noch, daß F *echte* Teilklasse von $\mathfrak{S}$ sein soll, dann folgt die Existenz zweier Klassen $\mathfrak{S}_0$, $\mathfrak{S}_1$ u. zwar auf folgende Weise:

Aus $\mathfrak{S} \neq F$ folgt zunächst, daß für kein Ding A aus $\mathfrak{S}$ A und $\overline{A}$ beide zu F gehören können. Denn für jedes Ding B aus $\mathfrak{S}$ gehört nach Ax. b)2. $A \rightarrow (\overline{A} \rightarrow B)$ zu F. Wenn nun A und $\overline{A}$ zu F gehören würden, so ergäbe sich nach Ax a), daß auch B zu F gehören würde u. zwar für jedes B aus $\mathfrak{S}$, entgegen der Vorauss. $\mathfrak{S} \neq F$.

Für das folgende unterscheide ich zwei Fälle, nämlich:

2 *Fall I*: *Für jedes Ding A aus $\mathfrak{S}$ gehört mindestens eines* | ~~*eines*~~ *der beiden*

[a]Cf. *Gödel 1932c*. In his memoir *1994* (p. 203) Menger recalled that he had mentioned the problem "in the letter of congratulation I wrote to Gödel expressing my admiration for his discovery" (item 1. above). If so, it must have been on a second sheet now lost.

somewhat with the pertinent literature. Your participation in the running of the seminar would of course be very valuable, not only to me but certainly also to Hahn.

I am leaving Texas at the end of May and then in all probability will travel westwards, that is home via California, Hawaii, Japan, China, and India and will arrive in Vienna just at the start of lectures. It will give me particular joy to see you then soon ⟦afterward⟧. In the meantime, with best wishes,

Yours,

Karl Menger

Unfortunately I have lost your address and am therefore sending this letter to you in care of Mr. Nöbeling.

2. Gödel to Menger

Vienna, 30 March 1931

Dear Professor,

I have thought about the question you posed[a] concerning the axiomatization of the propositional calculus and have obtained the following result: the existence of two classes $\mathfrak{S}_0, \mathfrak{S}_1$ with the required properties does not in fact follow from axioms a), b) 1–4 (which, as Mr. Nöbeling informed me, you yourself already established in the meantime);[b] but if, in addition to the five axioms, one postulates that F be a *proper* subclass of $\mathfrak{S}$, then the existence of two classes $\mathfrak{S}_0, \mathfrak{S}_1$ does follow, and indeed, in the following way:

First, from $\mathfrak{S} \neq F$ it follows that for no entity A from $\mathfrak{S}$ can A and $\overline{A}$ both belong to F. For by axiom b) 2., $A \rightarrow (\overline{A} \rightarrow B)$ belongs to F for every entity B from $\mathfrak{S}$. Now if A and $\overline{A}$ were to belong to F, then by axiom a) it would ensue that B would also belong to F for every B from $\mathfrak{S}$, contrary to the assumption that $\mathfrak{S} \neq F$.

In what follows I distinguish two cases, namely:

Case I: For every entity A from $\mathfrak{S}$, at least one of the two entities $A, \overline{A}$

[b]The five axioms referred to here are replaced in *Gödel 1932c* by four conditions, numbered I.(a)–(c) and II. Axiom a) herein corresponds to condition II. there, and axiom b) 2. to condition I.(b).

Dinge A, $\overline{A}$ *zu F.* Dann gehört nach dem eben bewiesenen genau eines zu F und eines zu $\mathfrak{S} - F$. In diesem Fall I genügen $\mathfrak{S}_1 = F$, $\mathfrak{S}_0 = \mathfrak{S} - F$ den verlangten Bedingungen, d. h. insbes: $X \to Y$ gehört dann u. nur dann zu $\mathfrak{S} - F$, wenn $X \epsilon F$ und $Y \epsilon \mathfrak{S} - F$. Bew:

Seien X, Y zwei beliebige Dinge aus $\mathfrak{S}$; dann gehören jedenfalls die folgenden 3 Dinge zu F:

$$\overline{X} \to (X \to Y) \tag{1}$$

$$Y \to (X \to Y) \tag{2}$$

$$X \to [\overline{Y} \to \overline{X \to Y}] \tag{3}$$

Denn die Formeln (1), (2), (3) sind allgemeingültige Aussageverbindungen und jede allgemeingültige Aussageformel ist aus den Axiomen b.)1–4 nach der Einsetzungsregel und dem Schlußschema (d. h. Axiom a.) ableitbar (vgl. Hilbert–Ackermann[c] Seite 22–23, 33).

α.) Gilt nun: $X \epsilon F$, $Y \epsilon \mathfrak{S} - F$, dann folgt $\overline{Y} \epsilon F$ und weiter aus (3): $\overline{X \to Y} \epsilon F$[,] das heißt: $X \to Y \epsilon \mathfrak{S} - F$.

β.) In jedem andern Fall ist entweder $\overline{X}$ oder Y Element von F und es folgt aus (1) bzw. (2): $(X \to Y) \epsilon F$.

Fall II: Es gibt Dinge A aus $\mathfrak{S}$*, so daß weder A noch* $\overline{A}$ *zu F gehört.*

In diesem Fall kann man die Klasse F successive solange erweitern, bis
3 schließlich doch Fall I eintritt u. zwar ge|schieht die Erweiterung so, daß immer die Axiome a), b)1–4 erfüllt bleiben. (Für b)1–4 ist dies selbstverständlich, sobald es sich um *Erweiterungen* von F handelt.) Zunächst definiere ich eine Klasse F' folgendermaßen. $X \epsilon F'$ soll gleichbedeutend sein mit: $(A \to X) \epsilon F$ (wobei A eines der Dinge ist, für welche~~s~~ $A \bar{\epsilon} F$, $\overline{A} \bar{\epsilon} F$). Wegen $(A \to A) \epsilon F$ gilt insbesondere $A \epsilon F'$. Ferner überzeugt man sich von folgendem:

1. F ist wirklich Teilklasse von F'. Denn aus $Y \epsilon F$ folgt $(A \to Y) \epsilon F$, weil ja nach ~~Ax. b 1~~ (2) $[Y \to (A \to Y)] \epsilon F$.
2. Axiom a) gilt auch für F' d. h.: Aus $X \epsilon F'$ und ~~$Y \epsilon$~~ $(X \to Y) \epsilon F'$ folgt $Y \epsilon F'$. Denn die Voraussetzung bedeutet:

$$(A \to X) \epsilon F \tag{4}$$

und

$$[A \to (X \to Y)] \epsilon F \tag{5}$$

[c] *Hilbert and Ackermann 1928.*

belongs to F. Then by what was just proven, exactly one of them belongs to F and one to $\mathfrak{S} - F$. In this first case $\mathfrak{S}_1 = F$, $\mathfrak{S}_0 = \mathfrak{S} - F$ satisfy the required conditions, that is, in particular: $X \to Y$ belongs to $\mathfrak{S} - F$ if and only if $X \epsilon F$ and $Y \epsilon \mathfrak{S} - F$.
Proof: Let X, Y be arbitrary entities from $\mathfrak{S}$; then in any case the following three entities belong to F:

$$\overline{X} \to (X \to Y) \tag{1}$$

$$Y \to (X \to Y) \tag{2}$$

$$X \to [\overline{Y} \to \overline{X \to Y}] \tag{3}$$

For the formulas (1), (2), (3) are logically valid propositional combinations, and every logically valid propositional formula is derivable from axioms b) 1–4 by means of the substitution rule and the inference schema (that is, axiom a.). (See Hilbert–Ackermann,[c] pages 22–23, 33.)

α.) Now if $X \epsilon F$ and $Y \epsilon \mathfrak{S} - F$, then it follows that $\overline{Y} \epsilon F$ and furthermore, from (3): $\overline{X \to Y} \epsilon F$, that is: $X \to Y \epsilon \mathfrak{S} - F$.

β.) In every other case, either $\overline{X}$ or Y is an element of F, and it follows from (1), or, respectively, (2), that $(X \to Y) \epsilon F$.

Case II: There are entities A from $\mathfrak{S}$ such that neither A nor $\overline{A}$ belongs to F.

In this case one can successively enlarge the class F until case I eventually obtains, and in fact the enlargement is done in such a way that axioms a) and b)1–4 always remain satisfied. (For b)1–4 this is immediate, as it is a question of *enlargements* of F.) I first define a class F' as follows: $X \epsilon F'$ is to be synonymous with $(A \to X) \epsilon F$ (where A is one of the entities for which $A \bar{\epsilon} F$ and $\overline{A} \bar{\epsilon} F$). Because $(A \to A) \epsilon F$, in particular $A \epsilon F'$ holds. One ⟦can⟧ furthermore convince oneself of the following:

1. F is actually a subclass of F'. For from $Y \epsilon F$ it follows that $(A \to Y) \epsilon F$, since according to (2) $[Y \to (A \to Y)] \epsilon F$.
2. Axiom a) also holds for F', that is: From $X \epsilon F'$ and $(X \to Y) \epsilon F'$ it follows that $Y \epsilon F'$. For the assumption means that:

$$(A \to X) \epsilon F \tag{4}$$

and

$$[A \to (X \to Y)] \epsilon F \tag{5}$$

Nun gehört sicher das Ding:

$$[A \to (X \to Y)] \to [(A \to X) \to (A \to Y)] \tag{6}$$

4 zu F (weil (6) eine allgemeingültige Aussageformel ist). | Daher gilt wegen (4), (5) $(A \to Y) \epsilon F$ und das heißt: $Y \epsilon F'$.

3. ~~A~~ F' ist *echte* Teilmenge von $\mathfrak{S}$. Es gehört nämlich $\overline{A}$ nicht zu F'. Denn $\overline{A} \epsilon F'$ würde bedeuten: $(A \to \overline{A}) \epsilon F$. Nun gilt: $[(A \to \overline{A}) \to \overline{A}] \epsilon F$, weil die Formel $(X \to \overline{X}) \to \overline{X}$ allgemeingültig ist. Daher würde aus $(A \to \overline{A}) \epsilon F$ folgen: $\overline{A} \epsilon F$ entgegen der Voraussetzung II.

Da die Klasse F' nach 1.–3. denselben Bedingungen genügt wie F, kann man auf F' dasselbe Verfahren anwenden, falls nicht etwa schon für F' der Fall I eintritt (womit man schon am Ziele wäre). Dieses Verfahren kann man wie leicht zu sehen ins Transfinite fortsetzen; denn ~~eine~~ ⟨die⟩ Vereinigungs⟨menge einer Folge⟩ monoton wachsender Klassen $F^{(i)}$, welche sämtlich dem Axiomen a.) b.)1–4 genügen und sämtlich echte Teilklasse von $\mathfrak{S}$ sind, ergibt wieder eine Klasse mit den genannten Eigenschaften (die Vereinigungsmenge kann nicht $= \mathfrak{S}$ sein, weil sonst X und $\overline{X}$ (bei beliebigen X) beide in der Vereinigungsmenge, folglich (wegen der Monotonie) auch beide in einem Summanden $F^{(i)}$ vorkommen müßten;
5 | dann wäre aber nach dem am Anfang Bewiesenem schon dieser Summand $F^{(i)} = \mathfrak{S}$). Auf Grund der Wohlordnungssatzes kann man daher exakt beweisen, daß man schließlich zu einer Klasse $F^{(\alpha)}$ kommt, für welche Fall I eintritt, und dann genügen, wie oben gezeigt, F^{α} und $\mathfrak{S} - F^{(\alpha)}$ den verlangten Bedingungen.

Übrigens glaube ich nicht, daß der eben bewiesene Satz für den Bernaysschen Widerspruchsfreiheitsbeweis erforderlich ist. Denn der bei Hilbert–Ackermann zu Grunde gelegte axiomatische Aufbau des Aussagenkalküls ist ein rein metamathematischer. D. h. es ist nicht von irgendwelchen Dingen die Rede, welche gewissen Axiomen zu genügen hätten, sondern Gegenstand der Betrachtung sind *Formeln gewisser Bauart* d. h. *Schriftzeichenkombinationen*, von deren Bedeutung abgesehen wird und deren Eigenschaften sich aus den anschaulichen Tatsachen der linearen räumlichen Anordnung ⟨von Zeichen⟩ (die als bekannt angenommen werden) ergeben. Daß beim Widerspruchsfreiheitsbeweis dann doch von einer "Interpretation" der Zeichen gesprochen wird, ist meiner Meinung nach bloß eine dem leichteren Verständnis dienende unpräzise Aus-
6 drucksweise. Streng genommen hätte | man zu sagen, daß gewissen Zeichenkombinationen nach bestimmten Vorschriften Zahlen ⟨⟨(bzw. Zahlenfunktionen)⟩⟩ zugeordnet werden. In diesem Sinn läßt sich, glaube ich, der Widerspruchsfreiheitsbeweis ganz exakt durchführen, wobei insbes.

Now certainly the entity:

$$[A \to (X \to Y)] \to [(A \to X) \to (A \to Y)] \tag{6}$$

belongs to F (because (6) is a logically valid propositional formula). Therefore $(A \to Y) \epsilon F$ holds on account of (4) and (5), and that means: $Y \epsilon F'$.

3. F' is a *proper* subset of $\mathfrak{S}$. That is to say, $\overline{A}$ does not belong to F'. For $\overline{A} \epsilon F'$ would mean: $(A \to \overline{A}) \epsilon F$. Now $[(A \to \overline{A}) \to \overline{A}] \epsilon F$ holds, because the formula $(X \to \overline{X}) \to \overline{X}$ is logically valid. Therefore $\overline{A} \epsilon F$ would follow from $(A \to \overline{A}) \epsilon F$, contrary to assumption II.

Since according to 1.–3. the class F' satisfies the same conditions as F, one can apply the same procedure to F', should case I not already obtain for F' (whereby the goal would already be reached). As one can easily see, this procedure can be continued into the transfinite; for the union of a sequence of monotone increasing classes $F^{(i)}$ that all satisfy axioms a.) and b.)1–4 and that are all proper subclasses of $\mathfrak{S}$ again yields a class with the stated properties (the union cannot be equal to $\mathfrak{S}$, because otherwise (for arbitrary X) both X and $\overline{X}$ would have to occur in the union set, and consequently (because of the monotonicity) both would also have to occur in a summand $F^{(i)}$; but then, by what was proven at the outset, that summand $F^{(i)}$ would already be equal to $\mathfrak{S}$). On the basis of the Well-ordering Theorem one can therefore prove precisely that one finally arrives at a class $F^{(\alpha)}$ for which case I obtains, and then, as was shown above, $F^{(\alpha)}$ and $\mathfrak{S} - F^{(\alpha)}$ satisfy the required conditions.

Anyway, I don't think that the theorem just proved is necessary for Bernays' consistency proof. For the axiomatic construction of the propositional calculus taken as the basis in Hilbert–Ackermann is a purely metamathematical one. That is, it is not a question of some sort of entities that have to satisfy certain axioms, but rather ⟦the⟧ object⟦s⟧ under consideration are *formulas of a certain structure*, that is, *combinations of signs*, whose meaning is disregarded and whose properties result from the concrete facts of the linear spatial ordering of the signs (which is assumed to be well known). That in the consistency proof one then nevertheless speaks of an "interpretation" of the signs is, in my opinion, merely an imprecise way of speaking that serves to make understanding easier. Strictly speaking, one would have to say that certain combinations of signs are correlated with numbers (or, respectively, numerical functions) according to definite prescriptions. In this sense I believe that the consistency proof may be carried out with complete precision, whereby it is

zu beachten ist, daß auch die zu beweisende Aussage der Widerspruchsfreiheit eine rein formale (metamathematische) Bedeutung hat⟦,⟧ d. h. sie besagt lediglich, daß man aus den 4 Ausgangsformeln a.)–d.) (Seite 22) durch Manipulieren nach den Regeln α), β) (Seite 23) niemals zwei Formeln erhalten kann, deren eine durch Überstreichen aus der andern hervorgeht.

~~Anbei~~ ⟨Gleichzeitig⟩ übersende ich Ihnen einige Separata meiner beiden Arbeiten und wäre Ihnen sehr dankbar, wenn Sie so freundlich wären, die überzähligen an eventuelle Interessenten Ihres Bekanntenkreises weiterzugeben.

Mit den besten Empfehlungen,

Ihr ergebener Kurt Gödel

3. Menger to Gödel

2./VI. 32

Lieber Herr Gödel, Die Notgemeinschaft deutscher Wissenschaft hat mir zur Fertigstellung & Herausgabe der Sammlung "Mengentheoretische Geometrie"[a] einen Geldbetrag bewilligt. Ich erlaube mir daraufhin vor allem an Sie die Frage zu richten, ob Sie bis Ende dieses Studienjahres durch genaue Lektüre des Manuskriptes des etwa 120 Druckseiten langes Einleitungsbandes über Geometrie mich unterstützen wollten. Die zur Verfügung gestellte Summe würde einstweilen nur die (natürlich in keinem Verhältnis zur Bedeutung Ihre Hilfe stehende) Remuneration von 250 Schillingen gestatten. Was Ihren (übrigens schon heute be-|2 trächtlichen) geistigen Anteil an dem Buche betrifft, so wird er,—dies bedarf wohl keines Bemerkung,—in demselben in volle Evidenz gesetzt werden. Mit der Bitte an baldige Antwort und besten Grüssen inzwischen

aufrichtig Ihr

Menger

[a]See *Menger and Nöbeling 1932*.

to be observed in particular that the assertion of consistency that is to be proved also has a purely formal (metamathematical) meaning, that is, it says only that from the four initial formulas a.)–d.) (page 22) one can never obtain, through manipulations according to the rules α) and β) (page 23), two formulas, one of which arises from the other through overlining.

I am sending you at the same time a few offprints of both my papers, and I would be very grateful if you would be so kind as to pass the extras on to those of your circle of acquaintances who might be interested.

With best wishes,

Yours,

Kurt Gödel

3. Menger to Gödel

2 June 1932

Dear Mr. Gödel, The Emergency Society for German Science has granted me a sum of money for the preparation and editing of the collection "Set-theoretic Geometry".[a] With that in mind I venture above all to ask you the question whether you would assist me up to the end of this academic year by a careful reading of the manuscript (about 120 printed pages in length) of the introductory volume on geometry. The amount made available (standing of course in no relation to the significance of your help) would for the present allow only the payment of 250 Schillings. As far as your intellectual share in the book is concerned (which, after all, is today already considerable), it will—as probably need not be remarked—be given full credit therein. With the request of a prompt answer and with best wishes in the meantime,

Yours sincerely,

Menger

4. Menger to Gödel

1932[a] Mönichkirchen N. Ö.
Villa Gisela

Lieber Herr Gödel

ich habe vergessen Ihnen zu sagen, daß ich vor meiner Abreise um Flüssigmachung der mir von der Österr. Deutschen Wissenschaftshilfe bewilligten kleinen Subvention ⟨und Übertragen derselben⟩ auf mein Kto bei der Österr. Creditanstalt angesucht habe & die letztere gleichzeitig beauftragt habe, Ihnen Ihren Anteil von 250 S. zu überweisen. Ich benütze den Anlass, um Ihnen nochmal für die Mühe, die Sie sich auf der Redaktion der Manuscriptes gegeben haben, bestens zu danken, während 2 ich ⟨die⟩ aus Ihrer Mitautor|schaft sich ergebenden Fragen mündlich mit Ihnen zu besprechen hoffe. Ich wäre Ihnen nun nur noch sehr dankbar, wenn Sie den Teil, der den formalen Aufbau der Mengenlehre betrifft, sich etwas durch den Kopf gehen lassen wollten, insbesondere auch ein Beispiel, etwa wie Endpkte eines topol. Raumes.

Hier ist es wie ein Paradis & ich bin schon in den zwei Tagen die ich der Wiener Hitze entflohen & faul bin, recht ausgeruht. Nächste Woche 3 will ich dann die Fertigstellung des noch zu |Erledigenden beginnen.

Schauen Sie nur auch, so bald als möglich aus Wien wegzukommen! Ich würde mich natürlich außerordentlich freuen, Sie in Laufe des Sommers einmal hier begrüßen zu können. Falls Sie ins Semmeringgebiet gehen, würde ich Ihnen gern auf halben Weg entgegenkommen & Sie dann herbegleiten. Ich muß aber noch die Karte näher ansehen. Halten Sie mich nur bitte über Ihre Adresse auf den Laufenden. Wenn Sie 4 von Wien | weggehen, vergessen Sie bitte nicht Ihre Notizen zu meinem Buch mitzunehmen, damit wenn wir uns treffen, wir die Sache in einem Nachmittag in Ordnung bringen können.

Are you already reading (and enjoying) "The house without key"?[b] I like this story very much. During our trip this summer, English will be our language of conversation if you like. I certainly hope to see you here. Meanwhile kindest regards and best wishes!

[a] This date appears to be an endorsement by Gödel.

4. Menger to Gödel

1932[a] Mönichkirchen N. Ö.
Villa Gisela

Dear Mr. Gödel,

I forgot to tell you that before my departure I requested the conversion to cash of the small subvention granted me by the Austro-German Scientific Aid ⟦Society⟧ and its transfer to my account with the Austrian Credit-Anstalt, and at the same time I directed the latter to remit to you your share of 250 Schillings. I take this opportunity to thank you very much once again for the trouble you have taken in the editing of the manuscript, while I hope to discuss with you orally the questions arising from your coauthorship. I would now be all the more grateful to you if you would be so kind as to mull over the part concerning the formal construction of set theory, especially also an example, perhaps like endpoints of a topological space.

It is like a paradise here, and in the two days that I have escaped the heat of Vienna and been lazy I have already fully rested. Next week I want to begin the preparation of what is still to be attended to.

Do take heed, too, to come away from Vienna as soon as possible! I would of course be exceedingly happy to be able to welcome you here sometime in the course of the summer. In case you go to the vicinity of Semmering, I would be glad to come to meet you halfway and then escort you here. But I must still look more closely at the map. Please do keep me informed of your address. If you go away from Vienna, please don't forget to take along my book, so that if we meet we can put the matter in order in an afternoon.

Are you already reading (and enjoying) "The house without key"?[b] I like this story very much. During our trip this summer, English will be our language of conversation, if you like. I certainly hope to see you here. Meanwhile kindest regards and best wishes!

---

[b] *Biggers 1925.*

Very cordially yours,

Karl Menger

Bitte um gelegentliche Returnierung beiliegender Beteiligung unter⟦??⟧ bei.

5. Gödel to Menger

Wien 4./VIII. 1932

Lieber Herr Menger,

vielen Dank für Ihren freundlich Brief und die 250 S. (Empfangsbestätigung liegt bei.) Es freut mich, daß Ihnen meine Mithilfe wertvoll war, und ich werde mir auch weiterhin die größte Mühe geben.—Augenblicklich habe ich mit dem Grundlagenbericht[a] viel zu tun und die Arbeit daran ist nicht immer sehr interessant. Daher bin ich auch in Ihrem englischen Buch noch nicht | über die ersten Seiten hinausgekommen; glaube übrigens ich sollte zunächst meine Englischkenntnisse anderweitig verbessern, da es sehr mühsam für mich zu lesen ist.—Neulich lernte ich einen Herrn Popper (Philosophen) kennen, der eine unendlich lange Arbeit geschrieben hat, in der, wie er sagt, sämtliche philos. Probleme gelöst sind. Er bemühte sich, mein Interesse dafür zu erwecken, und berief sich auch darauf, daß Sie ihn kennen und Ihnen seine Absicht, das "Sinn"problem | aus der Welt zu schaffen, sehr gut gefallen habe. Halten Sie etwas von ihm?—Ich bleibe vorläufig in Wien und werde voraussichtlich gegen Ende August für einige Tage nach Gösing fahren. Ich würde mich sehr freuen, wenn ich Sie bei dieser Gelegenheit irgendwo treffen könnte. Da aber meine Geographiekenntnisse schwach sind, muß ich die näheren Details Ihnen überlassen und kann nur hoffen, daß die Sache nicht den | Gesetzen der Geometrie widerspricht. Nochmals vielen Dank und die besten Grüsse

Ihr ergebener

Kurt Gödel

[a]The book that Gödel had agreed to coauthor with Heyting, from which he ultimately withdrew. For further details see the correspondence with Heyting in this volume or *Dawson 1997*, pp. 83–85.

Very cordially yours,

Karl Menger

At your convenience please return the enclosed ⟦??⟧.

5. Gödel to Menger

Vienna, 4 August 1932

Dear Mr. Menger,

Many thanks for your kind letter and the 250 Schillings (receipt enclosed). I'm happy that my assistance was valuable to you, and I will also take the greatest pains in the future.—At the moment I have much to do on the report on foundations[a] and the work on it is not always very interesting. Therefore I have also not yet gotten to the first pages in your English book; anyway, I think I should first improve my knowledge of English further, since it is very laborious for me to read.—Recently I became acquainted with a Mr. Popper (a philosopher), who has written an infinitely long paper in which, so he says, all philosophical problems are solved. He endeavored to awaken my interest in it, and also appealed to the fact that you know him and that his plan to remove the problem of "meaning" from the world has pleased you very much. Do you think anything of him?—I am staying for the time being in Vienna and will probably travel to Gösing for a few days toward the end of August. I would be very pleased if I could meet you somewhere on that occasion. But since my knowledge of geography is weak, I must omit more precise details and can only hope that the matter doesn't violate the laws of geometry. Once again, many thanks and best wishes

Yours,

Kurt Gödel

---

6. Menger to Gödel[a]

Lieber Herr Gödel,

ich habe den Brief von Bernays[b] genau gelesen. Aber besonders gefällt mir die Sache in dieser Form nicht. Nicht nur, daß sie von den tatsächlichen Anwendungen ⟨in⟩ der mengentheor. Geometrie entfernt ist, sondern ich habe den Eindruck, daß der Dualismus Menge-Klasse (Eigenschaft) entbehrlich sein muß. Zudem, wenn nun eine praktische Eigenschaft vorliegt—wann ist sie eine der als primitive Begriffe angenommen? Es ist mir nicht klar, wie man, wenn Mengen & Eigenschaften als prim. Begr. genommen werden, feststellt, daß die Menge aller Endpunkte einer Kurve eine Menge ist. Ich weiß: ~~Es ist~~ ⟨Wenn⟩ es eine ⟨durch eine Menge repräsentierte⟩ Eigenschaft eines Raumpktes ist, der Kurve anzugehören,
2 & wenn es eine | Teileigenschaft ist, der Kurve anzugehören & Endpkt zu sein, so wird auch diese Teileigenschaft durch eine Menge repräsentiert (die Menge aller Endpkte). Aber warum ist es eine Eigenschaft, Endpkt zu sein, da Eigenschaften doch nicht definiert wurden? Vermutlich übersehe ich etwas, aber daß die Duplizität Menge-Eigenschaft die Schwierigkeit des Aussonderungsaxioms lösen soll, verstehe ich nicht. Viel besser gefällt mir die Form, die Sie erwähnten, gewissen Quantifikatoren ⟨durch ein Axiom⟩ Mengen entsprechen zu lassen. Ferner hasse ich die
3 Form, | in der bei Bernays, wie üblich, unendl. Mengen eingeführt werden, aber das ist nebensächlich.

Aber ist es denn wirklich nötig, im Ausgangsbereich Mengen & Mengen von Mengen beisammen zu haben, so daß Summen einer Menge & eines Normalbereiches definiert sind, die doch kein Mensch verwendet? Das *muß* doch schöner zu machen gehen! Soll es nicht doch ⟦~~??~~⟧ möglich sein die Punktmengen (Teilmengen des Raumes, darunter auch die einpunktigen, d. h. jene welche die keine echten Teile enthalten) als un-
4 definiert einzuführen, in Verallge|meinerung der undefinierten Einführung gewisser Pktmengen (der linearen oder quadratischen) in der Elementargeometrie & für sie bloß Addition, Multiplik, Auswahlmengen etc. zu postulieren, allerding⟦s⟧ Summe, Durchschn & Auswahl gleich für *Mengen* solcher undef. Mengen. Beim *Operieren* mit undef. Mengen braucht man dann die abstrakte Mengenlehre (vermutlich übrigens in praxi nicht sehr viel von ihr). Aber die betrachteten Mengen sind "ir-
5 gendwelche Dinge, die gewissen Beziehungen genügen". Daß tatsäch|lich oft von "Punkt der Menge" gesprochen wird, spricht, glaube ich, nicht

[a]Although this letter is undated, it bears the endorsement "1932?", apparently in Gödel's hand; 1933, however, seems more likely.

6. Menger to Gödel[a]

Dear Mr. Gödel,

I have read Bernays' letter[b] carefully, but in this form the matter is not particularly pleasing. Not only is it remote from the actual applications in set-theoretic geometry, but I have the impression that the set-class (property) dualism must be dispensable. Moreover, if now a practical property is at hand—when is it one of the ones taken as a primitive concept? It is not clear to me how, if sets and properties are taken as primitive concepts, one establishes that the set of all endpoints of a curve is a set. I know: If lying on the curve is a property, represented by a set, of a point in space, and if lying on the curve and being an endpoint is a subproperty, then that subproperty will also be represented by a set (the set of all endpoints). But why is being an endpoint a property, since properties were, after all, not defined? Presumably I am overlooking something, but I do not understand how the set-property duplication would surmount the difficulties of the axiom of separation. I like much better the form that you mentioned, letting certain quantifiers correspond to sets by means of an axiom. Furthermore I hate the form in which, as is customary, infinite sets are introduced by Bernays; but that is beside the point.

But is it really necessary to have sets and sets of sets together in the domain one starts out with, so that unions of a set and of a normal domain are defined, which, however, no one uses? There *must* surely be a prettier way to do it. Shouldn't it be possible to introduce the point-sets (subsets of the space, including the singletons, that is, those which contain no proper part) as undefined ⟦notions⟧, as a generalization of the undefined introduction of certain (linear or quadratic) point-sets in elementary geometry, and to postulate for them only addition, multiplication, choice sets, and so on, and likewise of course ⟦axioms of⟧ union, intersection and choice for *sets* of such undefined sets? In *operating* with undefined sets one then uses abstract set theory (presumably anyway—in practice, not very much of it). But the sets considered are "arbitrary things that satisfy certain conditions". That one in fact often speaks of "point of the set" does not, I think, contradict this standpoint, for these

[b]Presumably that of 3 May 1931.

sehr gegen diesen Standpkt, denn diese “Punkte” lassen sich auf nicht allzukünstlichem (oft sogar fruchtbarem) Wege durch “einpunktige Mengen” ersetzen. Und so hat man mit einem Schlag 1) die Punktmengenlehre als naturgemäße Verallgemeinerung der Elementargeometrie & hat zugleich 2) die abstrakten Räume, ohne Typenvermischung! Ich wäre also doch für *diesen* Standpunkt, wenn ich nur wüßte was als Teilmenge
6 zulässig | ist, m. a. W. wenn das Aussonderung⟦s⟧axiom mir klar wäre. Hier handelt es sich natürlich um Festsetzungen über die inhaltlichen *Mengen* von undefin. Mengen ⟨m. a. W. um die Axiome der abstr. Mengenlehre⟩. Aber wenn Sie mir da helfen könnten, die Axiome so zu formulieren, daß die Menge aller Endpkte einer Kurve zulässig ist!!

Lieber Herr Gödel, lassen Sie sich diese mich furchtbar quälende Sache durch den Kopf gehen! Und lachen Sie mich nicht aus! Es kennt sich ja
7 auch außer mir kein Mensch | mehr aus. Sie sahen ja, daß Fränkel nicht wußte, daß man ohne Reduz. axiom das ganze Kontinuum herausbekommen könne. Wenn es Ihnen also auch als eine Trivialität erscheint unterziehen Sie sich doch dieser Mühe. M Die Sache muß endlich auch für Mathematiker, die nicht Speziallogiker sind, geklärt werden. Denken Sie nur an Weyls widerliches Geschwätz! Mein Einleitungsband muß einen die Sache vor meinem Gewissen rechtfertigenden logischen § enthalten
8 und ich brauche wohl nicht zu sagen, daß| wenn Sie sich der (für Sie noch dazu geringen) Mühe der Klarstellung unterziehen, die Sache als Ihre Arbeit in volle Evidenz gesetzt wird.

Ich komme am 1. April zurück & hoffe Sie dann bald zu sehen. Skilaufen müssen Sie doch noch einmal lernen. Es ist hier *so* schön. Ich gehe jetzt noch für ⟦ein⟧ paar Tage auf eine Hütte. Die Korrekturen des ersten Teiles von Heft 3 sind wohl schon in meiner Wohnung. Ich habe Frau Anna instruiert, Ihnen darüber telef. (A–12–4–93) Auskunft zu erteilen und wenn Sie gelegentlich im Vorbeikommen die Sendung ⟨ganz oder teilweise⟩ abholen wollen, sie auszufolgen. Gödelius, oder wenn Sie auf Titeln angesprochen werden wollen, Doctor mirabilis, ich grüße Sie

—Aufrichtig der Ihre

Karl Menger

"points" may be replaced in ways that are not too artificial (and are often even fruitful) by "singletons". And so in one stroke one has 1) point-set theory as a natural generalization of elementary geometry and at the same time has 2) abstract spaces, without mixing of types! I would thus really favor *this* standpoint if only I knew what is admissible as a subset, in other words, if the axiom of separation were clear to me. Here of course it is a question of stipulations concerning concrete *sets* of undefined sets, in other words, of the axioms of abstract set theory. If only you could help me to formulate the axioms so that the set of all end-points of a curve is admissible!!

Dear Mr. Gödel, do ponder these (for me) terribly tormenting matters! And don't laugh at me! No one besides me is more knowledgable about it. You saw, I'm sure, that Fraenkel didn't know that one can extract the entire continuum without the axiom of reducibility. So even if it looks like a triviality to you, do take the trouble to do this. In the end, the matter must be clarified for mathematicians who are not specialists in logic. Just think of Weyl's disgusting prattle! My introductory volume must contain a section logically justifying the matter confronting my conscience, and I probably need not say that if you undertake the task of clarification (a meager one at that, for you), that material will be fully acknowledged as your work.

I am coming back on 1 April and hope to see you then soon. You really must learn to ski again. It is *so* beautiful here. I am now going to a hut for a few more days. The proof sheets of the first part of book three are probably already at my house. I have instructed Mrs. Anna to give you information about it by telephone (A-12-4-93) and to hand it over to you if, while passing by, you occasionally want to pick up all or part of what has been sent. Gödelius, or, if you want to be addressed by title, Doctor mirabilis, I send you greetings

— Yours sincerely,

Karl Menger

7. Gödel to Menger

Wien 3./IV. 1933

Lieber Professor Menger,

besten Dank für Ihre freundliche Karte![a] Hoffentlich sind Sie mit Ihren Aufenthalt weiter zufrieden u. erholen sich gut—Ich sende heute die Korrekturen Ihres Vortrags[b] an Sie ab und möchte zur Verdeutlichung folgendes hinzufügen:
p. 3. Das erste ~~L~~ukasiewicz Axiom habe ich Ihnen, wie ich jetzt bemerke, Ł
⟨seinerzeit⟩ falsch angegeben. Daher muß auch das Beispiel für die Anwendung der Einsetzungsregel etwas abgeändert werden.
p. 4. zweiter Abs. Der Funktionenkalkül in der Frege–Russellschen Form, auf die Sie am Schluß des vorhergehenden Absatzes anspielen, enthält schon die Relationstheorie (wenigstens zum großen Teil), weil mehrstellige Funktionsvariable vorkommen.
p. 9. ⌐ Sobald man unter "Logik u. Mathm." ein bestimmtes formales System versteht, muß man die unter ⌐ eingeschalteten Worte hinzufü-
2 gen. "Logik u. Mathem." ohne | ~~ohne~~ nähere Präzisierung ist aber zu ungenau, um darüber irgend etwas behaupten zu können.
p. 9. unten. Das im Text stehende "kann man nicht" könnte vielleicht mißverstanden werden als "kann man derzeit nicht".

Zur Frage der unentscheidbaren Sätze (p. 10 oben) möchte ich noch bemerken, daß es in *jedem* formalen System, ⟨das die Arith. enthält,⟩ sogar unentsch. Sätze aus der Lehre von den natürlichen Zahlen gibt, die in höheren Systemen entscheidbar werden.

Die Gruppe der euklidischen Bewegungen + Ähnlichkeitstransformationen ist in der Ebene ~~z. B.~~ Schnitt der projektiven Abbildungen mit den Kreisverwandtschaften (analog im Raum), doch weiß ich natürlich nicht, ob Veblen das damals gemeint hat.

Von den Korrekturen des Kolloquiumberichtes (1. Teil) haben Sie mir, wie ich jetzt sehe, ein schon korrigiertes Exemplar gegeben. Ist das nicht ein Irrtum?

Kommen Sie bald zurück oder soll ich die Abzüge des Kolloquiumberichtes, bis ich sie gelesen habe, an Sie schicken?

Mit besten Grüßen

Ihr Kurt Gödel

---

[a]No longer extant.

7. Gödel to Menger

Vienna 3 April 1933

Dear Professor Menger,

Many thanks for your friendly postcard![a] Hopefully, you are further satisfied with your vacation and are making a good recovery.—I am sending the proof sheets of your lecture[b] off to you today and would like to add the following by way of elucidation:

p. 3. At the time, as I just now noticed, I misstated to you the first Łukasiewicz axiom. Therefore the example of the application of the substitution rule must also be altered somewhat.

p. 4. Second paragraph. The functional calculus in the Frege–Russell form, to which you allude at the end of the preceding paragraph, already contains the theory of relations (at least in large part), because many-place function variables occur.

p. 9. ⌟⌜ The moment one construes "logic and mathematics" as a definite formal system, one must add the words interpolated after the ⌟⌜ . Without more precise specification, however, "logic and mathematics" is too inexact to be able to assert anything about it.

p. 9. Bottom. The ⟦phrase⟧ "one cannot" that appears in the text could perhaps be misunderstood as "one cannot at present".

On the question of undecidable statements (top of p. 10) I would still like to remark that in *every* formal system that contains arithmetic there are even undecidable statements from the theory of the natural numbers that are decidable in higher systems.

In the plane the group of Euclidean motions + similarity transformations is the intersection of the projective mappings with the circular transformations ⟦those taking circles into circles⟧ (similarly in space), but of course I don't know whether Veblen meant that at that time.

As I now see, you gave me an already corrected copy of the proof sheets of the colloquium report (first part). Isn't that a mistake?

Are you coming back soon, or should I send you the galleys of the colloquium report before I have read them?

With best wishes,

Yours, Kurt Gödel

---

[b]Perhaps "Die neue Logik" (*Menger 1933*), one of a series of five public lectures by Menger and others, delivered at the University of Vienna during the academic year 1931–1932. On the basis of that lecture Menger subsequently claimed priority for having given the first "popular" exposition of the incompleteness theorems.

8. Gödel to Menger

Wien 6./IV. 1933

Lieber Professor Menger,

Das intuit.-form. Wörterbuch lautet folgendermaßen:

klass.	int.		
$\sim p$	$\neg p$	$p \to q$	$\neg(p \wedge \neg q)$
$p.q$	$p \wedge q$	$(x)F(x)$	$(x)F(x)$
$p \vee q$	$\neg(\neg p \wedge \neg q)$	$(Ex)F(x)$	$\neg(x)\neg F(x)$

Die arithmetischen Begriffe (=, +1, 0), sowie ⟨die⟩ daraus rekursiv definierten Funktionen ($a + b$, $a \times b$, ⟨$a < b$⟩, ... etc.) bleiben bei der Übersetzung unverändert.[a]

Eine Definition heißt imprädikativ, wenn in ihr auf eine Gesamtheit Bezug genommen wird, zu der das definierte Objekt selbst gehört. In Anwendung auf den Logikkalkül kann man exakter auch so sagen: Die Definition einer Klasse (Relation) k-ten Typs[1] ⟨heißt⟩ impräd, wenn in ihr ein Quantifikator vorkommt, der sich auf die Objekte (⟨d. h.⟩ Klassen oder Relationen) k-ten oder höheren Typs bezieht. Die intuit. Einwände richten sich nicht gegen solche Fälle von impräd Def., bei denen feststeht, daß das definierte Objekt auf andere Weise *auch* prädikativ definiert wer-
2 den kann (z. B. | die kleinste Zahl aus einer Menge von natürl. Zahlen; jede natürl. Zahl kann als Summe von Einsen prädikativ definiert werden!)

Mir ist noch eingefallen, daß Sie in der Formulierung meiner Resultate über Unentsch. und Widerspruchsfr. (Seite 9 und 10 gesperrt Gedrucktes) statt "Theorie" unbedingt sagen sollten "formale Theorie" oder "Formalismus" um anzudeuten, daß auch die Schlußregeln genau präzisiert sein müssen, was nach üblichem Sprachgebrauch von einer Theorie nicht verlangt wird.

Mit besten Grüßen

Ihr Kurt Gödel

[1] Eine Relation soll vom k-ten Typ heißen, wenn die Relationsglieder höchstens $k-1$-ten Typ haben, und mindestens eines wirklich $k-1$-Typ.

[a] The references here are evidently to Gödel's paper *1933e*.

8. Gödel to Menger

Vienna 6 April 1933

Dear Professor Menger,

The intuitionist-formalist lexicon reads as follows:

classical	intuitionistic	classical	intuitionistic
$\sim p$	$\neg p$	$p \rightarrow q$	$\neg(p \wedge \neg q)$
$p.q$	$p \wedge q$	$(x)F(x)$	$(x)F(x)$
$p \vee q$	$\neg(\neg p \wedge \neg q)$	$(Ex)F(x)$	$\neg(x)\neg F(x)$

The arithmetical notions ($=$, $+1$, 0) as well as the functions ($a + b$, $a \times b$, $a < b$, ... etc.) defined recursively from them remain unchanged in the translation.[a]

A definition is called impredicative if in it reference is made to a totality to which the defined object itself belongs. As applied to the logical calculus one can also say, more precisely: The definition of a class (relation) of the k-th type[1] is called impredicative if a quantifier occurs in it that refers to the objects (that is, classes or relations) of k-th or higher type. The intuitionistic objections are not directed against such cases of impredicative definitions in which it is established that the defined object can *also* be defined predicatively in another way (e.g., the least number in a set of natural numbers; every natural number can be defined predicatively as a sum of ones!)

It has occurred to me as well that in the formulation of my results about undecidability and consistency (printed in spaced type on pages 9 and 10), instead of "theory", unconditionally, you should say "formal theory" or "formalization" in order to make clear that the inference rules must also be precisely specified, which is not required of a theory according to ordinary linguistic usage.

With best wishes,

Yours, Kurt Gödel

[1] A relation is said to be of k-th type if the members of the relation have at most $k-1$-st type and at least one really is of type $k-1$.

9. Menger to Gödel

P. O. B. 206
Notre Dame, Ind.
U.S.A.
22/V/1937

Lieber Herr Gödel,

Wie gehts und was machen Sie immer? Uns geht es hier sehr gut. Die University of Notre Dame, eine liberale katholische Universität, ⟨$1\frac{1}{2}$ Bahnstunden östl.-Chicago,⟩ die ein~~e~~ erstklassiges chemistry-department hat. (Fr. Nieuwland hat hier den künstlichen Gummi erfunden), will jetzt ihre mathem. und physics departments ausgestalten. Im Zuge dieser Ausgestaltung hat sie ⟨den Physiker A.⟩ Haas und mich aus Wien herberufen. Ich erzählte neulich dem Presidenten von Ihnen und er bat mich, bei Ihnen anzufragen, ob Sie nicht auch herkommen wollen. Wenn Sie, was mich natürlich außerordentlich freuen würde~~m~~, im Prinzip einverstanden sein sollten, so lassen Sie mich es, bitte, ehestens wissen, damit die Details in Reine gebracht werden können.

2 | Ich würde natürlich meinen, daß Sie zunächst sich die ganze Situation ein Jahr lang ansehen sollen. Ich möchte Sie aber bitten, niemandem, der mit der Wiener Universität oder unserem Kolloquium zusammenhängt, vorläufig von meinem Schreiben irgendetwas mitzuteilen.

Lassen Sie bald von sich hören und seien Sie herzlichst gegrüßt,

Karl Menger

10. Gödel to Menger

Wien 3./VII. 1937

Lieber Professor Menger:

Vielen Dank für Ihren freundlichen Brief. Ich wäre im Prinzip einverstanden an die University of Notre Dame zu kommen; es würde mich sogar sehr interessieren den Betrieb an einer katholischen amerikanischen Universität kennen zu lernen. Die Bulletins, die mir zugeschickt wurden, haben mich sehr interessiert u. ich lasse mich bestens dafür bedanken. Ich ersehe aus Ihrem Schreiben nicht, wann die Sache aktuell werden würde. Meinerseits käme frühestens das Sommersemester 1938 in Betracht. Es hat mich zwar für dieses Semester Ph. Frank nach Prag ein-

9. Menger to Gödel

P.O. Box 206
Notre Dame, Ind.
U.S.A.
22 May 1937

Dear Mr. Gödel,

How are you, and what all are you up to? Things here are going very well. The University of Notre Dame, a liberal Catholic University ⟦lying⟧ east of Chicago one and a half hours by train, which has a first-class chemistry department (Father ⟦Julius⟧ Nieuwland discovered artificial rubber here), now wants to upgrade its mathematics and physics departments. In accordance with this upgrading, they called the physicist A. Haas and me here from Vienna. I recently told the president about you and he asked me to inquire of you whether you would not come here too. Should you agree in principle, which would of course make me exceedingly happy, please let me know as soon as possible, so that the details can be arranged.

Of course, I would think that you should first look at the whole situation for a year. But I would like to ask you, for the moment, to tell no one who is associated with the University of Vienna or with our colloquium anything of what I have written.

Let me hear from you soon, and accept my most cordial greetings,

Karl Menger

10. Gödel to Menger

Vienna, 3 July 1937

Dear Professor Menger:

Many thanks for your friendly letter. I would in principle agree to come to the University of Notre Dame; to become acquainted with the operation of a Catholic American university would even be very interesting to me. The Bulletins that were sent to me interested me very much and I wish to express my thanks for them. From what you wrote I did not learn when the ⟦appointment⟧ would become effective. On my side, the summer semester of 1938 would be the earliest that could be considered. Indeed, Ph⟦ilipp⟧ Frank has invited me to Prague for this semester,

geladen, doch ist durchaus noch nicht sicher, ob aus dieser Sache etwas wird. Im Wintersemester 1938/39 werde ich dann vermutlich in Princeton sein. Es wäre mir wesentlich, wenn ich mich nur für *ein* Semester verpflichten müsste. Wie Sie ja wissen, habe ich in Amerika mit meiner Gesundheit schlechte Erfahrungen gemacht u. möchte mich daher nicht im voraus für längere Zeit binden. Wenn aber die genannten Vorausset-
2 zungen zutreffen, würde ich | gerne akzeptieren, falls die übrigen Bedingungen (Bezahlung u. Verpflichtungen meinerseits) annehmbar sind.[a]

Bei mir gibt es nicht viel Neues. Seit ich aus Aflenz zurück bin, geht es mir gesundheitlich wieder schlechter, aber immer noch leidlich gut. Von meiner Vorlesung, in der ich am Schlusse die Widerspruchsfreheit des Auswahlaxioms im System der Mengenlehre bewiesen habe, hat Ihnen ja vielleicht Herr Wald geschrieben. Den analogen Beweis für das System der Principia Mathematica, sowie das Teilresultat über die Kontinuumhypothese, von dem ich Ihnen erzählte, habe ich im Kolloquium referiert.[b]

Augenblicklich überlege ich mir eben, ob ich im nächsten Semester etwas Elementares oder etwas Höheres lesen soll oder ob ich gar nicht lesen u. meine Zeit für eigene Arbeiten verwenden soll. Im zweiten Fall besteht die Gefahr⟦,⟧ dass ich keine Hörer unter den Studenten habe, weil keine genügende logistische bzw. mengentheoretische Vorbildung mehr vorhanden ist seit Carnap u. Hahn nicht mehr lesen.

Zum Schlusse möchte ich Ihnen noch zu Ihrer Berufung an die Notre Dame University, die ich aus dem Bulletin erfahren habe, herzlich gratulieren, wenn ich es auch sehr bedaure, dass ich damit wieder einen Freund in Wien verliere. Was sind Ihre Pläne betreffs des Wiener Kolloquiums?

Mit herzlichen Grüssen u. besten Wünschen für Sie u. Ihre werte Familie

Ihr Kurt Gödel

---

[a]Simultaneously, Gödel conveyed a similar tentative acceptance to Notre Dame president John O'Hara. That letter, however, was never received. (See Gödel's letter to Menger of 15 December 1937.)

though it is still not at all certain whether anything will come of that. In the winter semester 1938/39 I will then probably be in Princeton. It would be highly desirable to me if I had to be obligated for only *one* semester. As you know, I had a bad experience with my health in America and I would therefore not like to be bound in advance for a longer time. If, however, the aforementioned assumptions turn out to be right, I would gladly accept, assuming the remaining conditions (salary and duties on my part) are acceptable.[a]

There is not much new with me. Since I've been back from Aflenz my health has again been worse, but still tolerably good. Mr. Wald has perhaps written you about my lecture, at the end of which I proved the consistency of the axiom of choice in the system of set theory. In the colloquium I referred to the analogous result for the system of Principia Mathematica, as well as the partial result about the continuum hypothesis of which I told you.[b]

At the moment I am just pondering whether I should lecture on something elementary or something at a higher level next semester, or whether I should not lecture at all and use my time for my own work. In the second case there is the danger that I might have no listeners, because since Carnap and Hahn no longer lecture no suitable logical or set-theoretical preparation is available any more.

In conclusion I would still like to congratulate you heartily on your appointment at Notre Dame University, which I learned of from the Bulletin, even though I am also saddened that I am thereby again losing a friend in Vienna. What are your plans concerning the Vienna colloquium?

With cordial greetings and best wishes for you and your family

Yours, Kurt Gödel

[b]The date of this presentation is unknown, and no other reference to it has been found. The Menger colloquium ceased to meet not long afterward, and no reports of sessions held after the academic year 1935–1936 were ever published.

11. Menger to Gödel[a]

Hoffe Sie kommen erste~~n~~ Februar.[b] Artin wird hier sein und einige gute Studenten alle freuen sich auf Sie. Ruhige Arbeitsmöglichkeit billigere Lebenskosten als im Osten. Lehr~~e~~verpflichtung höchstens drei Wochenstunden und jede zweite Woche Seminar. Bitte baldige Antwort am besten telegraphisch.

n

Her⟦z⟧lichst

Menger

---

[a]Text of a telegram sent on 12 September 1937 from South Bend, Indiana, to Gödel at Josefstädterstrasse 43 in Vienna. For easier readability, we have transcribed the text into standard German and substituted periods for the stops in the telegram.

12. Menger to Gödel[a]

Notre Dame, Indiana
3. November 1937

Lieber Goedel,

in Princeton, wo ich einige Tage zu Gast war, aeusserte man die Absicht, Sie fuer dieses Studienjahr in irgend einer Ihnen zusagenden Form einzuladen. Die Einladung soll insbesondere Ihnen ermoeglichen, einen Besuch von Princeton mit einem Aufenthalt in Notre Dame zu verbinden. Praesident O'Hara, dem ich von dieser Besprechung Mitteilung machte, bat mich Sie wissen zu lassen, dass Ihre Einladung nach Notre Dame selbstverstaendlich auch fuer dieses Jahr aufrecht besteht, falls Sie einen Aufenthalt hier mit einem Besuche Princetons verbinden wollen. In diesem Falle bleibt es ganz Ihnen ueberlassen, in welcher Reihenfolge Sie die beiden Universitaeten besuchen wollen.

---

[a]On letterhead of the University of Notre Dame, Department of Mathematics, Notre Dame, Indiana.

11. Menger to Gödel[a]

Hope you are coming the first of February.[b] ⟦Emil⟧ Artin will be here and a few good students, all looking forward to you. Quiet opportunities for work, cheaper living costs than in the East. Teaching duties at most three hours per week and a seminar every second week. Please answer speedily, preferably telegraphically.

Most cordially

Menger

---

[b]In his reply of 15 December 1937 Gödel postponed his visit until the following year.

12. Menger to Gödel[a]

Notre Dame, Indiana
3 November 1937

Dear Gödel,

In Princeton, where I was a guest for a few days, the intention was expressed of inviting you for this academic year in some form congenial to you. In particular, the invitation should make it possible for you to combine a visit to Princeton with a stay in Notre Dame. President O'Hara, whom I informed of this discussion, asked me to let you know that of course your invitation to Notre Dame also holds good for this year, in case you want to combine a stay here with a visit to Princeton. In that case it is entirely up to you in what sequence you want to visit the two universities.

Es wird Sie in diesem Zusammenhange vielleicht interessieren, dass ⟨mich⟩ die University of Notre Dame bat, dieses Jahr wieder ein Symposium zu veranstalten, aehnlich dem ueber Variationsrechnung, dessen Programm ich beilege. Dasselbe wird an zwei Tagen zwischen dem 10. und 15. Februar stattfinden und die Algebra der Geometrie und ihre Beziehungen zur Booleschen Algebra, Gruppentheorie, Topologie und anderen Zweigen der Mathematik zum Gegenstand haben. Birkhoff jun., v. Neumann und Stone haben ihre Teilnahme bereits zugesagt. Ich brauche Ihnen nicht zu sagen, mit wie grosser Freude es allseits begruesst wuerde, wenn Sie um diese Zeit bereits hier waeren. Artin ist bereits in Amerika und auf dem Wege nach Notre Dame. Nach Beendigung des Symposiums wird Notre Dame wieder ein ruhiger Arbeitsplatz sein.

Ich hoffe von∧Ihrer Entscheidung, wie immer dieselbe ausfallen moege, ⊔ Nachricht zu erhalten. Es waere mir persoenlich natuerlich eine besondere Freude, wenn sich mir Gelegenheit bieten sollte, wieder mal einige Zeit mit Ihnen beisammen zu sein.

Mit herzlichen Gruessen, in alter Freundschaft

Karl Menger

13. Gödel to Menger

Wien 15./XII. 1937

Lieber Professor Menger:

Vielen Dank für Ihr Radiogramm u. Ihren Brief vom 3./XI. Ich habe mich doch entschlossen im laufenden akad. nicht nach Amerika zu kommen, wie ich ja auch schon in dem verloren gegangenen Brief an Präsidenten O'Hara[a] definitiv geschrieben habe. Auch für 1938/39 kann ich für Notre-Dame augenblicklich noch nicht definitiv annehmen, doch werde ich Ihnen darüber in ca 2 Monaten etwas Bestimmtes schreiben können. Ich würde mich jedenfalls sehr freuen, wieder ein Semester mit Ihnen beisammen sein zu können. Falls von dem Symposium über Algebra der Geometrie etwas im Druck erscheint, bitte ich Sie mir ein Separa-
2 tum zukommen | zu lassen. Das Programm des Symposiums über Variationsrechnung, von dem Sie schreiben, habe ich leider nicht erhalten. Es würde mich ebenfalls sehr interessieren.

[a]See note a to letter 10, 3 July 1937.

In this regard it will perhaps interest you that the University of Notre Dame asked me again this year to organize a symposium, similar to the one on the calculus of variations, whose program I am enclosing. It will take place on two days between the tenth and fifteenth of February and will have as its theme the algebra of geometry and its relations to Boolean algebra, group theory, topology and other branches of mathematics. ⟦Garrett⟧ Birkhoff, ⟦John⟧ von Neumann and ⟦Marshall⟧ Stone have already confirmed their participation. I needn't tell you with what great joy it would be greeted on all sides were you already here at that time. ⟦Emil⟧ Artin is already in America and on the way to Notre Dame. After the end of the symposium Notre Dame will again be a quiet place to work.

I hope to receive a report of your decision, however that may turn out. It would of course be a special joy to me personally should the opportunity present itself to me to be together with you for some time once again.

With cordial greetings, in old friendship,

Karl Menger

13. Gödel to Menger

Vienna, 15 December 1937

Dear Professor Menger:

Many thanks for your radiogram and your letter of 3 November. I have decided, however, not to come to America in the current academic ⟦year⟧, as I also already wrote definitely in the lost letter to President O'Hara.[a] For 1938/39 too I can for the moment still not definitely accept ⟦the offer from⟧ Notre Dame, though I will be able to write something more definite about it in about two months. In any case I would be very happy to be able to be together with you again for a semester. In case anything appears in print from the symposium on the algebra of geometry, I ask you to let me obtain an offprint. The program of the symposium on the calculus of variations, of which you wrote, I unfortunately did not receive. It would likewise interest me very much.

Ich habe im letzten Sommer meine Arbeit über das Kontinuumproblem fortgesetzt u. es ist mir schliesslich doch gelungen, die Widerspruchsfreiheit (sogar der verallgem. Kont. Hyp. $2^{\aleph_\alpha} = \aleph_{\alpha+1}$) in der gesamten Mengenlehre zu beweisen. Ich bitte Sie aber, einstweilen davon niemand etwas zu sagen. Ich habe bisher ausser Ihnen nur v. Neumann davon Mitteilung gemacht, den ich bei seinem letzten Aufenthalt in Wien auch den Beweis skizzierte. Gegenwärtig versuche ich, auch die Unabhängigkeit der Kont. Hyp. zu beweisen, weiss aber noch nicht ob ich damit durchkommen werde.

Bez. meines Kommens schreibe ich gleichzeitig an Präs. O'Hara. Mit besten Grüssen auch an Ihre Frau Gemahlin,

Ihr Kurt Gödel

Meine Adresse hat sich geändert.[b] Siehe Briefumschlag!

---

[b]In November 1937 Gödel moved from the apartment on Josefstädterstrasse that he had shared with his mother and brother since 1929 to an apartment on Himmelstrasse in the Viennese suburb of Grinzing.

14. Menger to Gödel

⟦Letter 14 of 20 May 1938 was written in English. See the opposing page.⟧

Last summer I continued my work on the continuum problem and finally succeeded in proving the consistency (even of the generalized continuum hypothesis $2^{\aleph_\alpha} = \aleph_{\alpha+1}$) with all of set theory. I ask you, however, for the time being to tell no one anything about it. Besides you I have up to now informed only von Neumann about it, ⟦to⟧ whom I also sketched the proof during his last sojourn in Vienna. Presently I am trying also to prove the independence of the continuum hypothesis, but don't yet know whether I will succeed with that.

Regarding my coming I am simultaneously writing President O'Hara. With best regards also to your wife,

Yours, Kurt Gödel

My address has changed.[b] See the envelope!

---

14. Menger to Gödel

May 20th, 1938

Dr. Kurt Gödel
Himmelstrasse
Wien XIX
Austria

Dear Gödel:

I still hope that we shall have the pleasure of seeing you here next year. Since Veblen told me that you suggested staying at Princeton for the six weeks before the Christmas vacation, I presume that you will prefer to visit Notre Dame during the spring term. As you know, the arrangements for lectures during the winter term have to be made a long time ahead. The spring term starts here in February, and ends at the beginning of June. The plan to spend the spring term here will thus give you a little rest, after you will be through at Princeton, which I presume will be welcome to you. I hardly need to say that I shall be glad if you spend this vacation month here at Notre Dame.

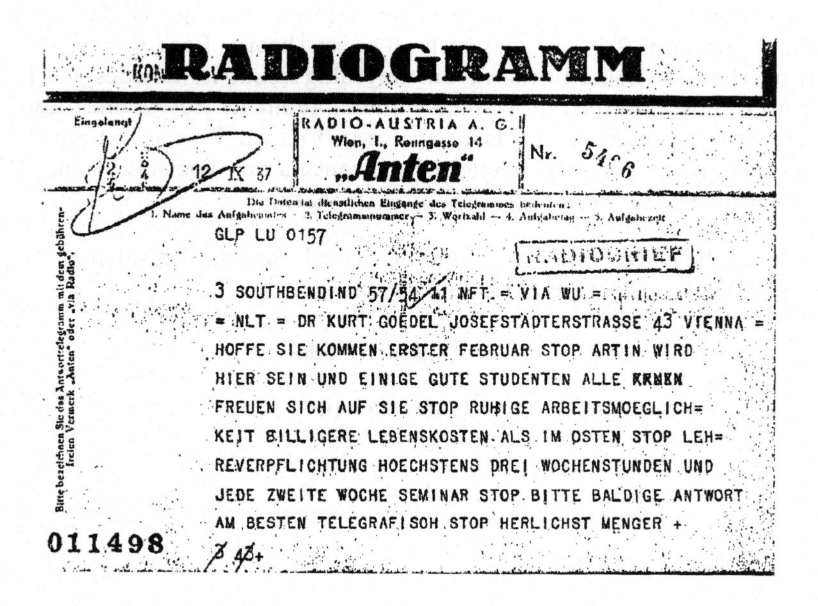

RADIOGRAMM

Eingelangt 12 IX 37

RADIO-AUSTRIA A. G.
Wien, I., Renngasse 14
„Anten"

Nr. 5406

Die Daten im dienstlichen Eingange des Telegrammes bedeuten:
1. Name des Aufgabeamtes — 2. Telegrammnummer — 3. Wortzahl — 4. Aufgabetag — 5. Aufgabezeit

Bitte bezeichnen Sie das Antworttelegramm mit dem gebührenfreien Vermerk „Anten" oder „via Radio".

GLP LU 0157

RADIOBRIEF

3 SOUTHBENDIND 57/54 41 NFT = VIA WU =
= NLT = DR KURT GOEDEL JOSEFSTADTERSTRASSE 43 VIENNA =
HOFFE SIE KOMMEN ERSTER FEBRUAR STOP ARTIN WIRD
HIER SEIN UND EINIGE GUTE STUDENTEN ALLE KRNKN
FREUEN SICH AUF SIE STOP RUHIGE ARBEITSMOEGLICH=
KEIT BILLIGERE LEBENSKOSTEN ALS IM OSTEN STOP LEH=
REVERPFLICHTUNG HOECHSTENS DREI WOCHENSTUNDEN UND
JEDE ZWEITE WOCHE SEMINAR STOP BITTE BALDIGE ANTWORT
AM BESTEN TELEGRAFISOH STOP HERLICHST MENGER +

011498

Radiogram, Menger to Gödel, 27 September 1937

15. Gödel to Menger

Wien 25./VI. 1938

Lieber Prof. Menger!

Besten Dank für Ihren Brief vom 20./V. Ich habe den Plan, im Herbst nach Notre-Dame to kommen, inzwischen schon selbst aufgegeben, weil es für mich doch zu anstrengend wäre. Ich komme also, falls nicht unvorhergesehene Umstände mich verhindern, im Feb. 1939.—

Was das Vorlesungsprogramm betrifft, so glaube ich, dass ich zum Halten einer elementaren Vorlesung wegen mangelnder Englischkennt-

Concerning your lectures, I should suggest that you give a very elementary introduction to mathematical logic (three hours a week), something of the Hilbert–Ackermann type. A course like that, given by an authority like you, is badly needed here, and will certainly be of great advantage to the university from many points of view. Besides, we expect you to give an advanced discussion either of the continuum problem, or of the Entscheidungs problem, for two hours once a week. You will have a small but good and interested group of listeners for this course. Please let me know at your earliest convenience whether this program is satisfactory to you.

At the end of the school year (that is, at the beginning of June) I intend to go with my family to California, where I am going to teach in summer school at Berkeley. Though it will be quite a trouble with so many, and so little babies,[a] we are looking forward with pleasure to making the trip west.

Hoping that you will have nice vacations too, and that I will hear from you soon, I am, with kind regards,

Cordially yours,

Karl Menger

[a] Menger's first child, Karl, Jr., was born in 1936, one year after Menger's marriage to Hilda Axamit, and twins Rosemary and Fred were born the following year.

15. Gödel to Menger

Vienna, 25 June 1938

Dear Prof. Menger,

Many thanks for your letter of 20 May. In the meantime I myself have already given up the plan of coming to Notre Dame in the fall, because it would really be too taxing for me. So, if unforeseen circumstances do not hinder me, I will come in February 1939.—

As concerns the program of lectures, I think that I am presently not very well suited to giving an elementary course of lectures, on account of

nisse, mangelnder Erfahrung in elem. Vorlesungen u. mangelnder Zeit zur Vorbereitung ⟨gegenwärtig⟩ nicht sehr geeignet bin. Andrerseits müsste ich in einem um 2-stündigen Sem. über das Kontinuum manches Interessante weglassen u. auch ziemlich hohe Anforderungen an die Hörer stellen. Ich würde es daher vorziehen, eine 3-stündige Vorlesung über Axiomatik der Mengenlehre zu halten, in welcher | ich auf die Ergebnisse von mir u. anderen über die Kontin.-Hyp. u. das Auswahlaxiom u. ihre Beziehungen zur Kardinalzahlarithmetik eingehen würde. Als Vorkenntnisse müsste ich nur die abstrakte Mengenlehre etwa im Umfang der ersten Kapitel von Hausdorff (2. Aufl.)[a] voraussetzen. Die elementarsten Tatsachen der Logistik müsste ich zu Beginn ohnehin bringen.

2

Mit besten Wünschen für einen schönen Sommer für Sie u. Ihre Familie u. herzlichen Grüssen,

Ihr Kurt Gödel

---

[a] *Hausdorff 1927.*

16. Gödel to Menger

Princeton 19./X. 1938

Lieber Professor Menger!

Ich bin seit 15./X hier in Princeton u. gedenke bis Ende Dezember hier zu bleiben. Ich hoffe Sie haben meinen Brief von Ende Juni, in dem ich Ihnen schrieb, dass ich das Sommersemester 1939 in Notre-Dame verbringen möchte, erhalten. Da jetzt manchmal Unregelmässigkeiten bei der Post vorzukommen scheinen, wäre es ja immerhin möglich, dass ein Brief von Ihnen oder von mir verloren gegangen ist. Von Veblen höre ich, dass Sie am 29./X. bei dem Meeting der Am. Math. Soc. in New York sein werden. Ich beabsichtige ebenfalls hinzukommen u. würde mich ausserordentlich freuen, wenn ich Sie dort treffen könnte. Wir könnten dann auch bezüglich meiner Vorlesungen in Notre-Dame alles Nähere besprechen. Wie ich Ihnen schon in meinem letzten Brief schrieb, befürchte ich, dass eine elementare Vorlesung aus | den dort genannten Gründen ein Misserfolg sein könnte. Wenn Sie aber glauben, dass ich damit Unrecht habe u. wenn die Universität einen grossen Wert darauf legt, so bin ich bereit, auch auf das vorgeschlagene Programm einzugehen.

2

Ich nehme an, Sie haben President O'Hara den beabsichtigen Zeitpunkt meines Kommens mitgeteilt. Ich selbst habe bisher an ihn nicht

insufficient knowledge of English, insufficient experience at elementary lectures and insufficient time for preparation. On the other hand, in about a two-hour seminar on the continuum I would have to omit much of interest and also put rather high demands on the listeners. I would therefore prefer to give a three-hour course of lectures on the axiomatization of set theory, in which I would go into the results of myself and others on the continuum hypothesis and the axiom of choice and their relations to cardinal arithmetic. As prior knowledge I would have to presume only abstract set theory about to the extent of the first chapter of Hausdorff⟦'s book⟧ (second edition).[a] I would have to mention the elementary facts of logic at the outset anyhow.

With best wishes for a lovely summer for you and your family, and with cordial greetings,

Yours, Kurt Gödel

---

16. Gödel to Menger

Princeton, 19 October 1938

Dear Professor Menger,

I've been here in Princeton since 15 October and am thinking of staying here until the end of December. I hope you received my letter from the end of June, in which I wrote that I would like to spend the summer semester of 1939 in Notre Dame. Since irregularities in the mails seem to occur now, it would after all be possible that a letter of yours or of mine has been lost. From Veblen I hear that on 29 October you will be at the meeting of the American Mathematical Society in New York. I intend to come there as well and would be very happy if I could meet you there. We could then also discuss all the details concerning my lectures at Notre Dame. As I already wrote you in my last letter, I fear that an elementary course of lectures could, for the aforementioned reasons, be a failure. But if you think that I am wrong and if the university places a great value on it, then I am ready to go on with the proposed program.

I presume you have informed President O'Hara of the intended time of my coming. I myself have not written him up to now, because under the

geschrieben, weil ja unter den gegenwärtigen Umständen nicht ⟨mit Sicherheit⟩ vorauszusehen war, ob ich überhaupt werde kommen können.

Ich hoffe Sie haben einen schönen Sommer gehabt u. es geht Ihnen u. Ihrer Familie recht gut u. bin mit besten Grüssen sowie Empfehlungen an Ihre Frau Gemahlin

Ihr Kurt Gödel

P.S. Hier in Princeton lese ich über das Kontinuumproblem u. beginne am 1. Nov.

17. Menger to Gödel

⟦Mid-October 1938⟧

Lieber Herr Gödel,

Ihr Brief kam, als ich eben im Begriffe war, Ihnen zu schreiben. Zunächst, um Ihnen meine allerherzlichsten Wünsche anläßlich Ihrer Vermählung auszusprechen. Hilda schließt sich mit den besten Glückswünschen an. Sodann, um Ihnen zu sagen, daß ich sehr froh bin, daß Sie gut in U. S. angekommen sind. Und schließlich, daß ich mich außerordentlich darauf freue, Sie bald zu sehen & hoffentlich ausführlich zu sprechen, da ich Ende nächster Woche für einige Tage nach dem Osten komme. Bei diesem Anlaß können wir dann auch die Details Ihres hiesigen Aufenthaltes besprechen, aber schon jetzt möchte ich Ihnen sagen, daß wir sicher ohne weiteres ein alle Teile befriedigendes Programm finden werden. Auch freue ich mich ganz besonders darauf, Näheres von Ihren fabelhaften neuen Arbeiten zu hören.

2 | Ich wäre sehr froh, wenn ich Ihnen auch von einer Kleinigkeit über Logik erzählen könnte, die ich selbst momentan in Arbeit habe. Vielleicht werden Sie einige geometrische & variationstheoretische Resultate interessieren, zu denen ich im letzten Jahre gelangt bin.

Für den Augenblick also nur nochmals die allerherzlichsten Wünsche und alles Nähere nächste Woche.

Mit vielen schönen Grüßen

Ihr

Karl Menger

present circumstances it is not foreseeable (with certainty) whether I really will be able to come.

I hope you have had a lovely summer and that you and your family are doing very well, and, with best wishes as well as compliments to your wife, am

Yours, Kurt Gödel

P.S. Here in Princeton I will be lecturing on the continuum problem and will begin on 1 November.

17. Menger to Gödel

⟦Mid-October 1938⟧

Dear Mr. Gödel,

Your letter came just as I was about to write to you—first of all, in order to express to you my very heartiest congratulations on the occasion of your marriage. Hilda joins in ⟦sending you⟧ all good wishes. Then, in order to say that I am very happy that you arrived in good health in the U.S. And finally, that I am exceptionally pleased at ⟦the prospect of⟧ seeing you soon, and hopefully speaking to you in detail, since I am coming to the East for a few days the end of next week. On that occasion we can then also discuss the details of your stay here, but I would already like to say now that we will certainly readily find a program pleasing to all parties. I am also most especially pleased to learn more about your fabulous new work.

I would be very happy if I could also tell you about a little thing concerning logic that I myself am just now working on. Perhaps you will be interested in some geometric and variation-theoretic results that I obtained during the past year.

For the moment, then, just very heartiest congratulations once more, and everything more particular next week.

With many beautiful wishes,

Yours,

Karl Menger

18. Gödel to Menger

Princeton 11./XI. 1938

Lieber Professor Menger:

Ich hoffe Sie haben das Wajsbergsche Separatum[a] u. den Abdruck meiner Vorlesungen[b] inzwischen erhalten. Ich habe mir die Frage meiner Vorlesungen in Notre-Dame nochmals überlegt u. wollte Ihnen noch folgendes darüber schreiben: Falls Sie lieber ein gemeinsames Seminar ankündigen wollen, so bin ich auch damit vollkommen einverstanden, unter der Voraussetzung dass nur solche Herrn sprechen, die den Gegenstand hinreichend kennen u. dass zweitens die elementare Einleitung in die Logik, welche ja einen grossen Teil des Semesters in Anspruch nehmen müsste, irgendwie einheitlich aufgebaut ist. (Z. B. dadurch dass ⟨nur⟩ ich oder Sie u. ich zusammen die entsprechenden Referate halten). 2 Nachher könnte man | dann vielleicht Referate anderer Herrn über etwas schwierigere Fragen einschalten. Ich überlasse, wie gesagt, die Entscheidung Ihnen, aber vielleicht wäre gerade aus den in meinen letzten beiden Briefen genannten Gründen ein Seminar einer Vorlesung vorzuziehen.

Bezüglich Ihrer logischen Frage glaube ich, es ist ganz sicher, dass durch die Stoneschen Resultate die genaue Korrespondenz zwischen den verschiedenen Arten von Satz-n-tupeln u. Klassen-n-tupeln erwiesen ist, wobei alle "bekannten" Sätze durch die Allklasse, alle "bekannt falschen" durch die leere Klasse u. die "fraglichen" durch alle übrigen Klassen repräsentiert werden. Die Wajsbergsche Arbeit ist nur dann von Interesse, wenn Sie die Operation "?" iterieren wollen.

Mit besten Grüssen, ⟨auch an alle Mathematiker von denen Sie mir erzählten⟩

Ihr Kurt Gödel

P.S. Bitte vergessen Sie nicht mir die Programme der Symposien u. das Bulletin der Univ. zuschicken zu lassen. An President O'Hara habe ich heute geschrieben.

---

[a]Probably *Wajsberg 1938*, which appeared earlier that year.

18. Gödel to Menger

Princeton, 11 November 1938

Dear Professor Menger:

I hope in the meantime you have received the Wajsberg offprint[a] and the copy of my lectures.[b] I have thought once again about the question of my lectures at Notre Dame and wanted to write you the following further about them: In case you want rather to announce a common seminar, I am also completely in agreement with that, under the assumption that only such persons speak who are sufficiently acquainted with the subject matter and that, secondly, the elementary introduction to logic, which must of course occupy a large part of the semester, somehow be built up uniformly. (E.g., by having only me, or you and me together, give the relevant talks.) Then afterwards talks by other persons concerning some more difficult questions could perhaps be brought in. As I said, I leave the decision to you, but perhaps, just for the reasons stated in my last two letters, a seminar would be preferable to a lecture course.

With regard to your logical question, I think it is quite certain that the precise correspondence between the various kinds of statement n-tuples and class n-tuples is established by Stone's results, whereby all "known" statements are represented by the universal class, all the ones "known to be false" by the empty class, and the "questionable" ones by all the rest of the classes. The Wajsberg paper is then only of interest if you want to iterate the "?" operation.

With best wishes, also to all the mathematicians of whom you spoke,

Yours, Kurt Gödel

P.S. Please don't forget to have the programs for the symposium and the Bulletin of the university sent to me. I wrote to President O'Hara today.

---

[b]The editors have been unable to identify these.

19. Menger to Gödel[a]

⟦December(?) 1938⟧

Lieber Herr Gödel,

ich hoffe Sie haben den Katalog der Univ. bekommen. Eine unangenehme Erkältung hinderte mich, Ihnen früher zu schreiben & für Brief & Sendungen zu danken. Beiliegend die 2 Programme.

Für eine Einführung in die math. Logik (sei es in Form eines Seminars oder einer Vorlesung) bekunden bereits jetzt einige Professoren großes Interesse. Ich bin ganz Ihrer Meinung, daß im Falle eines Seminars die Einführung | einheitlich (am besten natürlich von Ihnen) vorgetragen werden soll & nur mir wohlbekannte tüchtige Leute später einige wenige Referate halten sollen. Wir können alle Details in Jänner besprechen; es wird leicht sein, eine allseits befriedigende Lösung zu finden.

2

Mit den besten Grüßen & Wünschen

Ihr

Karl Menger

P.S. Warum lassen Sie nicht Ihre Frau herkommen? Das wäre doch viel netter für Sie. Und machen Sie sich doch wegen des nächsten Jahres (& der Zukunft überhaupt) nicht unnötige Sorgen!

Herzlichst

K. M.

---

[a]On letterhead of the University of Notre Dame Department of Physics.

20. Gödel to Menger

Wien 30./VIII. 1939.[a]

Lieber Herr Menger!

Ich bin seit Ende Juni wieder hier in Wien, u. hatte in den letzten Wochen eine Menge Laufereien, so dass es mir bisher leider nicht möglich

---

[a]Two days later the Nazis invaded Poland, prompting Menger to declare (*1994*, p. 225) that this letter "may well represent a record for unconcern on the threshold of world-shaking events".

19. Menger to Gödel[a]

⟦December(?) 1938⟧

Dear Mr. Gödel,

I hope you received the catalog of the university. A nasty cold prevented me from writing to you earlier and thanking you for ⟦your⟧ letter and parcels. I enclose the two programs.

Already now some professors have evinced great interest in an introduction to mathematical logic (be it in the form of a seminar or a lecture course). I am entirely of your opinion, that in the case of a seminar the introduction should be delivered uniformly (at best, of course, by you) and only competent people well known to me should later deliver a few little talks. We can discuss all the details in January; it will be easy to find a solution pleasing to all.

With greetings and best wishes,

Yours,

Karl Menger

P.S. Why didn't you let your wife come here? That would surely be much nicer for you. And don't trouble yourself unnecessarily on account of next year (and the future in general)!

Most cordially,

K. M.

---

20. Gödel to Menger

Vienna, 30 August 1939[a]

Dear Mr. Menger,

Since the end of June I've again been here in Vienna, and in the last weeks had a lot of running about, so that until now it was unfortunately

---

war, etwas für das Kolloquium zusammenzuschreiben. Wie sind die Prüfungen über meine Logikvorlesung ausgefallen? u. hat irgend jemand das Axiomensystem des Aussagekalküls, das ich zuletzt angab, weiterbearbeitet? Entschuldigen Sie bitte, dass ich ein Separatum von Łukasiewicz irrtümlich mitgenommen habe; ich sende es mit gleicher Post an Sie zurück. Wie geht es Ihnen u. Ihrer Familie? Im Herbst hoffe ich wieder in Princeton zu sein.

Mit besten Grüssen, sowie Empfehlungen an Ihre Frau

Ihr Kurt Gödel

P.S. Bitte richten Sie auch Grüsse von mir an die Mathematiker u. Physiker in Notre Dame aus.

21. Gödel to Menger

Princeton, 22./VI. 1942

Lieber Prof. Menger:

Darf ich fragen, was Sie in Ihrem Seminar behandeln wollen? Ich vermute die Widerspruchsfreiheit der Kontin.-Hyp., denn was ich sonst in meiner Notre-Dame Vorlesung gebracht habe, waren ja bloss Referate. Das einzige Originelle in diesem Teil war, soviel ich mich erinnere, die folgende triviale Folgerung aus der verallgem. Kontin. Hyp: Falls für jedes $x \in M$ m_x eine Kardinalzahl ist (wobei die m_x nicht von einander verschieden zu sein brauchen) so ist: $\prod_{x \in M} m_x = \left[\operatorname{Sup}_{x \in M} m_x\right]^{\overline{\overline{M}}}$ vorausgesetzt nur, dass unter den m_x kein grösstes existiert, was nebenbei bemerkt ein heuristisches Argument *gegen* die verallgem. Kontin. Hyp. ist weil *nur* die Konfinalität von M mit 1 u. mit keiner andern Ordinalzahl eine Ausnahmerolle spielt.

Was den Beweis für die Widerspruchsfreiheit der verallg. Kontin. Hyp. betrifft, so liegt ein in allen Details geführten | ja in den Princetoner Vorlesungen, die ich Ihnen schickte, vor. Ich glaube man braucht sich dabei nicht durch den Umfang u. die Verwendung der logischen Symbole abschrecken zu lassen, denn man kann die Sache ziemlich rasch lesen. Meine Notre-Dame Vorlesung (u. meine Princetoner Vorlesung von 1940, die dasselbe in etwas verbesserter Form enthält) sind nicht soweit in's Detail durchgeführt[1] u. ausserdem fürchte ich, dass sie wegen der zahlreichen

[1]Die Idee des Beweises ist allerdings etwas deutlicher sichtbar.

not possible for me to compile anything for the colloquium. How did the examinations over my logic course turn out? And has anyone further worked out the axiom system for the propositional calculus that I mentioned at the end? Please excuse that I mistakenly took with me an offprint by Łukasiewicz; I'll send it back to you with this same mailing. How are you and your family? In the autumn I hope to be in Princeton again.

With best wishes, as well as compliments to your wife,

Yours, Kurt Gödel

P.S. Please also convey greetings from me to the mathematicians and physicists at Notre Dame.

21. Gödel to Menger

Princeton, 22 June 1942

Dear Prof. Menger:

May I ask what you intend to treat in your seminar? I presume the consistency of the continuum hypothesis, for what I otherwise mentioned in my Notre Dame lecture course was merely a report. So far as I recall, the one original thing in that part was the following trivial consequence of the generalized continuum hypothesis: In case m_x is a cardinal number for every $x \in M$ (where the m_x need not differ from one another), then: $\prod_{x \epsilon M} m_x = \left[\operatorname{Sup}_{x \epsilon M} m_x\right]^{\overline{\overline{M}}}$, assuming only that among the m_x no largest exists, which by the way is a heuristic argument *against* the generalized continuum hypothesis, because *only* the cofinality of M with 1 and with no other ordinal number plays an assumptive role.

With regard to the proof of the consistency of the generalized continuum hypothesis, one carried out in all details is at hand in the Princeton lectures that I sent you. I think that in it one need not be intimidated by the extent and the use of the logical symbols, for one can get the gist of the matter rather quickly. My Notre Dame lecture (and my Princeton lecture of 1940, which contains the same material in somewhat improved form) are not carried through so far into the details,[1] and besides I'm afraid that they are difficult to read on account of the numerous subse-

[1] The idea of the proof is, I admit, manifestly somewhat more perspicuous.

nachträglichen Einschaltungen u. Änderungen[2] schwer lesbar sind. Wenn Sie aber wünschen kann ich sie gerne nochmals durchsehen (~~u. Ihnen~~ was nicht viel Zeit beanspruchen würde) u. Ihnen dann schicken.

Mit besten Grüssen u. Empfehlungen an Ihre Frau

Ihr Kurt Gödel

[2]u. der unleserlichen Schrift.

22. Gödel to Menger

Princeton, 14./IX. 1953

Lieber Prof. Menger!

Besten Dank für die Zusendung der neuen Einleitung zu Ihrem Buch über Calculus,[a] sowie der kurzen Note über Variable u. Funktion.[b] Über die erstere habe ich nichts weiter zu sagen als was ich Ihnen schon früher, teils mündlich, teils schriftlich, mitteilte. Was die letztere betrifft, möchte ich nur noch folgendes hinzufügen: Ich empfinde es als eine Härte, ja geradezu als wi~~e~~dersinnig, irgendein bestimmtes Objekt (sei es eine Klasse, oder eine Funktion, oder was immer) als "Variable" zu bezeichnen. Ich glaube, man sollte vielmehr sagen, dass jener Klasse etc. eine Variable zugeordnet ist.[1] Sie können natürlich sagen, dass die Terminologie willkürlich ist, aber ich glaube, man sollte doch darauf achten, dass die präzisierte Bedeutung | eines Wortes nicht *vollkommen* verschieden von der Bedeutung ist, die das Wort vorher hatte.[2]

2

Morgenstern ist gegenwärtig in Italien, kommt aber am 20. Sept. zurück, u. ich werde ihm dann die neue Einleitung zeigen.

Mit herzlichen Grüssen

Ihr Kurt Gödel

[1]oder noch besser, dass jene Klasse eine Variable definiert.
[2]u. ferner auch darauf, dass man nicht *vollkommen* verschiedene Dinge mit demselben Wort bezeichnet.

---

[a]*Menger 1953*.

quent insertions and changes.[2] But if you wish, I can willingly look them over once more (which would not require much time) and then send them to you.

With best wishes, and compliments to your wife,

Yours, Kurt Gödel

[2]and the illegible writing.

22. Gödel to Menger

Vienna, Princeton, 14 September 1953

Dear Prof. Menger,

Thanks very much for sending the new preface to your book on calculus,[a] as well as the short note on variable and function.[b] Concerning the first I have nothing further to say than what I imparted to you earlier, partly orally, partly in writing. As regards the latter, I would still like to add only the following: I feel it a hardship, indeed almost absurd, to designate some definite object (be it a class, or a function, or whatever) as a "variable". I think one should rather say that to that class, etc., a variable is assigned.[1] You can of course say that the terminology is arbitrary, but I think one should nevertheless see to it that the precisely defined meaning of a word is not *completely* different from the meaning the word had before.[2]

Morgenstern is presently in Italy, but is coming back on 20 September, and I will then show him the new preface.

With best wishes,

Yours, Kurt Gödel

[1]Or still better, that that class defines a variable.
[2]And furthermore that one not denote *completely* different things with the same word.

---

[b]*Menger 1953a.*

⟦After letter 22, the correspondence continues in English.⟧

23. Menger to Gödel

5506 N. Wayne Ave.
Chicago, Ill. 60640
Jan 2, 1968

Dear Gödel,

I have to write a report about the old papers on the foundation of projective geometry.[a] In looking through the Ergebnisse eines Mathematischen Kolloquiums I found in Heft 4, p. 14[b] a remark that you made in the discussion and I am enclosing a photocopy of the passage.

I must confess that I don't quite understand that remark. What is wrong with the following definition: P is a point if, for each element X, $X \subsetneq P$ implies either $X = P$ or $X = V$? I'd greatly appreciate it if you would drop me a line explaining what I am overlooking—especially, if I could get that line before Jan. 18. For on that day I shall leave for Turkey, where I shall teach at the Middle East Technical University in Ankara during the spring term.

It may interest you that in connection with that report I have been rethinking those old results and, jointly with an excellent student of mine, Helen Skala, have obtained rather unexpected results. The entire projective geometry of finite-dimensional spaces can be obtained from the assumption that we have a set with two operations, $\cup$ and $\cap$, and two elements V and U, such that $V \cup X = X$, $V \cap X = V$, $U \cap X = X$, $U \cup X = U$ for each X and satisfying one postulate:

$$X \cup ((X \cup Y) \cap Z) = X \cup ((X \cup Z) \cap Y)$$

and

$$X \cap ((X \cap Y) \cup Z) = X \cap ((X \cap Z) \cup Y)$$

for each X, Y, Z. Nothing else, not even associativity! Miss Skala has proved that in such a "projective structure" the elements that are joins of finite sets of points constitute a relatively complemented modular lattice. One can introduce dimension, etc.

---

[a] Perhaps *Menger 1970*.

[b] *Gödel 1933d*. The page number cited is incorrect; it should be 34.

How are you? Best wishes for 1968.

Sincerely,

Karl Menger

24. Gödel to Menger

Princeton, Jan. 22, 1968.

Dear Menger,

I am sorry I overlooked that the letter I received from you on Jan. 8 had a deadline for the reply. But what I have to say is quite trivial. If you define "point" as ⟨usual and as⟩ you suggest, then, if "x is a point" occurs as a hypothesis in a proposition, the whole proposition will contain existential quantifiers in the normal form. Of course, *some* propositions about points could be expressed in the restricted system.[1] ~~Incidentally, the same is true for definitions which contain universal quantifiers.~~

Many thanks for your New Years wishes. Axiomatics of geometry is rather far removed from my present interests.

Best wishes for your sojourn in Ankara.

Sincerely

Kurt Gödel

[1]This should have been mentioned in my remark. Also it should have been said that there is no substantial difference between existential and universal quantifiers in this respect. Only if both occur interlocked, *no* proposition about the defined term will be expressible in the restricted system.

25. Menger to Gödel

1/15/72

Dear Gödel,

I am writing a paper on the Wiener Kreis.[a] In connection with it, I'd greatly appreciate it if you would answer the following questions:

[a]Presumably incorporated as one of the chapters in the posthumously published *Menger 1994*.

1. Have you ever met Wittgenstein a) in Vienna? b) later in U. S. or England?
2. Does what Wittgenstein writes about your work in "Remarks on the Foundations of Mathematics" pp. 50–54 and pp. 176sq[b] give you the impression that he understood your work? (Personally, I have reservations.)
3. Did Schechter[c] (whom I don't remember well but whom you knew better) ever speak in the Kreis?

Best wishes and regards,

Sincerely

Karl Menger

---

[b] *Wittgenstein 1956.* Shorthand annotations at the top of the page read: "Hier sagt er ich hätte eine Ant. entdeckt." ⟦"Here he says I had discovered an Ant."⟧ "p. 172 21 1. Zeile" ⟦"First line"⟧
"p. 52e No. 14 letzte Zeile" ⟦"last line"⟧ "No. 16 erste Zeile" ⟦"first line"⟧
"52e Der Sinn eines Satzes ändert sich." ⟦"The sense of a statement changes."⟧
"p. 53e letzte Zeile—die Furcht der Mathematiker vor einem Widerspruch (Aberglaube)" ⟦"last line—the fear of mathematicians of a contradiction (superstition)"⟧

[c] Possibly the Viennese emigré Edmund Schechter, but, in the opinion of Friedrich Stadler, a leading authority on the Vienna Circle, the referent is more likely to have been Josef Schächter, a member of the Circle and student of Moritz Schlick.

26. Gödel to Menger

Princeton, April 20, 1972

Dear Professor Menger,

I am sorry my reply to your letter of Jan 15 comes so late. I hope it will be of some use to you. In consequence of frequent tiredness I hardly ever answer letters before a few weeks time. But in this case there was moreover a special reason, namely that I have always inhibitions to write about my relationship to the Vienna circle. I never was a logical positivist in the sense in which this term is commonly understood.[1] On the other hand, by some publications, the impression is created that I was.[2]

[1] and explained in Wissenschaftliche Weltauffassung, der Wiener Kreis, edited by Verein Ernst Mach, 1929.[a]

[2] probably, in part, due to my own fault.

---

[a] *Verein Ernst Mach 1929.*

As far as your specific questions are concerned I can see now that they have nothing to do with this problem. The answers are as follows:

1. I was never introduced to Wittgenstein and I have never spoken ⟨a
2 word⟩ with him. I only saw him once in my life | when he attended a lecture in Vienna. I think it was Brouwer's.[b]
2. As far as my theorem about undecidable propositions is concerned it is indeed clear from the passages you cite, that Wittgenstein did *not* understand it (or pretended not to understand it). He interprets it as a kind of logical paradox, while in fact it is just the opposite, namely a mathematical theorem within ~~the~~ an absolutely uncontroversial part of mathematics, namely finitary number theory or combinatorics.[c] Incidentally, the whole passage you cite seems non/sense to me. See, e.g. the "superstitious fear of mathematicians of a contradiction".
3. I don't remember any talk given by Schechter in the Vienna Circle. However, this means very little[3] because I am sure I did not enter the Circle before ~~1926/27~~ the academic year 1925/6 and very probably not before the calendar year 1926. Moreover I practically never attended after the spring term of 1933. I attended pretty regularly in 1926, 1927. I don't remember exactly how frequently I did 1928–1933.

Yours sincerely

Kurt Gödel

[3]Even if I was present I may have forgotten it. I remember only very few of the talks given. There were not many that really interested me. Was there anything special about Schechter's talk?

---

[b]Brouwer delivered two lectures in Vienna, on 10 and 14 March 1928; their published versions were entitled "Mathematik, Wissenschaft, und Sprache" ("Mathematics, science, and language") and "Die Struktur des Kontinuums" ("The structure of the continuum"). Their content is described in chapter X of *Menger 1994*, where it is noted that Wittgenstein attended the first, but not the second.

[c]For a contrary view of Wittgenstein's understanding of Gödel's work see *Floyd 1995*.

Ernest Nagel

Ernest Nagel

Ernest Nagel (1901–1985) was at the time of this correspondence the John Dewey Professor of Philosophy at Columbia University. During his long career, almost entirely at Columbia, he became one of the most eminent philosophers of science of his time. Logic was a continuing interest of his; for his generation, and certainly for him, logic and philosophy of science were often closely integrated.[a]

Nagel's correspondence with Gödel took place in 1957; it concerns the question whether to include one or both of *Gödel 1934* and (van Heijenoort's translation[b] of) *1931* as appendices to the book *Gödel's Proof* by Nagel and James R. Newman, to be published by New York University Press. This book was to be an expansion of two earlier essays by the authors, namely, their *1956* published in *Scientific American* and their *1956a* in Newman's anthology *The world of mathematics*. Their essays and book aimed to present informally, to a general audience, Gödel's incompleteness theorems and their logical and philosophical significance. The correspondence of Nagel and Gödel is part of a larger correspondence involving the two successive editors at NYU Press, Allan Angoff and Wilson Follett, as well as Newman.[c] Some of Gödel's correspondence with Angoff and Follett is included in volume IV of these *Works*.

Gödel's attitude toward this possible reprinting was complex and to some degree ambivalent, and he made one condition that seriously offended and angered Nagel, who found it totally unacceptable. For this and other reasons the project of including writings by Gödel was abandoned, and *Nagel and Newman 1958* consisted only of their expanded essay.

The larger correspondence begins with a letter of 17 January 1957 from Angoff to Gödel asking permission to include in Nagel and Newman's book "material from lectures you delivered at the Institute for Advanced Study during February–May 1934," i.e. from *Gödel 1934*. Gödel replied on 21 January, stating that the lectures were out of print but that Nagel could obtain them by inter-library loan from the Institute

[a]See for example the once very widely used textbook *Cohen and Nagel 1934*. *Nagel 1956* contains essays on philosophical issues about logic, and *Nagel 1979* contains reprints of historical essays on the development of formal logic.

[b]A version of this translation had been prepared by early in the correspondence. It was eventually published in *van Heijenoort 1967* after a process of revision that involved elaborate negotiation with Gödel. See the correspondence of Gödel and van Heijenoort in this volume.

[c]All citations and quotations in this note of letters not to or from Gödel are based on originals or copies in the Ernest Nagel papers in the Columbia University Library.

for Advanced Study. Subsequently he decided to send Nagel a copy and drafted a letter to Nagel dated 25 February (letter 1), which stated, however, that the lectures would only in part be useful for Nagel and Newman's purpose, because the proofs were in general less detailed than in *Gödel 1931*. The letter contains interesting comments on the relation of *1931* and *1934*, but it was marked "not sent." Possibly Gödel sent a brief note saying he would send a copy of the lectures; with not much delay he did.

Nagel acknowledged receipt on 9 March (letter 2) and said that it was planned to publish both van Heijenoort's translation of *Gödel 1931* and *Gödel 1934*. On 14 March Gödel drafted but did not send a reply (letter 3), which made two objections: Nagel and Newman's essay is intended for the general public, while his papers are not, and the reader "has a right to be informed about the present state of affairs," reflecting advances since 1934, in particular about Turing's analysis of computability which makes it clear that his proof "is applicable to *every* formal system containing arithmetic." The second point was stressed by Gödel in a number of his later writings, for example in the Postscriptum published in *Davis 1965* with *Gödel 1934* itself (see these *Works*, vol. I, pp. 369–370). Gödel's actual brief reply of 18 March (letter 4) only expresses (justified) surprise at the plan to publish both *1931* and *1934* in full, says he has not had time to consider the matter, and alludes to the second point above in a handwritten postscript.

Nagel replied on 21 March (letter 5) explaining the relation of the proposed book to the published articles. He welcomed the possibility that Gödel might add some comments to his papers and offered to come to Princeton in order to discuss any questions Gödel might have. But he also remarked that "time is of some importance here, for the book is to be published early next fall." This concern to keep to a schedule was common to Nagel, Newman and NYU Press, but Gödel showed no definite sign of sharing it. Gödel took up Nagel's offer to meet him in Princeton. Most of the correspondence leading up to their meeting deals with the arrangements of the meeting, but Gödel remarked on 25 March (letter 6) about *Nagel and Newman 1956*, "I liked this article very much as a non-technical introduction to the subject, although I don't agree in every respect with the formulation of the general meaning of my results."[d]

Nagel met Gödel at the latter's office at the Institute of Advanced Study on Saturday morning, 30 March. The meeting was lengthy, going on into the afternoon. One subject discussed was evidently the financial

[d]Gödel had clearly also read *Nagel and Newman 1956a*; see his letter to Angoff of April 9, remarked on below.

terms of Gödel's giving permission, since in a letter written later that day Nagel explained the terms of his and Newman's contract with NYU Press, correcting something he had said on the subject. Gödel must also have explained points on which he took exception to formulations of *Nagel and Newman 1956* and *1956a*. Nagel's remark on those issues is worth quoting for its cordial tone:

> But in any case, I want to thank you most warmly for your very great kindness to me this afternoon. You were most generous in giving me so much of your time, and I appreciate more than I can say in brief compass your comments on various points in the published forms of the paper by Mr. Newman and myself. I will, of course, check whether the blunders you noted are repeated in our longer version; and if they are (which I doubt), they will certainly be corrected.

This cordiality was soon to be undermined.

On 9 April Gödel wrote both to Nagel (letter 7) and to Angoff (letter 1), stating his conditions for agreeing to the reprinting of his own work. He is most explicit in the letter to Angoff. What proved to be the most contentious was the first condition:

> In view of the fact that giving my consent to this plan implies, in some sense, an approval of the book on my part, I would have to see the manuscript and the proof sheets of the book, including the appendix.

The second condition was that he should write an introduction to the appendix, dealing with advances since 1934 and "to supplement the considerations, given in the book, about the philosophical implications of my results." On this point he expressed dissatisfaction with Nagel and Newman's articles, and with reference to the first condition says that *1956a* "contains some very troublesome mistakes." The third condition was that if enough copies of the book are sold he should receive a share in the proceeds. Finally, he suggested reprinting only *1934* and the first chapter of *1931*, to avoid duplication, and changing the title of the book to *Gödel's theorems on the principles of mathematics*. Apparently referring to the three conditions just mentioned, Gödel wrote in his letter to Nagel that he had made to Angoff "the same suggestions I mentioned to you in our conversation in Princeton".

Nagel did not respond at this point, but Angoff's brief reply of 12 April expressed confidence that matters could be worked out. He evidently expected his fuller reply of 22 April (letter 2) to be satisfactory to Gödel, since he proposed to have their business manager draw up a contract. Concerning the first condition he wrote:

> We are glad to accede to your request that you examine the manuscript and galley proofs of the book prior to publication.

He accepted the proposal of an introduction by Gödel, offered him a share of the royalties, agreed to Gödel's suggestion about the title, but saw no need to omit most of *1931*.

In his reply of 6 May (letter 3), Gödel modifies Angoff's proposal as to sharing royalties: he resigns his share entirely, if fewer than 6,000 copies are sold, but requests a larger share in case, "contrary to expectations, more copies should be sold." Concerning the first condition formulated in his letter of 9 April he writes now:

> Of course I shall have to see the manuscript of the book before I sign the contract, so that I can make sure that I am in agreement with its content, or that passages with which I don't agree can be eliminated, or that I may express my view about the questions concerned in the introduction.

Angoff states in his reply of 15 May simply that he will have to discuss this point and the request concerning royalties with Nagel and Newman. He wrote on the same day to Nagel, included a copy of Gödel's letter, and mentioned Gödel's financial demand, noting that it would cause the Press to increase the price of the book. But "in the interest of expeditious publishing" he did not advocate asking Gödel to agree to a reduced share, even though he thought his demand excessive. He also favored agreeing to Gödel's other requests.

Angoff evidently did not anticipate how Nagel would react to the demand expressed in the last quotation.[e] In his reply of 21 May, Nagel wrote, "I have been reflecting on the copy of Gödel's letter you sent me with mounting amazement and anger." His reply reflected a discussion of the issues on the telephone with Newman, but although he found them in agreement he did not presume to speak except for himself. He thought Gödel's position concerning royalties "grasping and unreasonable" but was prepared to swallow it. Clearly that was not a deal-breaker either for Nagel and Newman or for NYU Press, but it may have generated some ill-will that colored their reactions in the other matters.

About the demand to see the manuscript before signing the contract Nagel wrote:

> ...I could scarcely believe my eyes when I read his ultimatum that he is not only to see the manuscript of our essay *before* signing the contract with you, but that he is to have the right to eliminate anything in the essay of which he disapproves. In short, he stipulates as a condition of signing the contract the right of censorship.
>
> This seems to me just insulting, and I decline to be a party to any such agreement with Gödel. I am utterly opposed to his seeing the manuscript before he signs a contract with you; and I cannot consent to giving him the right to veto anything Jim and I have written. Quite apart from hurt vanity, I believe I have good reasons for taking this stand. If Gödel's conditions were granted, Jim and I would be compelled

[e]Indeed, Nagel's reply to him of 21 May indicates that he had not reacted in anything like the same way to the earlier formulations of the same condition: what Gödel might have said in the conversation in Princeton on 30 March and what he wrote in his letter to Angoff of 9 April.

> to make any alterations Gödel might dictate, and we would be at the mercy of his tastes and procrastination for a period without foreseeable end. Now I for one am not prepared to give more of my time to revising the manuscript. Gödel is of course a great man; but I decline to be his slave.

Gödel's condition, quoted above from his letter of 6 May to Angoff, might have allowed Nagel and Newman's essay still to contain statements he took exception to, since he mentioned that he could express his view in the introduction to his own writings. Nagel passed over this and thus may have interpreted Gödel as claiming a more final "right of censorship" than he had intended. But he was surely reasonable in anticipating that Gödel's demand would lead to difficult and prolonged negotiation. Although he says he is not sure what the next move should be, he does say that it may be necessary to abandon the plan of including Gödel's papers.

If the disagreement had been faced at this point, there might have been hope of resolving it. But Nagel did not respond to Gödel directly, and Angoff's letter of 28 May (letter 4) papers over the disagreement. The letter mentions only the practical difficulty of sending the manuscript before printing because of the delay it would cause; it assures Gödel that he will be sent the galleys and that Nagel and Newman "would be delighted to consider all your suggestions." Gödel's reply of 3 June (letter 5) sets him straight rather brutally, pointing out that his consent to the reprinting of his papers was "conditioned *on my seeing the manuscript of the book and approving of its content.* In view of the fact that I may be in substantial disagreement with some parts of it it does not seem advisable to set the book in type before I had a chance to see it."

Angoff wrote on 4 June to Nagel and Newman, still hopeful and declining to take literally Gödel's statement that he must approve of the content of Nagel and Newman's book. But he thought they should agree to Gödel's request to see the manuscript before printing. He also suggested that they might reprint *1931* without Gödel's permission, saying that the paper was probably in the public domain.[f] But he was reluctant about this idea because of the contention to which it might give rise.

In his reply of 5 June, Nagel wrote that Angoff had "misunderstood the main point" of his previous letter:

> I there stated that I was opposed to Gödel seeing the manuscript of the Nagel–Newman book before giving his consent to the inclusion of his papers in the book. I really meant what I said; and despite your quite charitable interpretation of Gödel's latest communication, I do not find

[f] Angoff is silent at this point about *Gödel 1934*. Later the translation of *1931* by B. Meltzer (*Gödel 1962a*) was published without consulting Gödel or obtaining his permission.

> that he has withdrawn what I regard as an insolent demand that he must approve the contents of the manuscript as a condition of his granting the right to reprint his materials. And so I am compelled to repeat to you that UNDER NO CIRCUMSTANCES WILL I AGREE TO GÖDEL'S DEMAND. Indeed, after talking to Jim ⟦Newman⟧ this morning, I must go further; I DECLINE TO GIVE MY PERMISSION TO GÖDEL'S HAVING ACCESS TO THE *MANUSCRIPT*, except possibly in your office.

In writing to Gödel on 6 June (letter 6), Angoff still does not mention the dispute, and even finesses the fact that Gödel had previously stated that he had given his permission only conditionally. Angoff says it will be impossible to send Gödel the manuscript but invites him to look at it in the office before it goes to the printer "a week or two from now"; he again assures him that he will be sent galleys and that his suggestions will be considered. In his reply of 25 June (letter 7), Gödel softens his position somewhat; he accepts being sent the galleys "if you are willing to bear the costs in case changes should be necessary." He will not give permission for reprinting his papers until any mistakes (such as he found in *Nagel and Newman 1956a*) have been corrected.

Angoff probably never saw this reply, since his employment with NYU Press had probably already ended. The problem was taken up in July by another editor, Wilson Follett.[g] In a letter to Nagel of 9 July, he too mentions the option of printing the Gödel material without his consent and rejects it. He says he has consulted a copyright lawyer: "His advice, like your own conviction, is uncompromisingly against doing that, whether the material is in the public domain or not; and we cannot very well go ahead against such a consensus." He proposes printing the book without the Gödel material if Nagel and Newman should agree.

A letter of Newman to Nagel of 8 July shows extreme reluctance to give up the Gödel material, at least the translation of *1931*, and advocates publishing the latter if necessary without Gödel's consent.[h]

[g]Wilson Follett (1887–1963), evidently the senior of the two editors, had come to NYU Press at the end of a long career, first as an English professor and then as an editor, for some time with Alfred A. Knopf. He published several books on English and American literature. His posthumously published guide to American usage, *Follett 1966*, was widely used by editors and was recently reissued in revised form *Follett 1998*.

In his letter to Nagel of 9 July, Follett wrote that Angoff's connection with the Press "ended on the day of his latest letter to you." This would probably be a note, not preserved, informing Nagel of his letter to Gödel of 6 June. That is likely to be the date Follett refers to, but there is no hint in Angoff's letter to Gödel that he is about to leave the Press. That he was fired is hinted in a letter of Newman to Nagel of 12 July.

[h]The letter also expresses discontent with NYU Press, not only (quite understandably) about Angoff's handling of communication with Gödel but also about delay.

From Nagel's letter to Gödel of 22 August (letter 9) it appears that they proposed to NYU Press that a last effort to reach an understanding with Gödel be made, and that they ask for permission only to print the translation of *1931*, with a disclaimer that his permission did not signify "approval in the slightest degree" of Nagel and Newman's essay. Nagel reports that since writing to Follett to this effect, he has had no further word.

Follett did write to Gödel to make this proposal, but by 16 August Gödel had not received his letter.[i] On that day he wrote to Nagel (letter 8) inquiring where the publication of the book stood and briefly reiterating his previous position. Nagel's letter of 22 August was actually a response to this inquiry of Gödel. In addition to reporting the above to Gödel, Nagel expressed directly his response to Gödel's demand to see the manuscript of his and Newman's essay and approve it in some way. He did so more diplomatically than he had done to Angoff and with regret that he had not written to Gödel about the matter much earlier. He then made on his own the proposal for publishing only the translation of *1931*.[j]

Gödel responded to Nagel on 29 August (letter 10) shifting the ground and saying that the lectures *1934* would be more appropriate. He enclosed a copy of his reply to Follett (letter 1) stating his position more fully, saying that *1931* "is not well suited to serve, by itself, as an appendix to a book which is destined for the general public." He again insisted that neither his paper nor his lectures should be reprinted without mentioning "certain advances made since" and that he could not leave it to others' judgment what should be added. He also wrote:

> All difficulties that have arisen could easily have been avoided if the authors had discussed with me the plan of the book beforehand.

This seems disingenuous, since he and Nagel had had an extensive discussion in March. From Gödel's postscript, it is also evident that the proposed disclaimer of his approval for what Nagel and Newman wrote did not satisfy him at all; he maintains that even his limited cooperation would signify approval.

At this point it was clear that no plan for including writing by Gödel in the book could succeed, but it took until 25 September for Follett to reach Nagel and Newman to confirm this. Follett replied to Gödel on

[i]This letter is lost, but Gödel's letter 1 to Follett is evidently a reply to it. Since he says he received Follett's letter "with some delay", it may well have been sent before his letter to Nagel of 16 August.

[j]Curiously, this letter is not preserved in Gödel's papers, although much less significant ones are.

26 September (letter 2) and informed him about this conclusion. He added, speaking "for myself only and without prompting from any source," that he considered Gödel's position "a deeply shocking contravention of the generally recognized canons of free inquiry." The extended criticism of Gödel's attitude that follows is the only document known to us in which someone really chews Gödel out.

Why did this history unfold as it did? In other documented cases of publication of writings by Gödel with the latter's approval, in *van Heijenoort 1967* and *Wang 1974*, the others' attitude was more patient and deferential than that of Ernest Nagel. Between Gödel and Nagel, there was also a serious failure of communication at a decisive point: When Gödel offended Nagel by asserting what Nagel took to be a "right of censorship", Nagel himself did not communicate with Gödel directly to raise the issue, and Allan Angoff made matters worse by his attempts to paper the matter over. Another factor was the strong desire of Nagel, Newman, and NYU Press not to delay publication by a long negotiation with Gödel.

The failure of communication concerned, however, not only the issue of "censorship". It is puzzling that Nagel did not ask Gödel to read the manuscript of the book shortly after their Princeton conversation; why not make sure that the "blunders" Gödel had pointed to had not been repeated in the book? It would be possible to identify a number of difficulties in the articles that might have been the "blunders" Nagel refers to or the "troublesome and serious mistakes" mentioned in Gödel's later remarks. Indeed, in *Nagel and Newman 1956* and *1956a* the very statement of the first incompleteness theorem is not correct.[k] The problem is inadequately dealt with in the book.[l] The articles contain other local errors that could have been eliminated quite easily, but some such errors persist in the book.[m]

[k]The authors consistently attribute to Gödel the proof that if a system S containing arithmetic is consistent, then the standard Gödel sentence G is not only not demonstrable but not refutable either. Of course, to show G non-refutable one needs a stronger assumption; or, to show that if S is consistent then some F is neither provable nor refutable, one needs a different example, e.g. a Rosser sentence. Cf. these *Works*, vol. I, pp. 133–134, 140.

[l]In the text of *Nagel and Newman 1958* the same incorrect claim about G is made (p. 91), but in a footnote the authors write that "for simplicity of exposition" they have given an "adaptation" of the theorem of *Rosser 1936*, apparently missing the second of the points made in note k. Cf. *Putnam 1960a*, p. 205. This problem largely remains after the editor's revisions of the text in *Nagel and Newman 2001*; see pp. 99–100. However, the statement at the end of note 28 of *1958* (note 35 of *2001*) is corrected.

[m]Cf. *Putnam 1960a*, pp. 206–207. One might reasonably ask how much time Nagel and Newman devoted to revision of their manuscript after Nagel's 30 March meeting with Gödel. On 15 May Angoff wrote to Nagel and Newman that the copy-edited manuscript was in his office. That leaves one and a half months for revision, typing of the revised manuscript, and copy editing.

Gödel had broader misgivings with the presentation from the very beginning, and it is equally puzzling that he did not convey them openly to Nagel at the outset. They are expressed in Gödel's first two letters to Nagel (of 25 February and 14 March); but neither letter was actually sent.[n] They are connected, as we pointed out already, to the proper explication of "formal system" achieved by Turing and Post.[o] That explication allows, on the one hand, the formulation of the incompleteness theorems in their proper generality and serves, on the other hand, as the background for the philosophical interpretation of the results. The former point is made most extensively in the unsent letters and is alluded to, for example, in Gödel's letter to Angoff of 9 April. There he says that in an introduction to the appendix with his papers he would like "to mention advances that have been made after the publication of my papers." As to the philosophical interpretation of his results he writes in the same letter to Angoff:

> I am not very pleased with the treatment of these questions in the articles that came out in the Scientific American and in the "World of Mathematics". This, for the most part, is not the fault of the authors, because almost nothing has been published on this subject, while I have been thinking about it in the past few years.

It is likely that Gödel refers to such reflections on philosophical implications of his theorems as he presented in the Gibbs Lecture **1951*.

These remarks were made shortly after Nagel's visit in Princeton on 30 March. It is possible that Nagel was less grateful for the criticism he received than he let on in his letter later that day, and his later reactions may have reflected resentment or defensiveness. It might also be that his responses did not inspire confidence in Gödel. Nagel's own views in the philosophy of logic and mathematics show some influence of logical positivism. One can detect this influence in some of the general remarks about mathematics near the beginning of both articles with Newman, and it is not unlikely that Gödel was familiar with some of Nagel's articles, especially the well-known *1944*. He thus could well have

[n]It is possible that these matters were among those Gödel brought up in his meeting with Nagel on 30 March, but the later correspondence gives no indication to that effect. Gödel seems to have envisaged dealing with issues of this kind in his own introduction to his papers.

[o]The necessity of having a sharply formulated notion of "mechanical procedure" is not emphasized by Nagel and Newman in either of their publications, the central mathematical fact of the representability of (primitive) recursive relations is not made explicit in their exposition, and the equivalence of recursiveness and Turing computability is not mentioned. On account of this lack, there is no mathematical connection between formalisms and machines, thus, no firm grounding for their reflection on machines, for example, on p. 1695 of *1956a*.

expected views he disagreed with to be expressed in the longer version of the essay.

Finally, there is a reason why one might think that the project was misconceived from the beginning. In his unsent letter of 14 March, Gödel offered the objection that the book was intended for "the general public," while his own papers were not. One can agree that they did not fit well with Nagel and Newman's semi-popular exposition. If Gödel had pressed this objection, the ultimate conclusion, that the book would appear without his papers, might have been reached more quickly and amicably.[p] However, some comments in Nagel's letters might be offered as replies to this objection. On 21 March (letter 5) Nagel wrote:

> Mr. Newman and I believe we have done something to fill this need ⟦of giving an account of Gödel's work for a general audience⟧; and the N.Y.U. press agrees with us that we have written a fairly intelligible introduction to a major intellectual achievement of our times, and that our book may stimulate some of its readers to go deeper into the subject. It is for this last reason that the Press proposed adding your papers as appendices—so that those with sufficient interest in the subject might have the materials for further study readily available.

In the letter of 22 August Nagel reiterates his interest in making *Gödel 1931* available to students.

That Gödel did not press the practical objection and apparently even suppressed his substantive misgivings in his direct communications with Nagel probably shows that he *was* genuinely interested in having his papers reprinted, in spite of the questions he did raise. Indeed, once a more appropriate setting was offered, *Gödel 1931* and *1934* appeared with Gödel's approval in *Davis 1965*, and *1931* again, in a translation gone over carefully by Gödel, in *van Heijenoort 1967*. From a broader perspective, it seems unfortunate that the proposed cooperative venture between Gödel and Nagel foundered. It is perhaps indicative of the emerging separation of (mathematical) logic and philosophy of science.

Charles Parsons and Wilfried Sieg[q]

A complete calendar of the correspondence with Nagel appears on p. 451 of this volume.

[p]Nonetheless, an ironic postscript to this history is offered by the publication in *Nagel et alii 1989* of a French translation of *Nagel and Newman 1958*, together with a French translation of *Gödel 1931*!

[q]We are indebted to Leigh S. Cauman and Alexander Nagel for sharing their recollections of the episode narrated in this correspondence. Professor Cauman was manuscript editor for *Nagel and Newman 1958*, and the book was designed by her husband, the late Samuel Cauman. We thank Paul Benacerraf, John Dawson and Warren Goldfarb for information and comments.

1. Gödel to Nagel[a]

Princeton, Feb. 25, 1957

Dear Professor Nagel:

Mr. Angoff has informed me that you and Mr. James R. Newman are planning to write a book on my results about undecidable propositions and that you would like to use material contained in a course of lectures I gave in Princeton in 1934.[b] I am enclosing herewith a copy of mimeographed notes of these lectures. However, I am afraid that, except for the items mentioned below, they won't be of much use for your purpose. I had planned originally to give the proofs in more detail in these lectures than I had done it in my paper in Monatshefte f. Math. u. Phys. 38, 1931,[c] but I underestimated the time necessary. As a result the proofs, on the whole, do not contain more, but rather less detail than those given in my paper.

The only exception is that the formulas by which recursive functions are expressed in the system are constructed explicitly (see bottom of p. 16[d]). This really is all that is needed of Satz V (stated on p. 186 of my paper in Monatsh.) for the existence of undecidable propositions. For, if for the formulas thus constructed Satz V does not hold, the existence of undecidable propositions follows trivially, provided no propositions refutable in finitary arithmetic are demonstrable in the system. That the first one of the conditions under which the proof goes through (stated on p. 190 of my paper and on p. 19 of the lectures) is satisfied by *all* formal systems (or, to be more exact, that each formal system can easily be so transformed that it is satisfied) could not be proved ⟨rigorously⟩ before the ⟨general⟩ concept of "formalism" had been clarified, which was done by Turing and Post.

| What also might be of interest to you is ch. 7 of my lectures. It has some philosophical significance and, moreover, sketches the way in which I arrived at my results. The Diophantine problem constructed in ch. 8 can be considerably simplified, namely to a proposition containing only *one* change between universal and existential quantifiers in the prefix.[e] 2

[a]On the back of this sheet this letter is marked "nicht abgesch" (not sent). There is also a largely shorthand annotation, "Über Korrekturen in meinen lectures 1934" ("On errata in my lectures 1934"). It is not clear how and exactly when a copy of *Gödel 1934* was sent to Nagel; he acknowledges its receipt in letter 2 below.

[b]*Gödel 1934*. References to "the lectures" are to this work.

[c]*Gödel 1931*. References to "my paper" or "my paper in Monatshefte" are to this work.

[d]of *1934*.

[e]Cf. *Gödel *193?*.

As to a further elaboration of ch. 9 see S. C. Kleene's paper in Math. Ann. 112, p. 727,[f] where general recursiveness as defined in my lectures is proved equivalent with a considerably simpler concept.

I am sorry that, due to a bad cold and other circumstances, I was not able to send you this letter sooner.

Sincerely yours

Kurt Gödel

P.S. Unfortunately I have to ask you to return the lecture notes when you are through with them, because this is my last copy.

---

[f]*Kleene 1936.*

2. Nagel to Gödel[a]

March 9, 1957.

Dear Professor Gödel

Thank you most warmly for the copy you sent me of your ON UNDECIDABLE PROPOSITIONS OF FORMAL MATHEMATICAL SYSTEMS.[b] It is very kind of you to let me have it, and I am deeply appreciative of your generosity.

As you doubtless know, a somewhat more expanded version of the article by Mr. James Newman and myself (on your incompleteness theorem) which appeared in SCIENTIFIC AMERICAN last year,[c] will be published in book form by the New York University Press. We are planning to include in an Appendix a translation of your 1931 paper in the *Monatshefte*[d] (the translation was made by Professor John van Heijenoort of the department of mathematics in Washington Square College, and has been checked by me), as well as the paper you just sent me. I

---

[a]On letterhead of Columbia University in the City of New York, New York 27, N.Y., Department of Philosophy.

[b]*Gödel 1934.*

[c]*Nagel and Newman 1956.*

[d]*Gödel 1931.*

assume that all this is familiar to you, and that you have no objections to our including this material in the Appendix. I hope also that you will not mind if the corrections are made in the text of your 1934 Princeton Lectures of the *Errata* which are listed on pgs. 29–30 of the paper. If you do object to these alterations being made, would you be good enough to advise me at your earliest convenience.

With all good wishes and personal regards,

Sincerely yours,

Ernest Nagel

3. Gödel to Nagel[a]

Princeton, March 14, 1957.

Dear Professor Nagel:

Thank you for your letter of March 9. I am sorry to say that the plan of the book as the publishers apparently have it in mind seems rather bad to me, for the following reasons:

1. While the book evidently is intended for the general public, my papers are not. Therefore they take certain things for granted, in particular the possibility of completely formalizing ⟨all⟩ mathematics as far as it has been developed.[1] Special instances of this fact are even used in certain passages of the proof.

2. Considerable advances have been made in these questions since 1934. Not to mention other things, it was only by Turing's work that it became completely clear, that my proof is applicable to *every* formal system containing arithmetic. I think the reader has a right to be informed about the present state of affairs.

[1]and, more specifically, of formalizing it in the systems mentioned in the beginning of my paper in Monatsh. f. Math. u. Phys.[b]

---

[a]This letter was appparently not sent.

[b]*Gödel 1931*.

In view of all this the least that would have to be done is to reprint also extracts from papers by Turing, Kleene, etc. and to add an introduction commenting on this material and quoting other relevant work.

Sincerely yours

Kurt Gödel

4. Gödel to Nagel

Princeton, March 18, 1957.

Dear Professor Nagel:

Thank you for your letter of March 9. I had no idea that it is planned to publish complete reprints of my paper of 1931 and of my lectures of 1934.[a] I am sorry I had no time yet to give careful consideration to this proposal. But I am going to write you about it soon.

Sincerely yours

Kurt Gödel

P.S.[b] It seems to me that, if this plan is carried through, some explanations for the reader and quotations of other relevant work will be necessary.

[a] *Gödel 1931* and *1934*.
[b] The P.S. is handwritten.

5. Nagel to Gödel[a]

March 21, 1957.

Dear Dr. Gödel,

Thank you most warmly for your kind letter.
When I last wrote you, I had assumed you were familiar with the plan

---

[a] On letterhead of Columbia University in the City of New York, New York 27, N.Y. Department of Philosophy.

of the New York University Press to reprint your 1931 paper in English translation as well as your 1934 Princeton Lectures,[b] as appendices to a book which will contain an expanded version of the essay on your work by Mr. James R. Newman and myself in the June 1956 issue of *Scientific American.*[c] Mr. Denis Flanagan, editor of the magazine, told me that you had read our article and gave it your general approval.

You may perhaps recall that our essay was addressed to the general reader, in the hope of making clear to those without specialized knowledge the basic ideas underlying your 1931 paper and the significance of your discoveries. If one may judge by the favorable comments Mr. Newman and I have received, we have been at least partially successful in our effort; and we thought that our essay might reach a wider class of readers if it were made available in book form. The N.Y.U. Press has accepted it for publication in an enlarged version. In this version we are introducing a number of detailed explanations for which there was no room in the magazine article; and we also try to anticipate various difficulties the reader may experience in understanding the problems under discussion, by supplying a fuller background for crucial steps in the exposition.

I have long felt the need for an account of your work which could be placed into the hands of the intelligent layman as well as of university students whose primary concern is not in the foundations of mathematics. Mr. Newman and I believe we have done something to fulfill this need; and the N.Y.U. Press agrees with us that we have written a fairly intelligible introduction to a major intellectual achievement of our times, and that our book may stimulate some of its readers to go deeper into the subject. It is for this last reason that the Press proposed adding your papers as appendices—so that those with sufficient interest in the subject might have the materials for further study readily available.

It goes without saying that the value of the proposed appendices would be even greater should you add some further comments to your papers—as you suggested you might. I venture to note, however, that time is of some importance here, for the book is to be published| early next fall. 2

I shall of course be happy to answer, as far as I am able, any questions you may wish to raise. If you think it might help to expedite matters, I would be glad to visit you in Princeton, at some time convenient to both of us.

Sincerely yours,

Ernest Nagel

[b] *Gödel 1931* and *1934*.

[c] *Nagel and Newman 1956*.

6. Gödel to Nagel

Princeton, March 25, 1957.

Dear Professor Nagel:

Thank you for your letter of March 21. Your suggestion of discussing orally the question of publishing reprints of my papers in the proposed book seems very good to me. Could you perhaps come to Princeton this weekend (Friday or Saturday)? ør some day of the subsequent week?

Incidentally, how much do you expect to enlarge the article published in the Scientific American?[a] I liked this article very much as a non-technical introduction to the subject, although I don't agree in every respect with the formulation of the general meaning of my results.

Hoping to hear from you soon about our proposed meeting in Princeton, I remain

sincerely yours

Kurt Gödel

[a] *Nagel and Newman 1956.*

7. Gödel to Nagel

Princeton, April 9, 1957.

Dear Professor Nagel:

Thank you very much for your letter of March 30. The stipulations about sale abroad and sales of unbound sheets seem very unfair to me, since the authors get only a very small fraction of the income derived from them and, moreover, the "income" is not so precisely defined as the sales∧price of the book. The 5 years clause also is unfavourable to the authors, because it leaves them at the mercy of the publishers after 5 years.

I am writing to Mr. Angoff today making the same suggestions I mentioned to you in our conversation in Princeton.[a]

[a] See letter 1 to Allan Angoff in these *Works*, vol. IV.

Sincerely yours,

Kurt Gödel

P.S.[b] I made two more suggestions to Mr. Angoff, namely 1. To reprint only my lectures and the 1[rst] chapter of my paper (because of the duplication which would otherwise result). 2. To change the title of the book to: "Gödel's Theorems on the Principles of Mathematics".

---

[b]The P.S. is handwritten.

8. Gödel to Nagel

129 Linden Lane
Princeton, August 16, 1957.

Dear Professor Nagel:

I wonder in what stage the publication of the book on my proof is. The last I heard from Mr. Angoff was that he would send me galley proofs in a few weeks. That was on June 6. Of course I cannot give my permission to the reprinting of my papers, or make the necessary additions about more recent developments, before I have seen the text of the book.[1] Mr. Angoff seemed to be very anxious to speed matters up, but I am not willing to do anything hastily.

Sincerely yours

Kurt Gödel

[1]I have stated so in three letters to Mr. Angoff. Has it perhaps been decided to publish the book without the appendix?

9. Nagel to Gödel

South Wardsboro, Vermont.
August 22, 1957.

Dear Professor Gödel,

Thank you for your letter, which has just reached me at my summer address. I am afraid I know little more than you do about the present publication plans for the book on your work by Mr. Newman and myself; and the following few items will bring you up-to-date as far as my own knowledge of the matter is concerned.

More than six weeks ago I received a letter from Mr. Wilson Follett of the New York University Press, informing me that Mr. Angoff's connections with the Press were ended, and that he was replacing Mr. Angoff. Mr. Follett also wrote that negotiations with you concerning permission to reprint your papers seemed to him to have reached an impasse, and that he recommended publishing the Newman–Nagel essay without your material. Mr. Newman and I thereupon suggested to Mr. Follett that before the Press adopted this plan, he should make one final effort to reach an understanding with you. More specifically, we proposed to Mr. Follett that he ask for permission to include *only the English translation* of your 1931 paper,[a] *without* further addenda by you; and that he should assure you that a prominently placed statement in the book would make absolutely clear that your permission to do so does not signify your approval in the slightest degree of our expository essay. Since that letter to Mr. Follett was sent I have heard nothing further about the matter; and I gather from your own letter to me that Mr. Follett has not communicated with you either.

May I revert for a moment to your correspondence with Mr. Angoff? According to the information Mr. Angoff sent me last spring, you stated two conditions for giving your consent to reprinting your papers in our book: the first related to financial matters; the second required that the *manuscript* of our expository essay be submitted to you *for your approval*. Now as far as I know, your financial conditions never constituted serious obstacles, for both Mr. Newman and I made perfectly clear to the Press that any financial arrangements which were satisfactory to you (and of course to the Press) would be entirely agreeable to us. The real hitch arose over your second condition. I must say, quite frankly, that your second stipulation was a shocking surprise to me, since you were ostensibly asking for the *right to censor* anything of which you disapproved

[a] *Gödel 1931*.

in our essay. Neither Mr. Newman nor I felt we could concur in such a demand with⟨out⟩ a complete loss of self-respect. I made all this plain to Mr. Angoff when I wrote him last spring, though it seems he never stated our case to you. I regret now that I did not write you myself, for I believe you would have immediately recognized the justice of our demurrer to your second condition.

| It is probably too late now to carry through the original plan of the Press to include both of your papers (with such further additions as you wished to make). I venture to make a plea for the emended plan to include, with your permission, only the English translation of your 1931 paper.[b] The *sole* reason why I should still like to include the translation is that its publication in our book would make your paper available to students who otherwise have no access to it and would never see it. (Most college libraries do not have files of the *Monatshefte*, and according to one reliable informant not even the Library of Congress in Washington has it.) It seems to me a pity, moreover, that since an English translation of your paper has already been made (in my judgement the translation is an accurate one), the translation will otherwise not be likely to be published. I fully appreciate your hesitation to grant permission to include even the English version of your paper in our book, for your consent might be construed as an approval on your part of our expository essay. But I think, nevertheless, that the chances of such miscontrual can be reduced to a minimum, by inserting a conspicuously placed statement in the book explicitly disavowing such an interpretation. I hope you will see your way to giving the Press your consent to print the English translation as an appendix in our book, for I believe that by doing so you will be rendering a great service to many students. 2

I am leaving Vermont tomorrow to attend some meetings in the midwest, and I will have no fixed mailing address for some time. If you care to write to me, mail will reach me from Sept. 6 to Sept. 14 c/o Professor Herbert Feigl, Dept of Philosophy, University of Minnesota, Minneapolis, Minnesota.

With all good wishes and my pr⟨o⟩found personal esteem,

Sincerely yours,

Ernest Nagel

---

[b] *Gödel 1931.*

10. Gödel to Nagel

Princeton, Aug. 29, 1957.

Dear Professor Nagel:

Thank you for your letter. I am enclosing herewith a copy of a letter I am sending to the New York University Press in reply to their latest proposal to reprint only my paper of 1931. I think you will agree with me that my lectures would fit much better to the book than my paper. More‿over they doubtless are still less available than my paper.

Sincerely yours

Kurt Gödel

Donald Perlis

This item of correspondence from Gödel, dated 20 December 1968, was in response to an inquiry from Donald Perlis—at that time a Ph.D. student of Martin Davis[a]—as to notions of absolute demonstrability and definability that had been suggested by Gödel in his remarks for the Princeton Bicentennial Conference (*1946*). Those remarks had first been published in Davis' collection *The undecidable* (*1965*), which is where Perlis had seen them; he asked (in a letter dated 28 October 1968) whether any further work had been done about the notions in question.

In his response, Gödel began by distinguishing the idea of absolute demonstrability in set theory from that in number theory. As to the former he referred to "rather sketchy" work by Takeuti; perhaps Gödel had in mind here the work *Takeuti 1961* on "Cantor's absolute". As to absolute number-theoretic demonstrability, he pointed to the attempts of Turing (*1939*) and the undersigned (*1962*) via ordinal logics, or transfinite progressions of formal axiomatic systems. Completeness of a progression based on the iteration of a uniform reflection principle along a suitable path P recursive in the class O of Church–Kleene notations for ordinals was established in the latter. But it was shown in *Feferman and Spector 1962* (cited by Gödel) that there is incompleteness even for Π^0_1 statements along paths which are many-one reducible to O (i.e., are Π^1_1 definable). This sensitivity of completeness questions for ordinal logics to the kind of path through O along which a progression might be taken is the reason for Gödel's statement that "[t]he results show that a distinguished ordinal notation would be necessary to carry through this idea."

Concerning the notion of absolute definability in set theory, Gödel pointed to the work of Myhill and Scott (*1971*) on ordinal definability. As detailed in the introductory note to *1946* in volume II of these *Works*, Gödel had explained this notion informally there, and, somewhat later, others—including Myhill and Scott—independently rediscovered it. Gödel's response to Perlis concludes with a pointed mention of the result in *McAloon 1966* that the assumption HOD (that all sets are hereditarily ordinal definable) is consistent with the negation of the continuum hypothesis CH and even with the assumption that the power

[a]Perlis is presently Professor of Computer Science at the University of Maryland, specializing in artificial intelligence.

of the continuum "is much greater than $\aleph_1$." Apparently, Gödel thought this supported his view (reiterated since *1947*) that CH is false.

Solomon Feferman

1. Gödel to Perlis

December 20, 1968

Mr. Donald Perlis
18 Blenheim Gardens
London, NW2, England

Dear Mr. Perlis:

Some (rather sketchy) work on absolute demonstrability in the theory of *constructible* sets was done by Professor G. Takeuti (University of Illinois). Turing's ordinal logics, of course, are an attempt at absolute *number theoretic* demonstrability. Se⟦e⟧ also: Feferman, J. S. L. 27, p. 259,[a] p. 383.[b] The results show that a distinguished ordinal notation would be necessary to carry through this idea.

Absolute definability (under the name of "ordinal definability") was treated by Professor John Myhill (New York State University at Buffalo) and by Professor Dana Scott (Stanford University).[c]

I would suggest that you inquire with the authors mentioned about the literature.

As to the conjecture (2) of my Bicentennial Lecture I was recently informed that it has been verified by Kenneth McAloon in a dissertation at the University of California in Berkeley.[d] To be more precise, Dr. McAloon, using Cohen's method, has proved the consistency (with the Zermelo–Fraenkel axioms of set theory) of the assumption that all sets are "ordinally definable" and that $2^{\aleph_0}$ is much greater than $\aleph_1$.

Sincerely,

Kurt Gödel

[a] *Feferman 1962*.

[b] *Feferman and Spector 1962*.

[c] See *Myhill and Scott 1971*.

[d] *McAloon 1966*.

Walter Pitts

Walter Harry Pitts, Jr. (1923–1969) was a mathematician and co-author of some well-known papers, in particular *McCulloch and Pitts 1943*, a pioneering paper in mathematical modeling of neural functioning.[a] However, he is almost an unknown figure.[b] This must be due largely to his very untraditional academic career. He never received any higher academic degree—he did not even finish college—, never held any important academic posts, and published only scantily. Pitts was born of working class parents in Detroit, and probably at the age of fifteen he ran away from home. Already at that young age Pitts displayed amazing intellectual powers, having taught himself several languages, including Latin, Greek and Sanskrit. He had also studied logic, and in connection with this he had a short correspondence with Russell about *Principia mathematica*, which gained him an invitation to come to England as a graduate student. From 1938 to 1943 he studied at the University of Chicago but earned only the degree of Associate of Arts, which he was awarded because of his achievements in *McCulloch and Pitts 1943*.

In 1943, apparently at the urging of Norbert Wiener, Pitts moved to the Massachusetts Institute of Technology, where he was to spend the rest of his life with the exception of war work in New York beginning in 1944, part of the Manhattan Project. In his first decade at MIT he worked as an unofficial graduate student with Wiener. *Wiener 1948* identifies him as one of the early workers in cybernetics.[c] But during this period he registered as a student only for one academic year, namely in physics in 1943–1944.[d] In 1952 there was a sudden split between Pitts and Wiener, and their relationship came to an end. Pitts remained at MIT, continuing in the position he had held since 1951 as a research associate in the Research Laboratory of Electronics. He retained that position until his death in 1969.

Pitts had a lifelong interest in Leibniz, which started in his early teens. He was particularly impressed by Leibniz's logical theory. In the 1950s Pitts was working on a reading of Leibniz's metaphysics which aimed to be better informed by Leibniz's work in engineering than pre-

[a]Of the 21 papers collected in *McCulloch 1965*, Pitts is a co-author of five, one the well-known *Lettvin et alii 1959*. There are a few papers of the 1940s by Pitts alone or with other authors.

[b]But see *Smalheiser 2000*, which appeared at an advanced stage of our work but prompted some changes in the present note.

[c]See *Wiener 1961*, pp. 13–14.

[d]He was again registered as a student at MIT in 1956–1958, this time in electrical engineering and computation.

vious studies. The guiding idea was that there are notions and examples in Leibniz's metaphysics which draw on Leibniz's more practical work but which the published translations fail to capture. During the period from 1953 to 1960 Pitts gave lectures on Leibniz at the Bell Telephone Laboratories in New Jersey. We do not know of any surviving written record of these lectures.[e] The occasion for Gödel's letter to Pitts is not definitely known, but it was very likely one or more of these lectures. No letter from Pitts is to be found among Gödel's papers. So far as is known Gödel did not attend any of the lectures, but others from Princeton did. Furthermore, Gödel addressed his letter to Pitts at Bell Laboratories.

Gödel's engagement with Leibniz's work was of quite long standing. According to Karl Menger, already in the early 1930s Gödel was intensely interested in Leibniz and desired to see his unpublished manuscripts.[f] The suggestion in the letter about the value that unpublished writings of Leibniz might have echoes the remarks at the end of *Gödel 1944*, where Gödel seems to regard the hopes that Leibniz expressed for a universal characteristic as realistic. In the present letter Gödel goes further, saying that Leibniz made definite claims of discoveries of which it is not definitely known whether he wrote them down. It is not clear what Gödel may have had in mind more specifically. The suggestion that important parts of Leibniz's work were lost or concealed had been expressed by Gödel in conversation with Karl Menger and Oskar Morgenstern.[g]

The last sentence of the letter alludes to an effort of Gödel and Morgenstern, beginning in 1949, to have a microfilm of Leibniz's papers made and deposited in the Princeton University Library.[h] That effort ultimately failed, but a parallel effort by Paul Schrecker of the University of Pennsylvania was successful. Gödel's remark indicates that he thought that whatever Schrecker had obtained was short of being complete.[i] It is natural to conjecture that the new effort he refers to was made by Professor Giorgio de Santillana of MIT, possibly in collaboration with Pitts or at his suggestion. The evidence of this is an agreement dated 7 April 1959 between the Niedersächsische Landesbibliothek and MIT, signed for MIT by de Santillana, by which MIT would obtain a copy of the microfilm already in the possession of the University of Pennsylvania.[j] However, we have not succeeded in locating

[e]However, Jerome Lettvin has a tape of them.

[f]*Menger 1994*, p. 210.

[g]*Wang 1987*, pp. 103–104.

[h]See *Dawson 1997*, pp. 189–190.

[i]Dr. Michael T. Ryan, Director of Special Collections of the University of Pennsylvania Library, informs us that Schrecker obtained "a selection on 33 reels of film" (communication of 29 January 1999).

[j]Communication of Anke Holzer of the Niedersächsische Landesbibliothek, 6 May 1999.

such a microfilm in the MIT Library, and in any event it would not have fulfilled what seems to have been Gödel's hope for a more comprehensive microfilming of Leibniz's papers.

Øystein Linnebo and Charles Parsons[k]

[k]We are greatly indebted to Jerome Lettvin, who is the principal source of our biographical information about Pitts, and who commented on an earlier draft. We also thank John Dawson and Neil Smalheiser for comments.

1. Gödel to Pitts

September 23, 1958

Dr. Walter Pitts
Bell Laboratories
Murray Hill, N.J.

Dear Dr. Pitts

I was very much interested to hear that you are considering applications of Leibnizian ideas to modern physics. Are you publishing something in print, or does there exist a manuscript about it?

I do think that the possible applications of Leibniz's work to modern science are far from being exhausted. In particular the unpublished manuscripts of Hannover may contain invaluable ideas as to the systematic solution of mathematical, as well as other scientific, problems. In fact this must be so, if Leibniz ever put down on paper what he definitely claimed to have discovered, and if the manuscripts concerned were not lost in the subsequent centuries. I am, therefore, very glad to hear that the plan of microfilming the Leibniz papers is being taken up again.

Sincerely yours,

Kurt Gödel

David F. Plummer

The following letter is Gödel's response to an inquiry of 10 July 1967 from Mr. David Plummer of Bellevue, Washington, a non-mathematician who had become interested in the incompleteness theorems through his reading of Ernest Nagel's and James R. Newman's essay *1956a*. Plummer wrote Gödel to request further references on the subject and to pose two questions concerning the import of those theorems. Having gained the impression that the undecidability results had "cast certain gloomy ... epistemological shadows", he asked whether Gödel thought the "overriding ... philosophical implication" of his theorems was "basically nihilistic or destructive". He opined to the contrary that perhaps the theorems could be invoked to refute determinism and justify the view expressed by F. A. von Hayek (in his *1960*) that "the advance and even preservation of civilization are dependent upon a maximum of opportunity for accidents to happen."

Unsurprisingly, Gödel disagreed with both suggestions. His response to Plummer should be compared with the disjunction he propounded in his Gibbs Lecture (**1951*, p. 310) and with his replies to similar inquiries that he received from Georg Brutian and Leon Rappaport.

John W. Dawson, Jr.

1. Gödel to Plummer[a]

Mr. David F. Plummer
14414 NE 14th Place
Bellevue, Washington 98004

July 31, 1967

Dear Mr. Plummer:

In my opinion the philosophical implications of my theorems are anything but nihilistic or destructive. Nor do I see any connection with liberty or ac-

[a]In addition to the footnotes there is an annotation saying "number theoretic statt arithmetic an 2 stellen" ("number-theoretic instead of arithmetic in 2 places").

cident. What has been proved is only that the kind of reasoning necessary[1] in mathematics cannot be completely mechanized. Rather[2] constantly renewed appeals to mathematical intuition are necessary. The decision of my "undecidable" proposition (see p. 93 of Nagel's and Newman's book, 2nd ed., 1960)[b] results from such an appeal. Other instances are explained in the first passage cited below. Whether every arithmetical yes or no question can be decided with the help of some chain of mathematical intuitions is not known. At any rate it has not been proved that there are arithmetical questions undecidable by the human mind. Rather what has been proved is only this: Either there are such questions or the human mind is more than a machine. In my opinion the second alternative is much more likely.

I have briefly explained the situation in the following books:

1. Philosophy of Mathematics, ed. by Paul Benacerraf and Hilary Putnam, Prentice Hall, Inc. 1964, p. 264–265 and p. 269–272.[c]
2. The Undecidable, ed. by Martin Davis, New York, Raven Press, 1965, p. 71–73.[d]

A slightly different version of my proof is given on p. 59–63 of "The Undecidable" and brought up to date on p. 72.

Sincerely yours,

Kurt Gödel

[1](to be more precise, for ⟨answering⟩ ~~the solution of~~ of all number theor. yes or no questions of a rather simple structure).

[2](if the human mind is at all able to give the answers).

---

[b]The editors know of no 1960 edition; Gödel's reference would fit *Nagel and Newman 1959*.

[c]*Benacerraf and Putnam 1964*, i.e., *Gödel 1964*.

[d]*Davis 1965*, Postscriptum to *Gödel 1934*.

Karl Popper

Karl Popper (1902–1994) was a noted philosopher, born and educated in Vienna, who from 1946 on taught at the London School of Economics. He first met Gödel in 1935 when he gave a presentation to the Menger Colloquium. Another meeting took place in 1950, when, on a visit to Princeton, Popper went to see Gödel to discuss *Gödel 1949a*.[a] On 24 March 1964, Popper wrote Gödel two letters. The first invited Gödel to a meeting of the British Society for Philosophy of Science in London, planned for July 1965, which was to include a session on foundations of mathematics. Popper elaborated:

> Our plan is ... to discuss questions such as:
> (1) Is it true that very few competent people continue to think or to write about the status of mathematics?
> (2) If so, why? Have the old great problems been tacitly dropped? If so, why? Have they turned out to be pseudo-problems? If so, in which sense? And why not say so?
> (3) Can we uphold the distinction between pure mathematics and pure science? If so, what are the characteristic differences? If not, how can we explain the mistake made by those who say, or who said, there was a distinction?

The second letter, written in German, concerned Paul Schilpp's invitation to Gödel to contribute a paper to his planned volume on Popper in the *Library of living philosophers* series.[b] Popper wrote, "I really don't need to tell you how highly I would value such a contribution."

In his reply Gödel declines both invitations. Some of his reasons are of interest. When he notes that "mathematicians and philosophers, for the most part, are empiricists", he means logical empiricists or logical positivists (and not post-positivist empiricists, like W. V. Quine). The attempt "to subsume mathematics satisfactorily under this viewpoint" that Gödel has chiefly in mind, it is plausible to suppose, is the positivists' linguistic or syntactical account of logic and mathematics. As he expressed in **1953/9*, he thought any such view could be conclusively refuted.

Gödel's experience in writing **1953/9* is reflected in his remarks about not writing for the Schilpp volume on Popper. **1953/9* was intended for the Carnap volume *Schilpp 1963*. Gödel had worked on it over a stretch of years, but in the end did not arrive at a version with

[a] *Popper 1974*, pp. 80 and 105.

[b] *Schilpp 1974*. Schilpp's invitation to Gödel is letter 38 of the Schilpp correspondence in this volume.

which he was satisfied. In his letter to Schilpp of 3 February 1959 explaining that he would not be submitting anything, he admitted that "a complete elucidation of the situation turned out to be more difficult than I had anticipated, in consequence of the fact that the subject matter is closely related to, and in part identical with, one of the basic problems of philosophy, namely the question of the objective reality of concepts and their relations."[c]

Warren Goldfarb

[c]Letter 36 of the Schilpp correspondence in this volume.

1. Gödel to Popper

April 10, 1964

Dear Professor Popper:

I am terribly sorry I cannot take part in the prospective conference in 1965 of the British Society for the Philosophy of Science. The fact that, for reasons of health, I have, for many years, avoided any travelling except for short trips to nearby places would by itself be decisive, even if I had no other reasons. Incidentally, I believe it will be very hard, if not impossible, to get the kind of discussion underway, which you have in mind. You must not forget that mathematicians and philosophers of science, for the most part, are empiricists, while all attempts to subsume mathematics satisfactorily under this viewpoint have failed, and very conspicuously in recent times.

Unfortunately, I am not in a position either to write a contribution to the Schilpp volume devoted to you as a philosopher.[a] Here the question of expenditure of time, about which I have some experience from previous contributions, plays an important part. But still more important is the fact that, as of now, I don't like particularly to write or speak about philosophical questions, because I have not yet developed sufficiently my own point of view.

I hope you won't take my negative response amiss.

Very sincerely yours,

Kurt Gödel

[a]*Schilpp 1974*.

Emil L. Post

Shortly after his return from Europe on 15 October 1938, Gödel attended a regional meeting of the American Mathematical Society in New York City.[a] During the meeting, on October 29, he made the acquaintance of Emil L. Post. Post wrote a brief, moving note to Gödel the very same day and an extended letter on the following day. Post reflected on their meeting and, relatedly, on his own work on absolutely undecidable problems that had started in the early 1920s and "anticipated" Gödel's first incompleteness theorem. It had been highly emotional for Post to meet the man "chiefly responsible for the vanishing of that dream", i.e., the dream of astounding the mathematical world by his "unorthodox ideas" and by establishing the existence of unsolvable problems. "Needless to say", he emphasized at the end of his second letter, "I have the greatest admiration for your work, and after all it is not ideas but the execution of ideas that constitutes the mark of greatness." Post wrote a third letter on 12 March 1939 after he had read Gödel's abstract *1939* on the relative consistency of the continuum hypothesis. Gödel had sent Post a "sheaf of reprints" in the fall of 1938, but responded to Post's letters only on 20 March 1939.

At the time of his meeting with Gödel, Post was a faculty member at the City College of New York. He had been appointed there in 1935 and remained at the institution until his death in 1954. His education was also deeply connected with City College. Born in the Polish town of Augustow on 11 February 1897, Post emigrated with his parents to New York in May of 1904. He attended Townsend Harris High School, a free secondary school for gifted students located on the campus of City College, and received his B.S. from City College in 1917.

From 1917 to 1920 he was a graduate student at Columbia University, finishing with a thesis under the direction of Cassius Keyser. The thesis was published as *Post 1921* and concerned the propositional calculus in Whitehead and Russell's *Principia mathematica*. Post established the

[a]For a fuller discussion of Gödel's circumstances at the time, see *Dawson 1997*; the meeting between Post and Gödel is described on pp. 130–132. *Post 1994* provides information on Post's life and work, in particular about the enormous difficulties Post had to face, from his severe mental illness to the restrictive working conditions under which his scientific endeavors had to be pursued.—The biographical facts in this Note stem from these two sources.

Special thanks to Martin Davis, John Dawson, Solomon Feferman, and Charles Parsons for helpful e-correspondence concerning a draft of this Note; John Dawson helped me to locate information on the mathematician Jesse Douglas mentioned in the postscript of Post's letter of 29 October 1938.

Emil Post

semantic completeness[b] of the calculus; he went on to generalize the "postulational method" and the "truth-table development" for finitely many, arbitrary propositional connectives. The former generalization was the starting point for Post's investigation of symbolic logics that led to the anticipation of Gödel's result. Post's *1941* gives an account of this work; the paper was rejected for publication and appeared only in *Davis 1965*.

The central new mathematical result of *1941*—the reduction of arbitrary canonical systems to systems in normal form—was presented in *Post 1943*. Post's Thesis[c] secures then the reduction for all symbolic logics. Post describes in his letter how this reductive result led him first to the discovery of an absolutely unsolvable problem and then to the realization of the incompleteness of all symbolic logics. Post emphasizes that a particular proposition can be seen to be undecidable, i.e., "...a particular enunciation of the logic, determined by the logic, and of course the entscheidung problem ⟦sic⟧ and the method of proving the above contradiction, was such that neither it nor its negative was asserted in the logic." These very sketchy considerations are detailed in section 2 of *Post 1944*; that section is entitled "A form of Gödel's theorem".

Post formulates then in his letter what he takes to be the main point of Gödel's Theorem: "...the existence of an undecidable proposition in each logic sufficiently general and yet a 'symbolic logic'." This is a formulation of sufficient generality, such as Gödel himself had been aiming for. Its rigorous mathematical version depends for its adequacy on Post's Thesis.[d] Indeed, Post's proof of the incompleteness result relies on it by identifying symbolic logics with normal systems. The remainder of the letter is devoted to the underlying methodological problem and how its difficulty, together with the personal and professional reasons hinted at in Note 1, prevented Post from publishing his considerations.

The thesis has, according to Post, "but a basis in the nature of physical induction". He believes that that is true not only for his own work, but for "any work". Such a perspective had been taken in *1936* where

[b]This is the Fundamental Theorem in section 3 of *Post 1921*.—Hilbert and Bernays established the (Post) completeness of the propositional calculus in lectures of 1917–1918 and Bernays's *Habilitationsschrift* of 1918 (later published in abridged form as *Bernays 1926*). Some of the results of the latter were published only in *Bernays 1926*; cf. *Sieg 1999* and *Zach 1999*.

[c]In *Davis 1982*, p. 21, Post's Thesis is formulated roughly as follows: any set of strings on some alphabet that can be generated by a finite process (thus any symbolic logic) can be generated by canonical productions and, using the reductive result, by normal productions.

[d]Gödel emphasized in the Postscriptum to his *1934* the need of Turing's penetrating analysis for being able to formulate his incompleteness theorems for all formal systems (that are consistent and include a modicum of elementary number theory).

Post presents his formulation 1, a model of computation essentially identical with Turing's. On p. 105 of that paper he asserts:

> The writer expects the present formulation to turn out to be logically equivalent to recursiveness in the sense of the Gödel-Church development. Its purpose, however, is not only to present a system of a certain logical potency but also, in its restricted field, of psychological fidelity. In the latter sense wider and wider formulations are contemplated. On the other hand, our aim will be to show that all such are logically reducible to formulation 1. We offer this conclusion at the present moment as a working hypothesis. And to our mind such is Church's identification of effective calculability with recursiveness.

In a footnote to the last sentence, Post claims that the actual work of Church and others has carried "this identification considerably beyond the working hypothesis stage". He warns, however, that calling the identification a definition may blind us "to the need of its continual verification" by considering, quasi-empirically, wider and wider formulations and reducing them to formulation 1. The success of this research program, he says in the main text, would "change the hypothesis not so much to a definition or an axiom but to a natural law".

Post, in his letter to Gödel, states that the quasi-empirical work supporting the "induction" could be extended to cover the system of *Gödel 1931* and to obtain, in this way, the incompleteness theorem specifically for that system without appeal to the thesis. That this could be done for the system of *Principia mathematica* itself Post claims to have seen in the 1920s. Post did not then pursue the inductive avenue, because he thought he saw a way of properly analyzing "all finite processes of the human mind" and thus a possibility of establishing the theorem "in general and not just for Principia mathematica". Post adds in parentheses after "human mind" in the last quote "something of the sort of thing Turing does in his computable number paper".[e] How closely related Post's foundational considerations were to those of Turing can be seen from later developments reflected in *Post 1947*, *Turing 1950* and *Turing 1954*. Post used in *1947* a description of Turing machines by production systems to show the unsolvability of the Thue problem (established independently in *Markov 1947* using quite directly Post's normal systems). Turing employed in *1950* the same techniques to extend Post's result. In his semi-popular *1954*, he formulated a version

[e]The reference is of course to *Turing 1937* and, more particularly, to Turing's argument in section 9 of that paper; as to the analysis of that argument, see *Sieg 2002* and literature quoted there. In his own *1943* Post writes in note 18: "Since the earlier formal work made it seem obvious that the actual details of the outline ⟦of the proof of his version of the incompleteness theorem⟧ could be supplied, the further efforts of the writer were directed towards establishing the universal validity of the basic identification of generated sets with normal sets."

of his thesis in Post's way: all puzzles (i.e., combinatory problems) can be transformed into substitution puzzles (i.e., Post's normal systems); then he gave a perspicuous presentation of solvable and unsolvable problems via substitution puzzles.

Wilfried Sieg

1. Post to Gödel

1253 St. Nicholas Ave
New York, N.Y.
Oct. 29, 1938

My Dear Prof. Gödel,

I am afraid that I took advantage of you on this, I hope but our first meeting. But for fifteen years I had carried around the thought of astounding the mathematical world with my unorthodox ideas, and meeting the man chiefly responsible for the vanishing of that dream rather carried me away.

Since you seemed interested in my way of arriving at these new developments perhaps Church can show you a long letter I wrote him about them. As for any claims I might make perhaps the best I can say is that I would have *proved* Gödel's Theorem in 1921—had I been Gödel.

Emil L. Post

P.S. Douglas,[a] now at Princeton, might recall some talks I had with him on these very matters. E. L. P.

[a]Post refers presumably to the mathematicican Jesse Douglas (1897–1965). Douglas had also attended City College and obtained his Ph.D. at Columbia in 1920. During the academic year 1938–1939 he was a member of the Institute for Advanced Study.

2. Post to Gödel

1253 St. Nicholas Ave
New York, N.Y.
Oct. 30, 1938

My dear Dr. Gödel,

In our conversation of yesterday each time a comparison of your Theorem and absolutely unsolvable problems arose you kept emphasising that in the former a particular proposition appeared as undecidable. I therefore want to point out that that is exactly what arose in my procedure of getting your Theorem as a corollary of the existence of absolutely unsolvable problems.

The problem in question was an entscheidung problem. I therefore considered any logic in which that problem could be formulated. On the basis of the reductions I refer⟦r⟧ed to I assumed that that logic was likewise reducible to 'normal form' (equivalent I am sure to general recursiveness). Then the assumption that that logic assigned a unique yes or no answer to each particular problem forming the entscheidung problem led by a Cantor diagonal argument to a contradiction. Hence my conclusion of the unsolvability of that problem by any 'symbolic logic'⟦,⟧ and hence its absolute unsolvability.

But furthermore I examined the source of the contradiction and found that it led to a particular problem of the class of problems forming the entscheidung problem receiving neither a yes or no answer—assuming that the logic was consistent, i⟦.⟧e. didn't give both a yes and no answer. That is that a *particular* enunciation of the logic, *determined* by the logic, and of course the entscheidungsproblem and the method of proving the above contradiction, was such that neither it nor its negative
2 was asserted in the logic. Hence my conclusion | that a complete symbolic logic does not exist—not in the obvious sense that one may add to its concepts and hence enunciations, e.g. the Principia with its restricted theory of types gives but a partial theory of particular transfinites, but that relative to a fixed and indeed simple set of enunciations, those of the particular entscheidungsproblem, it fails to include a yes or no answer for each of those enunciations. (It may be that with my eye fixed on the entscheidung problem in question I failed to pause and state your theorem as you give it as a final theorem. It is so stated though as a preliminary to the above statement.)

I may further mention that the above reexamination of the reductio ad absurdum proof showed that granting the ~~the~~ consistency of the logic, and I might say the meaning of the entscheidung problem the analysis did lead to a definite yes or no answer to the enunciation that the assumed logic failed to so decide. I therefore concluded that mathematical proof was essentially creative in that once having set up a formal system relative to say the above entscheidungsproblem we could then always transcend that system, i⟦.⟧e. add to the set of its assertions relative to the same body of enunciations—a conclusion I believe also reached in your work.

Of course your particular undecidable proposition for a logic has intrinsic interest in its reinterpretation as a statement of the consistency of the formal system. I'm afraid its very interest has led both to misinterpretations of the meaning of your theorem in its relation to possible proof of consistency and to neglect of what I think is the main point of your theorem, the existence of an undecidable proposition in each logic sufficiently general and yet ⟨a⟩ 'symbolic logic'.

| May I finally say that nothing that I had done could have replaced 3 the splendid actuality of your proof. For while corollary your theorem may be of the existence of an absolutely unsolvable problem, the absolute unsolvability of that problem has but a basis in the nature of physical induction at least in my work and I still think in any work. Of course with sufficient labor that induction could have gone far enough to include your particular system theorematically. That that could be done for Principia Mathematica I saw then. My only excuse for not doing so—well there are many and having written so much I might add them. Chiefly I thought I saw a way of so analyzing "all finite processes of the human mind" [something of the sort of thing Turing does in his computable number paper] that I could establish the above conclusions in general and not just for Principia Mathematica. Secondly that the absolute unsolvability of my problem would not achieve much recognition from others merely on the basis of an incompleteness proof for Principia Mathematica. And lastly while the above general analysis enticed me it seemed foolish to do all the labor involved in the more special Principia Theorem. As for not publishing my work as outlined above, or rather in Church's letter, the whole force of my argument depended on identifying my 'normal systems' with any symbolic logic and my sole basis for that were the reductions I mentioned. And with the difficulty I had in getting my elementary propositions paper[a] published at all, and the rejection | outright of its 4 companion piece[b] I saw no hope of getting those reductions into print. And then came illness, and ⟨then⟩ a sort of preparatory regime for work

[a] *Post 1921*.

[b] See *Post 1965*; in note 19 Post refers to an "abstract of this yet unpublished paper", i.e., *Post 1921a*. In that paper Post gave a decision procedure for canonical systems in which the primitive functions are all unary. The same reference is also given in *Post 1943*, note 11.

which would gradually make me less excited by those general ideas of mine. And that regime took, ⟨and is taking,⟩ so much more time than I had planned for it that finally you and Church and even Turing sped by me.

I hadn't meant this letter to be the sort of thing it has turned out to be. Please forgive my taxing your patience this way. And of course any resentment I may have is at the Fates if ⟨not⟩ myself. It was a real pleasure to meet you and I hope my egotistical outbursts have spent themselves with that first meeting and this letter. Needless to say I have the greatest admiration for your work, and after all it is not ideas but the execution of ideas that constitute⟦s⟧ a mark of greatness.

Sincerely yours

Emil L. Post

3. Post to Gödel

1253 St. Nicholas Ave
New York, N.Y.
March 12, 1939

My dear Dr. Gödel,

Your very interesting abstract in the Bulletin[a] reminded me that I failed to thank you for your very generous sheaf of reprints you sent me last semester. I do appreciate them and hope you will send me a reprint of this latest and very important paper.

I was going to ask if you had tackled certain other outstanding questions by your methods. But it might be unfair to so intrude. May I suggest that the reason for 'your success' (if I may use a phrase of Wiener's) is your genius for producing | a simplified logical system which has the earmarks of generality and yet is manageable. In the case of your undecidable proposition paper any loss of generality thus *possibly* incurred is unimportant since the method obviously would ~~obviously~~ generalize. But I wonder if the same is true of the continuum hypothesis paper. Is the system of axioms as general as what Principia Mathematica in its three volume published form would be if it were completely formalized.

[a] *1939*.

Then again your very undecidable proposition paper limits the significance of such a result gotten from any one formal system. For that paper does give a new method of proof for any formal | logic, not present in that logic. And so what guarantee of the consistency of the continuum hypothesis under such a prolongation of the logic you analyse.

However the main thing to do about an important problem is to do something about it. And your abstract shows that your methods can do more than just something with the significant problems of today.

Sincerely yours

Emil L. Post

4. Gödel to Post

Notre-Dame, March 20, 1939

My dear Dr. Post,

First of all I have to apologize to you for not yet having replied to your letters of Oct. 29 and 30. One of the reasons for the delay was, that I wanted first to read in detail your letters to Church, which you mentioned to me. Your method of treating formal systems, which you sketch there, is certainly very interesting and worth while to follow up in its consequences.

As to the personal matters which you touch in your letters, I can assure you, that I have noticed nothing of what you call egotistical outbursts in your letters or in the talk I had with you in New York; on the contrary it was a pleasure to speak with you. I hope to see you again when I come back to the East in May and shall certainly be very interested to hear more | about your way of getting at undecidable propositions.

The remarks which you make in your letter of March 12 about my latest paper, I am sorry I could not quite understand, because you apparently failed to enclose some pages of your letter in the envelope. The guarantee for the consistency of the Continuum-Hypothesis consists in this, that a contradiction derived from it in any of the known formal systems could be transformed into a contradiction obtained ~~fro~~ without it in the same system (as stated in the abstract of which I enclose a reprint). A sketch of the proof is to come out in April.

Sincerely yours

Kurt Gödel

Math. Dep.
University of Notre Dame, Ind.

Leon Rappaport

On several occasions Gödel received letters requesting his own assessment of the significance of the incompleteness theorems, and in at least four instances he took the time to reply—most extensively to Leon Rappaport, a Swede who had studied mathematics at the University of Warsaw in the years prior to World War II and had participated in seminars there conducted by Leśniewski, Łukasiewicz and Tarski.[a] (See also Gödel's letters to Yossef Balas, Georg Brutian and David Plummer.)

Rappaport's purpose in writing to Gödel was to verify the accuracy of three remarks he had made about the incompleteness theorems in a book manuscript he was preparing for submission, an extract of which he enclosed with his letter of 27 June 1962. The assertions in question, labeled E1–E3 by Gödel in his reply, were (E1) that the incompleteness theorems "confronted the entire, meticulously built-up structure ⟦of axiomatized mathematics⟧ with the danger of breaking down"; (E2) that "we still do not know what is the basis of our belief in the possibility of distinguishing truth from falsity" in logic and mathematics; and (E3) that if a mathematical theory used to describe physical phenomena were found to contain undecidable propositions, then the aspects of those phenomena corresponding to such propositions must in principle be unpredictable.

In the letter itself Rappaport added the further two-part question (cited by Gödel as L3): "a) Is there a proof...that by increasing the number of axioms" of the theory in a suitable manner an undecidable statement can always be decided in the extended theory?, and b) Is the number of undecidable statements in a "non-elementary deductive theory" known to be infinite? With regard to a), Gödel noted that "there are, for any S and T, infinitely many P for which it is undecidable in S, whether P is decidable in T." For if formal *un*decidability in T could be decided in S for all but finitely many statements of T, formal undecidability in T would be a recursively enumerable notion; but since formal decidability in T *is* recursively enumerable, it would then follow that formal undecidability in T, and hence provability in T, would be recursive.

John W. Dawson, Jr.

[a]Gödel also addressed the significance of his incompleteness theorems in several lectures and manuscripts that remained unpublished during his lifetime. See in particular items **1933o*, **1951* and **1953/9* in volume III of these *Works*.

1. Gödel to Rappaport

Dr. Leon Rappaport
Saltsjöbadsvägen 35
Saltsjö-Duvnäs, Stockholm
Sweden

August 2, 1962

Dear Dr. Rappaport:

Replying to your letter of June 27, 1962, 1 would like to say this:

Nothing has been changed lately in my results or their philosophical consequences, but perhaps some misconceptions of them have been dispelled or weakened. My theorems only show that the *mechanization* of mathematics, i.e. the elimination of the *mind* and of *abstract* entities, is impossible, if one wants to have a satisfactory foundation and system of mathematics.

I have not proved that there are mathematical questions undecidable for the human mind, but only that there is no *machine* (or *blind formalism*) that can decide all number theoretical questions (even of a certain very special kind).

Likewise it does not follow from my theorems that there are no *convincing* consistency proofs for the usual mathematical formalisms, notwithstanding that such proofs must use modes of reasoning not contained in those formalisms. What is practically certain[1] is that there are, for the classical formalisms, no conclusive *combinatorial* consistency proofs (such as Hilbert expected to give), i.e. no consistency proofs that use only concepts referring to finite combinations of symbols and not referring to any infinite totality of such combinations.

I have published lately (see Dial., vol. 12 (1958)[a] p. 280) a consistency proof for number theory which probably for many mathematicians is just as convincing as would be a combinatorial consistency proof, which however uses certain abstract concepts (in the sense explained in this paper).

2 | Your formulation E1[2] is overdramatized and not true as it stands. It is not the structure itself of the deductive systems which is being threatened with a break/down, but only a certain *interpretation* of it, namely its interpretation as a blind formalism.

[1]No formal proof has yet been given, because the concept of a combinatorial proof, although intuitively clear, has not yet been precisely defined.

[2]I am referring to your letter by L and to the excerpt by E.

---

[a]*Gödel 1958.*

As far as L3b is concerned, the existence of an infinity of such theorems can be proved even for the system of number theory.

E2 is substantially correct, because in contradistinction to the amazing progress of the experimental sciences and, to a lesser degree, also of mathematics, our knowledge of the abstract mathematical entities themselves (as opposed to the *formalisms* corresponding to them) is in a deplorable state. This is not surprising in view of the fact that the prevailing bias even denies their existence.

The answer to L3a is this: If you can prove in some formalism S that a proposition P is undecidable in the formalism T you can, by adding P or $\sim P$, construct a demonstrably consistent system in which P is decidable. But there are, for any S and T, infinitely many P for which it is undecidable in S, whether P is decidable in T. Therefore, as long as you confine yourself to definite formalisms T, S for mathematics and metamathematics, you cannot construct for these P any demonstrably consistent extension of T in which they are decidable.

As to E3 it should be noted that it depends on the structure of the physical theory concerned, whether the incompleteness of the systems of mathematics implies the existence of unpredictable physical phenomena. If the mathematical problems involved in the derivation of individual phenomena from the laws of physics are sufficiently complicated, specific physical questions (such as the occurrence or non-occurrence of a phenomenon under specified conditions) *can* become undecidable. I don't think it has ever been investigated whether the physics of today has reached this degree of complication. Of course, if it has, the same restriction would apply that I mentioned about mathematics in the beginning of this letter. On the other hand, if it has not, this would not answer your question in the negative, because of the unfinished state of physics and, to an even higher degree, of biology.

Sincerely yours,

Kurt Gödel

Wolfgang Rautenberg

In the period leading up to Paul Cohen's proof of the independence of the axiom of choice and the continuum hypothesis, rumors circulated to the effect that Gödel had already obtained these results, although he had done nothing to publish or even announce them.[a]

Shortly after Cohen's result became known, the well-known Polish logician Andrzej Mostowski (1913–1975), who had earlier worked on the independence of the axiom of choice, endorsed these reports in a rather extreme form, stating in a short expository article that it had been "known since 1938" that Gödel had proofs of the independence of the continuum hypothesis and of the axiom of constructibility but "in spite of many inquiries, he never betrayed his secret".[b] Evidently it was this claim that prompted Church and Gödel to set the record straight in Church's presentation on Cohen's work at the International Congress of Mathematicians in 1966; see Gödel's letters to Church of 10 August and 29 September 1966 in volume IV of these *Works*, and the remark on the subject in *Church 1968*, p. 17.

In the spring of 1967, before the congress proceedings containing Church's paper appeared, Wolfgang Rautenberg, a logician then at the Humboldt University in East Berlin, wrote to Gödel quoting the same remark of Mostowski and inquiring about its truth, in connection with his work on an article on the history of the continuum problem (letter 1).[c] It is interesting that according to Rautenberg, none of the colleagues he consulted thought the claim improbable. However, he notes that *Gödel 1947* (or *1964*) gives no indication that he had such a result.

Gödel's reply of 30 June 1967 (letter 2) refers Rautenberg to Church's forthcoming paper but goes over the matter once again. A slight differ-

[a]My own memory (as well as that of Solomon Feferman) is primarily of rumors concerning the independence of the axiom of choice. But the statement of Mostowski that prompted the present correspondence indicates that such rumors existed also concerning the independence of the continuum hypothesis. In fact Mostowski had heard as early as the fall of 1938 a report that Gödel had proved something on the independence of the continuum hypothesis, as Bernays reports in his letter to Gödel of 21 June 1939, in volume IV of these *Works*.

[b]*Mostowski 1964*, p. 124, quoted in Rautenberg's letter below. It seems likely that Mostowski thought better of this claim soon afterward, since it is not repeated in his extensive survey *1965*, based on lectures in the summer of 1964. In the chapter in which he presents Gödel's consistency results, he discusses the axiom of constructibility and remarks that "it was shown quite recently" that it is not derivable from the other axioms (p. 87). Cohen's results are described later (pp. 144–149) with no mention of any possible anticipation by Gödel.

[c]Rautenberg's letter is undated. But in a communication of 13 July 2000, he states that, as far as he can remember, he got a reply from Gödel within two months.

ence between what he says here and his final position in the correspondence with Church may show some uncertainty on his part on whether to claim that he had in 1942 a proof of the independence of the axiom of choice in type theory. For publication in *Church 1968*, he would only say that he believed that his independence proof for the axiom of constructibility could be extended to one for the axiom of choice, although he writes to Church that he believed that he could carry out this extension "without serious difficulty". To Rautenberg, however, he writes that he was "in possession of" a proof of the independence of both the axiom of constructibility and the axiom of choice in type theory. What he writes to Rautenberg is in fact very close to the statement attached to the letter to Church of 10 August 1966. In both he denies that he had a proof of the independence of the continuum hypothesis; about this he is a little more emphatic in writing to Rautenberg. He adds in a footnote that his method was closer to Scott's (i.e. Boolean-valued

1. Rautenberg to Gödel

Dr. Wolfgang Rautenberg
Institut für mathem. Logik
Humboldt Universität
108 Berlin

Hochverehrter Herr Professor Gödel!

Bitte erlauben Sie mir, bezüglich des Kontinuumproblems, über dessen Geschichte ich einen Aufsatz verfertigen muß, eine Frage an Sie zu richten. Professor A. Mostowski schreibt in einem Artikel in den "Elementen der Mathematik" (1964)

"Es ist seit 1938 bekannt, daß Gödel einen Unabhängigkeitsbeweis dieser Hypothesen [gemeint sind die Kontinuumhypothese und das Konstruktibilitätsaxiom] besitzt; trotz vieler Anfragen verriet er nie sein Geheimnis."[a]

Bisher hat niemand aus dem mir bekannten Fachkollegenkreis diese Behauptung für unwahrscheinlich gehalten, da ja die Methode von P. Cohen offenbar auf den Ideen Ihrer Arbeit "Consistency ..."[b] beruht. Andererseits finden sich in einem Aufsatz "What is Cantor's Continuum Problem?"[c] von Ihnen keinerlei Andeutungen.

[a] *Mostowski 1964*, p. 124.
[b] *Gödel 1940*.

models, more accurately attributed to Scott and Solovay) than to Cohen's (i.e. forcing).

In his article *1968*, Rautenberg quoted from Gödel's letter the first sentence of the second paragraph and its accompanying note (p. 20). There is no correspondence in Gödel's papers indicating that he asked Gödel's permission to do this. After Gödel's death, almost the whole body of the letter appeared in English translation in the introductory note to *1947* and *1964* in these *Works*, vol. II, p. 159.

Charles Parsons

The translation of Rautenberg's letter is by Charles Parsons, that of Gödel's letter by Gregory H. Moore and Charles Parsons.

1. Rautenberg to Gödel

Dr. Wolfgang Rautenberg
Institut für mathem. Logik
Humboldt Universität
108 Berlin

Most honored Professor Gödel,

Please allow me to put to you a question relating to the continuum problem, about whose history I must prepare an essay. Professor A. Mostowski writes in an article in "Elemente der Mathematik" (1964):

"It has been known since 1938 that Gödel has a proof of the independence of these hypotheses [meant are the continuum hypothesis and the axiom of constructibility]; in spite of many inquiries, he never betrayed his secret."[a]

Up to now no one in the circle of colleagues in the field that I know has considered this claim improbable, since the method of P. Cohen obviously rests on the ideas of your monograph "Consistency ...".[b] On the other hand, no hints at all are to be found in an essay of yours "What is Cantor's Continuum Problem?"[c]

[c] *Gödel 1947* or *1964*.

Über eine kurze Erklärung Ihrerseits zu der Äußerung Mostowskis wären ich und die Leser meines Aufsatzes Ihnen gewiß sehr dankbar. Bitte aber schreiben Sie mir, wenn es Zeit und Umstände Ihnen erlauben.

Hochachtungsvoll

Wolfgang Rautenberg

2. Gödel to Rautenberg

Princeton, 30. Juni 1967

Sehr geehrter Herr Doktor Rautenberg!

In Beantwortung Ihrer Anfrage moechte ich Sie auf die Darstellung des Sachverhaltes verweisen, die Prof. A. Church in seinem Vortrag auf dem letzten internationalen Mathematikerkongress gegeben hat.[a]

Die Mostowskische Behauptung is insofern unrichtig, als ich bloss im Besitze gewisser Teilresultate war, naemlich von Beweisen fuer die Unabhaengigkeit des Konstruktibilitaets- und Auswahlaxioms in der Typentheorie.[1] Da aber bald darauf mein Interesse sich der Philosophie, und spaeter auch der Kosmologie zuwandte, wurden diese Beweise niemals im Detail ausgearbeitet und die Untersuchungen nicht fortgesetzt. Bis zu einem Beweis der Unabhaengigkeit der Kontinuumhypothese vom Auswahlaxiom bin ich nicht gelangt und es war mir zweifelhaft, ob die Methode, die ich verwendete, ueberhaupt dazu ausreichen wuerde.

mit besten Gruessen

Kurt Gödel

[1] Auf Grund meiner hoechst unvollstaendigen Aufzeichnungen von damals koennte ich heute ohne Schwierigkeit nur den ersten dieser Beweise rekonstruieren. Meine Methode hatte eine viel naehere Verwandschaft mit den neuerdings von Dana Scott entwickelten als mit der Cohenschen.

---

[a] Cf. *Church 1968*, p. 17.

I and the readers of my essay would certainly be very grateful for a short explanation on your part concerning Mostowski's statement. But please write to me when time and circumstances permit you to.

Respectfully yours,

Wolfgang Rautenberg

2. Gödel to Rautenberg

Princeton, 30 June 1967

Dear Doctor Rautenberg,

In reply to your inquiry, I would like to refer you to the presentation of the facts that Professor A. Church gave in his lecture at the last International Congress of Mathematicians.[a]

Mostowski's claim is incorrect, insofar as I was merely in possession of certain partial results, namely proofs for the independence of the axiom of constructibility and the axiom of choice in type theory.[1] Since, however, soon thereafter my interest turned towards philosophy and later also ⟦towards⟧ cosmology, these proofs were never worked out in detail, and the investigation was not continued. I never got so far as a proof of the independence of the continuum hypothesis from the axiom of choice, and I found it doubtful that the method that I applied would be at all sufficient for such a result.

With cordial greetings,

Kurt Gödel

[1]On the basis of my highly incomplete notes of that time I could now reconstruct only the first of these proofs without difficulty. My method had a much closer connection with that developed recently by Dana Scott than with Cohen's ⟦method⟧.

Constance Reid

Constance Reid

Constance Reid (1918–) is a well-known biographer of a number of notable mathematicians, including David Hilbert, Richard Courant, Jerzy Neyman, Eric Temple Bell and last, but not least, (Reid's sister) Julia B. Robinson; her eminently readable biographies were initially intended for a general audience. The correspondence with Gödel was begun by Reid in September 1965 (letter 1) when she wrote asking about his personal contacts, if any, with Hilbert, for the benefit of the biography *Reid 1970*, then in progress. Gödel answered in March of the following year (letter 2), saying that he had never met Hilbert nor did he have any correspondence with him. However, in his reply, Gödel went beyond the request, to volunteer a number of very clear and interesting comments on the significance of his incompleteness theorems for Hilbert's consistency program. He added that he thought to publish such remarks "on some suitable occasion," but nothing per se did subsequently appear; perhaps, though, the work on *1972a* was an attempt in that direction.

When she was closer to seeing *Hilbert* into print, in February 1969 Reid wrote Gödel once more, this time at the suggestion of John Addison, who was one of the readers of her manuscript. Addison recalled that Gödel had once mentioned having given another proof of the consistency of the continuum hypothesis that was closer to Hilbert's attempt (*1926*) than the one that he had published, and Reid asked Gödel if he would comment on the value of Hilbert's much criticized effort. In his June 1969 response (letter 4), Gödel again went beyond what was requested, into the question of the relationship of Hilbert's "scheme" to the unpublished version of his own proof. He said that he had explained that in his note published on p. 368 of the introduction in *van Heijenoort 1967* to a translation of *Hilbert 1926*. But now he added that in the cited note he had "somewhat understated the great similarity in the outward structure." By way of redress, Gödel explained what might be considered a closer connection. Interestingly, a month after this letter, Gödel went over the same issues in correspondence with Paul Bernays. (See letters 68b and 70 of that correspondence, as well as the introductory note to the Bernays–Gödel correspondence in these *Works*, vol. IV.)

Reid made good use of Gödel's answers, quoting judiciously from both of them on pp. 217–218 of *Hilbert.* It was good that much of his first response could be used in that way and thus be made available to a general audience. And from the second response she made use of almost the whole of the last paragraph (the only part of that letter which does not involve technical notions) concerning the *general* value of Hilbert's attempt.

Solomon Feferman

1. Reid to Gödel

70 Piedmont Street
San Francisco, California 94117
September 1, 1965

Dear Professor Gödel:

In spite of David Hilbert's great importance for twentieth century mathematics, there is very little personal or biographical material about him in English. For this reason, I am now in the process of writing a short, essentially non-technical biography of Hilbert,[a] to be published by Thomas Y. Crowell Company next year.

The dramatic conclusion to Hilbert's career is of course the publication of your own paper of 1931[b] and its effect on his program of Formalism. In writing about this, I would like to give also something of the personal background for this event.

I am wondering if you ever met Hilbert personally or had any sort of discussion with him about your work. If you did, I would appreciate your sending me a description of the meeting and of your direct impression of Hilbert himself.

I realize that providing this information will require time and effort on your part. Please, be assured that I would not make the request if I did not so completely believe that this information should be put in permanent form—and that Hilbert and his conception of mathematics should be preserved for students of today and tomorrow.

I am enclosing a stamped self-addressed envelope for your convenience.

Sincerely yours,

Constance Reid
Mrs. Neil D. Reid

[a]The work described here became *Reid 1970*.
[b]*1931*.

2. Gödel to Reid

March 22, 1966

Mrs. Neil D. Reid
70 Piedmont Street
San Francisco, California 94117

Dear Mrs. Reid:

I am sorry I am rather late in replying to your inquiry about my relations with Hilbert. However, I presume it won't be too late for you to take this letter into account.

I would like to say first that I never met Hilbert; nor did I ever have any correspondence with him.

Next I would like to call your attention to a frequently neglected point, namely the fact that Hilbert's scheme for the foundations of mathematics remains highly interesting and important in spite of my negative results.

What has been proved is only that the *specific epistemological* objective which Hilbert had in mind cannot be obtained. This objective was to prove the consistency of the axioms of classical mathematics on the basis of evidence just as concrete and immediately convincing as elementary arithmetic.

However, viewing the situation from a purely *mathematical* point of view, consistency proofs on the basis of suitably chosen stronger metamathematical presuppositions (as have been given by Gentzen and others) are just as interesting, and they lead to highly important insights into the proof theoretic structure of mathematics. Moreover, the question remains open whether, or to what extent, it is possible, on the basis of the formalistic approach, to prove "constructively" the consistency of classical mathematics, i.e., to replace its axioms about abstract entities of an objective Platonic realm by insights about the given operations of our mind.

As far as my negative results are concerned, apart from the philosophical consequences mentioned before, I would see their importance primarily in the fact that in many cases they make it possible to judge, or to guess, whether some specific part of Hilbert's program can be carried through on the basis of given metamathematical presuppositions.

| I think, on some suitable occasion, I shall publish this, or some similar account of the relationship between my work and Hilbert's program. 2

I would like to add that, in my opinion, the formulation of my result given above is far preferable to saying it has been shown that "there are no absolute consistency proofs for the systems of classical mathematics", because I believe epistemology has not yet reached the stage of development where it could make final statements about such concepts as "absolute proof".

The proposition under quotation marks has been proved only if "absolute consistency proof" is interpreted according to Hilbert's philosophical point of view.

Sincerely yours,

Kurt Gödel

3. Reid to Gödel

70 Piedmont Street
San Francisco, California
February 10, 1969

Dear Professor Gödel:

Since I last wrote to you, I have completed the book on Hilbert. The manuscript has now been read by various mathematicians. One of these was Prof. John Addison at Berkeley.

Professor Addison recalls that at one time you mentioned to him that you had carried through a proof more along the idea of Hilbert's sketch of a proof of the continuum hypothesis although your own original proof had not been motivated by Hilbert's.

I am wondering if you would be willing to comment on the value of this particular work of Hilbert's. He seemed to think a great deal of it and was most upset by the criticism which it received from other mathematicians at the time.

May I thank you again for your assistance? I have found your letter of March 6, 1966, most valuable.

Sincerely yours,

Constance Reid

4. Gödel to Reid

June 25, 1969

Dear Miss Reid:

I have explained the relationship between my consistency proof of the Continuum Hypothesis and Hilbert's scheme in a note published on p.

368 of "From Frege to Gödel",[a] ed. by Jean van Heijenoort, Harvard University Press, 1967. The unpublished version of my proof, which uses recursive functions of ordinals, does not approximate Hilbert any more closely, because I am freely using bound variables and possibly undefinable ordinals in the recursive definitions.

However, it seems to me now that, in rightly stressing the great differences in the heuristic ideas and the epistemological outlook between my proof and Hilbert's scheme, I have somewhat understated the great similarity in the outward structure. It also extends to Hilbert's Lemma I, although I don't *need* anything like Lemma I, because I can prove stronger lemmas. Note that also Hilbert, before stating Lemma I, hints at a stronger lemma, which perhaps is an analogue of the consistency of $V = L$.

All in all, I think, it may be said that in general outline my proof agrees with Hilbert's scheme, except that Hilbert's transfinite recursive functions (or, which is the same, functions defined without using Hilbert's ϵ) are replaced in my proof by the "constructible" functions.

Of course the constructible functions are a far cry from Hilbert's recursive functions. However, it is not impossible that a closer approximation to Hilbert's scheme could be achieved, although *some* generalizations of Hilbert's concepts are no doubt *necessary*. For the consistency proof of the Continuum Hypothesis such a closer approximation to Hilbert would only introduce unnecessary complications. But Hilbert had the more ambitious aim to prove the "truth" of the Continuum Hypothesis. This, for him, could only mean its truth for finitary (today one would have to say for constructivistic) mathematics. Its "formal truth" in classical mathematics then follows from Hilbert's philosophical idea that the latter is only a "completion" of the former, hence cannot contradict it in any way.

Finally I would like to say that, in judging the value of Hilbert's work on the Continuum Problem, it is frequently overlooked that, disregarding | questions of detail, one quite important *general* idea of his has proved perfectly correct, namely that the Continuum Problem will require for its solution entirely new methods deriving from the foundations of mathematics. This, in particular, would seem to imply (although Hilbert did not say so explicitly) that the Continuum Hypothesis is undecidable from the usual axioms of set theory. 2

Sincerely yours,

Kurt Gödel

---

[a] *van Heijenoort 1967.*

Abraham Robinson

Abraham Robinson

Abraham Robinson (1918–1974), one of the most eminent logicians of the post-war period, will be known to readers of these *Works* for his considerable work in model theory and perhaps especially for his development of nonstandard analysis. The biography *Dauben 1995* contains background material on the interaction between Gödel and Robinson.[a] The later phases of their relationship are of particular significance in view of Gödel's apparent wish that Robinson should succeed him at the Institute for Advanced Study.[b]

The two met for the first time early in the academic year 1960/61. Robinson was then on sabbatical leave from Jerusalem, which he spent as visiting professor at Princeton University's department of mathematics. There he replaced Alonzo Church who, in turn, was on sabbatical leave in California. That year was pivotal in Robinson's scientific career: the idea of nonstandard analysis—his most notable contribution to mathematics, using logic as an indispensable tool—occurred to him in the fall of 1960, and was developed soon thereafter.[c]

On 29 September 1960, a few days after arriving at Princeton, Robinson invited Gödel to participate in a regular logic colloquium he was organizing at Princeton University.[d] The invitation was not taken up, but Gödel agreed to receive Robinson privately. It is likely that during their meeting the latter mentioned his recent discovery. If so, then Gödel was one of the first to learn about it.

On 28 August 1961, shortly before leaving the U.S., Robinson sent Gödel a note from Berkeley,[e] in which he explained that two months earlier, on his departure from Princeton, he did not wish to impose on Gödel with another visit, because he had heard that the latter was "not feeling too well". The note also mentions that Robinson's first paper on nonstandard analysis[f] ("which you have in typescript") is about to be published. On the reverse of the note Gödel jotted down that he had seen Robinson "only once, and this was evidently in fall 1960" and that he had a manuscript of Robinson's paper.

[a]Excerpts from their correspondence have been published there, p. 398, p. 457ff., p. 468f., p. 485f.

[b]*Dauben 1995*, p. 458. The correspondence itself contains some evidence for this.

[c]See Preface to *Robinson 1966*.

[d]Letter of 29 September 1960.

[e]Letter of 28 August 1961.

[f]*Robinson 1961*.

The next written communication between the two preserved in the *Nachlaß* dates from 1967. Robinson, who held a chair in the UCLA philosophy department from 1962 to spring 1967, co-organized with Paul Cohen and Dana Scott a summer institute on axiomatic set theory (July–August 1967). Early that year, he sent Gödel an invitation to attend the institute,[g] which the latter declined, while expressing an interest "in seeing any material on the conference that may be available, such as a program or abstracts of papers."[h]

The organizers decided to dedicate the published proceedings of the institute to Gödel, and on 4 May Robinson wrote to Gödel, asking him to accept the dedication.[i] The letter of 7 July 1967 (letter 1 in this volume) is Gödel's response, which speaks for itself. However, three remarks may be added in this connection. First, a brief handwritten draft[j] contains the significant statement: "As you know, my present interests are widely distant from set theory. But I hope, some time, to return to this subject." Gödel indeed returned to set theory a few years later.[k] Second, a longer handwritten draft,[l] which was evidently the one used by the typist, mentions, in addition to Azriel Lévy, also William Hanf; but the latter mention is crossed out and replaced by a general reference to "the Tarski school". Third, in the published proceedings of the 1967 UCLA summer institute[m] the dedication to Gödel does not appear after all; it was probably simply forgotten.

Gödel kept up an interest in Robinson's work, for which he had high regard. In response to an enquiry by Michael F. Atiyah,[n] Gödel replied:

> I think very highly of Abraham Robinson as a mathematical logician. In fact, in the field of applications of mathematical logic to mathematics proper (excluding abstract set theory) he is probably the most outstanding among living logicians. In particular, his theory of infinitesimals and its application for the solution of analytical problems seems to me of greatest importance. I believe that once this theory will have been more fully developed, it will play a major role in mathematics.[o]

[g]Letter of 16 January 1967.

[h]Letter of 27 April 1967.

[i]Letter of 4 May 1967.

[j]Document 011942.

[k]See *Gödel *1970a*.

[l]Document 011943.

[m]*Scott 1971* and *Jech 1974*.

[n]Document 011946. The enquiry was made on behalf of the Royal Society, London, in connection with consideration of Robinson for election to the Society.

[o]Document 011948. Cf. also document 011983, written later in the same connection, which mentions also Robinson's method of forcing in model theory, developed since 1969.

Gödel's high opinion of nonstandard analysis did not diminish with time; on the contrary, his remarks in *Gödel 1974* and his letter to Dan Mostow of 21 May 1974 (the text of which follows the correspondence between Gödel and Robinson) are, if anything, more laudatory. However, his sanguine expectations regarding the acceptance of nonstandard methods by mathematicians at large have so far not been vindicated. Perhaps it is too soon to pronounce the final verdict.

After the spring term of 1967 Robinson moved to the mathematics department at Yale, where he was to remain until his death. By 1971, Gödel seems to have worked out a plan to attract him for extended visits to the I.A.S., with a view to Robinson's eventually becoming Gödel's successor there. In this connection Robinson sent Gödel, at the latter's request, a full bibliography of his publications, as well as reprints of his recent "and, where relevant, some earlier papers".[p] Letter 2 in this volume, from March 1971, is Gödel's response, and Letter 3, 14 April 1971, is Robinson's reply. One point in this exchange calls for some clarification. Gödel states that "the correct framework for number-theoretical problems will in many cases be highly transfinite systems because highly transf⟦inite⟧ axioms are often needed for their solution". This no doubt alludes to the fact that large-cardinal axioms of set theory have number-theoretic "diophantine" consequences.[q] It is a profound insight, which inspires much current work in set-theoretic foundations. However, Gödel's further remark that Robinson's nonstandard infinite numbers would perhaps "come in here" is somewhat misleading. Whereas the highly transfinite axioms mentioned earlier are not conservative with respect to the "ordinary" axioms of set theory, Robinson's nonstandard analysis (and hence the use of his nonstandard integers) *is* conservative. Robinson himself stresses this very clearly in a subsequent letter of 3 January 1972. But in the letter of 14 April 1971 his response is more diplomatic. Without contradicting Gödel directly, he points out that the large-cardinal axioms, being non-conservative, go beyond the "results and methods that I (for example) have succeeded in developing." Nonstandard analysis has in fact been used, by Robinson and others, to prove number-theoretic results. Significantly, Robinson kept Gödel informed of such achievements. In this connection note letter 10, dated 16 October 1973, which is probably the last letter from the former to the latter, and the latter's reply, letter 11, dated 22 November.

[p]Robinson to Gödel, 8 February 1971.

[q]This point was made by Gödel before; see, for example, *Gödel *1951*, p. 193. It was further strengthened by the Matiyasevich–Robinson–Davis–Putnam theorem (*Matiyasevich 1970*), which reduced the number m of parameters in the "diophantine problems" of *Gödel *1951* to 0.

But this sort of work has nothing to do with the use of large-cardinal axioms to solve "diophantine" problems.

The letters of 22 November 1971, 3 January 1972, 29 December 1972, 4 January 1973, 23 April 1973, 2 July 1973 and 18 July 1973 call for little detailed comment. They date from the period 1971–73, during which the relation between Gödel and Robinson became closer, and they met relatively frequently. These letters contain several references to Gödel's plans to bring Robinson over to the Institute. There is also some evidence of friendly disagreements between the two on methodological and philosophical issues. Thus, in one of his last letters to Gödel, dated 23 August 1973, Robinson says:

> I am distressed to think that you consider my emphasis on the model theoretic aspect of Non Standard Analysis wrongheaded. As a "good formalist" I am quite willing to write a book on the assumption that the reals are unique, and in fact have done so. However, even on this basis the present evidence for the uniqueness of a non standard ω-ultrapower of the reals is not strong. In addition it would not make any difference to my arguments, although I realize its foundational significance.

And referring to their philosophical disagreements, manifested in the paper he enclosed with that letter,[r] Robinson adds:

> But as far as the main thrust of my paper is concerned, with which you are bound to disagree, I can only quote Martin Luther: "Hier stehe ich, ich kann nicht anders."

In the fall of 1973 Robinson complained of abdominal pains. An exploratory operation revealed that he was suffering from terminal pancreatic cancer. Thus Gödel's intended successor, twelve years his junior, was to pre-decease him by four years: he died on April 11 of the following year.

The letter draft of 20 March 1974, written a few days before Robinson's death, is a desperate attempt to comfort a friend. It is ironic that the mathematical statement at the end of this letter, no doubt intended to try to take Robinson's mind off his illness, relates to the very topic that had occupied Gödel's mind when he himself was afraid he was about to die: the subject of *Gödel *1970a*.[s]

Moshé Machover

A complete calendar of the correspondence with Robinson appears on pp. 452–453.

[r] *Robinson 1975.*

[s] The result stated in the letter is evidently a weakening of Axiom 4.

1. Gödel to Robinson[a]

July 7, 1967

Professor Abraham Robinson
Department of Mathematics
University of California
Los Angeles, California 90024

Dear Professor Robinson:

Thank you very much for your letter of May 4. I am very honoured by the fact that the Organizing Committee of the Institute on Axiomatic Set Theory wishes to dedicate the proceedings of this conference to me. Please give my best thanks to the Committee for this expression of high regard for my work. I accept gladly, with one reservation however, namely that the wrong impression should be avoided as if these highly interesting recent developments were only an elaboration of my ideas. I really took only the first step. Moreover, I perhaps stimulated work in set theory by my epistemological attitude toward it, and by giving some indications as to the further developments, in my opinion, to be expected and to be aimed at. I did not, however, give any clues as to how these aims were to be attained. This has become possible only due to the entirely new ideas, primarily of Paul J. Cohen and, in the area of axioms of infinity, of the Tarski school and of Azriel Levy. I request that this should be mentioned in the introduction to the book.

Sincerely yours,

Kurt Gödel

[a]Retained copy.

2. Gödel to Robinson[a]

Dear P. Rob.

I thank you very much for your letters of *Jan 29 & Feb 8* ~~and~~ for the bibl. of your papers & the ~~rep~~ two folders of reprints. It seems to me ~~your endeavor to~~ the problem raised by you of find⟨ing⟩ the correct mathematical framework for a given math. [entity] ⟨subject⟩ is ~~quite highly~~ ⟨one of the most⟩ important ⟨of ⟨all⟩ math. qu.⟩ ~~&~~ should (taken in a ⟨sufficiently⟩ general sense) ⟨it/'s correct treatment⟩ should lead to ~~the~~ systematic methods for the solution of math. problems. ~~For~~ E.g. one may say that the number field $K(\sqrt[n]{1})$ ⟨probably⟩ is the correct framework for the ~~solut~~ ~~treatment~~ ⟨solution⟩ of Fermat's problem ⟨with exp. n⟩. ~~I don't~~ ⟨ ~~Evidently~~ ⟩ ~~know if~~ ⟨~~This ex (It is true) cannot perhaps somehow directly~~ ⟩ ~~But perhaps~~ ~~This could probably not be sub~~ ~~Evidently this completion of~~ ~~(it is true)~~ ⟨This ex., it is true, does not come⟩ ~~should~~ under your concept of model completion but it is ⟨certainly⟩ something [related to it] ⟨similar in nature⟩. Generally speaking I think that the correct framework for number⟦-⟧theoretical problems will ⟨in many cases⟩ be highly transfinite ⟨systems⟩ because highly transf ax. are often needed ~~nec~~ for their solution. Perhaps your ⟨the⟩ ⟨~~non-standard~~⟩ ~~in~~finite | ~~numbers~~ ⟨integers of your nonstand. analysis⟩ will come in here.

Your work is probably the first attempt of systematically applying math-logic to the sol of math. problems i.e. to use i~~s~~ for the ~~find~~ purpose for which i~~s~~ was originally conceived by mathematicians like Peano.

Many thanks again,

Yours very sinc.

⟦No signature⟧

[a]A draft only, the letter has written at the top "1971" and, mostly in shorthand, "nicht abgeschickt, statt des Disk." ("not sent instead of Disc⟦ussion⟧), followed by "ca III 71.

3. Robinson[a] to Gödel

April 14, 1971

Dear Professor Gödel:

I enjoyed our conversation and wish to thank you for giving me so much of your time. Let me first deal with some practical points that you raised during my visit.

As I mentioned, it would be difficult for me to obtain leave of absence for a whole year. However, I shall have half a year's sabbatical (so-called Triennial) leave due to me in 1972–73. Accordingly, I shall be happy to spend the first semester of that academic year (until the end of January 1973) at the Institute.[b] This would not require any financial support from the Institute except, possibly, for the additional cost of having to rent an apartment in Princeton. In any case, this point is inessential.

You also mentioned the possibility that I might come to the Institute for five years, and I greatly appreciate your suggestion. It would give me a chance to spend a considerable period at the center of the mathematical universe, where my contacts with you and with some of your colleagues might be crucial for my thinking. On the other hand, such a step would probably force me to dissolve my ties with Yale and would lead to some uncertainty for the future. However, having thought the matter over, I wish to say that the importance of your suggestion seems to me so great that I would give very serious consideration to a concrete proposal in this direction.

Your point that even seemingly remote questions concerning the existence of large cardinals react on ordinary arithmetic is certainly highly relevant in the context of the application of logic to mathematics. And inasmuch as any concrete steps in this direction would provide the working mathematician with an effective (non-conservative) extension of his tools, it certainly goes beyond the results and methods that I (for example) have succeeded in developing. But although some of the questions that interest me | seem rather special, I believe that a satisfactory answer 2 to them also has provided, or may provide, a genuine widening of the framework within which the mathematician carries out his work. And to the extent to which these questions obtain their natural answer within

[a]All of Robinson's letters to Gödel are on the letterhead of Yale University, New Haven, Connecticut 06520, Department of Mathematics, Box 2155 Yale Station.

[b]According to *Mitchell 1980*, Robinson was a member of the I.A.S. only in the spring of 1973.

the framework of formal languages and of general syntactical or semantical considerations, they constitute a genuine contribution of logic to mathematics.

I have given some thought to your suggestion that I prepare a survey of the existing achievements in this area and I certainly intend to comply with your wishes. However, the task is, by now, quite complex and it may take some time before I can do the thing properly.

Yours very sincerely,

Abraham Robinson

4. Gödel to Robinson

Princeton, Nov. 22, 1971

Dear Professor Robinson,

I was delighted to hear that you wish to spend one term of 1972/73 at the Institute. I am sorry there was some delay in my proposing you for membership because of the absence of a faculty member and other circumstances. But now that I have initiated the matter it will probably be settled within a week.

Professors on sabbatical leave usually receive some grant from the Institute in order to make up for the~~ir~~ reduction of their salary. On the basis of your letter, I was assuming that you will receive your full salary or some equivalent of it. Please, let me know in case I am mistaken.

2 Thank you very much for sending me the survey of | mathematical results obtained with the help of mathematical logic. I wonder if there are some more analytical results first obtained by using infinitesimals, and if there aren't some striking examples of known theorems whose proofs are *much* simpler in nonstandard analysis. Also it occurred to me: Didn't Novikov disprove the Burnside Conjecture by means of recursion theory?

I hope very much that your visit to the Institute will materialize, and I am looking forward to hearing more from you about your work.

Very sincerely yours

Kurt Gödel

5. Gödel to Robinson

Princeton, Dec. 29, 1972

Dear Professor Robinson,

I thank you very much for the list of results of Nonstandard Analysis which you sent me some time ago. In my opinion Nonstandard Analysis (perhaps in some non-conservative version) will become increasingly important in the future development of Analysis *and* Number theory.

The same seems likely to me, with regard to all of mathematics, for the idea of constructing "complete models", in various senses, depending on the nature of the problem under discussion. Has the concept of forcing completion borne any concrete fruits yet?

I am looking forward to seeing you soon.

Seasons greetings and best wishes.

Yours sincerely

Kurt Gödel

6. Robinson to Gödel

January 4, 1973

Dear Professor Gödel,

I greatly appreciated your kind letter. As for the particular points raised in it, I certainly hope that you will have time to discuss them with me further. At the moment, let me say only that I quite agree with you that it would be desirable to strengthen the deductive powers of classical mathematics by means of some of the higher axioms of infinity suggested by set theory, either by means of Nonstandard Analysis or in some other way. (I do not believe in mathematical monopolies.) So I hope this will be done sooner or later but I cannot tell how soon.

We expect to arrive in Princeton next Monday and I shall call you either on Monday or on Tuesday in order to time an appointment.

With best wishes for the New Year

Yours sincerely

Abraham Robinson

7. Robinson to Gödel

April 23, 1973

Dear Professor Gödel

I enclose a preprint of the paper "Nonstandard Arithmetic and Generic Arithmetic,"[a] which I read at the Bucharest Conference. It looks rather shabby because it is one of my last two copies.

During one of our last meetings you raised the question whether generic arithmetic is nonconservative in the sense that it leads to theorems not provable in Peano arithmetic, and I replied that I had not considered the question. In fact, it now seems to me that in order to give a precise meaning to this question one would first have to state the axiomatic framework (presumably set-theoretic) within which the construction of the class G is carried out.

To clarify another point, the "resultant" of a predicate $A(x_1, \ldots, x_n)$ can be written as $A'(x_1, \ldots, x_n) = \bigwedge_\nu \bigwedge_\mu A_{\nu\mu}(x_1, \ldots, x_n)$ where all $A_{\nu\mu}$ are existential. That is to say, if $M \in \Sigma$ and $a_1, \ldots, a_n$ denote elements of M then $M \models A'(a_1, \ldots, a_n)$ if and only if $A(a_1, \ldots, a_n)$ holds in all generic structures that are extensions of M. This expresses at least a good part of the meaning of the notion of a resultant in algebraic field theory. The elimination of all quantifiers from A' is possible if any two structures of Σ can be embedded in a structure that belongs to Σ. This is *not* the case if Σ is the class of all models of the universal sentences true in N (with $+, \cdot, =$).

I gather that Henkin sent you a copy of Bertrand Russell's letter. Henkin agrees with my view that Russell's remarks on your incompleteness theorem are misleading. It is hard to say whether at the time he wrote this letter, Russell still had a coherent philosophy of Mathematics.

I shall leave for Europe tomorrow and I hope to see you in Princeton in August.

Yours very sincerely,

Abraham Robinson

---

[a] *Robinson 1973.*

8. Gödel to Robinson

July 2, 1973

Professor Abraham Robinson
Department of Mathematics
Yale University
Box 2155, Yale Station
New Haven, CT 06520

Dear Professor Robinson:

I thank you very much for your letter of April 23 and the preprint on generic arithmetic. I also thank you for asking Henkin to send me Russell's letter. Russell evidently misinterprets my result; however, he does so in a very interesting manner, which has a bearing on some of the questions we discussed a few months ago. In contradistinction Wittgenstein, in his posthumous book, advances a completely trivial and uninteresting misinterpretation.

I am looking forward to your visit in August and hope very much that you will stay for more than a few days. I am sure payment of your expenses and housing in the Institute can be provided for any period of time you wish to spend here, except that, to my knowledge, all apartments have to be vacated for cleaning purposes by August 24. Please let me know the exact date of your visit and whether you will be accompanied by Mrs. Robinson.

Sacerdote is working hard on the problems you discussed with him. The other day he sent me a manuscript of 33 pages. Unfortunately I am not sufficiently acquainted with the subject matter to give him any guidance.

I hope you had a pleasant and interesting trip abroad and that I shall hear from you about your experiences. Our discussions last spring interested me very much and I am looking forward to seeing you again soon.

Very sincerely yours,

Kurt Gödel

9. Gödel to Robinson[a]

July 18, 1973

Professor Abraham Robinson
Department of Mathematics
Yale University
New Haven, CT 06520

Dear Professor Robinson:

Thank you very much for your letter of July 6. Arrangements for your stay in Princeton have been made already. I was trying to get you the same apartment and office you had before, but unfortunately that was not possible. But they will be of the same type and in the same neighborhood.

I am sorry you won't stay for more than a few days. But I would like, at any rate, to have several conversations with you. I'll be glad to see you in my office on Wednesday at 4:00 p.m., or at any other time that may be more convenient for you.

I don't think I am overestimating the importance of non-standard analysis. Your generic arithmetic, too, is most interesting. In particular the definability of the standard integers is almost unbelievable. Of course the true value of such constructions must be judged by successful applications to concrete problems. But I have no doubt that such will be forthcoming.

Looking forward to seeing you soon,

Yours sincerely,

Kurt Gödel

---

[a]Retained copy.

10. Robinson[a] to Gödel

October 16, 1973

Dear Professor Gödel,

Professor Stephan Körner, of the Department of Philosophy at Yale University and of the University of Bristol has expressed his strong desire to visit you at the Institute. Professor Körner, whose many writings include a book on the Philosophy of Mathematics⟦,⟧ is not only well versed in contemporary Philosophy of Science but also in classical Philosophy, particularly in Kant and Leibniz. During the present term he and I jointly are giving a course on the Philosophical Foundations of Mathematics. I have no doubt that you also would find it interesting to talk to Körner.

You may be interested to hear that, in recent weeks, Peter Roquette and I have produced a nonstandard proof of Siegel's theorem on integer points on curves. This theorem is regarded as one of the high points of 20th⟦-⟧century number theory. We also believe that in one direction we can go slightly beyond Siegel's result. As you may know, Roquette, of Heidelberg University, is a leading "standard" expert in this area. He was visiting here for a month and will come back later to continue our collaboration.

Yours cordially,

Abraham Robinson

[a]Retained copy.

11. Gödel to Robinson

Princeton, Nov. 22, 1973

Dear Professor Robinson,

Thank you very much for the copy of your Bristol talk and your letters of Aug 23 and Oct 16. I am sorry I am so slow in answering.

I agree with you that the epistemological investigation of applied mathematics has been neglected.

That Siegel's theorem on rational points has been proven and strengthened in non-standard analysis is quite interesting. May I assume that the proof also is shorter?

I shall be very glad to have a talk with Prof. Körner. If he calls me at home any day after 7 PM we can make an appointment at his convenience. My Tel. No. is 609.924.0569.

What do you think of Sacerdote's latest work?

With best regards to Mrs. Robinson

Yours sincerely

Kurt Gödel

12. Gödel to Robinson

Princeton, March 20, 1974

Dear Professor Robinson,

In view of what I said in our discussions last year you can imagine how very sorry I am about your illness, not only from a personal point of view, but also as far as logic and the Institute for Advanced Study are concerned.

As you know I have unorthodox views about many things. Two of them would apply here: 1. I don't believe that any medical prognosis is 100% certain, 2. The assertion that our ego consists of protein molecules seems to me one of the most ridiculous ever made. I hope you are sharing at least the second opinion with me.

I am glad to hear that, in spite of your illness, you are able to spend some time in the Mathematics Department. I am sure this will provide some welcome diversion for you. Perhaps the following (surprising, but very easily proved) result, which I obtained recently, will interest you: Hausdorff proved that the existence of a "continuous" system of orders of growth | (i.e. one where every decreasing ω_1-sequence of closed intervals has a non-empty intersection) is incompatible with Cantor's Continuum Hypothesis. Surprisingly the same is true even for a "dense" system, i.e., one where every decreasing ω_1-sequence of closed intervals, *all of which are larger than some fixed interval I*, has a non⟦-⟧empty intersection. I think many mathematicians will consider this to be a strong argument against the Continuum Hypothesis.

With my very best wishes, in every respect

Yours sincerely

Kurt Gödel

P.S. Please, give my regards to Mrs. Robinson.

⟦After Robinson's death, Professor Dan Mostow, then chairman of the department of mathematics at Yale, wrote to ask Gödel if he would speak at the memorial service. Gödel replied on 21 May 1974 as follows:⟧

Dear Professor Mostow:

Thank you for your letter of May 3. I am very sorry that, in view of my state of health, I cannot attend any memorial service for Professor Robinson. I am happy to hear that my letter brought some comfort to him and Mrs. Robinson.

As far as my statement after his Princeton lecture is concerned it seems too long and too technical to me for being quoted in a memorial service. If it seems appropriate it might be mentioned that in my opinion "Abraham Robinson will always live in the memory of mathematicians as the man who first developed a satisfactory theory of infinitesimals, a discovery whose fundamental importance for the development of analysis and number theory is not yet sufficiently recognized at the present time."

Yours sincerely,

Kurt Gödel

⟦In May 1974 Gödel wrote Renée Robinson a letter of condolence. A tortured draft survives in Gödel's *Nachlaß*, with an annotation reading "10./V.74 abgeschickt" ("sent 10 May 1974"); the draft has been reconstructed insofar as possible to read as follows:⟧

Dear Mrs. Robinson

As you ~~probably~~ ⟨may⟩ have heard I was sick in a hosp. ⟨myself at the time of⟩ your husband's ~~death~~ ⟨decease⟩ ~~reached me with great delay, because I ⟦was?⟧ in the a hospital myself at that time. My reason that I~~

~~was ⟨for this reason it was⟩ not possible for me~~ ⟨I am very sorry that I was not able⟩ to send you a timely letter of condolence.

So ⟨please⟩ let me ~~please~~ say now how deeply I regret the ~~⟦???⟧ly~~ premature death of your husband whom I ~~considered~~ ⟨valued very highly indeed⟩ not only as a personal friend, but also ~~as the one logician math. log. who ⟨was more succ. than anyb. else⟩ really succ. in making this~~ 2 ~~science fruitful for math ⟨though had he⟩~~ | ~~as the one ⟨math⟩ log. who pursued~~ ~~logic primarily for the sake of it's appl. in math & was quite succ. in this endeavour.~~ ⟨as the one ⟨math.⟩ log. who acc⟨omplished⟩ ~~inc~~ ⟨incomparably⟩ more than anybody else in making ~~math log. fruitful~~ this sci. fruitful for math. I am sure his name will be rem. by math for centuries.

I can only hope that modern med. succ⟨eeded⟩ in making the last few ~~weeks~~ ⟨months⟩ of his life tolerable for him & that moreover his phil. or 3 | rel. views let him bear his fate with equanimity. ⟨~~I am sure ⟦he⟧ will be rem. by math. for centuries.~~⟩

~~With the expression of~~ ⟨Allow me please to convey to you⟩ my deepest sympathy.

Yours sincerely

Bertrand Russell

The one letter below is the only known correspondence between Gödel and Bertrand Russell.[a] It concerns *Schilpp 1944*, the volume for which *Gödel 1944* was written, in particular the question of whether Russell will write a reply, as he had done for the other papers in the volume. Why this issue arose, and its resolution, is related in the introductory note to the correspondence with Paul Schilpp in this volume. The letter exists only in a handwritten version in the *Nachlaß*, in a form that Gödel would ordinarily give to a typist. However, in a letter to Paul Schilpp of 5 October 1943, Russell reports having received a letter from Gödel, and thus it is pretty certain that this text was in fact sent.

Warren Goldfarb

[a]In later years, Gödel apparently forgot he had ever written to Russell. See the correspondence with Kenneth Blackwell in these *Works*, vol. IV.

---

1. Gödel to Russell

Princeton, Sept. 28, 1943

Dear Prof. Russell,

I regret ~~that~~ very much ⟨that⟩ ⟨(as Prof. Schilpp ⟨has⟩ informed me) ~~that~~ you don't intend to write a reply to my article[a] in the Library of ⟦Living Philosophers.⟧[b]⟩ and am very sorry that this is apparently due to my sending in the manuscript so late. On the other hand I still hope that you may perhaps change your mind, since your decision ~~apparently was~~ ⟨seems to have been⟩ based on the wrong assumption that my article will not be controversial. This is by no means so. I am advocating ⟨in some respects⟩ the exact oppposite of the development inaugurated by Witt-

---

[a]*1944*.

[b]The letter originally began "Professor Schilpp has informed me that you don't intend to write a reply to my article in the Library of Living Philosophers. I regret that very much and..." Gödel revised this by a series of arrows that resulted in the insertions shown, but failed to include the last two words of his first sentence in the material to be moved.

genstein and therefore suspect that many passages will contradict directly your present opinion. Furthermore I am criticizing the vicious circle principle and appendix B of Principia, which I believe con|tains formal mistakes that make the proof invalid. The reader would probably find it very strange if there is no reply, not to mention that the readers of this volume will ~~probably~~ ⟨naturally⟩ be more interested in your opinions than in mine. ~~Frankly speaking~~ ⟨In fact⟩ ~~I am in favor of not publishing~~ ⟨I doubt very much if⟩ my article ⟨should be published⟩ at all without a reply ~~from you~~ ⟨at least to the main items⟩ and ~~believe that~~ you will ~~come to the same conclusion~~ ⟨~~may~~ perhaps agree with me⟩ after having read it. 2

Very sincerely yours,

Kurt Gödel

P.S. I have sent in the manuscript to/day. I did not ~~want to~~ send it ⟨directly⟩ to you because it is not very easily readable in it/s present condition.

Frederick W. Sawyer, III

While a graduate student in history and philosophy of science at the University of Pittsburgh, in 1974 Frederick W. Sawyer III wrote Gödel with questions about *Gödel 1958*. In particular, he asks after the revised English version of that paper, of which he had seen a mention. This version was first published in these *Works*, vol. II, as *Gödel 1972*. Gödel drafted a reply, but very probably did not send it. In it, he minimizes the importance of his revisions and added notes, mentioning as important only the elaboration published as footnote h (these *Works*, vol. II, p. 275), and giving a clear summary of its motivation. As we know, Gödel had been dissatisfied with the notes he added, and did not allow the revised version to be published, even though it had been sent to the printer in 1970 and Gödel had continued to make revisions on the proof sheets until 1972. (See the introductory note to *Gödel 1958* and *1972* in these *Works*, vol. II, p. 219.) Finally, Gödel admits he had presented the material in *Gödel 1958* years earlier—the draft lacks dates, but the lectures in Princeton were in the spring of 1941, and the lecture at Yale was on 15 April 1941. (The Yale lecture appears as *Gödel *1941* in these *Works*, vol. III.) He adds that he did not publish the material at the time because his own interests shifted (indeed, this was the last work Gödel did on intuitionism and constructivity) and because there was not too much interest in Hilbert's Program at that time.

Warren Goldfarb

1. Sawyer to Gödel

University of Pittsburgh
Pittsburgh, Pennsylvania 15213
February 1, 1974

Dear Professor Gödel,

I am a Pitt graduate student interested in the history and current development of attempts to prove the consistency of number theory and analysis. I am currently doing some preliminary reading in this general area in preparation for my proposed thesis, so I have naturally been drawn to the study of various of your papers.

Most recently I have been considering your 1958 Dialectica article[a] (and the literature which has grown up around it). This paper was listed in the bibliography of your published work [in Foundations of Mathematics-Symposium Papers Commemorating the Sixtieth Birthday of Kurt Gödel] with the notation that a revised English version was to appear in a future issue of Dialectica. If I am not presuming too much I would greatly appreciate receiving a copy of this revision if it currently exists in circulable form.

Since I am also concerned with the history of the topic, I was interested to read Kreisel's remark [JSL, 33:333][b] to the effect that you had incorporated the Dialectica interpretation into your lectures at Princeton as early as 1941. I would be pleased if you would verify this statement for me and perhaps elaborate on it.

Needless to say, there are a number of other questions I would like to pose relating both to your views concerning the philosophical significance of consistency proofs and to the manner in which the Dialectica interpretation grew out of your earlier work; however, given the current preliminary nature of my study, I suspect that to do so would unduly impose upon your time. Of course, I would hope that I might have the opportunity to pose such questions when I am better able to appreciate your responses. In the meantime, I hope that you will be able to provide me with a copy of the revised and translated article.

Thank you for your time and attention.

Sincerely,

Frederick W. Sawyer, III

---

[a] *1958*.

[b] *Kreisel 1968*.

2. Gödel to Sawyer[a]

Dear Mr. Sawyer,

I don't think the translation of my Dialectica paper is relevant for your purpose, since in the "revisions" I made it a point not to change

---

[a] Undated draft or retained copy, unsigned.

the meaning, but rather confine myself to "rewordings". I added some notes, but most of them only serve the convenience of the reader. The only one of some importance concerns the second half of footnote 1 on p. 283. There I tried to show that the most direct way of arriving at an intuitionistic interpretation of T, starting with the truly primitive (i.e. irreducible) terms and evidences of Intuitionism, does *not* pass through Heyting's logic, or the general intuit. concept of proof or implication, but rather through much narrower (and in principle decidable) concepts of "provable" and "implies". Thus the implicit use of "implication" and "demonstrability" occurring (through the words "immer ausfuehrbare"[1] and "erkennbare") in the definition of "computable function of finite type" on p. 282–283 does *not* give rise to any circularity. However, to carry that out in detail is rather cumbersome, and the matter, probably, cannot be explained convincingly in a footnote.

| It is true that I first presented the content of my Dialectica paper in a course of lectures at the Institute in Spring   and in a talk at Yale in   . There were several reasons why I did not publish it then. One was that my interest shifted to other problems, another that there was not too much interest in Hilbert's Program at that time. 2

Sincerely,

⟦No signature⟧

[1] I.e., *if* the arguments are computable

Paul Arthur Schilpp

Paul Arthur Schilpp

Paul A. Schilpp (1897–1993) was professor of philosophy at Northwestern University and subsequently at Southern Illinois University at Carbondale. He was founder and, until 1981, editor of *The library of living philosophers*, a series of volumes on major philosophical figures. The General Introduction to the series states that each volume is to contain

> *First*, a series of expository and critical articles written by the leading exponents and opponents of the philosopher's thought;
> *Second*, the reply to the critics and commentators by the philosopher himself;
> *Third*, an intellectual autobiography of the thinker whenever this can be secured...
> *Fourth*, a bibliography of writings of the philosopher...

Schilpp placed great emphasis on the second component. The General Introduction speaks of the pervasive problem of deciding among differing interpretations of a philosopher's works, and the difficulty of adjudicating among the contenders. It continues

> One effective way of meeting the problem at least partially is to put these varying interpretations and critiques before the philosopher while he is still alive and to ask him to act at one and the same time as both defendant and judge. If the world's great living philosophers can be induced to co-operate...they will have taken a long step toward making their intentions clearly comprehensible.

The Gödel–Schilpp correspondence concerns Gödel's contributions to various of these volumes, namely, those on Bertrand Russell (*Schilpp 1944*), Albert Einstein (*Schilpp 1949*) and Rudolf Carnap (*Schilpp 1963*), although in the last case Gödel wound up not submitting an essay. The correspondence is voluminous, containing more than 100 items over a span of 30 years. It shows Gödel to have been rather difficult to deal with on editorial matters: unresponsive to deadlines, finicky about editorial suggestions and inclined to ask for minor changes of his own even very late in the editorial process. Schilpp generally dealt with such problems diplomatically, and occasionally firmly.

The correspondence opens with Schilpp's invitation in November 1942 to Gödel to contribute a paper to the Russell volume, setting the rather astonishing deadline for submission of four months later, March 1943. Gödel accepted the invitation, wrote twice in the spring to ask for more time, and sent Schilpp a handwritten manuscript in May. Schilpp returned the edited manuscript and a typescript by early July. By this time Russell was already drafting his Reply to the other essays in the volume, and Schilpp urged on Gödel the importance of getting his essay in some form to Russell quickly, lest it come too late for Russell to reply to it. Gödel pleaded for more time to assess Schilpp's editorial changes and to make changes of his own, and refused to send the typescript as is to Russell. On 10 August 1943 Gödel wrote "I am terribly sorry I was not able to decide sending you the manuscript for print, because whenever I read it I find some improvements possible." Despite Schilpp's warning that Russell would not respond if the submission came any later, Gödel did not send the corrected typescript to Schilpp until the end of September. The next day he wrote Russell urging him to compose a Reply.[a] Russell did not respond to Gödel; rather, Schilpp relayed to Gödel Russell's decision not to write a reply. With a further letter of 14 October 1943 (not reproduced here) Schilpp enclosed a letter from Russell on the matter, along with Russell's note that was to be published at the end of his "Reply to Criticisms":

> Note: Dr. Gödel's most interesting paper on my mathematical logic came into my hands after my replies had been completed, and at a time when I had no leisure to work on it. As it is now about eighteen years since I last worked on mathematical logic, it would have taken me a long time to form a critical estimate of Dr. Gödel's opinions. His great ability, as shown in his previous work, makes me think it highly probable that many of his criticisms of me are justified. The writing of Principia Mathematica was completed thirty-three years ago, and obviously, in view of subsequent advances in the subject, it needs amending in various ways. If I had the leisure, I should be glad to attempt a revision of its introductory portions, but external circumstances make this impossible. I must therefore ask the reader to give Dr. Gödel's work the attention that it deserves, and to form his own critical judgment on it. (*Schilpp 1944*, p. 741)

The remainder of the correspondence about *Gödel 1944* centers on the provision of corrections and on difficulties the printers were having, due to wartime conditions, in obtaining the special symbols used in the article, such as the Hebrew aleph and circumflexed variables. Gödel responded with paraphrases of his text that avoid the use of such notation, so that, for example, in his mention of the continuum hypothesis on page 147 of *Gödel 1944* the use of alephs is avoided. In these months Gödel also continued to suggest some emendations and additions to his text; on 7 December Schilpp wrote Gödel that further changes are not possible, but Gödel persisted in asking for some slight alterations in letters of 19 December and 22 December. Schilpp apparently allowed them to be made.

Schilpp first met Gödel in person in Princeton, on 29 May 1946. At that occasion, he broached the idea of Gödel's contributing to *The*

[a]The letter to Russell is printed in this volume, above.

Library of Living Philosophers volume on Einstein. Gödel apparently was interested, and after a letter from Schilpp, he agreed to do so, but says he will contribute only a short note on the relation of relativity theory and Kant's doctrines. Schilpp found this acceptable, and set a deadline of 1 March 1947 (letters 19–23). Gödel asked for more time in letters of February and July 1947. After several more proddings from Schilpp, Gödel wrote on 11 May, 1948, claiming to have finished a draft and to have given a copy to Einstein. He sent nothing to Schilpp, however, until March 1949; but in this case what he submitted is indeed the final version, and no further emendations were offered.

As we now know, Gödel wrote several manuscripts that are preparatory to the essay *Gödel 1949a* that actually appeared in the Schilpp volume, but differ from it considerably. Two of these manuscripts are published as **1946/9* in these *Works*, vol. III, pp. 230–259. It is plausible to suppose that the draft to which Gödel referred in his letter of 15 July 1947 was one of these preparatory manuscripts; this would help explain the length of time between this letter and his final submission. Which document he gave to Einstein in May 1948 as a draft of his contribution is not possible to ascertain.

The most protracted correspondence between Gödel and Schilpp concerns Gödel's never-submitted contribution to the volume on Carnap. The invitation was offered on 15 May 1953 and accepted on 2 July, with Gödel again saying he would write only a short note (letters 27–28). At nearly a year off, Schilpp's deadline of April 1954 is more generous than in the earlier cases. In March 1954, Gödel asked for a postponement "by a few weeks"; in May, that he hoped "to be finished soon"; on 28 December 1954 that "the manuscript for this article is ready in handwriting. It has only to be typed." At that point he claimed he would send it in January. On 31 January 1955 he sent Schilpp a telegram, "manuscript Carnap volume in process of typing sending it in few days."

Gödel preserved six drafts of his essay for the Carnap volume, labeling them Fassung I–Fassung VI.[b] It is likely that the manuscript that Gödel refers to in December 1954 is version I. Version II is a typed copy of that manuscript, with numerous revisions and additions in Gödel's hand, and several additional typed pages. Clearly, when Gödel read the typescript he was not satisfied with it, and so began to revise it rather than send it to Schilpp. Nearly a year later, in November 1955, Gödel claimed that the essay "has been finished for quite some time now", and that he needed only to reformulate two footnotes, and make minor corrections to the whole. He promised it for early December (letter 32). There is some sporadic prodding from Schilpp over the next year. Gödel did not

[b] *Nachlaß* items 040431-040446. Versions III and V were first published in these *Works*, vol. III, pp. 334–362. Versions II and VI were published in *Rodríguez-Consuegra 1995*, pp. 171–222.

write again until July 1957, when he says the essay was in a state that could be finished in two or three weeks, and promises it by September 1 (letter 33). In September he drafted a letter saying that he plans to reduce the size of the paper. This is probably the point at which he abandoned version III, and wrote out a new manuscript, version IV, which is about 40% shorter. Version IV was subsequently condensed into the two very short final versions, each about one quarter the length of version III. Finally, after another inquiry from Schilpp, in February 1959 Gödel wrote that he would not be submitting anything, explained some of the difficulties he found in arriving at a version that would satisfy him, and concluded "because of widely held prejudices, it may do more harm than good to publish half done work" (letter 36). By "widely held prejudices", it appears, Gödel meant positivist, antimetaphysical ones. Schilpp responded with a sympathetic note, taking Gödel and himself to have similar philosophical outlooks (letter 37).

Unfazed by the experience with the Carnap volume, Schilpp wrote Gödel in 1964 inviting him to write an essay for the volume on Popper (*Schilpp 1974*). Gödel declined this invitation, while indicating a more positive view of Popper than of the other Vienna positivists (letter 39).[c]

Some final correspondence between Schilpp and Gödel (not reproduced here) took place in 1971: it is concerned with the reprinting of *Godel 1944* in *Pears 1972* and Gödel's desire to add new material, namely, an expansion of an initial footnote that had first appeared in the reprint of the paper in *Benacerraf and Putnam 1964*. He sent the new text to Schilpp on 14 May 1971, but subsequently altered it in galley proof in two places. The text of 14 May speaks of "an a priori, and in part abstract, intuition" as the basis for the constructions of both Hilbert-School constructivists and intuitionists. In the revised text, this becomes simply "a mathematical intuition".[d] The earlier text says, "What can be obtained by Russell's constructivism is the system of finite orders of the ramified hierarchy without the axiom of infinity for individuals"; the revision replaces the first seven words with "What, in Russell's own opinion, can be obtained by his constructivism." Gödel justified these changes in a letter of 18 October 1971, which accompanied the corrected galleys, to an editorial assistant named William Strachan at Doubleday and Company, the publisher of the Pears volume. Of the first change, Gödel said, "This little change is important, because otherwise Intuitionists may accuse me of misrepresenting their views, since they generally refuse to call their intuition 'priori'." Of the

[c]Gödel elaborates his reasons for not contributing to the Popper volume in a letter to Popper, which appears in this volume, above.

[d]The final version of this footnote, as it appeared in *Pears 1972*, can be found in these *Works*, vol. II, p. 119.

second, he said simply, "the objective truth of the last but one sentence [i.e., the original sentence beginning 'what can be obtained by Russell's constructivism'] is questionable." These changes provide another example of Gödel's caution and great concern not to leave himself open to criticism.

Warren Goldfarb

A complete calendar of the correspondence with Schilpp appears on pp. 454–458 of this volume.

1. Schilpp[a] to Gödel

November 18, 1942

My dear Professor Gödel:–

I do not know whether or not you are at all familiar with our *Library of Living Philosophers*. I am, therefore, taking the liberty of sending you, under separate cover, several explanatory announcements concerning the nature, intent, and success of our project—thus far.

However, the particular cause of my venturing to address you is the following. We are, at present, planning to publish the volume on *The Philosophy of Bertrand Russell* some time during the fall or early winter 1943 as Volume V in our *Library*.

I consider it a very great honor and privilege to invite you herewith to contribute an essay on "Russell's Mathematical Logic" to our Russell volume. Your essay might range from 25 to 40 (or 45) typewritten (double-spaced) pages, and should be in my possession (in duplicate) on or about March 15, 1943.

In talking the matter over last night with Lord Russell in person, I learned that he too would not only very greatly appreciate your participation in this project, but that he considers you the scholar *par excellence* in this field. Since I completely share Lord Russell's opinion on this point, you will realize that I consider it of the utmost importance that we should be able to secure your kind and greatly valued co-operation. I trust, therefore, that I may have the very great pleasure

[a]On letterhead of *The Library of Living Philosophers*. With the exception of a few postcards and two letters of 1959, all of Schilpp's letters are on this letterhead.

and privilege of receiving an early and, I hope, favorable reply from you to this sincere invitation and urgent request.

You will, I am sure, be interested to know that Professor Albert Einstein has already kindly consented to contribute the essay on "Russell's Epistemology" to our Russell volume. I mention this fact merely in order to indicate to you that we are getting, for our Russell volume, the very first minds in America and England.

Thanking you in advance for your very kind consideration of this invitation, and with the expression of the hope that I may have the privilege of hearing favorably from you at your earliest convenience, I beg to remain,

most sincerely and respectfully yours,

Paul A. Schilpp
Associate Professor of Philosophy

2. Gödel to Schilpp

Princeton, Nov. 30, 1942

Dear Prof. Schilpp:

Excuse me please for not having replied earlier, but unfortunately the promised announcements concerning the nature & purpose of the "Library of living philosophers" arrived only last Friday, so that I had to obtain some information on my own account. The project of your library is certainly a fascinating one and I am quite willing to collaborate. I am aware that what you want is chiefly a critical attitude and also, I presume, a presentation of the subject in the light of preceding and subsequent historic developments will be welcome to you; both of which suit me very well.

Of course I would confine myself to strictly logical problems, so that e.g. the greater part of "The philosophy of logical atomism" (Monist, vol 28)[a] would not come under my treatment. Furthermore I have noticed | that M[r] Quine in the Whitehead volume[b] has already given a thorough review of the formal part of the "Principia Mathematica", irre-

[a] *Russell 1918.*

[b] *Schilpp 1941.*

spective of what is due to Whitehead and what to Russell, so that I shall
e have to deal chiefly with the controversial items and the epist/mological background. This lies somewhat outside my own field of work. Therefore I have hesitated ~~to~~ at first, but finally I yielded, in view of the fact, that the meddling of scientist⟦s⟧ into philosophy has so often proved useful for both.

Please give my regards to Lord Russell. I should appreciate very much if I could have a complete list of his publications (not only those bearing on logics).

I regret that the time given to me is so short. I shall do my best but it might be that it will take me a little longer.

Very sincerely and respectfully yours

Kurt Gödel

3. Gödel to Schilpp

March 27, 1943

Dear Prof. Schilpp:

I am very sorry I cannot complete the article about Russell's logic quite as promptly as you wish. It will however not take me more than two or at least three weeks from now.

I had considerable trouble with the expression in English since I never wrote in English except about pure mathematics.

Very sincerely yours

Kurt Gödel

4. Gödel to Schilpp

April 18, 1943

Dear Prof. Schilpp:

I am extremely sorry I was somewhat delayed in the writing of my article by a bad attack of my stomach trouble. But still I hope to be able to complete the handwritten manuscript in a few days. I would however prefer to keep it for another week, because I might be able to find clearer formulations for certain passages. If it is important for you to get the article soon, I can send you the handwritten manuscript, because typewriting it here might cause a delay of another week.

Very sincerely yours

Kurt Gödel

5. Gödel to Schilpp

Princeton, May 18, 1943

Dear Prof. Schilpp:

I am sending you herewith the handwritten manuscript about Russell's logic and hope you will kindly excuse the long delay. I made quite a number of additions and changes, so that the manuscript is now perhaps not quite as easily readable as I had expected it would be. So if you prefer to get it in typewritten form, please send it back at once. In the opposite case I should ~~be~~ appreciate very much if you would call my attention to mistakes in the English expression and suggest the appropriate words in cases where I failed to find them. The footnotes are to be renumbered as indicated on the last page of the book containing them.

Very sincerely yours

Kurt Gödel

6. Schilpp to Gödel

May 22, 1943

My dear Professor Gödel:–

Thank you very much indeed for your kind letter of the 18th and also for your manuscript for our Russell volume. Both reached me safely this morning.

Although you are quite right that the manuscript is not as easily read as are your personal letters, my secretary has already made some "Stichproben" from the same, and I think we shall probably manage to get your manuscript into typescript in a reasonably accurate way.

Since—as you yourself suggested in your letter—I have already found that it will be necessary to re-word your English expressions in quite a few places (in order to have the article appear as nearly idiomatically correct as possible), we are making three (instead of merely two) copies of the typescript. One of the three copies I shall have submitted to you for final corrections before your essay goes to the printer. In this way you will need to return to us only such pages as you find still standing in need of correction. Also, in order to facilitate such further corrections on your part, we are typing the entire manuscript in triple (instead of merely in double) spacing. This will allow you very easily to make the corrections right in between the lines.

Thanking you again for your very great kindness in coöperating with us in this enterprise and especially for the excellent paper which you have written for us, I beg to remain, as ever,

most sincerely and gratefully yours,

Paul A. Schilpp

7. Gödel to Schilpp

Princeton, May 26, 1943

Dear Prof. Schilpp:

Many thanks for your letter of May 22. I would suggest that you confine changes in the text of my paper to an absolutely necessary minimum, because it seems to me that in such matters stylistic elegance is of no importance as conpared with a precise and unambiguous formulation of what one wants to say and I spent much work for this purpose. If you should have made a typescript of my manuscript also in it/s original form I should appreciate if you send it to me too, because some corrections ~~might~~ may have become necessary only through mistakes in copying my original manuscripts.

Very sincerely yours

Kurt Gödel

8. Schilpp to Gödel

May 31, 1943

Dear Professor Gödel:-

Thank you very much for your so very kind letter of the 26th. I hasten to reply.

Please rest assured that I shall make only absolutely necessary changes in your manuscript. I have always made it a very definite editorial policy of mine, in so far as at all possible, never to interfere with the intentional meaning and therefore with the preferred expression of any particular author. In editing the volumes in our *Library*, I have actually leaned over backwards in sticking to this rule.

However, I am sure that you will not take it amiss, if I say to you that your manuscript gives evidence that you have not yet fully caught the idiom of American English and that, as a consequence, many of your sentences would not at all be clear to American readers. Although I quite agree with you, that, in any scientific or philosophical treatise, "stylistic elegance is of no importance as compared with a precise and unambiguous formulation of what one wants to say," it is just in the interest of "precise and unambiguous formulation" that most—if not indeed all—of my corrections and changes will be made. I have no interest in "stylistic elegance," probably because I am not capable of using an elegant style myself. But I am greatly concerned with making meanings as clear as possible, and therefore find it necessary at times to re-word a sentence or sentences.

However, you will, in any case, be the final judge of all corrections and changes; since, as I wrote you before, I expect to send you the revised typescript for your own inspection and final corrections. I shall be happy, at the same time, to return your handwritten manuscript to you so as to enable you to judge all changes for yourself.

I am herewith enclosing a sheet containing some questions which I am unable to answer myself, and a decision on which will make a difference in the typing of the sentence(s). I imagine you will be able to answer those questions at once; please send us the answer on the same sheet(s) on which we have typed the questions. Thank you in advance for this help.

Thanking you again for your so very kind, helpful, and greatly appreciated co-operation, I remain, with every good wish, as ever,

most sincerely and gratefully yours,

Paul A. Schilpp

9. Gödel to Schilpp

Princeton, June 7, 1943

Dear Prof. Schilpp:

Thank you very much for your kind letter of May 31. I should appreciate if you would send me as soon as possible the portion of the corrected manuscript you have finished so far. I am somewhat worried about the whole matter, because in the questions I treated much depends on the wording chosen, so that it may become necessary to send the manuscript to and fro several times. I did of course expect that some improvements of my English might be necessary or desirable, but by no means did I expect that "many of my sentences would not at all be clear to American readers", since when some years ago Prof. Church of Princeton University was kind enough to correct the manuscripts of some lectures I gave in this country, he never found it necessary to make any essential changes. May I repeat my question whether in making your corrections you had before you my handwritten manuscript or a typewritten copy made by somebody else? In the latter case I would | urgently request to send me this copy too, especially because you said in one of your letters that you hope to be able to get my manuscript into typescript "in a reasonably accurate way". If you had any real difficulties in reading my manuscript you better send it back ⟨here⟩ for typewriting ~~here~~. 2

As to your question about "his" I belive it is perfectly clear both from the meaning and the grammatical structure of the sentence that only Hilbert can be the person meant.

Very sincerely yours

Kurt Gödel

10. Schilpp[a] to Gödel

June 27, 1943

My dear Professor Gödel:–

Please accept my sincerest apologies for the long delay in my reply to your kind letter of June 7th. Your letter reached me in the middle of the

[a] A handwritten note at the side of the page begs pardon for a smudge "caused by heat and glue on the envelope."

final examinations here at the University, when I was unable—for a couple of weeks—to do much of anything aside from the necessary chores connected with reading and grading examinations and getting final grades into the Registrar's offices.

However, during the last few days I was able to get back at your MS. again. Let me assure you, in the first place, that I have done all editorial work directly on *your own* hand-written manuscript; *not* on a secretarial typescript of the same. This you can see for yourself on the enclosed first "Heft", which I return to you herewith, together with the first carbon-copy of the typescript I am having made. I am sending you herewith the first 13 typewritten pages of the new typescript, which represent about the first 23 handwritten pages (together with the resp. footnotes) of your own manuscript. This material should give you a fair sample of what I am doing. I feel quite confident that you will not find a single one of your ideas either changed or in the least misrepresented. But: you are, of course, the final judge of that. For I am sending you *both* your own MS. and our typesript precisely in order that you may be able to check everything as carefully as possible and then return the original as well as the typescript to me *with your corrections.*

Before the end of this next week I hope to be able to send you the balance of your paper (both your original and our typewritten revision). However, I would appreciate it if you would be so kind as to send me the material I am returning to you today just as soon as you have carefully checked it and have made all the necessary corrections on the typescript, *without waiting* for the balance of your essay.

Thanking you again for your kind and most valued co-operation, and hoping that I may have the privilege of hearing from you soon concerning the enclosed material, I beg to remain, as ever,

most sincerely, gratefully and respectfully yours,

Paul A. Schilpp

P.S. Please note specifically the questions—marked in red pencil—on pp. 6 and 10 ⟨and 13⟩ of the typed MS.

⟦Schilpp's letter of 27 June was followed by another on 5 July transmitting the rest of the typescript. Gödel answered that letter on 12 July, asking for more time on account of illness, objecting to some of Schilpp's corrections, and expressing his desire to improve formulations.⟧

11. Schilpp to Gödel

July 14, 1943

My dear Professor Gödel:

Thank you very kindly for your letter of July 12th, which has just arrived. I hasten to reply.

First of all, let me say that I would by no means expect you to agree with me in all instances where I found it necessary to make changes in your manuscript. Also I feel equally confident that, if you will have one of your American colleagues go over such instances, you will discover that you will not be able to leave it as it was originally written. However, I obviously do not have the slightest objection to your making changes at any point where you think such are called for; especially in view of the fact that we both seem to quite agree upon the proposition that we do not want the article to be printed in bad English.

I am also, of course, glad to permit you a few days extra time for the purpose of "improving your original formulations by slight changes or additions."

On the other hand, since Lord Russell now has all the other essays except your own in his possession, and, since moreover, he desires to write his "Reply" in the order in which the essays are to appear in the book, and since your essay is marked for the third place in the book, I am wondering whether you would at least permit me to send a copy of the typescript as we have thus far made it to Lord Russell—*provided* I would tell Lord Russell at the same time that there would be some minor changes made by yourself in the article. This would at least give Lord Russell a general idea of what to expect from your essay, and therewith an over-all view of all the contributors' papers. If you will agree to this procedure,—and *I hope very much* that you will,—I would appreciate it greatly if you would send me a telegram to that effect immediately upon receipt of this letter; thereby enabling me to send the typescript (of which we still have two copies left here) to Lord Russell at once.

Thanking you again for your kind cooperation, and hoping that I will receive such telegraphic permission as I have requested, I beg to remain, as ever,

most sincerely and gratefully yours,

Paul A. Schilpp

⟦Schilpp's papers contain a note on his desk calendar leaf of 14 July 1943, possibly the transcription of a telegram, that Gödel would not agree to his suggestion of sending the typescript to Russell, but would be sending a list of "important corrections" soon. On 16 July, Schilpp replied saying that Russell had almost completed his Reply and that he (Schilpp) was sending an extra copy of the manuscript that he hoped Gödel would mark and send directly to Russell.⟧

12. Schilpp to Gödel

July 28, 1943

My dear Professor Gödel:–

I am terribly sorry to have to trouble you again. However, it is now $3\frac{1}{2}$ weeks since I sent you the last portions of your essay and two weeks since I received your telegram, in which latter you said that you would send me a list of important corrections *V E R Y S O O N.* Yet, since that time I have heard nothing, and I have certainly not received that list of important corrections.

May I merely remind you that yours is the only contributed essay to our Russell volume which Lord Russell has still not seen? And also: that Lord Russell has practically completed his formal Reply to all other essays?

In other words, it is *not* for my—or the printer's sake—that I am urging haste; rather it is first of all for your own sake: for I take it for granted that you would like to have Lord Russell reply to your essay in his formal "Reply." And, secondly, it is for Lord Russell's sake, who—at this same time—has not merely been writing his Reply to our volume, but is also finishing a large work on the History of Philosophy[a] (from earliest times to the present), a work which is scheduled to be in the printer's hands in September.

I am, of course, still in hopes that your corrections (or completely corrected set of your MS.) is already in the mails before this letter can reach you; and also, that another set is on its way directly to Lord Russell. However, if this should not be the case, would it be possible for you to let me know by return mail just exactly when we might expect your essay?

---

[a] *Russell 1945.*

Thanking you again for your kind cooperation, and in hope of hearing from you by return mail, I beg to remain, as ever,

most sincerely and gratefully yours,

Paul A. Schilpp

⟦August was filled with pleas from Schilpp, excuses from Gödel (including a medical exam for military service that he did not pass) and attempts by Schilpp to get the manuscript while he was on a lecture tour.⟧

13. Schilpp to Gödel

August 25, 1943

My Dear Professor Gödel:

Your letter of August 10 finally reached me here in Oregon on the five weeks lecture tour, concerning which I had written you earlier. The same mail which brought me your letter also brought me one from Lord Russell which made it pretty obvious (and I should say practically final) that he does not intend to reply to your essay at this late date. Personally I think this is very unfortunate indeed. And, if you had permitted me to follow my original suggestion of letting me send to Lord Russell at least the first typewritten tentative draft of your paper (unfortunately I did not get permission from you) Lord Russell's reply could have easily been given. But since you refused to give me your permission to send that draft to Lord Russell, and since Lord Russell has obligated himself to finish his History of Philosophy by this fall and also to write a series of lectures which he is to deliver in October and since, moreover, on top of all this, he is planning to leave the United States for Great Britain in November I simply do not see how I can use any more pressure than I have already done to try to get him to reply to your essay at this late date. I still feel that this is most unfortunate indeed; but I must remind you that it is certainly not my fault.

As you will have noted on the itinerary sent to you by my secretary, I am not staying at any one place on this tour very long. However, I am strongly in hopes that your essay will reach me certainly not later than during my stay at the camp in Santa Barbara, California. I shall be at that camp on September 5, 6, and 7. But in order to make very sure that it reaches me there before I leave, I hope you will make very sure to get the essay to Santa Barbara by September 5 at the very latest. We are already becoming quite inconvenienced at our printers by the absence of your essay. The address at Santa Barbara at which you can reach me will be as follows:

c/o C.P.S. Camp No. 36
Star Route
Santa Barbara, California

Thanking you again for you⟦r⟧ cooperation and hoping most sincerely that when I reach Santa Barbara I shall find your essay awaiting, I beg to remain,

Most sincerely and respectfully yours,

Paul A. Schilpp

14. Gödel to Schilpp

Princeton, Sept. 13, 1943

Dear Prof. Schilpp:

Excuse me please that I didn't send in the manuscript during your lecturing tour. I was unfortunately unable to concentrate on any work during the last few weeks. My wife was sick in the hospital after an operation and I had to move to another apartment just at that time. In addition my own condition of health was (and is) extremely ~~bad~~ poor too. Since however the manuscript was almost ready when I stopped working on it, I hope that (unless I should be prevented by some other unexpected circumstances) it won't take me very long from now on.

Very sincerely yours,

Kurt Gödel

15. Gödel to Schilpp

Princeton, Sept. 20, 1943

Dear Prof. Schilpp:

Your letter of Sept. 16 reached me on Saturday. The manuscript is now ready except that I wanted to speak about the last few changes and additions I made with Prof. Church who was kind enough to look over the whole article with reference to the English language. I am going to send only one copy because the manuscript in it's present form would not be convenient to read for Lord Russell.

Concerning Russell's reply I must say that ~~I find~~ it seems to me extremely unfair toward Lord Russell to publish an article in which he is pretty severely criticized, when he is prevented by other obligations from writing a reply. I can therefore agree to a publication under these circumstances only if this should be Lord Russell's | express wish. In addition a few words of explanation for the reader would seem desirable ⟨in this case⟩. I am personally in favor of not publishing my article at all without a reply from Russell, since I wrote it largely for the purpose of a discussion.

Very sincerely yours,

Kurt Gödel

16. Schilpp to Gödel

September 22, 1943

Special Delivery Airmail

My dear Professor Gödel:

This morning's mail brought me your letter of September 20, but still no essay from your pen! And this is the middle of the week in which I had asked to receive your essay at the beginning of it! Nevertheless, I am profoundly relieved to note from your letter that your manuscript is ready; but I do wish you would send it without *any further delay*.

I note what you have to say about the absence of a reply from Lord Russell to your essay. Nor can I blame you for feeling that it would be "extremely unfair toward Lord Russell to publish an article in which he is pretty severely criticized, when he is prevented...from writing a reply." I quite agree. However, I must remind you that this ⟦is⟧ neither Lord Russell's fault nor my own. If you had not sent me your telegram of July 14 or 15 in which you refused to let me send your original copy to Lord Russell, we could at least have had a reply from him to your major contentions. As it was, Lord Russell wrote me on August 8 as follows: "I think I will say nothing about Gödel except a postscript that it came too late. In any case it is not likely to be controversial." When, after that letter, I still raised the possibility of his writing a reply at least as a postcript to his major "Reply", he replied to me under date of August 14 as follows: "As for Gödel, you could insert a note saying his essay came too late."

As you can see from these quotations from Lord Russell's letters, it was not I but Lord Russell who has decided against replying to your essay. Nobody would be any happier than I if, even at this late date Lord Russell could be gotten to change his mind as to that. However, his own statements to me on this matter have so much the tone of finality that I do not believe the way is open for me to raise the question again with him. On the other hand, I not only have no objection to your raising the question with him directly, if you care to do so, but I should frankly be happy if you would. You may address him c/o Bell & Clapper, Phoenixville, R. D., Pennsylvania. I ought, however, to warn you in advance that it is most likely that he will refuse to reconsider; not merely because it is so late, but even more so because he has a series of lectures to deliver this month and writes me that he is leaving for England next month which, as he writes, "involves masses of red tape." In spite of all that, I repeat that I should be happy to have you try.

Let me express the hope once more that by the time this letter reaches you, your essay will be on its way here.

With every good wish,

Most sincerely yours,

Paul A. Schilpp

17. Gödel to Schilpp

Princton, Sept. 27, 1943

Dear Prof. Schilpp:

I am sending you herewith the article about Russell's logic. I am very sorry I had to keep you waiting so long but I do think that the changes and additions I made were very much in the interest of the cause. Since Prof. Church has looked over the whole manuscript[1] I hardly think that any corrections of the English text will be necessary. If some mistakes should have been overlooked, please call my attention to them.

[1]except the additions (on the attached sheets) to p. 36 and p. 37, which I wrote afterwards, and the footnote on p. 1.

From your letter of Sept. 22 I gather that Lord Russell's decision not to write a reply was based on the entirely wrong assumption that my article will not be controversial. | I therefore take it for granted that my manuscript won't go into print ~~until~~ ⟨before⟩ Russell has read it and either decides to write a reply or declares himself in favor of a publication without a reply. I am going to write to Lord Russell to-day. 2

Very sincerely yours

Kurt Gödel

18. Schilpp to Gödel

October 7, 1943

My dear Professor Gödel:–

On October 2nd we sent *your own* corrected copy of your essay on "Russell's Mathematical Logic" directly to Lord Russell (after we had transferred all of your corrections, changes, and additions to our own original typescript of your essay, which latter we sent on the same day to the printer). Together with your typescript I sent a letter to Lord Russell, in which I told him that we could still insert any reply he cared to make to your essay in his formal "Reply" in our volume.

Yesterday afternoon, I received a letter from Lord Russell, dated October 5th, in which, among other things, he writes as follows:

> "I have received Gödel's article, and a letter from him, urging me to answer it. It is quite impossible for me to make a detailed answer. I have not worked at mathematical logic since 1927 and it would take me at least a month's work. I am prepared to write a short paragraph saying I am unable to form a critical estimate of his article, but I think it quite probable that most of his criticisms are justified. I hope this will satisfy him and you."

At this same time I am writing to Lord Russell and am asking him to send me that "short paragraph" he has promised in the above, and to send it to me at the earliest possible moment. As you will note from the content of his letter, there is nothing further that I can ask him to do—under the circumstances.

Within about two weeks you may expect to receive galley-proof of your essay, which, I trust, you will check at once and return the corrected galleys to me just as quickly as possible.

Although you have probably gotten to be quite out of patience with me, due to the fact that I *had* to keep urging your getting your typescript back to me, I would like to take this opportunity to say two things:

1) Your essay, as you sent it to me under date of Sept.$_{\wedge}$27th, was—stylistically—a tremendous improvement over the original manuscript you had sent me last spring; and
2) I consider the actual content of your essay to be one of the most important contributions to our Russell volume (which is, of course, precisely the reason why I did not want our volume to have to go to press without your contribution).

Again accept my sincere thanks. Most sincerely yours,

Paul A. Schilpp

19. Schilpp to Gödel

July 10, 1946

My dear Dr. Gödel,

After all the correspondence we had with each other in connection with your highly valued contribution to our Russell volume, it was a real privilege, at last, to have the opportunity of making your personal acquaintance during my brief visit to Princeton on May 29th.

Ever since that visit I have been in hopes of hearing from you since you had so kindly promised to let me know on what aspect of Professor Einstein's work you might wish to write your essay for our proposed Einstein volume. However, not having heard from you in the past six weeks, I wonder if you would permit me to make a suggestion of my own. How would the subject, "The Realistic Standpoint in Physics and Mathematics" suit you as a title for your essay? Of course, if you have something else in mind, I hope you will feel entirely free to suggest another possible title to me (although, in that case, I would hardly know to whom else to turn for the writing of the above suggested subject).

Thanking you again for your kind former co-operation, expressing to you also once more my sincere pleasure in having met you personally, and looking forward with interest not merely to your early reply to this letter, but to your further collaboration with us, I remain, as ever,

most sincerely and gratefully yours,

Paul A. Schilpp

20. Gödel to Schilpp

Princeton, July 25, 1946

Dear Prof. Schilpp:

I regret that after careful consideration I have come to the conclusion, that I am not in a position to write an article of any considerable length for the Einstein/volume. I could, however, if you wish, write about 3 pages under the title: "Some remarks about the relation between the theory of relativity and Kant",[1] where I would treat this question in a positive sense pointing out the similarities, which seem to me incomparably more important than the differences. I am doubtful, however, if such a contribution would be of any use, for, since this question touches the very essence of "relativity", namely the role of the observer, I presume it will be treated in one or more of the longer articles already. Please let me know your opinion.

Sincerely yours

Kurt Gödel

[1] or perhaps: "Some remarks about the philosophical significance of relativity/theory".

21. Schilpp to Gödel

July 30, 1946

My dear Doctor Gödel:–

Thank you very kindly for your letter of the 25th. I had begun to worry about not hearing from you.

Although I am, in a way, sorry that you did not see your way clear to accept our first invitation and to write on "The Realistic Standpont in Physics and Mathematics,"I do not feel so badly, in the light of the kind offer which your letter contains.

If I have any question at all on your kind offer, it merely concerns the words "Some Remarks," on the one hand, and the problem of saying what I am sure you will be wanting to say under your suggested title in "3 pages" or, for that matter, even in 30 pages. Of course, I wouldn't care, if your paper should run to much more than 30 typewritten pages: not merely *even* under the head of your own suggested contribution, but *precisely because* of this new heading. For, in some ways, I am even more enthusiastic about your suggestion than I was about the original subject on which I asked you to write.

As a matter of fact, I am so very enthusiastic about your suggested topic—in either formulation in which you have mentioned it, i.e., either under the title “The Relation Between The Theory of Relativity and Kant” or under the title “The Philosophical Significance of Relativity-Theory” or, for that matter, under any other title under which you might care to formulate the problem of ‘the role of the observer’ in Kant and in Einstein—that I do hope most earnestly and want to urge you most sincerely to treat this subject as fully as it needs and des⟦e⟧rves to be treated even if, by doing so, you⟦r⟧ essay should run to 50 instead of to 3 typewritten pages. For I am, of course, in entire agreement with you that this problem touches the very essence of “relativity,” and therefore deserves the fullest kind of treatment. Frankly, I rather like your first wording of the paper best, namely: “The Theory of Relativity and Kant,” and I wish very much, therefore, that you might see fit to treat this very important subject just as fully as possible. I hope you will consent to this and that you will be so kind as to send me your kind acceptance of this assignment by return mail. You would make me a very happy man, if you would be so kind as to accept this and let me know at once.

With every good wish, I remain, as ever,

most sincerely and gratefully yours,

Paul A. Schilpp

22. Gödel to Schilpp

Princeton, Aug. 12, 1946

Dear Prof. Schilpp:

I am sorry that, in reply to your letter of July 30, I can only repeat, what I wrote before, that it is not possible for me to write more than a few (i.e., 3 to 5) pages for the Einstein-/volume. The reasons are 1.) that ⌴ I cannot afford to spend so much time on more or less expository articles, especially on a subject which, strictly speaking, lies outside my own field of work ⟦and⟧ 2.) that what I have to say on the subject I chose, and what of it seems to me of importance and general interest does not require more than a few pages. I hope very much that such a short contribution will fit in your scheme of the book, because otherwise I am afraid I shall not be able to contribute anything at all.

Most sincerely yours

Kurt Gödel

P.S. ~~Please~~ Excuse me, please, that I did not reply earlier, but unfortunately your letter was delivered to me belatedly.

23. Schilpp to Gödel

August 17, 1946

My dear Dr. Gödel:–

Thank you for your letter of the 12th, which I found awaiting me today, when I returned from a week's lecture-tour in the Middle West.

Let me say at once that it goes almost without saying that I shall much prefer to have a paper of from three to five pages from you than to have nothing at all from your pen. I merely wish to emphasize once more that, in case—when you actually start writing—your paper should be longer than you had anticipated, that also will be quite alright with us. In any case, I shall expect to receive a paper on

"The Theory of Relativity and Kant"

from your pen as soon as you have it ready, and certainly no later than next March 1st (1947).

Thanking you again for previous favors as well as for your kind agreement to contribute to our Einstein/volume, and with every good wish, I remain, as ever,

most sincerely and gratefully yours,

Paul A. Schilpp

⟦In February 1947 Gödel writes, asking for an additional month to complete his essay. In June, Schilpp writes to ask whether he might soon receive the promised essay.⟧

24. Gödel to Schilpp

Princeton, July 15, 1947

Dear Prof. Schilpp:

As for my article on Einstein and Kant I have essentially finished a draft of it more than a month ago. There is however one important point, which has turned out to depend on the solution of a mathematical problem, at which I am working now. You will understand that under these circumstances it is hard for me to promise a definite date, but I hope I shall have solved the problem in question soon and if I should not succeed in a reasonable time I intend, of course, to send you the article anyway. I hope you won't object to some further delay in this situation.

Sincerely yours

Kurt Gödel

⟦In the following year, Schilpp inquires twice more—at one point even resorting to a postcard in German—as to when he might see Gödel's essay and receives more excuses as Gödel revises further.⟧

25. Schilpp to Gödel

March 9, 1949

My dear Professor Gödel:

After talking with Professor Einstein this afternoon by long distance telephone, I venture to write to you once more about your contribution to our Einstein volume.

As you may recall, you had promised this to me many months ago; yet, up to now, I have not received it. Fortunately, Professor Einstein tells me that you had promised him to bring the finally completed and revised manuscript to him *today.* If this is the case, I trust that you will also be so kind as to send a copy to me, if you have not already done so.

Chagrined as I am over the fact that it proved to be impossible to get this volume off the press by Professor Einstein's 70th birthday (March 14, 1949), I am nevertheless very anxious that it should at least appear during the calendar year of his 70th birthday, i.e., during 1949. However, this is obviously only possible if all the essays which are still outstanding come into my possession at once. I therefore beg you once more to please send me a copy of your essay for our Einstein volume by return mail, if at all possible.

Thanking you for past favors, and hoping that I may have the pleasure of hearing from you by return mail, I beg to remain, as ever,

Most sincerely and gratefully yours,

Paul A. Schilpp

⟦In March Gödel finally sends off the manuscript, and in July he writes Schilpp to inquire about proof sheets. In December, he writes Schilpp again saying he has never received proof sheets, worrying about corrections he had sent in after sending the manuscript, and asking for a copy of his article.⟧

26. Gödel to Schilpp

Princeton, Dec. 21, 1949

Dear Prof. Schilpp:

I am sorry I bothered you about the proof-sheets. I remember now that I received them exactly at the time when we were moving to our new house we had just bought. The excitement connected with this event was no doubt responsible for my forgetfulness.—Thank you for the author⟦'⟧s copy of the book you sent me. I have found one rather annoying misprint in my article. In footnote 11 it should read: $t \ll 10^{11}$ instead of: $t \ll 10''$. The expression $10''$ might be understood to mean 10 seconds, whereas the intended meaning is 100 billion years. Fortunately it is clear from the context (at least for an attentive reader) that 10 seconds cannot be the intended meaning, but it is not so clear how the misprint is to be corrected. ~~In case~~ I presume that the printing of | the first edition is finished, but I hope this misprint will be corrected in the later editions, and perhaps in the reprints.—Shall I receive any reprints of my article? If so I should appreciate if ~~I~~ you would send them as early as possible, so that I can send them out together with the reprints of my mathematical paper about the same subject. 2

Sincerely yours

Kurt Gödel

27. Schilpp to Gödel

May 15, 1953

My dear Professor Gödel:

I do not know whether you have heard that we are planning Vol. X in our LIBRARY OF LIVING PHILOSOPHERS on *The Philosophy of Rudolf Carnap.*

It is with a great deal of personal pleasure that I am hereby inviting you to contribute an essay—of, say, somewhere between 25 and 40 double⟦-⟧spaced typewritten pages—on

"Carnap and the Ontology of Mathematics"

to this our projected Carnap volume. As you know, we shall need two copies of your paper (of which one may, of course, be a carbon copy); one of these is for editorial purposes and for the printer, and the other will go to Professor Carnap himself to aid him in formulating his "Reply".

Although I know how terrribly busy you are, I trust that it will be possible for you to let me have your essay by, say, April 2, 1954. As you will note, this will allow you almost a full year for the composing of your essay.

Moreover, the very fact that Professor Carnap himself is with you at the Institute at present, should make this undertaking a labor of love on your part. At least, so I hope.

Thanking you in advance for the courtesy of what I hope will not merely be an early but also a favorable reply to this sincere and urgent request, thanking you again for many past favors and kindnesses, and with every personal good wish, I beg to remain, as ever,

Most sincerely and gratefully yours,

Paul A. Schilpp
Professor of Philosophy

P.S. You will, I am sure, be interested to know that I am inviting Professor Einstein to contribute an essay on "Logical Positivism and Physical Reality" to this, our Carnap volume.

P.A.S.

28. Gödel to Schilpp

Princeton, July 2, 1953

Dear Prof. Schilpp:

I regret that I shall not be able to contribute a long article to the Carnap/Volume of the Library of Living Philosophers. However, if that serves your purpose, I shall be glad to ~~send you~~ write a few pages under the title:

Some observations about[a] the nominalistic view on[b] the nature of mathematics.

I am purposely avoiding to mention Carnap's name in the title, because I don't know to which extent the view I am going to criticize is (or still is) his. I want to leave that for him to decide in his reply. I don't know if you are aware that I have discussed exactly the same question in my Gibbs/Lecture of 1951, which will be published soon.
| My paper may, therefore, in part, literally agree with what I said there. 2

Sincerely yours

Kurt Gödel

P.S. Please excuse me for the long delay in replying to your letter.

---

[a]In a different handwriting, above "about", the word "on" appears.

[b]In a different handwriting above "on", the word "of" appears.

29. Schilpp to Gödel

July 6, 1953

Dear Professor Gödel:

Thank you very much, indeed, for your kind lettter of July 2nd, just received, to which I hasten to reply.

You need not at all apologize for offering to write a shorter paper for our Carnap/volume. After all, it is not quantity that matters, but quality. And, on the matter of quality, there can never be any doubt, when a paper is written by Professor Gödel.

I am most grateful to you, therefore, for agreeing to write a brief paper on:

"Some Observations on the Nominalistic View of the Nature of Mathematics"

for the projected volume on *The Philosophy of Rudolf Carnap.* Under the circumstances I think you are quite wise in leaving Carnap's name out in the title of your paper. But, by discussing the problem itself Carnap will be able to say whether or not he still holds to a nominalistic view of the nature of mathematics; or, if he has changed his position on this point, how he has changed and in which direction. This should make not only interesting but important reading, from any point of view.

I am very happy to be able once more to welcome you to the group of contributors of still another volume in our series. It is always an honor to have you among the contributors, and I am delighted with your response.

Thanking you for your kind co-operation, and with the very best of good wishes for a pleasant and enjoyable summer, I remain, as ever,

most sincerely and gratefully yours,

Paul A. Schilpp

30. Gödel to Schilpp

Princeton, March 28, 1954[a]

Dear Prof. Schilpp:

I am writing you in order to ask for a postponement, by a few weeks, of the date of delivery of my contribution to the Carnap Volume. I am substantially finished with the first draft; it will fill about 7 pages in print. However, I still have to translate it into English. Moreover I hope I shall be able to improve the formulations, if I devote a little more time to it.

I realize, of course, that a year is plenty of time for writing a few pages. But I have on purpose been waiting till now, because my work during the past year was in part concerned with the question of realism versus

[a]A note under the date says "repl. by postcard March 31, 1954. P.A.S."

nominalism, so that a postponement of the time of composition could only be to the advantage of | the paper. 2

Sincerely yours

Kurt Gödel

⟦The ensuing year and a half has sporadic communication between Schilpp and Gödel, with requests for postponements and some correspondence about the German translation of *1949a*.⟧

31. Schilpp to Gödel

November 2, 1955

Dear Professor Gödel:–

This is a word of inquiry in order to find out as to just when I may ␣ expect your essay, which you have promised us for the CARNAP/volume of our "Library".

I think you will grant that I have not troubled you for a long time with this question. But, in the light of the last information I had from you about this, I think I am justified in asking for this information—and this all the more so in view of the fact that most of the essays for our ␣ CARNAP/volume are now in my hands. I would deeply appreciate it, therefore, if you could let me know how soon I may expect your essay.—With all good wishes, as always, I remain,

most sincerely and gratefully yours,

Paul A. Schilpp

32. Gödel to Schilpp

Princeton, Nov. 14, 1955

Dear Prof. Schilpp:

My essay for the Carnap Volume, up to two footnotes, has been finished for quite some time now. I only have to reformulate these and once more read through the whole paper for minor corrections. I was about to do this at the beginning of this term, but was prevented by some unusual and important faculty matters, which demanded my whole attention. As soon as I shall be through with them, which, I hope, will be early in December, I'll be able to send the manuscript within two weeks.

I apologize for having delayed the matter so long. One of the reasons was that I had ⟨heard⟩ from friends of Carnap that the publication of the volume had been postponed for about a year.

Sincerely yours Kurt Gödel

33. Gödel to Schilpp

Princeton, July 11, 1957

Dear Prof. Schilpp:

I hasten to reply to your letter of July 8. I am very sorry I was not able yet to send you the manuscript. It is by now in a stage where I could finish it in two or three weeks, if I could concentrate on it but, I was, after the close of the term in May, prevented from ⟨doing⟩ ~~it~~ ⟨so⟩ by a number of urgent matters partly of a professional & partly of a personal nature. I certainly do hope I shall be able to send it by ~~Mo~~ Sep. 1. I shall write more definitely soon.

Sincerely yours,

Kurt Gödel

34. Gödel to Schilpp[a]

Princeton, Sep. 9, 1957

Dear Professor Schilpp:

I am sorry I could not send the article for the Carnap Volume by the 1^{rst} of Sep. In addition to the circumstances mentioned in my preceeding

---

[a]Marked by Gödel "nicht abgesch⟦ickt⟧ ⟦"not sent"⟧.

letter I caught a very bad influenza about the middle of August and am not quite over it yet. However, I have now decided to reduce the size of the paper to $\frac{1}{3}$ or $\frac{1}{4}$, i.e., to about ten typewritten pages. This should enable me to get done very soon, and I think the paper will even win in readability and usefulness by this procedure. Please let me know what the present plans about the date of publication are.

Sincerely yours

Kurt Gödel

35. Schilpp[a] to Gödel

January 24, 1959

My dear Professor Gödel:

Although I have not heard from you for over a year now, and although Professor Carnap's "Reply to My Critics" is not merely completed but in my hands, I venture to write to you once more, asking you whether it will not be possible for you to get your own essay for our CARNAP/
⊔ volume into my hands before the completed (and edited) manuscript will be sent off by us to the printers.

Whereas it is already too late to get a reply (to your essay) from Professor Carnap, I would still very much like to include your essay in our
⊔ CARNAP/volume—if this is at all possible.

Won't you be so kind as to reply to this letter of inquiry—as well as of entreaty, please?

With all personal good wishes, I remain, as ever,

Cordially and gratefully yours,

Paul A. Schilpp

---

[a]This and letter 37 are on the letterhead of the American Philosophical Association, Western Division.

36. Gödel to Schilpp

Princeton, Feb. 3, 1959

Dear Professor Schilpp:

I am extremely sorry I cannot give an aff~~e~~rmative answer to your inquiry of Jan. 24. In view of the fact that my article would ~~be~~ severely criticize some of Carnap's statements, it does not seem fair to publish it without a reply by Carnap. Nor would this be conducive to an elucidation of the situation. i

However, I feel I owe you an explanation why I did not send in my paper earlier. The fact is that I have completed several different versions, but none of them satisfies me. It is easy to allege very weighty and striking arguments in favor of my views, but a complete elucidation of the situation turned out to be more difficult than I had anticipated, doubtless in consequence of the fact that the subject matter is closely related to, and in part identical with, one of the basic problems of philosophy, namely the question of the objective reality of concepts and their relations. On the other hand, ~~in view~~ because of widely held prejudices, it may do more harm than good to publish half done work.

I hope that in view of the reasons stated, and also in view of the fact that I was considerably hampered in my work by illness and other difficulties,∧ you will kindly excuse it that I could not carry out my original intention. ⊔

Very sincerely yours

Kurt Gödel

37. Schilpp to Gödel

February 6, 1959

Dear Professor Gödel:

Thank you very much for your letter of February 3rd, just received. I hasten to reply.

Whereas I must admit that, without Carnap's having an oppportunity to reply to your criticisms, much of the value of your critical remarks might be lost (on the readers of the book), I am, of course, terribly sorry that—under these circumstances—you have decided that you should not contribute to our CARNAP/volume. For, I do think that an exceedingly ⊔

critical essay from your distinguished pen would have been a most valuable asset in our book. And, for this reason I am very sorry, indeed, for your decision. I only wish that your health might have been better and that, in that case, it would have been possible for you to get your essay in to us before Carnap had completed the writing of his "Reply." However, all of these regrets now seem to be "water over the dam," as an American colloquialism has it.

As far as the "widely held prejudices," to which you refer in your letter, are concerned, you may be interested to know that I myself expect to blast away at some of them in my presidential address on May 1st, under the title: "The Abdication of Philosophy."[a] (I am sure you must have seen—and read—Bertrand Russell's scathing attack on J. O. Urmson's book, *Philosophical Analysis*[b] in the January–March 1958 issue of the *Zeitschrift für Philosophische Forschung*, pp. 3–16.[c] Whether one agrees wholly with Russell or not, it certainly cannot be denied that, in this critical review, he is doing something far more than just "tilting at windmills." But, quite frankly, I have little hope that Russell's (or my own) attack will bear much fruit. They are too definitely and dogmatically ensconced in their (from their own point of view) supposedly "impregnable" position to let themselves even ⟦be⟧ so much as disturbed by adverse criticism, let alone to let themselves be dislodged from it. 'Nough said.

Sincerely, even though regretfully yours,

Paul A. Schilpp

---

[a] *Schilpp 1959.*

[b] *Urmson 1956.*

[c] *Russell 1958.*

38. Schilpp to Gödel

February 16, 1964

My dear Professor Gödel:

As a greatly valued former contributor to one of the volumes of our LIBRARY OF LIVING PHILOSOPHERS, I venture to come to you once more with and earnest and urgent request for your kind and valued cooperation.

As will note (from this letterhead), we are planning a volume on *The Philosophy of Karl R. Popper* as Vol. XIV in our LIBRARY.

I would like, herewith, to invite you most sincerely to contribute an essay—topic of your own choosing—to this our projected *Popper/* volume. ⊔ The length of your essay also may be of your own choosing. And, what is best of all, I am able to allow you all of 12 or 13 months for the composing and writing of your essay, that is to say until the end of March or April 1965.

I understand that you are quite (negatively) critical of some of Popper's ideas and work. This is precisely why I would welcome a contribution from your pen all the more. As you know perhaps better than almost anyone else, it is precisely by threshing such matters out among scholars that real scientific progress is being made.

For all of these reasons I hope most earnestly that you will find it possible to accept th/is sincere and urgent invitation and that you will be so kind as to send me an early and, I trust, favorable reply. i

With all personal good wishes, I remain, as ever,

Most cordially and gratefully yours,

Paul A. Schilpp
Professor of Philosophy
Northwestern University

39. Gödel to Schilpp

Princeton, July 23, 1964

Dear Professor Schilpp:

I assume you have heard from Professor Popper already that I am not in a position to write an article for the Popper Volume of the Library of Living Philosophers.

As far as my attitude toward Popper's ideas and work is concerned, you apparently have been misinformed. I am, in the main, *less* critical of them than of positivistic philosophy in general.

Sincerely yours,

⟦No signature on retained copy⟧

Carl Seelig

Carl Seelig (1894–1962) was a Swiss editor, critic, journalist, novelist and poet who wrote and edited several volumes about Einstein.[a] Gödel's letters to Seelig, apparently, were written in response to an initial inquiry from the writer about Gödel's associations with Einstein, and a subsequent invitation to expand on his remarks. (Seelig's half of the correspondence is no longer extant.) Seelig was planning a new third edition of his "documentary biography" of Einstein at the time, and long quotations from both of Gödel's letters appear in it (*Seelig 1960*, pp. 421–423).

The first letter is of special interest because it gives some indication of the character of Gödel's conversations with Einstein, and gives evidence of the range of his interests within physics. It suggests that Gödel was in a position to discuss with Einstein the latter's search for a unified field theory, and was himself "very skeptical" about the prospects for such a theory. (It would be interesting to know the basis for Gödel's doubts. Of course, most physicists at the time were skeptical, but it would seem out of character for Gödel to simply defer to received opinion.)

The two letters, jointly, are also noteworthy because of Gödel's comments about the "unfinished state" of relativity theory, and his proposals for further work. He suggests that a mathematical analysis of Einstein's equation might make it possible to properly attack the problem of finding its solutions and determining their general properties. He further proposes that it should be possible to prove "analogues of the fundamental integral theorems of Newtonian physics", and that these results would provide significant insight into the theory.[b]

David B. Malament

The translation is by John Dawson, revised in consultation with David Malament.

[a]For further information about Seelig, and a complete list of his publications, see *Weinzierl 1982*.

[b]It is not entirely clear what Gödel has in mind here. He may be thinking of theorems that would capture a sense in which, for example, (total) energy or angular momentum is conserved in general relativity. (A good deal of work was, in fact, devoted to (attempting to prove) theorems of this sort at just about the time Gödel wrote his letters.) If so, his claim is problematic. Today it is widely agreed that it does *not* make sense to speak of total energy or angular momentum (or their conservation) in general relativity except in the presence of very special conditions (e.g., asymptotic flatness). The problem is that, except under those conditions, one does not have a clear notion of the energy or angular momentum content of the gravitational field itself. For a discussion of the issues involved, and further references, see, for example, *Geroch 1973* and *Trautman 1962*.

1. Gödel to Seelig

Princeton, Sep. 7, 1955

Sehr geehrter Herr Seelig:

In Beantwortung Ihres Schreibens moechte ich Ihnen folgendes mitteilen:

Ich lernte Einstein im Jahre 1933 bei meinem ersten Besuch in Princeton kennen. Aber erst nachdem ich mich dauernd in Princeton niedergelassen hatte, kam ich (1942) in naeheren Kontakt mit ihm. Seitdem erfreute ich mich sehr haeufig seiner Gesellschaft. Unsere Diskussionen bezogen sich hauptsaechlich auf Philosophie, Physik und Politik. Einstein erzaehlte mir regelmaessig ueber die Fortschritte (oder Modifikationen) der unifizierten Feldtheorie. Ferner sprachen wir des oefteren ueber Gegenstaende, ueber die Einstein's Meinung in publizierter Form vorliegt, und schliesslich auch ueber eine Reihe von Fragen, die durch meine Arbeiten und Ansichten angeregt wurden. Ich habe oft darueber nachgedacht, warum wohl Einstein an den Gespraechen mit mir Gefallen fand, und glaube eine der Ursachen darin gefunden zu haben, dass ich haeufig der entgegengesetzten Ansicht war und kein Hehl daraus machte. Ein "Mitarbeiter" Einstein's bin ich nie gewesen, da ich der unifizierten Feldtheorie sehr skeptisch gegenueberstehe. Meine eigene Arbeiten ueber die Relativitaetstheorie beziehen sich auf die 1916 veroeffentlichte reine Gravitationstheorie, von der ich glaube, dass sie, sowohl von Einstein selbst, als auch von der ganzen zeitgenoessischen Physikergeneration als Torso stehen gelassen wurde, und zwar in jeder Hinsicht: physikalisch, mathematisch und hinsichtlich ihrer Anwendungen (Kosmologie).
2 | Ich hoffe, dass diese Mitteilung fuer Ihre Zwecke genuegt. Andernfalls stehe ich Ihnen gerne mit Ergaenzungen zur Verfuegung. Entschuldigen Sie bitte die verspaetete Beantwortung Ihres Briefes. Fuer ein Exemplar der 3. Auflage Ihrer Einsteinbiographie[a] waere Ich Ihnen sehr dankbar. Ich besitze uebrigens auch die frueheren Auflagen nicht.

Mit besten Gruessen

Kurt Gödel

---

[a]Presumably *Seelig 1960*, then in preparation.

1. Gödel to Seelig

Princeton, September 7, 1955

Dear Mr. Seelig,

In answer to your letter I would like to inform you as follows:

I became acquainted with Einstein in the year 1933 during my first visit to Princeton. But only after I had settled permanently in Princeton did I come into closer contact with him (1942). Since then I have quite often enjoyed his company. Our discussions principally related to philosophy, physics and politics. Einstein regularly told me about the progress (or modifications) of the unified field theory. Furthermore we spoke on many occasions about matters on which Einstein's opinion is available in published form, and finally also about a series of questions that were stimulated by my own works and views. I have often pondered why Einstein took pleasure in his conversations with me, and I believe one of the causes is to be found in the fact that I frequently was of the contrary opinion and made no secret about it. I was never a "collaborator" of Einstein's, since I was very skeptically opposed to the unified field theory. My own papers on relativity theory are concerned with the pure theory of gravitation published in 1916, which I think was left to stand by Einstein himself, as well as by the whole contemporary generation of physicists, as an untouched torso, and indeed, in every respect: physical, mathematical, and with regard to its applications (cosmology).

I hope that this information suffices for your purposes. On the other hand, I will be glad to provide you with supplementary details. Please pardon this delayed reply to your letter. I would be very grateful for a copy of the third edition of your biography of Einstein.[a] By the way, I also do not own the earlier editions.

With best wishes,

Kurt Gödel

2. Gödel to Seelig

Princeton, Nov. 18, 1955

Sehr geehrter Herr Seelig,

Besten Dank fuer die Zusendung der Einsteinbiographie.[a] Ich hatte leider noch keine Zeit, sie ganz zu lesen, aber auf Grund der Stellen, die ich gesehen habe, kann ich Ihnen das Kompliment machen, dass man sich bei der Lektuere Einstein irgendwie persoenlich nahe fuehlt, in viel hoeherem Grade als bei den anderen Biographien, die ich bisher gesehen habe.

Was den weiteren Ausbau der reinen Gravitationstheorie betrifft, von dem ich in meinem letzten Brief sprach, so meine ich damit nicht eine Erweiterung in dem Sinn, dass die Theorie einen groesseren Tatsachenbereich umfassen wuerde, sondern eine mathematische Analyse der Gleichungen, welche es ermoeglichen wuerde, ihre Loesung systematisch in Angriff zu nehmen und allgemeine Eigenschaften der Loesungen zu erkennen. Bisher kennt man ja nicht einmal die Analoga zu den fundamentalsten Integralsaetzen der Newtonschen Theorie, die meiner Meinung nach unbedingt existieren muessen. Da solche Integralsaetze[1] auch eine direkte physikalische Bedeutung haetten, so wuerde dadurch auch das physikalische Verstaendnis der Theorie wesentlich vertieft werden. Umgekehrt koennte eine genauere Analyse des physikalischen Inhalts der Theorie zu solchen mathematischen Theoremen fuehren. Einstein stand solchen Bestrebungen, zumindest den zuerst genannten, nicht geradezu ablehnend gegenueber, aber er war diesbezueglich wenig optimistisch.

2 | Der Grund, weswegen ich diese Fragen nicht selbst in Angriff genommen habe, ist einfach der, dass sie sehr weit von meinem eigenen Arbeitsgebiet, der Logik und Grundlagenforschung, abliegen und andrerseits so schwierig sind, dass sie zweifellos die ganze Arbeitskraft eines Mathematikers in Anspruch nehmen wuerden. Ich habe mich jedoch, im Zusammenhang mit gewissen philosophischen Problemen, mit einem weniger schwierigen Fragenkomplex aus der allgemeinen Relativitaetstheorie, ⟨nämlich⟩ der Kosmologie, eine Zeit lang beschaeftigt. Die Tatsache, dass ich hier als Neuling auf dem Gebiet der Relativitaetstheorie sofort wesentlich neue Ergebnisse erzielen konnte, scheint mir den unfertigen Zustand der Theorie hinreichend zu beweisen.

Verbesserungsvorschlaege fuer die 3. Auflage der Einsteinbiographie werde ich wohl kaum machen koennen, da ich ja ueber Einstein's persoenliche Verhaeltnisse fast gar nicht unterrichtet bin und unsere Gespraeche fast ausschliesslich wissenschaftliche und andere allgemeine Fragen zum

[1]und ebenso andere mathematische Saetze

[a]Probably *Seelig 1952*, which was in Gödel's library at one time and, along with *Seelig 1960*, was donated to Princeton's Firestone Library after Adele's death.

2. Gödel to Seelig

Princeton, November 18, 1955

Dear Mr. Seelig,

Thanks very much for sending the biography of Einstein.[a] Unfortunately I still haven't had the time to read the whole of it, but on the basis of the passages I have seen I can pay you the compliment that in reading it one somehow feels personally close to Einstein in a much higher degree than in the other biographies I have seen up to now.

As to the further development of pure gravitation theory, of which I spoke in my last letter, I do not thereby mean an extension in the sense that the theory would encompass a greater range of facts, but rather a mathematical analysis of the equations that would make it possible to tackle their solution systematically and to recognize general properties of the solutions. Up to now we don't even know the analogues of the most fundamental integral theorems of the Newtonian theory, which in my opinion must certainly exist. Since such integral theorems[1] would also have a direct physical significance, the physical understanding of the theory would thereby be essentially deepened too. Conversely, a more precise analysis of the physical content of the theory could lead to such mathematical theorems. Einstein did not reject such efforts, at least the first mentioned, out of hand, but he was not very optimistic in that regard.

—The reason why I didn't tackle these questions myself is simply that they are remote from my own area of specialization, logic and foundational research, and on the other hand are so difficult that they would undoubtedly engage the full working power of a mathematician. I have, however, in connection with certain philosophical problems, devoted myself for some time to a less difficult complex of questions from general relativity theory, namely cosmology. The fact that here I, as a newcomer to the field of relativity theory, could immediately obtain essentially new results seems to me sufficient to demonstrate the unfinished state of the theory.

I will scarcely be able to make any suggestions for the improvement of the third edition of your biography of Einstein, since I am hardly informed at all about Einstein's personal relationships, and our conversations almost exclusively had scientific and other general questions as their

[1]and other mathematical theorems as well

Gegenstand hatten. Das einzige, was ich vorschlagen moechte, ist ein ausfuehrliches Sachregister, das nach meiner Meinung in keiner Biographie fehlen sollte. Aufgefallen ist mir auch, dass Sie nichts ueber die naeheren Umstaende von Einstein's Austritt aus der israelitischen Glaubensgemeinschaft und ueberhaupt ueber sein Verhaeltnis zu den konfessionellen Religionen ⟨in seiner fruehen Jugend⟩ sagen. Philipp Frank scheint diesen Dingen doch einen nicht zu vernachlaessigenden Einfluss zu Beginn der geistigen Entwicklung Einstein's zuzuschreiben. Vergl. Biogr. Deutsche Ausg. 1949,[b] p. 21–23.

Ich danke Ihnen bestens fuer Ihre freudliche Einladung, Sie in der Schweiz zu besuchen. Leider werde ich davon kaum jemals Gebrauch machen koennen.

Mit herzlichen Gruessen

Ihr Kurt Gödel

---

[b] *Frank 1949.*

subject. The one thing that I would suggest is a detailed subject index, which in my opinion no biography should lack. It also struck me that you say nothing about the more particular circumstances of Einstein's withdrawal from the Zionist community and in general about his relationship to denominational religions in his early youth. Phillip Frank seems, however, to ascribe to these things an influence at the beginning of Einstein's intellectual development that is not to be ignored. See Biogr. Deutsche Ausg. 1949,[b] pp. 21–23.

I thank you very much for your friendly invitation to visit you in Switzerland. Unfortunately I will hardly ever be able to make use of it.

With hearty greetings,

Yours truly, Kurt Gödel

Thoralf Skolem

This item is a handwritten letter, dated 1 November 1931, addressed to a "Sehr geehrter Herr Professor." From the contents of the letter, in particular from the reference to "Ihrer Arbeit 'Über einige Satzfunktionen in der Arithmetik'", it is clear that the "Herr Professor" is Thoralf Skolem[a] and that the paper is *Skolem 1931*. The actual missive comes from a private source, and it is not known whether it was ever received by Skolem. The letter is a second and private response to *Skolem 1931*. The public response was *Gödel 1932d*, which is a very careful and strictly "mathematical" review of that article. In this private response, Gödel offers hints and suggestions related to Skolem's views on the issue of set-theoretic relativism. These suggestions are based on the results of his incompleteness paper *1931*, which had appeared the previous January.

The 1931 article by Skolem in question was an intermediate work in his program of translating the original and rather general results on the existence of countable models and the set-theoretic relativism that Skolem inferred from it (see his *1923a* and *1929*) into the actual construction of nonstandard models (see his *1933a* and *1934*); for a careful discussion of this development see the remarks by Hao Wang in his survey of Skolem's works (*Wang 1970*, p. 41). In the two papers referred to on nonstandard models for arithmetic, Skolem does not seem to have taken any notice of the private letter, and there is no mention of *Gödel 1931*. The papers *Skolem 1933a* and *1934* were reviewed in *Gödel 1934c* and *1935*, respectively; see the careful analysis by Robert Vaught in the introductory note to them in volume I of these *Works*, pp. 376–379; note, in particular, the remarks on the top of p. 377 op. cit., where Vaught discusses in detail the relationship between these papers of Skolem and some immediate consequences of *Gödel 1931*.

It is perhaps of interest to note that Skolem refers to *Gödel 1931* on many later occasions, for example in *Skolem 1937* and *1940*, but with main attention to Gödel's result (Theorem VII, pp. 191–192) on the arithmetical definability of (primitive) recursive functions. Interestingly, in *1940*, Skolem gives a simple proof of the undecidability of arithmetical problems that is remarkably similar in mathematical content to *Gödel *193?*.

Jens Erik Fenstad[b]

The translation is by Gabriel Sabbagh, revised by John Dawson.

[a]For a biographical sketch of Skolem (1887–1963) by the undersigned, see *Skolem 1970*.

[b]I wish to thank John W. Dawson, Jr., Solomon Feferman and Charles Parsons for helpful comments on a draft of this note.

1. Gödel to Skolem

Wien 1./XI.1931

Sehr geehrter Herr Professor!

Besten Dank für Ihre freundliche Karte und die Sonderdrucke Ihrer Arbeiten. Ich übersende Ihnen gleichzeitig Separata meiner beiden Abhandlungen über die Grundlagen; ihr Inhalt berührt sich an manchen Stellen mit den von Ihnen erzielten Resultaten.

Z.B. liefert meine Arbeit "Über formal unentscheidbare Sätze etc."[a] ~~einen B~~ auch einen Beitrag zu dem von Ihnen vertretenen mengentheoretischen Relativismus. Aus den Seite 190 betrachteten | widerspruchsfreien, aber nicht ω-widerspruchsfreien Systemen ergibt sich nämlich durch eine einfache Überlegung,[1] daß für die Axiomensysteme der Mengenlehre Realisierungen existieren, in denen gewisse im absoluten Sinn unendliche Mengen innerhalb des Systems "endlich" sind. D.h. dasselbe, was Sie für den Begriff "unabzählbare Menge" gezeigt haben, gilt auch für den Begriff "endliche Menge", daß er sich nämlich axiomatisch (durch Zählaussagen) nicht charakerisieren läßt.—Da sie in |Ihrer Arbeit "Über einige Satzfunktionen in der Arithmetik"[b] eine Andeutung gemacht haben, die in diese Richtung weist, so glaube ich, daß Sie dies besonders interessieren wird.

Mit vorzüglicher Hochactung

Ihr ergebener

Kurt Gödel

[1]die ich in meiner Arbeit nicht durchgeführt habe

---

[a]*1931*.

1. Gödel to Skolem

Vienna
1 November 1931

Dear Professor,

Thank you very much for your friendly card and the offprints of your papers. I am concurrently sending you reprints of my two essays on the foundations; their content relates in many places to the results that you obtained.

For example, my paper "On formally undecidable propositions, etc."[a] also provides a contribution to the set-theoretical relativism expounded by you. Namely, from the consistent but not ω-consistent systems considered on page 190, it follows by a simple consideration[1] that there exist realizations for axiom systems of set theory in which certain sets that are infinite from an absolute standpoint are "finite" within the system. In other words, what you showed for the concept "uncountable set" also holds for the concept "finite set", namely that it cannot be axiomatically characterized (by means of first-order formulas). Since you made a suggestion in your paper "On some propositional functions in arithmetic"[b] that points in this direction, I think you will find this particularly interesting.

With special esteem,

Yours sincerely,

Kurt Gödel

[1]which I did not carry out in my paper

---

[b]*Skolem 1931.*

Patrick Suppes

Born in 1922, Patrick Suppes is the Lucie Stern Professor of Philosophy, Emeritus, at Stanford University, where he has been on the faculty since 1950. His work has ranged quite broadly over the philosophy of science, especially that of physics, as well as the theory of measurement and the philosophy of economics and psychology. In the early 1950s he carried out work with J. C. C. McKinsey and others on the set-theoretic axiomatization of various branches of physics, and through McKinsey was further influenced by Alfred Tarski in adopting more generally the set-theoretic and model-theoretic viewpoints in his philosophical work. In addition to that, Suppes has maintained strong pedagogical interests over a long period of time; toward the end of the 1950s he published two widely used introductory texts on logic and set theory, and he pioneered (and continues to direct) computer-assisted instruction in these and other parts of mathematics.

There is a single item of correspondence from Gödel to Suppes, dated 22 July 1974, in response to an inquiry from Suppes of 26 June. The latter has disappeared, but in a personal communication[a] Suppes explained that he had proposed to the Sloan Foundation to "make a series of one-hour films of the then-living famous contributors to the foundations of mathematics in this century" and that he had approached Church, Gödel, Kleene and Tarski (among others) with this idea. Gödel turned down Suppes in his response, whose interest lies primarily in the variety of reasons adduced for not participating, including that he is not a very good lecturer. After suggesting that "[s]omebody else could probably establish better contact with the audience," Gödel's letter concluded with the demand that "[a]t any rate I ought to see a copy of the lecture beforehand so that the impression can be avoided that I agree with everything that is said, in case I don't."

According to Suppes, Kleene also declined to participate "because he felt it would interfere with the current demand for him to give lectures in many places." So, after that, Suppes gave up on the project and nothing came of it.

Solomon Feferman

[a] Via e-mail, 12 July 1995.

1. Gödel to Suppes

July 22, 1974

Professor Patrick Suppes
Institute for Mathematical Studies
in the Social Sciences
Stanford University
Stanford, California 94305

Dear Professor Suppes:

I am sorry I am not in a position to take part in the project outlined in your letter of June 26. I have been asked several times already to have a film of this kind made, but have always declined. There is more than one reason for that.

The first one is that the preparation of such a film would be rather time consuming and, in view of my poor health, I feel obliged to use the working power that is left to me for research and other activities at the Institute.

Another reason is that my philosophical views on the foundations of mathematics and my interpretation of my results are rather unorthodox, namely Platonistic. Therefore my lecture would probably arouse antagonism and involve me in further discussions that would again be a drain on my time.

I believe, moreover, a presentation of the mathematical part of my work could be done just as well or even better by some other logician well acquainted with it, e.g., Feferman. I don't think I am a good lecturer. I have very little practice in lecturing. Somebody else probably could establish much better contact with the audience.

As far as my philosophical views and heuristic principles are concerned I would suggest reading to the audience some letters and remarks of mine that were published recently in the book: "From Mathematics to Philosophy" by Hao Wang[a] (cnf. in particular pp. 8–12, 324–326).

At any rate I ought to see a copy of the lecture beforehand so that the impression can be avoided that I agree with everything that is said, in case I don't.

Sincerely yours,

K. Gödel

P.S. As far as Kreisel is concerned I am doubtful he is good at lecturing to undergraduates.

---

[a] *Wang 1974*.

Alfred Tarski

Alfred Tarski

Alfred Tarski first met Kurt Gödel on the occasion of his visit to Vienna early in 1930, at the invitation of Karl Menger. Their subsequent contact, both personal and by mail, which begins with a letter to Tarski from Gödel in 1931, extended at least to 1970; the relationship between them over this entire period is traced in *S. Feferman 1999*. The entire collection, whose sole source is the Gödel *Nachlaß*, consists of 22 items, of which five are from Gödel to Tarski (not all of which were sent). Following the first letter, there is nothing more until 1942; from then until 1944, all but one of the letters and cards from Tarski are of a personal nature, and so are not included here. Moreover, up until that date all the correspondence is in German; from 1945 it is all in English. The complete correspondence from Tarski to Gödel from 1942 to 1947 has been translated and edited by Jan Tarski, and has been published as *Tarski 1999*. After 1947, there were only three letters between them. The first, from Gödel to Tarski in 1961, is reproduced below. The second and third, from May 1970, concern the manuscript **1970a* that Gödel sent to Tarski for submission to the *Proceedings of the National Academy of Sciences*; the history of that is described in the introductory note to **1970a, b* and *c* in these *Works*, vol. III. The last of these items, **1970c*, is a letter from Gödel to Tarski, which was probably not sent, in response to Tarski's letter to him of 19 May, in which his manuscript was returned without comment. Neither of these letters is included below.

An account of Tarski's visit to Vienna in February 1930 is to be found in *Menger 1994*, pp. 143–157. Tarski, then aged 29 and five years senior to Gödel, was already an established researcher with some two dozen articles in print. In Vienna, he gave three lectures to Menger's Colloquium, one on set theory, and the other two on logic. It is likely that Gödel attended all three; at any rate, Gödel requested a personal meeting with Tarski, in order to explain to him his own recently completed dissertation work (*Gödel 1929*). We can guess that this impressed Tarski due to the intrinsic interest of the completeness theorem established there, and the elegance with which it was proved.

The first item of correspondence between the two which is included below is a letter from Gödel to Tarski, dated 20 January 1931, saying that in the previous summer he had arrived at some new results in metamathematics that he thought should interest Tarski. This indirect reference to Gödel's stunning incompleteness theorems is the understatement of the century! While it took the logical world at large about a year to absorb the significance of *Gödel 1931*, a few individuals were quick to appreciate their importance—Tarski among them. In Poland he was the first to explain Gödel's accomplishment, in a report to the Warsaw Philosophical Society on 15 April 1931. More importantly, Tarski was quick to apply the method of arithmetization that Gödel had used to construct self-referential statements, in order to cap his theory of truth by showing that for suitably expressive languages L, one cannot define the notion of truth-in-L within L itself. In a brief announcement of his work on the concept of truth to the Vienna Academy of Sciences (*1932*), Tarski explicitly credited Gödel for providing the "Gedankengang" needed to establish this. It appears that the two had only one other meeting in the decade of the 1930s, during a visit to Vienna in the winter and spring of 1935.[a]

An invitation from W. V. Quine to speak at a Unity of Science conference at Harvard University in September 1939 in effect saved Tarski's life; he made a last-minute decision to attend and left Warsaw just three weeks before Germany invaded Poland and World War II broke out. Distressingly, his wife and their two children were left behind in Warsaw, and could not be brought out to join him. Stranded in the United States, Tarski spent three difficult years going from one temporary position to another until 1942 when he finally obtained a position at the University of California, Berkeley as a lecturer in the Mathematics Department.[b]

Meanwhile (as described in the opening essay on Gödel's life and work in volume I of these *Works*), Gödel emigrated with his wife Adele from Austria to the United States early in 1940 and became established as a member of the Institute for Advanced Study in Princeton. Within two years, Tarski and Gödel renewed their personal contact, since Tarski was able to spend part of 1941–1942 as a Guggenheim Fellow at the Institute. The subsequent correspondence, *Tarski 1999*, not reproduced here, shows that Tarski enjoyed considerable personal contact with Gödel and his wife during that time. Tarski's letters to Gödel from California in the period 1942–1944 speak of his personal and scientific isolation and burdens, and his wishes that he were in the East again with his friends and colleagues. An effort was made in the spring of 1944 to arrange such a visit to Princeton, but the requisite invitation from the Institute was not forthcoming in time in order to secure funds to support the travel. Instead, Tarski spent his break from teaching with his friend and co-worker, J. C. C. McKinsey, then at Montana State College in Bozeman.

[a]This meeting is commemorated in the photo of the two of them together, to be found in vol. I of these *Works*, following p. 15.

[b]See *A. B. Feferman 1999* for fuller information concerning Tarski's difficulties during these trying years.

This is the personal background to the next two items of correspondence included below. The scientific background is as follows. The paper *Tarski 1938a* had initiated a study of the relationship between the classical and intuitionistic propositional calculi and topology. There, topological spaces S are taken to be specified by a closure operation C applicable to arbitrary subsets of S satisfying the usual conditions. Every formula A with propositional variables $X, Y, Z, \ldots$ has associated with it an operation $\varphi_{\mathrm{A}}(X, Y, Z, \ldots)$ on subsets of S via the interpretations of $\mathrm{X} \vee \mathrm{Y}$ by $X \cup Y$ and $\neg\mathrm{X}$ by $S - C(X)$. The main results of *Tarski 1938a* were that A is provable in the classical propositional calculus if and only if $\varphi_{\mathrm{A}}(X, Y, Z, \ldots)$ is dense in S for all subsets $X, Y, Z, \ldots$ of S, and that A is provable in the intuitionistic propositional calculus if and only if $\varphi_{\mathrm{A}}(X, Y, Z, \ldots) = S$ for all open subsets $X, Y, Z, \ldots$ of S. In their first collaboration McKinsey and Tarski (*1944*) introduced abstract algebras of topology, which are Boolean algebras with an additional "closure" operation C satisfying the general conditions of closure operations in topology. An evident project of further joint work during Tarski's visit to Montana in 1944 was the algebraization of the work on intuitionistic calculi and topology of *Tarski 1938a*, the results of which were to appear as *McKinsey and Tarski 1946*. The basic ones connecting closure algebra with the algebra of intuitionistic logic were that the set of all closed elements of a closure algebra forms (in their terminology) a Brouwerian algebra[c] and that, conversely, every Brouwerian algebra can be obtained in this way. For inclusion in that publication, Tarski asked permission of Gödel in letter 2[d] to quote a result (and its proof) about Brouwerian operations in Euclidean spaces that Gödel had communicated to him when they were together in Princeton. Tarski pointed out that it would follow from Gödel's result and his own paper (*1938a*) that infinitely many closed sets can be constructed by Brouwerian operations in every Euclidean space. Gödel obliged in his response (letter 3); he states there that his construction implies the existence of infinitely many operations of one argument in Heyting's intuitionistic propositional calculus, without having to use Tarski's "(much deeper) result about the equivalence of topology and Heyting calculus"—only the easy direction for that equivalence (in the sense of *Tarski 1938a*) being needed. (A crossed-out passage in the copy of this letter found in Gödel's *Nachlaß* refers to *Gödel 1933f*, where an interpretation—formally equivalent to the topological interpretation—had been given of the intuitionistic propositional calculus IPC in the modal calculus S4.)

[c]Heyting algebra in current terminology.

[d]A postcard dated 27 April 1944.

The first meeting of Tarski and Gödel following the war was on the occasion of the Princeton Bicentennial Conference on the Problems of Mathematics in 1946. By then, to Tarski's great relief, he was reunited with his wife and children, who had managed to survive the war after six and a half harrowing years. In the meantime, he had become tenured at U.C. Berkeley, and rose to the position of full professor there in 1946. Tarski's letter to Gödel a week in advance of the Princeton meeting (item 4 below) begins by outlining the topic of the presentation he is to make on that occasion. Evidently Gödel had explained his own plans in an earlier letter (not found in the *Nachlaß*), and Tarski was concerned to relate the two. At the conference, Tarski was designated as the discussion leader for the section on logic.[e] His talk for that was about decidable, undecidable and unsolvable problems; it surveyed previous work by himself and others and offered suggestions for directions of further work.

The proceedings of the conference were never published. Gödel's lecture (concerning the notion of "absolute" or ordinal definability in set theory) eventually found its way into print in the collection *Davis 1965*; it appears in these *Works* as *Gödel 1946*. On the other hand, Tarski's lecture never appeared in print during his lifetime and languished in the Tarski archives at the Bancroft Library of the University of California in Berkeley for the remainder of the twentieth century following his death (in 1983). An edited version has finally appeared in *Sinaceur 2000*. In this lecture text, as he wrote Gödel he would, Tarski made the suggestion that certain propositions of set theory such as Cantor's continuum problem might in some sense be absolutely undecidable, in contrast to the number-theoretic propositions shown independent by *Gödel 1931*, which are only relatively undecidable. Of course, this contrasts with the program advanced in *Gödel 1947* to settle the continuum problem by suitable new "large cardinal" axioms for set theory.

There is a distant echo of this difference of views in the final item of correspondence included below, from Gödel to Tarski in August 1961. Gödel refers there to the result of *Scott 1961* that the negation of the axiom of constructibility, V = L, follows from the existence of a measurable cardinal, MC. Gödel had expressed the belief in his 1947 article that the continuum hypothesis CH is false, and hence that large cardinal axioms of Mahlo type, which are consistent with V = L, would not settle the continuum problem. Thus Scott's result opened up the possibility of finding a stronger kind of large cardinal axiom, such as MC, which *would* settle CH. Subsequently Lévy and Solovay (*1967*), using Cohen's

[e]Cf. *Sinaceur 2000*, p. 1.

method of forcing, showed that CH is independent of MC, and their argument extends to all large cardinal axioms that have been proposed to date and that are assumed to be consistent. To that extent, at least, Tarski's suggestion that CH may be absolutely undecidable has been supported by subsequent developments, but of course the matter remains controversial. For further discussion of these matters, see the introductory note to *Gödel 1938*, *1939*, *1939a* and *1940*, in volume II of these *Works*, p. 19ff.

In his letter to Tarski, Gödel described Scott's result as "beautiful" but "not surpris⟦ing⟧". What did surprise him, though, was the result that he had heard ascribed to Tarski that the first inaccessible cardinal is not measurable. This question was a long-standing one dating back to *Ulam 1930*, where the notion of measurable cardinal was introduced and it was shown that any such cardinal must be inaccessible.[f] Then in the late 1950s, when questions concerning the compactness of certain infinitary languages $L_{\kappa,\kappa}$ were being considered in Berkeley, Tarski suggested to his student William Hanf that he consider the related question of possible weak compactness of inaccessible cardinals. In 1960 (see *Hanf 1964*), Hanf showed by use of suitable $L_{\kappa,\kappa}$ sentences that there are "many" inaccessible cardinals below the first weakly compact cardinal. Tarski (*1962*) followed that up with a proof that the least measurable cardinal (if it exists) is greater than the least weakly compact cardinal. It is interesting that Gödel was so impressed with this; he said "I still can't believe that this is true", but added that he didn't have time to check the proof because he was "working mainly on phil⟦osophy⟧."[g]

Solomon Feferman[h]

A complete calendar of the correspondence with Tarski appears on p. 459 of this volume.

[f]As mentioned by Ulam (*1930*, p. 146), the inaccessibility of measurable cardinals was also obtained independently by Tarski.

[g]See, in this respect, Gödel's revised note 20 to *1964* appearing on p. 260 of volume II of these *Works*.

[h]I have benefited from useful comments by John W. Dawson, Jr., Aki Kanamori and Charles Parsons on a draft of this note.

1. Gödel to Tarski

Wien 20./I. 1931

Sehr geehrter Herr Dozent!

Herzlichen Dank für Ihre Karte vom 2./IX. 30 sowie die in liebenswürdiger Weise nach Königsberg gesandten Separata. Anbei übersende ich Ihnen 5 Sonderdrucke meiner Arbeit über den Funktionenkalkül, die Sie bitte an eventuelle Interessenten verteilen wollen. Im letzten Sommer ist es mir gelungen, einige neue metamathematische Sätze zu beweisen, von denen ich annehme, daß sie Sie interessieren werden. Die Resultate (ohne Beweis) habe
2 ich in einer Mitteilung | an die Wiener Akad. d. Wiss.[a] zusammengefaßt, von der ich 2 Separata beilege. Die Beweise erscheinen in wenigen Wochen in den Monatsh. f. Math u. Phys.[b] und ich werde Ihnen, sobald es mir möglich ist, Sep. zuschicken. Ich sehe mit großem Interesse der Fortsetzung[c] Ihrer in den Monath. erscheinenden Arbeit[d] entgegen und bin sehr gerne bereit Korrekturen mitzulesen. Sehr dankbar wäre ich Ihnen, wenn Sie mir Separata der in Warschau erscheinenden metamathem. Arbeiten[e] versorgen könnten. Ist die exakte Formalisierung der Principia Mathematica schon erschienen? Diese
3 würde mich ganz besonders inter|essieren.—Die Abschrift meiner Dissertation[f] brauche ich nicht und überlasse ich sie Ihnen sehr gerne.

Mit besten Grüßen

Ihr ergebener Kurt Gödel

---

[a] *1930b*.

[b] *1931*.

[c] *Tarski 1933*.

1. Gödel to Tarski

Vienna, 20 January 1931

Dear docent,

Cordial thanks for your card of 2 September 1930, as well as the offprints you so kindly sent to Königsberg. With this letter I am sending you five offprints of my paper on the functional calculus, which you might want to distribute to potentially interested parties. Last summer I succeeded in proving some new metamathematical theorems that I presume will interest you. I have collected the results (without proof) in a contribution to the Vienna Academy of Sciences,[a] of which I enclose two offprints. The proofs are to appear in a few weeks in the Monatshefte für Mathematik und Physik,[b] and I will send offprints to you as soon as possible. I look forward with great interest to the continuation[c] of your paper that is to appear in the Monatshefte[d] and would be very glad to look over the proof sheets. I would be very grateful to you if you could provide me with offprints of ⟦your⟧ metamathematical papers that are to appear in Warsaw.[e] Has the precise formalization of Principia Mathematica already appeared? That would especially interest me.

I don't need the copy of my dissertation[f] and am happy to send it to you.

With best wishes,

Sincerely yours,

Kurt Gödel

---

[d] *Tarski 1930a.*

[e] Presumably *Tarski 1924* and *1930.*

[f] *1929.*

2. Tarski to Gödel

April 27, 1944

Lieber Kurt!

Du hast mir in Princeton ein Ergebnis mitgeteilt das die Brouwerschen Algebra (Logik) betrifft: man kann im Eukl. Raume R_n eine ⟨abgeschlossene⟩ Punktmenge konstruieren aus deren man mittels 'Brouwerschen Operationen' n verschiedene abgeschl. Mengen erhalten. Daraus und aus meinen Ergebnissen im 'Aussagenkalkül und Topologie'[a] kann man erschließen daß es in jedem Eukl. Raume abgeschl. Mengen gibt aus denen man unendlich viele abgeschl. Mengen bekommen kann. Ich beabsichtige jetzt zusammen mit McKinsey eine Arbeit über Brouwersche Algebra zu veröffentlichen[b] (eine Art der Fortsetzung unserer 'Algebra of Topology'), und wir würden uns sehr freuen wenn Du uns erlaubst Dein Ergebniss (zusammen mit dem Beweis) anzugeben. Schreibe mir, bitte, ob Du nichts dagegen hast.—Ich bin schon zurück in Berkeley und beginne die Vorlesungen in einigen Tagen. Was hört man bei Dir? Beste Grüße für Dich und Adele.

Dein

Alfred

---

[a] *Tarski 1938a.*

3. Gödel to Tarski

Princeton, 12./V. 1944

Lieber Alfred!

Vielen Dank für Deine beiden Karten u. Deinen Brief aus Montana. Ich hoffe dass Du Dich in Montana gut erholt hast. Landschaftlich u. klimatisch war es ja bestimmt schöner als Princeton, aber Adele u. mir tat es wirklich sehr leid dass Du nicht kamst. Besteht denn gar keine Aussicht, dass Du die Reise nach dem Osten im Herbst (oder Sommer) doch noch machst? Ich habe noch keine Sommerpläne u. werde wahrscheinlich bis auf ein Intervall von 2–3 Wochen in Princeton bleiben.

2. Tarski to Gödel

April 27, 1944

Dear Kurt,

In Princeton you told me about a result that concerns Brouwer's ⟦intuitionistic⟧ algebra (logic): in the Euclidean spaces R_n one can construct a closed point set from which n different closed sets result by means of the 'Brouwerian operations'. From that, and from my results in 'Aussagenkalkül und Topologie',[a] one can conclude that in every Euclidean space there are closed sets from which one can obtain infinitely many closed sets. Together with McKinsey I am now planning to publish[b] a paper on Brouwer's algebra (sort of a continuation of our 'Algebra of Topology'), and we would be very pleased if you would allow us to refer to your result (together with the proof). Please write me whether you have any objection to that. I am already back in Berkeley and begin my lectures in a few days. What are people talking about where you are? Best wishes to you and Adele.

Your

Alfred

---

[b] *McKinsey and Tarski 1946.*

3. Gödel to Tarski

Princeton, 12 May 1944

Dear Alfred,

Many thanks for your two cards and your letter from Montana. I hope that you kept well in Montana. In terms of the landscape and climate it was decidedly more beautiful ⟦there⟧ than Princeton, but Adele and I were really very sorry that you did not come. Is there then no prospect at all that you ⟦might⟧ still make the trip East in the fall (or summer)? I still have no plans for the summer and will probably stay in Princeton except for an interval of two to three weeks.

Bezüglich des Resultats über den Heyting-Kalkül, habe ich natürlich nicht das geringste dagegen, wenn Du es publizierst (ganz im Gegenteil). Da die Konstruktion ja sicher auf mehrere Arten möglich ist, ist es vielleicht gut wenn ich das Wesentliche ganz kurz wiederhole:

Es sei H = Hülle, I = Inneres, $\phi(A) = \text{Rand}(HI(A))$

2 $\psi(A) = HI(A).H(A - HI(A))$, ϕ_R, ψ_R die auf R | relativierten Operationen, R_n der n-dimens. Eukl. Raum, dann kann man (induktiv nach i) eine Folge abgeschlossener Menge A_i ($A_i \subseteq R_i$) definieren, so dass:

$$A_i = \psi_{R_{i+1}}(A_{i+1}), R_i = \phi_{R_{i+1}}(A_{i+1}).$$

—Daraus folgt sofort die Existenz unendlich vieler nicht-äquivalenter einstelliger Aussagenoperationen des Heytingkalküls u. zwar ohne Dein (viel tiefer liegendes) Resultat über die Äquivalenz von Topologie u. Heyting-kalkül zu benutzen. Man ~~offenbar~~ braucht nur das Enthaltensein in einer Richtung, welches sich ⟨ja⟩ ganz leicht ~~⟨[z.B. auch⟩ aus meiner Bemerkung in Erg. Math. Koll. Heft 4, p 39[a] ergibt, da ja das dort angegebene Axiomensystem für den Begriff der Beweisbarkeit offenbar mit den topologischen Axiomen äquivalent ist.]~~ ⟨ergibt⟩.[b]

Dein KG

---

[a] *1933f.*

[b] A shorthand note at the bottom of the page continues:
Und zwar⟦?⟧ unter alleiniger Verwendung der Tatsache, daß der Heyt. Ax. aus den Topol folgt, was ganz leicht beweisbar ist—Die viel tiefer liegende Äquivalenz zwischen H und Topol ist ⟦?⟧. Ich glaube, das sollte man in einer Fußnote erwähnen, weil das Th. über Heyt.kalkül doch an sich interessant ist—Meine Arbeit über Russell noch nicht erschienen obwohl Korrekturen vor 4 Monaten.

⟦From here the correspondence continues in English.⟧

With respect to the result about the Heyting calculus, of course I have not the slightest objection to your publishing it (quite the contrary). Since the construction is certainly possible in several ways, it is perhaps good that I repeat the essentials very briefly:

Let H stand for the hull, I for the interior, $\phi(A)$ = the boundary of $HI(A), \psi(A) = HI(A).H(A - HI(A))$, ϕ_R and ψ_R be the operations relativised to R and R_n be n-dimensional Euclidean space. Then one can (by induction on i) define a sequence A_i of closed sets ($A_i \subset R_i$) such that

$$A_i = \psi_{R_{i+1}}(A_{i+1}), R_i = \phi_{R_{i+1}}(A_{i+1}).$$

From that the existence of infinitely many inequivalent one-place propositional operations of the Heyting calculus follows immediately, and in fact, without using your (much more profound) result about the equivalence of topology and the Heyting calculus. One only needs the containment in one direction, which follows quite easily.[c]

Your KG

---

[c]⟦Translation of shorthand note at bottom:⟧ And in fact, with sole use of the fact that Heyting's axiomatization follows from the topology, which is quite easily provable, the much more profound equivalence between H and topology is ⟦?⟧. I think that should be mentioned in a footnote, because the theorem about the Heyting calculus is interesting in and of itself. My paper on Russell has still not appeared, although ⟦I received⟧ proof sheets four months ago.

4. Tarski to Gödel[a]

1001 Cragmont Ave.
Berkeley 8, Calif.
Dec. 10, 1946

Dear Kurt,

Your letter of Dec. 2 reached me yesterday.

---

[a]On letterhead of the University of California, Department of Mathematics, Berkeley 4, California.

I am so glad to hear that you are planning to talk at the Princeton Conference about the notions of absolute provability, definability, etc.; I hope to learn much from your remarks. As you see from the summary of my talk, my program is rather ambitious and probably too comprehensive; and you can easily realize that within a 30-minutes talk I shall merely be able to touch upon most of the problems involved. As regards the question in which you are interested, I don't think that I can do anything else but to emphasize the fundamental difference between all the undecidable statements known at present in elementary number theory ~~one~~ on the one hand and some undecidable statements (like ⟦the⟧ continuum hypothesis) in analysis and set theory; the statements of the first kind being clearly undecidable in a relative sense while those of the second seem to be undecidable in some absolute sense. And in this connection I shall raise the problems (1) whether ⟨and how⟩ the notions of relative and absolute undecidability can be made precise and (2) whether, on the basis of some adequate definition of these notions, it will be possible to show that a number-theoretical problem can be undecidable only in a relative sense. Perhaps I shall make some additional remarks | emphasizing the difficulty of these problems; and, of course, since now I know what you are planning to discuss, I shall refer to your subsequent remarks. Unfortunately, I cannot send you a copy of my talk since I have none and I doubt whether the text will be written down before my talk.

Mostowski wrote me of the following result which he found during the war and which may interest you. Consider a consistent ⟨axiom⟩ system S of set theory; Consider also a property P, say, of binary relations. The property P is called absolute if it does not depend on the model of S to which the notions 'set', 'class', etc. are related; ⟨thus, every property expressible in the lower predicate calculus is absolute.⟩ All the binary relations R between integers can of course be enumerated by means of real numbers; let x_R be the number correlated with R. Now, it turns out that a necessary condition for P to be absolute is that the set of all x_R for which R has the property P be a Borelian set. If, e.g., P is the property of being a well-ordering relation, then the set of x_R is not Borelian, and hence P is not absolute. Mostowski doubts whether the condition is also sufficient.

I am so glad that in a few days I shall see you and Adele again. I hope to spend much time with you during these few days. But it would make no sense to bother you with my staying overnight since the Princeton University offers its hospitality and reserves rooms for the outside members of the conference in Berkeley Inn. I don't have to add that your invitation was in itself something very pleasant.

Best greetings to both of you and "auf Wiedersehen"

Alfred

P. S. I am leaving Berkeley tomorrow by plane, since I have promised to give a talk in Minneapolis. My address on Dec. 12–15 will be: c\o Herbert Feigl, 5601 Dupont Ave So. Minneapolis 9, Minn.

5. Gödel to Tarski[a]

ca. Aug. 61

Dear Alfred,

I have heard it has been proved that there is no two valued denumerably additive measure for the first inacc. number. I still can't believe that this is true ~~at~~, but don't have the time to check it because I am working mainly on phil. I understand the proof is based on some work of yours? You probably have heard ~~about~~ of Scott's beautiful result that $V \neq L$ follows from the existence of any such measure for any set. I have not checked this proof either | but the result does *not* surprise me. 2

~~We were very s~~ Adele & I were very sorry you could not come to see us in Pr when you were in New York. Adele is in Italy at the present time & won't come back before the middle of Sep.

Sincerely yours —

⟦Gödel's letter to Tarski concerning his submission of **1970a* and **1970b* for publication in the *Proceedings of the National Academy of Sciences* was published as **1970c* in these *Works*, vol. III, pp. 424–425.⟧

[a] A note in shorthand beside the salutation reads: wahrscheinlich nicht geschickt (probably not sent).

Hans Thirring

Hans Thirring (1888–1976) was one of the professors with whom Gödel studied physics during his early years at the University of Vienna. Apart from the exchange reproduced here no other correspondence between the two men is known. Evidently, though, despite their lack

1. Thirring to Gödel[a]

Wien, 31. Mai 1972

Lieber Kollege Gödel!

Ich bin jetzt 84 Jahre und seit vier Jahren leide ich an Aphasie und bin sehr sprachbehindert. Ich war sechsmal in USA und das letzte Mal im Jahr 1968 fuhr ich per Schiff die Südroute und erlitt einen Gehirnschlag und seit damals kann ich nicht mehr richtig sprechen und auch schlecht schreiben. Ich bin aber absolut nicht gelähmt und auch meine Gehirntätigkeit ist vollkommen normal.

Ich erinnere mich deutlich an Ihren Besuch im Jahr 1940 als Sie sich vor Ihrer abenteuerlichen Flucht vor Hitler über Rußland und China von mir verabschiedeten und schließlich in USA landeten. Ich habe Sie damals auf den Artikel in den Naturwissenschaften vom 9. Juni 1939, Bd. 27,
4 S. ~~1~~02 von Flügge,[b] einem Mitarbeiter Hahns, aufmerksam gemacht, dessen Titel lautete: “Kann der Energieinhalt technisch nutzbar gemacht werden.” Ich bat Sie damals sich mit Einstein in Verbindung zu setzen und ihn dringend zu bitten, Präsident Roosevelt auf die große Gefahr aufmerksam zu machen, die entstehen würde, wenn Hitler-Deutschland die Bombe früher fertig hätte als USA. Ich sprach damals mit drei Leuten, die aus Deutschland emigrierten und bat sie um dasselbe. Ich

[a]On letterhead reading: Prof. Hans Thirring
Wien
IX., Boltzmanngasse 5
Tel. 34 22 48

of contact following Gödel's emigration, they continued to regard each other with esteem.

Because Einstein's famous letter to Roosevelt was sent well before Gödel left Austria, the question to which Thirring so urgently sought an answer from Gödel is of little historical moment. Gödel's reply, however, is of interest for the Spenglerian view of history it implicitly embraces.

John W. Dawson, Jr.

---

1. Thirring to Gödel[a]

Vienna, 31 May 1972

Dear colleague Godel!

I am now 84 years old and for the past four years have been suffering from aphasia and have had great difficulty speaking. I was in the U.S.A. six times, and the last time, in the year 1968, I was travelling by ship via the southern route and suffered a stroke. Since then I can no longer speak properly, and I also write badly. However, I am not at all crippled and my mental activity is completely normal.

I remember distinctly your visit in the year 1940, as you bade good-bye to me before your adventurous escape from Hitler across Russia and China and landed finally in the U.S.A. At that time I drew your atten-
4 tion to the article by Flügge,[b] one of Hahn's co-workers, on p. ~~1~~02, vol. 27 of Naturwissenschaten (9 June 1939), whose title read: "Can nuclear energy be made technologically usable?" I asked you then to get in contact with Einstein and urge him to make President Roosevelt aware of the great danger that would arise if Hitler's Germany had the bomb ready earlier than the U.S.A. I spoke at that time with three people who were emigrating from Germany and asked them ⟦to do⟧ the same.

---

[b] *Flügge 1939.*

weiß nicht, ob einer dieser Männer jemals mit Einstein darüber gesprochen hat. Diese Kollegen waren Prof. Karl Prizibram, der nach Belgien ging, Henry Hausner fuhr nach New York und Sie flohen nach China. Es kam mir im Jahr 1940 sehr wichtig vor, Einstein zu informieren. Es ist bekannt, daß Einstein mit Szil/ard im Jahr 1942 bei Präsident Roosevelt war und dieser sofort alle Mittel zur Verfügung stellte und aus aller Welt die besten verfügbaren Physiker nach Amerika holte.

2 | Wie gut befreundet ich mit Albert Einstein war, ersehen Sie daraus, daß er meinem Sohn Walter als er Einstein in Princeton besuchte eine Photographie für mich mitgab mit folgender Widmung: "Meinem guten alten Freund und Gesinnungsgenossen Hans Thirring.

Albert Einstein, 1953."

Das war das letzte Lebenszeichen, das ich persönlich von unserem großem Kollegen erhielt, denn ich Jahr 1958 als ich das erstemal nach dem Krieg in USA war, weilte er nicht mehr unter den Lebenden. Ich möchte nun sehr gerne wissen, ob Sie Einstein damals auf meine Befürchtungen aufmerksam machten.

Ich nahm an fast allen Pugwash-Konferenzen teil und es wurde das Einstein–Russell-Manifest ausgearbeitet. Ei⟨n⟩er der d/e/r/ ersten Wissenschafter, der damit einverstanden war, war Albert Einstein, welcher zwei Tage vor seinem Tod das Manifest unterschrieb. Fast alle prominenten Wissenschafter unterschrieben das Manifest und es wurde, wie Sie ja wissen werden, bei der Presse Konferenz in London am 9. Juli 1955 von Bertrand Russell vorgelesen.

Mit diesen Gedanken beschäftige ich mic/h/ noch immer und ich liefere auch jedes Jahr einen Beitrag zu der Konferenz; heuer findet eine große Pugwash-Konferenz im September in Oxford statt, an der ich leider aus gesundheitlichen Gründen nicht teilnehmen kann. h

Das Wesentliche, das ich von Ihnen wissen wollte, was geschah von 1940–1942.

Mit vielen herzlichen Grüßen

Ihr alter

Hans Thirring

I do not know whether one of those men ever spoke with Einstein about the matter. Those colleagues were Prof. Karl Prizibram, who was going to Belgium; Henry Hausner, ⟦who⟧ was travelling to New York; and you, who were fleeing to China. In the year 1940 it seemed to me very important that Einstein be informed. It is well known that in the year 1942 Einstein and Szilard met with Roosevelt, who immediately put all means at ⟦their⟧ disposal and brought the best available physicists from all over the world to America.

How good a friend I was of Albert Einstein you will learn from the fact that when my son Walter visited him in Princeton Einstein gave him a photograph for me, accompanied by the following dedication: "To my good old friend and soulmate Hans Thirring.

Albert Einstein, 1953."

That was the last token of ⟦his⟧ existence that I personally received from our great colleague, for in the year 1958, the first time after the war that I was in the U.S.A., he was no longer living. I would now very much like to know whether you made Einstein aware of my fears at the time. I took part in almost all the Pugwash Conferences, ⟦at which⟧ the Einstein–Russell manifesto was drawn up. One of the first scientists who expressed agreement with it was Albert Einstein, who signed the manifesto two days before his death. Almost all prominent scientists signed the manifesto, and it was, as I'm sure you know, read aloud by Bertrand Russell at the press conference in London on 9 July 1955.

I still busy myself with those ideas, and every year I also furnish a contribution to the conference; this year a large Pugwash Conference is scheduled for September in Oxford, in which, for reasons of health, I unfortunately cannot take part.

The essential thing that I want to know from you ⟦is⟧, what happened from 1940 to 1942?

With many heartfelt greetings,

Your old ⟦friend⟧,

Hans Thirring

2. Gödel to Thirring

Princeton, 27/VI. 1972.

Lieber Professor Thirring!

Es tat mir sehr leid, aus Ihrem Briefe zu erfahren, dass Sie einen Schlaganfall hatten, der einen dauernd Schaden verursachte. Ich wünsche Ihnen das best-mögliche für die weitere Entwicklung Ihres Zustandes.

Was Ihre Frage betrifft, so erinnere ich mich nur daran, dass ich Einstein Grüsse von Ihnen überbrachte. Ich hatte damals seit etwa 10 Jahren jeden Kontakt mit der Physik u. mit Physikern verloren. Ich wusste nicht, dass an der Herstellung einer Kettenreaktion gearbeitet wurde. Als ich später von diesen Dingen hörte, war ich sehr skeptisch, nicht aus physikalischen, sondern aus soziologischen Gründen, weil ich glaubte, dass diese Entwicklung erst gegen das Ende unserer Kulturperiode erfolgen wird, die vermutlich noch in ferner Zukunft liegt. In den Jahren 1940 u. 1941 sah ich Einstein *sehr* selten. | Von 1942 an sah ich ihn sehr oft. Aber es ist sehr unwahrscheinlich, dass das irgend etwas mit seiner Tätigkeit oder der Atombombe zu tun hatte, da eine hinreichende andere Erklärung dafür vorhanden ist. Über die Atombombe sprachen wir vor ihrer Fertigstellung nur ein einziges mal u. ganz kurz. Leider kann ich Ihnen weiter nichts über die Sie interessierende Frage berichten.

Mit herzlichen Grüssen u. besten Wünschen

Ihr Kurt Gödel

2. Gödel to Thirring

Princeton, 27 June 1972

Dear Professor Thirring,

I was very sorry to learn from your letter that you had suffered a stroke which caused lasting damage. I wish you the best possible for the further improvement of your condition.

As concerns your question, I only recall that I conveyed greetings from you to Einstein. At that time I had had no contact with physics and physicists for some 10 years. I didn't know that work was going on toward the production of a chain reaction. When I later heard of such things I was very skeptical, not on physical, but rather on sociological grounds, because I thought that that development would only happen toward the end of our culture period, which presumably still lay in the distant future. In the years 1940 and 1941 I saw Einstein very seldom. From 1942 on I saw him quite often. But it is very unlikely that that had anything to do with his action or with the atom bomb, since another sufficient explanation for that is at hand. We spoke about the atom bomb only a single time prior to its completion, and only quite briefly.

Unfortunately I can report nothing further to you about the question that interests you.

With heartfelt greetings and best wishes,

Yours, Kurt Godel

Stanisław Ulam

Stanisław Ulam

Stanisław Marcin Ulam (1909–1984) was a brilliant, pioneering and eclectic mathematician, very much the fox rather than the hedgehog. In his early years in Poland he provided a number of seminal results in the budding fields of set theory, topology, measure theory and functional analysis. Established at Los Alamos in 1944, he soon made a crucial contribution to the development of the hydrogen bomb and, with the development of computers, inspired and named the Monte Carlo method of statistical sampling. In later years he made important contributions to the study of fluid dynamics and non-linear and ergodic systems.[a]

The Ulam–Gödel correspondence is sparse and sporadic, with Ulam raising mathematical issues or making inquiries and Gödel writing always in reply. The correspondence does cast some light, albeit from the periphery, on several issues in the historical development of set theory in its formative period.

The first several letters have as a common thread possibilities for having measures on collections of sets. For present purposes, a *measure* on a collection $\mathcal{F}$ is a function $m: \mathcal{F} \longrightarrow [0,1]$, the unit interval of reals, such that m is not identically zero, $m(\{x\}) = 0$ for any singletons $\{x\} \in \mathcal{F}$, and m is countably additive, i.e. for pairwise disjoint $\{X_n \mid n \in \omega\} \subseteq \mathcal{F}$, $m(\bigcup_n X_n) = \sum_n m(X_n)$. In the fundamental paper *Ulam 1930*, drawn from his dissertation, Ulam had investigated the possibility of having a measure on the full power set $\mathcal{P}(X)$ of some set X, and the concept of a *measurable cardinal*, the most important of all concepts in the theory of large cardinals in set theory, emanates directly from this paper. A cardinal κ is *measurable iff* there is a measure on $\mathcal{P}(\kappa)$ with range $\{0,1\}$, i.e. the measure is two-valued.

In letter 1, the initial letter of 16 January 1939, and with more detail in letter 2, the succeeding letter of three years later, Ulam refers to their "joint remark" that it is not possible to find a measure on the collection of projective sets of reals. The reconstruction is straightforward: *Ulam 1930* established that there is no measure on the power set $\mathcal{P}(\aleph_1)$ by using a doubly indexed system of sets now known as an *Ulam matrix.* Gödel established that the restriction of the canonical well-ordering of L to its reals is a projective, in fact $\mathbf{\Sigma}^1_2$, well-ordering. He derived from this that if $V = L$, then there is a projective, in fact $\mathbf{\Delta}^1_2$, set of reals that is not Lebesgue measurable. Using Gödel's projective well-ordering one can in fact generate an Ulam matrix consisting of projective sets

[a] *Ulam 1974* is a collection of selected papers; *Ulam 1976* is an autobiography; and *Cooper 1989* is a festschrift.

and thereby draw the following conclusion generalizing Gödel's non-Lebesgue-measurability result: If $V = L$, then there is no measure at all on the collection of projective sets.[b]

Ulam in letter 1 writes that he thinks that their joint remark makes it "imperative to change the problem of measure!" and asks whether the remark is worth publishing. In letter 2 Ulam inquires "whether you ever published your results on the non-measurable projective sets or our joint remark." In letter 3, his undated response, Gödel somewhat cryptically says no. In fact, only in a footnote of his later, 1951 edition of *Gödel 1940*, his monograph on L, did he sketch his result that if $V = L$, then there is a Σ^1_2 well-ordering of the reals.[c] In any case, Ulam's inquiries tend to undermine Kreisel's eye-catching remark in his memoir on Gödel that "...according to Gödel's notes, not he, but S. Ulam, steeped in the Polish tradition of descriptive set theory, noticed that the definition of the well-ordering ... of subsets of ω was so simple that it supplied a non-measurable PCA [i.e. Σ^1_2] set of real numbers ..."[d]

Ulam's letter of 14 August 1942, letter 2, actually begins with an inquiry about an idea for establishing that the existence of a non-Lebesgue-measurable projective set of reals already follows from the continuum hypothesis (CH), or possibly from the weaker hypothesis that the power of the continuum is less than the least weakly inaccessible cardinal.[e] Ulam includes a sketch (Gödel *Nachlaß* document 012880) of his idea, which turns on another use of Ulam matrices in a projective setting. If successful, the idea would replace Gödel's $V = L$ assumption for getting a non-Lebesgue-measurable projective set simply by a cardinal hypothesis. In his reply, letter 3, Gödel points out a problem already at the second step of the sketch. He also points out that because Ulam's proposed construction is effective, if successful, it would establish the existence of a non-Lebesgue-measurable projective set without appealing to the axiom of choice, "which is unlikely." On the margin he writes "I have a feeling that the measurability of all proj. sets is consist. log. with the ax. of choice."

Ulam later published the idea sketched in 012880 in his collection of problems, *Ulam 1960*, p. 17ff. With the advent of forcing, Gödel's "feeling" turned out to be vindicated: *Solovay 1970* established that if ZFC + "there is an inaccessible cardinal" is consistent, then so is ZFC

[b]For the projective sets of reals and their properties in L see Solovay's note in these *Works*, vol. II, p. 13ff., or *Kanamori 1997*, §§12 and 13. See also the latter, §2, p. 24, for the Ulam matrices argument.

[c]See these *Works*, vol. II, p. 33.

[d]*Kreisel 1980*, p. 197.

[e]A *weakly inaccessible* cardinal is an uncountable regular limit cardinal.

+ CH + "all projective sets are Lebesgue measurable". Moreover, CH here can be replaced by $2^{\aleph_0} = \aleph_\alpha$ for a wide range of α's.[f]

Ulam's letter of 14 August 1942 is significant in other respects. In it is also a remarkable anticipation of effective descriptive set theory. Alluding to some remarks of Nikolai Luzin about "really" constructive Borel sets, Ulam formulates topological concepts in terms of the recently developed mathematical concept of recursive set of integers. This is a very early example of the formalization of effectiveness in terms of recursiveness. Ulam describes the open, closed and G_δ sets of reals with recursive codes and notes that one can proceed into the finite Borel classes and "even some transfinite ordinal classes." He then inquires whether, as with the classical Borel hierarchy, new sets appear at each level of the effective version and also of the effective versions, analogously defined, of the projective hierarchy. In effect, Ulam is asking whether what are now known as the "lightface" arithmetical and analytical hierarchies are proper.[g]

Gödel in letter 3 astutely points out, in effect, that the universal sets of classical descriptive set theory used to establish the properness of hierarchies are "lightface"—the implicit reply to Ulam's question is therefore in the affirmative. This exhibits a comparative expertise in descriptive set theory on Gödel's part, a circumstance that further militates against Kreisel's remark quoted above.

The actual mathematical development would proceed in reverse order: The arithmetical hierarchy was developed in *Kleene 1943* and the analytical hierarchy in *Kleene 1955*, *1955a* and *1955b* in terms of formula complexity and shown to be proper. The connection through effective topology with the hierarchies of classical descriptive set theory was only made afterwards in *Addison 1958*.

Finally, Ulam in letter 2 discusses incorporating projection as an operation into Boolean algebras involving Cartesian products, as a way of treating quantification on an algebraic basis. This would be developed in his study of *projective algebras* with his friend and collaborator C. J. Everett in *Everett and Ulam 1945*.[h] Gödel in his undated reply sagaciously makes a connection to the work of Ernst Schröder and moreover raises "the problem of decision". This is an early speculation for algebraic theories; the theory of Boolean algebras itself was shown to be

[f]See *Kanamori 1997*, §11 for an exposition of Solovay's result. The hypothesis that there is an inaccessible cardinal is nowadays considered a very mild assumption. In any case, *Shelah 1984* established that it is necessary in Solovay's result, by showing that if ZFC + "every $\mathbf{\Delta}^1_3$ set of reals is Lebesgue measurable" is consistent, then so is ZFC + "there is an inaccessible cardinal".

[g]See *Kanamori 1997*, §12 for effective descriptive set theory.

[h]Projective algebras are precursors of the better known *cylindric algebras* of *Henkin, Monk and Tarski 1971* and *1985*.

decidable by *Tarski 1949a*.[i] Overall, Gödel's reply is remarkable for addressing each of Ulam's concerns by getting to the heart of the matter with an economy of words.

Ulam in letter 4, dated 6 December 1947, revisits his analysis of measure in *Ulam 1930* in connection with the power of the continuum. He first praises Gödel's recent article on the continuum problem, *Gödel 1947*; notes that he (Ulam) too believes in the falsity of CH in the sense that new axioms will be added that, e.g,, will entail that $2^{\aleph_0}$ is the least weakly inaccessible cardinal; and recalls discussing possibilities of such axioms with Gödel in 1938 and later.

Ulam then proceeds to sketch an approach via measure. He first notes that he had shown in *Ulam 1930* that for no cardinal λ less than the least weakly inaccessible cardinal is there a measure on the power set $\mathcal{P}(\lambda)$. He then writes "I believe that it is possible" for the least weakly inaccessible cardinal κ that there *is* a measure on $\mathcal{P}(\kappa)$. For this he envisions a transfinite construction procedure similar to Gödel's for his L and suggests the likelihood of postulating an axiom asserting the existence of more and more sets of reals "definable by transfinite induction," entailing that $2^{\aleph_0}$ is the least weakly inaccessible cardinal.[j] Ulam concludes his letter with speculations about certain collections of sets of reals and a possible axiom about such collections never being the full power set $\mathcal{P}(\mathbb{R})$.

Whether there could be a measure on $\mathcal{P}(\kappa)$ for κ the least weakly inaccessible cardinal would remain an open problem for several decades after *Ulam 1930* until Solovay in 1966 (see *Solovay 1971*) established that this is false in a strong sense.[k] Ulam's letter of 25 October 1957, letter 5,[l] is part of an interchange, most of it unpublished here, regarding his memoir on his recently deceased friend, John von Neumann. Ulam writes to Gödel somewhat flippantly: "I have always regarded as the principal value of von Neumann's articles on axiomatization, the fact that you used his system or a similar one for some of your work!" Gödel

[i]Tarski does state at the end of the abstract that its results were obtained in 1940.

[j]Note that this measure approach to the size of the continuum complements and yet is consistent with his speculations at the beginning of letter 2 about the existence of non-Lebesgue-measurable sets of reals under the assumption that $2^{\aleph_0}$ is less than the least weakly inaccessible cardinal.

[k]The least cardinal κ for which there is a measure on $\mathcal{P}(\kappa)$ is the least "real-valued measurable cardinal", and Solovay established just from a "saturated ideal" property of such a cardinal κ that it must be the κth weakly inaccessible cardinal and a fixed point in the sequence of cardinals in various strong senses. See *Kanamori 1997*, §§2 and 16, particularly 16.8.

[l]Archives of the American Philosophical Society.

in his reply of 8 November 1957, letter 6,[m] warmly demurs and proceeds to make several remarks of considerable import on von Neumann's work on the axiomatization of set theory. First, Gödel considers that von Neumann's "necessary and sufficient condition which a property must satisfy, in order to define a set, is of great interest because it clarifies the relationship of axiomatic set theory to the paradoxes." Presumably, Gödel is referring to Axiom IV2 in *von Neumann 1925*, to the effect that a class is a set exactly when there is no surjection from that class onto the universe of sets. Gödel continues, "That this condition really gets at the essence of things is seen from the fact that it implies the axiom of choice, which formerly stood quite apart from other existential principles."[n] Gödel regards von Neumann's inferences as "not only very elegant, but also very interesting from the logical point of view." He continues, "Moreover I believe that only by going farther in *this* direction, i.e., in the direction opposite to constructivism, will the basic problems of abstract set theory be solved."

Ulam in letter 8, dated 6 January 1963, asks Gödel of his opinion of Paul Cohen's recent work on the independence of the continuum hypothesis, which introduced the forcing method. In a postscript, Ulam notes that he had in print expressed how he shared Gödel's belief in the falsity of CH, and inquires whether properties of sets of power $\aleph_1$ "seem now to be perhaps more interesting."

Gödel in his response, letter 9 dated 2 February 196[4], writes significantly that Cohen's work is a milestone "since it introduces for the first time a general method for independence proofs." He continues, "This is indispensable in set theory because the present axioms simply do not determine some of the most important properties of sets." These remarks gain more significance when it is pointed out that Gödel around 1942 had partial results toward the independence of the Axiom of Choice but neither pursued nor published this work.[o]

Gödel then voices agreement with Ulam about properties of sets of power $\aleph_1$ having become much more interesting, alludes to Hausdorff's "Pantachie Problem," and proceeds to describe it. This merits some discussion.

[m]Archives of the American Philosophical Society.

[n]Von Neumann's Axiom IV2 implies that there is a surjection F from the class ON of ordinals onto the class V of sets. But then, F induces a well-ordering $\prec$ of V defined by: $x \prec y$ exactly when the least member of the preimage $F^{-1}(\{x\})$ is less than the least member of the preimage $F^{-1}(\{y\})$. Hence, the Axiom of Choice in a strong, global sense holds. Also, by considering whether a class, once well-ordered by $\prec$, is bijective with ON or not, one sees that Axiom IV2 admits the following self-refinement: A class is a set exactly when there is no bijection from that class onto the universe of sets. This is how von Neumannn's approach to classes is most often characterized.

[o]For Gödel's partial results see *Moore 1988*, p. 130ff. and the correspondence with Rautenberg in this volume.

Most of *Hausdorff 1907* is devoted to the analysis of pantachies, and the last subsection is entitled "The Pantachie Problem." The term 'pantachie' derives from its initial use by *Du Bois Reymond 1880* to denote (everywhere) dense subsets of the real line and then to various notions connected with his work on rates of growth of real-valued functions and on infinitesimals. Discussing subsets of ${}^{\omega}\mathbb{R}$, the collection of (countable) sequences of reals, Hausdorff argued (p. 107) that "an infinite pantachie in the sense of Du Bois Reymond does not exist." Hausdorff then went on to redefine 'pantachie' as a subset of ${}^{\omega}\mathbb{R}$ maximal with respect to being linearly ordered under the eventual dominance ordering. (For $f, g \in {}^{\omega}\mathbb{R}$, f *eventually dominates* g *iff* $\exists m \in \omega \forall n \in \omega (m \leq n \to g(n) < f(n))$.) This anticipated Hausdorff's later work on maximal principles, principles equivalent to the Axiom of Choice. For an ordered set $\langle X, < \rangle$, an (ω_1, ω_1^*)-*gap* is a set $\{x_\alpha \mid \alpha < \omega_1\} \cup \{y_\alpha \mid \alpha < \omega_1\} \subseteq X$ such that $x_\alpha < x_\beta < y_\beta < y_\alpha$ for $\alpha < \beta < \omega_1$, yet there is no $z \in X$ such that $x_\alpha < z < y_\alpha$ for $\alpha < \omega_1$.

Hausdorff's subsection "The Pantachie Problem" lists several problems, the first of which (p. 151) is whether there is a pantachie (in his sense) with no (ω_1, ω_1^*)-gaps. What Gödel describes in letter 8 as "what Hausdorff calls the 'Pantachie Problem' " is actually the following variant of Hausdorff's "Scale Problem" (p. 152): whether there is a subset of ${}^{\omega}\omega$, the collection of (countable) sequences of natural numbers, cofinal in the eventual dominance ordering and of ordertype ω_1. This problem would be among the first adequately analyzed by the emerging method of forcing: Solovay observed in the mid-1960s that it is consistent to have such an ω_1 cofinal subset of ${}^{\omega}\omega$ with $2^{\aleph_0}$ arbitrarily large.[p] Then Stephen Hechler in his dissertation of 1967 (see *Hechler 1974*) would prove that it is consistent that there is no such ω_1 cofinal subset of ${}^{\omega}\omega$.[q]

It is a testament to Gödel's continuity of thought that in 1970 he would frame axioms (see these *Works*, vol. III, p. 405ff.) about the existence of certain collections of functions linearly ordered by eventual dominance, axioms that were intended to imply that CH is false in a specific way. Among his postulations was a pantachie with no (ω_1, ω_1^*)-gaps, the focus of Hausdorff's "The Pantachie Problem". Solovay in

[p]In modern parlance, Solovay had observed that the forcing for adjoining any number of random reals is ${}^{\omega}\omega$ bounding, i.e. any member of ${}^{\omega}\omega$ in the generic extension is eventually dominated by a member of ${}^{\omega}\omega$ in the ground model. The argument is given by Solovay in these *Works*, vol. III, §6.5 on p. 414.

[q]Hechler introduced a notion of forcing for adjoining a function in ${}^{\omega}\omega$ eventually dominating all ground model functions in ${}^{\omega}\omega$. With repeated applications he was able to establish the general assertion that if, in the sense of the ground model, κ and λ are cardinals of uncountable cofinality such that $2^{\aleph_0} \leq \kappa$ and $\lambda \leq \kappa$, then there is a cardinal-preserving generic extension in which $2^{\aleph_0} = \kappa$ and there is cofinal subset of ${}^{\omega}\omega$ of ordertype λ.

his commentary in these *Works*, vol. III, p. 409, provides in effect a proof that the existence of such a pantachie implies $2^{\aleph_0} = 2^{\aleph_1}$. This result actually appears in *Hausdorff 1907* (Theorem V, p. 128), which was presumably the route to Gödel's invocation (p. 421) of $2^{\aleph_0} \geq \aleph_2$.

Returning to the correspondence, Ulam in his final letter of significance, letter 10 (dated 17 February 1966), discusses possibilities for adapting notions like Borel and projective set to subsets of a set E of power $\aleph_1$. This is in the spirit of his question, raised two years earlier in letter 8 (6 January 196[4]), of interesting properties of sets of power $\aleph_1$ independent of CH.

In retrospect, what is remarkable about this Ulam–Gödel correspondence is the lack of any mention at its end of two major developments in set theory in the 1960s, developments having to do with both Ulam's fruitful speculations from the 1930s and Gödel's interest in new axioms for set theory. The first development is Dana Scott's 1960 result (see *Scott 1961*) that if there is a measurable cardinal, then $V \neq L$, and the result of William Hanf and Alfred Tarski around the same time (see *Hanf 1964* and *Tarski 1962*) that a measurable cardinal κ is "much larger" than the least strongly inaccessible cardinal.[r] Measurable cardinals, as was written above, had emanated directly from Ulam's work on measure in *Ulam 1930*, and Gödel pointed out Scott's result in *Scott 1961* and referred to *Tarski 1962* in footnote 20 of his 1964 revision of his article *Gödel 1947* on the continuum problem (see these *Works*, vol. II, pp. 260–261). Moreover, measures had figured prominently in the early correspondence, and Solovay in 1966 (see *Solovay 1971*) had established the equiconsistency of the following two theories: (a) ZFC + "there is a measurable cardinal", and (b) ZFC + "there is a measure on $\mathcal{P}(\mathbb{R})$".[s] The second development left unmentioned in the correspondence is the investigation of determinacy, which was getting into full swing in the 1960s. The determinacy of infinite games is now a mainstream of set theory that brings together both descriptive set theory and large cardinals in the analysis of sets of reals.[t] Ulam had essentially asked in the 1930s, in the famous book of problems kept at the Scottish Cafe in Lwów (see *Mauldin 1981*, p. 113ff.), when a specific player has a winning strategy in a certain two-player infinite game. The related question of when such a game is *determined*, i.e. when there is a winning strategy for one or the other player, would become the focus of subsequent

[r]A *strongly inaccessible* cardinal is an uncountable regular cardinal κ such that whenever $\lambda < \kappa$, then $2^\lambda < \kappa$. The Hanf–Tarski argument shows that a measurable cardinal κ is in fact the κth strongly inaccessible cardinal and a fixed point in the cardinal sequence in various strong senses.

[s]See *Kanamori 1997*, §16.

[t]See *Kanamori 1997*, chapter 6.

investigations. Ulam's compatriots Jan Mycielski and Hugo Steinhaus proposed in *Mycielski and Steinhaus 1962* what is now known as the Axiom of Determinacy, and Mycielski himself would become a close colleague of Ulam's, but perhaps not until after Mycielski emigrated to Boulder, Colorado in 1969.[u] With infinite games becoming a rich paradigm for the articulation of dichotomies across the breadth of set theory, it is intriguing to consider what Gödel would have made of the developments in this direction and how he would have regarded determinacy axioms for set theory.

Akihiro Kanamori

A complete calendar of the correspondence with Ulam appears on p. 460 of this volume.

[u]It is noteworthy that in his contribution to the Ulam festschrift *Cooper 1989*, Mycielski does not mention determinacy.

1. Ulam to Gödel

Cambridge
Jan. 16, 1939

Dear Dr. Gödel,

I would be very grateful to you for a reprint of your paper in the Proceedings.[a] Will your full paper appear soon? If possible, please send me a copy of your Princeton lectures.[b]—I think the remark we made in Williamsburg: non-existence of any completely additive measure in the class of projective sets makes it imperative to change the problem of measure! Do you think it is worth publishing?—In connexion with a possibility of generalizing the problem of measure this question occurred to me: Let us consider the Boolean algebra of all subsets of the interval $(0,1)$ modulo countable sets. Can this algebra which has 2^c elements be mapped homomorphically on an algebra with only c elements?

[a]*Gödel 1938.*

[b]*Gödel 1940.*

I intend to write you soon on some subjects which we discussed in Williamsburg. Hoping that this letter reaches you (please send me your address in Notre-Dame) and expecting to hear from you soon I am

Yours sincerely

S. Ulam

2. Ulam to Gödel[a]

August 14, 1942

Dear Gödel:

I should like very much to have your opinion about an idea which seems to me to hold out some promise of establishing the existence of non-measureable (Lebesgue) projective sets from the assumption of the continuum hypothesis or even, perhaps, a weaker hypothesis—namely, that the power of the continuum is smaller than that of the first inaccessible aleph. It may be distin⁄quished from your result in that it would hold, presumably without the independent axiom of yours about constructible sets. I enclose a few pages where this possibility is sketched.

Also, I wanted to ask you whether you ever published your results on the non-measurable projective sets or our joint remark that from the construction used by me in Fundamenta Mathematica⟦e⟧, vol. 16,[b] it would follow, through the use of your method, that it is not possible to find any measure at all for the projective sets that would be zero for sets consisting of single points, countably additive and not identically zero.

Also, I should appreciate very much your comment on the following question: Luzin wrote in several papers and in his book rather vaguely

[a]On letterhead of the Department of Mathematics, 306 North Hall, The University of Wisconsin, Madison. Gödel appears to have numbered the first three paragraphs of this letter.

[b]*Ulam 1930.*

about "really" constructive Borel sets.[c] He formulated the problem of existence of such sets of the fourth class or higher classes. (Zero-class are open and closed sets. First-class, F_σ and G_δ sets, etc.) His lack of precision in formulating what constitutes the "really" constructive sets makes his problems mathematically meaningless. It seems to me that one way to define his notion is to try to define recursive sets of points on the line or recursive functions of the real variable as follows: Let us order, once for all, the sequence of intervals with rational endpoints $\omega_1, \omega_2, \ldots,$ $\omega_n \ldots$. By a recursive open set we shall mean a set which can be represented in the form $Z = \sum_{k=1}^{\infty} \omega_{n_k}$ where the sequence of integers $n_1, n_2 \ldots$ $n_k \ldots$ is recursive in the accepted sense. A recursive closed set would be a complement of a recursive open set. A recursive G_δ-set would be one that can be written as $Z = \prod_l \sum_k \omega_{n_{k,l}}$, where the double sequence of integers $n_{k,l}$ is recursive. One can proceed in this fashion and define Borel sets that are recursive for any finite class and even some transfinite ordinal classes. Of course, one can start with various definitions of recursiveness for sequences or multiple sequences of integers.

The first problem that arises is this: Do there exist recursive Borel sets of the n-th ⟦class⟧ that do not belong to any lower class? An analogous question for recursive analytic or projective sets, if one defines a recursive analytical set as a projection on the line of a recursive plane G_δ-set. Do you have theorems in this direction? By the way, it seems that a possible definition of a recursive continuous (or more generally Baire) function of a real variable x would be this: For every rational interval the set $f^{-1}(\omega_n)$ is a recursive open (or Borel) set. [$f^{-1}(\omega_n)$ is the set of all x such that $f(x) \in \omega_n$.]

2 | I have studied the following situation which generalizes algebraically the idea of Boolean algebra and gives one posssibility, it seems to me, of a treatment of quantifiers on an algebraical basis: Suppose that we have a set E, its product with itself $E \times E = E^2$. (We single out a point p_0

---

[c]Luzin's book is *Luzin 1930.* In it Luzin discussed (pp. 89–104) the "constructive existence" of Borel sets in the first four classes. (The Borel sets are ramified into a hierarchy consisting of classes K_α for $\alpha < \omega_1$.) Luzin called (p. 89) 'canonical' those members of K_α given by "particularly simple" properties, stipulating that every member of K_α is to be a countable union of canonical members. In a footnote Luzin observed that the definition of 'canonical' was vague since when a property is "particularly simple" was not specified. Luzin went on to describe the known "arithmetical definitions" of sets in the first four classes. See also *Luzin 1930a* where Luzin states (p. 65) as a problem to demonstrate the "constructive existence" of a Baire function of class 5. The Baire hierarchy of real functions has a direct correlation with the Borel hierarchy of sets of reals.

in E and consider the subsets $p_0 \times E$ and $E \times p_0$ of E^2, as representing E itself.) Let us consider two given classes of subsets of E^2: $\mathfrak{A}$ ~~called~~ $\mathfrak{B}$ ~~and~~ . We shall call them projectively isomorphic if there exists a one-to-one transformation T of sets from $\mathfrak{A}$ to $\mathfrak{B}$ satisfying the following:

1. $T(A - B) = T(A) - T(B)$
2. $T(A \times B) = T(A) \times T(B)$
3. $T(\mathrm{Proj}_E(A)) = \mathrm{Proj}_E(T(A))$, (if all the sets on the left belong to the class $\mathfrak{A}$).

I have not completed any systematic investigation of such systems under the notion of projective isomorphism, but have noticed various facts about them. This set-up permits the making of a mathematical investigation of logic more completely, it seems to me, than the mere treatment through Boolean algebras. I studied in particular the question of closure of a finite number of sets under the Boolean operations and the operations of product and projection. The interesting feature of it is that one can obtain an infinite number of sets starting from only two sets given in E^2. Do you think that it is worthwhile undertaking a systematic investigation in this direction and if there is, perhaps, something in the logical literature with which I am not familiar, covering this already? As an example of theorems that I have obtained I may say that they arise from the fact that there exist systems which are isomorphic as Boolean algebras but not projectively isomorphic. The model on which I studied this situation is the case where E is a set of integers.—I enclose a few pages on which I sketched a possibility of obtaining constructive nonmeasurable sets.[d] For the understanding of it I may add that, as you know, it is sufficient to obtain $\aleph_1$ sets of measure zero ⟨whose sum would be the whole interval⟩ such that the system itself would be constructive, e.g. in the sense of Kuratowski's papers in Fundamenta M. vols. 27, 28, 29.[e] The property (w) ⟨used in the enclosed pages⟩ is the following: If a set Z contains a number x, it contains all numbers differing from it by a rational number.

I am extremely interested in knowing on what you are working now and to learn your results about the intuitionistic logic and the continuum hypothesis. Please write me if you find time.

---

[d] These pages are included in the *Nachlaß* as document number 012880. See also the introductory note.

[e] The relevant papers are *Kuratowski 1936, 1937, 1937a, 1937b* and *1937c*. *Kuratowski 1937* introduces the concept of an 'elementary' projective set, which is defined through relativization. In the document 012880 Ulam describes constructions similar to Kuratowski's, constructions that do not appeal to the Axiom of Choice. See also the introductory note.

With best regards and greetings,

Yours sincerely,

S. Ulam

3. Gödel to Ulam[a]

Dear Ulam

Excuse me please for not having ~~answered~~ ⟨replied earlier to⟩ your letter of Aug 14. The ~~first~~ sec. of your questions whether I publ. my results on non⟦-⟧meas. proj. set was I think answered by this very fact. As to your quest. about constr. ~~projective~~ ⟨Borel⟩ sets it seems to me that the univ. sets constr. in the usual manner are recursively proj. in your sense; ~~but~~ ⟨What⟩ Luzin ~~seems to have something else in~~ had in mind ~~is hard to see because I ⟨really⟩ don't know~~ when he considers the Baire-set ⟨of order 3⟩ as more constructive than the univ. sets. ~~I rea~~ is ~~hard to see~~ ⟨I don't know at all⟩ unless he meant something like "'"perspicuity" of the construction.—Your generalized Boole⟦an⟧ algebras, with the op. $\times$, proj would amount to a calculus of relations which is still more restricted | than Schröder⟦'⟧s [which already does not allow to express all statements of the first order.] It would be interesting whether the problem of decision could be solved for this system. As to your ⟨interesting⟩ sketch[b] of a proof for the exist. of non⟦-⟧meas. proj. sets I am sorry that already the sec. step is not clear to me. Furthermore there is no reason why the proj. character should be preserved in sums of $\aleph_1$ summands. Finally it seems to me that if the arg went through it would give ~~a pro~~ ~~the constr.~~ non⟦-⟧measurable sets without the ax. of choice which is unlikely. Or do you have a diff. opinion about this question. ~~As for me~~ ⟨On ~~all~~ these grounds⟩ I am ~~more than~~ ⟨extremely⟩ doubtful about your whole scheme of proof.

With best reg. & greetings,

Your ——

---

[a] Undated letter draft.

[b] Gödel is referring to *Nachlaß* document 012880, enclosed with letter 2 by Ulam.

⟦Written vertically in the right margin:⟧ and also bec. I have a feeling that the measurability of all proj. sets is consist. log. with the ax. of choice.

4. Ulam to Gödel

Dec. 6, 1947

Dear Gödel,

Have just read your wonderful article in the Monthly on the Continuum hypothesis.[a] Would be very grateful to you for a reprint.

As you know, I believe in the falsity of the cont. hypothesis in the same sense as you state it in your predictions. i.e. that new axioms, as unavoidable as the axiom of choice will be added— from which it will follow that c is, for example the first inaccessible aleph. (I think that we discussed such possibilities in 1938 and later.)

May I indicate some more grounds for this feeling?

In a paper in Fundamenta[b] ⟨...Theorie der Mass. in allg. Mengenlehre⟩ (I think vol. 16 ⟨*or 15*⟩) I proved the impossibility of an absolutely additive measure function for *all* subsets of a set of power $\aleph_1$, then the same, without any hypotheses for sets of power $\aleph_2 \ldots$ all alephs less than the first inaccessible one. Now I believe that it *is* possible to have such a measure for the first inaccessible aleph, call it $\aleph_i'$. The reason is that probably one can "construct" all its subsets, somewhat as you construct "all" subsets of the continuum in your book in the Princeton series;[c] *and at the same time* construct a measure of the sets as one goes along with transfinite procedure. This means, in particular, the *im*possibility of a class of decompositions of $\aleph_i'$ with the properties given in my paper in Fund. 16. Of this, I am fairly certain.

Now, in addition, it seems very likely to me that some axiom, asserting the possibility | of existence of "more and more" sets definable by transfinite induction can be postulated within the class of all subsets of the continuum, in which case it would be impossible to reject the conclusion that ⟦the⟧ continuum is $\aleph_i'$. ($\aleph_i^0 = \kappa^0$.) 2

---

[a] *Gödel 1947.*

[b] *Ulam 1930.*

[c] *Gödel 1940.*

At present I cannot find a sufficiently general and natural statement for an axiom of this sort.

I was writing you, for the last few months a longish letter on such set-theoretical questions but it is still unfinished.—Plan to be in Princeton towards the end of January and would like very much if I could see you and have a discussion on such problems.

As I think I mentioned to you once, I tried to examine properties of a class $\langle\mathcal{R}\rangle$ of sets (say of real numbers) with the properties:

(1.) $\mathcal{R}$ is a Borel field.

(2.) Given an arbitrary division of the interval I into disjoint sets, consisting of at least two elements each: $I = A_1 + A_2 + \ldots + A_\xi + \ldots$, then there exists in $\mathcal{R}$ a set having one element in common with *each* of the A_ξ: $(Z \varepsilon \mathcal{R},\ Z \cdot A_\xi = \text{one pt.})$.

The first question was whether $\mathcal{R}$ must be the class of *all* subsets of I—up to countable sets. Now the axiom I had in mind would assert the possibility of constructing classes of sets—which ~~do not have this~~ satisfy 1, 2 and are *not* the full class, in fact of course more generally—any non-contradictory set of properties—and different from the class of *all* subsets. Somehow the axiom should perhaps assert a *freedom of construction* beyond that given by the axiom of choice—this still *within* a given system of axioms and not as a "meta-system" statement.

Please send me a reprint then; would be very grateful to you if you let me know whether you will be in Princeton towards end of January.

With cordial greetings,

Yours as ever

S. Ulam

5. Ulam to Gödel

October 25, 1957

Dear Gödel:

Many thanks for your letter and the copy of your letter to Price.[a]

---

[a]G. B. Price, professor of mathematics at the University of Kansas, was managing editor for the *Bulletin* for volume 64.

It was my intention, of course, to discuss, in an article on von Neumann, his work on logic, foundations, and set theory. This will probably occupy several pages. It would be wonderful, of course, if you were willing to write a special article on it since it would be much more authoritative and, to mention just one thing, I have always regarded as the principal value of von Neumann's articles on axiomatization, the fact that you used his system or a similar one for some of your work!

I was in Princeton for half a day a week ago, and dropped into your office, but you were not there. I hope to be in Princeton some time this winter and would like very much to have a chance again to continue our discussions of this spring. When I know more precisely the date, I will let you know in advance.

Cordially yours,

S. Ulam

P.S. I would like to have the chance of showing you the manuscript of the article on von Neumann, when I have it finished—probably within the next two months.

6. Gödel to Ulam

Princeton, Nov. 8, 1957

Dear Ulam:

Thank you for your letter of Oct. 25. You are doing me too much honour by seeing the principal value of von Neumann's papers on axiomatics of set theory in the fact that I used his, or a similar, system of axioms! His results, it is true, are mathematically not very intricate, but nevertheless, in my opinion, quite important for the foundations of set theory. In particular I believe that his necessary and sufficient condition which a property must satisfy, in order to define a set, is of great interest, because it clarifies the relationship of axiomatic set theory to the paradoxes. That this condition really gets at the essence of things is seen from the fact that it implies the axiom of choice, which formerly stood quite apart from other existential principles. The inferences, bordering on~~n~~ the paradoxes, which are made possible by this way of looking at things, seem to me, not only very elegant, but also very interesting from the logical point of view. Moreover I believe that only by going farther in *this* direction, i.e., in the direction opposite to constructivism, will the basic problems of ~~set~~ abstract set theory be solved.

I don't know what von Neumann's attitude toward his work on the foundations of set theory was in his later years. Was he perhaps prejudiced against investigations of this kind because the consistency of the axioms cannot be proved?

Morgenstern told me that you may come to Princeton in the near future. It would be very nice if I could see you in that case.

Cordially yours,

Kurt Gödel

7. Gödel to Ulam[a]

Princeton, Jan. 28, 1958.

Dear Ulam:

I am sending you herewith the changes ⟨in⟩, and additions to, your report about v. Neumann's work on the the foundations of mathematics, which I would suggest. I also have reformulated my remark which you quote on p. 18. Incidentally I believe that this remark would better be quoted in connection with insertion 4 at the place marked by a star.

Concerning the relationship of v. Neumann's work to mine and Turing's I would like to add the following: ⟨To my knowledge⟩ the only passage in von Neumann's writings ~~I know of~~ which ⟨~~perhaps~~ ⟩ can be interpreted to be (as you rightly say) a "vague forecast" of the existence of undecidable propositions in any formal system (although certainly not of the unprovability of consistency) is the one you ~~quote~~ ⟨mention⟩ on p. 15. But this *one* ⟨doubtful⟩ passage (which, incidentally, has very little to do with the main purpose of the paper concerned) certainly does not justify what you say on p. 17, line 8 and line 17.

~~Of course~~ I do not want to deny that v. Neumann, by his virtuosity in using formalism,[1] ~~stimulated research [?]~~ [b] ~~clarify~~ ⟨~~on, and~~ contribu-

[1]and also by his general considerations about the actually occurring mathematical formalisms, although these form only a *very* small part of all possible formalisms

[a]Retained copy.

[b]The words here are illegible, but may say "and attempted to".

ted toward a clarification of,⟩ the concept and limits of ⟦?⟧ ⟨formalism⟩. However ~~I don't think~~ the words "anticipated" or "forerunner" are ~~suitable~~ ⟨much too strong⟩ for describing th~~is~~⟨e⟩ situation.[2]

Thank you for sending me your collection of problems. I should be delighted to see you, if you came to Princeton sometime in the near future.

Sincerely yours

Kurt Gödel

[2] "prepares the grounds", which you say on p. 17, line 18, is a much more suitable expression.

8. Ulam to Gödel[a]

January 6, 1963[b]

Dear Gödel:

Last August I had to give a lecture, the so-called von Neumann lecture, which is given annually. The subject was "Combinatorial Analysis in Infinite Sets and Some Physical Theories."[c] In the first part of it I discussed your results on the continuum hypothesis and I also alluded to the results of Cohen, which I heard about but I have not seen his proof. I still do not know the details of it.

What do you think of these results? It seems to me that yours and Cohen's work really initiates, so to say, "non-Cantorian" set theories—really in analogy to non-euclidean geometries.

I am now writing up my lecture—it should be ready in a few days and I will permit myself to send you a preprint.

With best regards and greetings and wishes for a Happy New Year,

Yours, as ever,

S. M. Ulam.

[a] On letterhead of University of California, Los Alamos Scientific Laboratory, P.O. Box 1663, Los Alamos, New Mexico.

[b] Internal evidence shows the year should be 1964.

[c] *Ulam 1964*.

P.S. In my little book, "A Collection of Mathematical Problems",[d] published four years ago, I have written, in the Introduction on page one, that your impression was that "in a suitably large and 'free' axiomatic system for set theory, the continuum hypothesis is false. This feeling, based on indications provided by results on projective sets and the abstract theory of measure, has been shared by the author for many years."[e] I am glad that this proved to be correct. In general, results on properties of sets of power, $\aleph_1$, etc., which used no hypothesis whatsoever, seem now to be perhaps more interesting?

S. M. U.

[d] *Ulam 1960.*

[e] This entire quote was set off by Gödel in brackets.

9. Gödel to Ulam

February 10, 1964

Professor S. M. Ulam
Los Alamos Scientific Laboratory
P. O. Box 1663
Los Alamos, New Mexico

Dear Professor Ulam:

Many thanks for your letter of January 6.

A sketch of Cohen's proof of the independence of the continuum hypothesis has just come out in the PROCEEDINGS OF THE NATIONAL ACADEMY OF SCIENCES.[a] I think very highly of Cohen's work. It seems to me to mark a milestone in the history of set theory, since it introduces for the first time a general method for independence proofs. This is indispensable in set theory because the present axioms simply do not determine some of the most important properties of sets.

I perfectly agree with you that properties of subsets of power $\aleph_1$ have now become much more interesting. In particular what Hausdorff calls the "Pantachie Problem", i.e., whether there exist $\aleph_1$ sequences of integers whose orders of growth (Wachstumsordnungen) surpass any given

[a] *Cohen 1963* and *1964*.

order, seems to me a basic question about the structure of the continuum, once the continuum hypothesis is dropped.

I am looking forward to the preprint of your v. Neumann Lecture.

Sincerely yours,

Kurt Gödel

10. Ulam to Gödel[a]

February 17, 1964

Dear Professor Gödel:

Please find enclosed two copies of a lecture given in 1963. At the time it was given Paul Cohen's paper had not yet been published. I am sending *two* copies in case the library in the Institute or somebody else would want to have one.

It was very nice to see you at Morgenstern's last week and I only wish I could see you again soon—to discuss some of the many problems in set theory which now appear to me even more important. In particular, it seems to me, given a set E of power $\aleph_1$, it might be interesting to consider a class of subsets of it which would contain "elementary" subsets and also all sets obtained by, say, Borel operations on them. This in analogy to studying on the interval a class of sets which contains the elementary subsets, e.g., the subintervals, and then building new sets from them. The question is: what should we consider as an elementary subset of the set of all ordinals up to Ω? One could include among them all sets which form arithmetical progressions in this set (always including also limit ordinals of the sequence of such). Or perhaps one could consider as elementary the "recursive" subsets. Then if we add all countable sums of such to our class, complements, etc., that is to say the analogue of the Borel class, some questions arise. For example, can one define a countably additive measure function for all such sets? One could, of course, go farther and define projective operations on such sets by considering the direct products of the set E, again in analogy with the case of the continuum of real numbers. This would be a wider class of sets and I must say

[a]On letterhead of University of California, Los Alamos Scientific Laboratory, P.O. Box 1663, Los Alamos, New Mexico.

that I do not know whether there were studies of such in recent literature. The questions of existence of various "paradoxical" decompositions seem to me of interest. Can one consider the set E as a group in some natural way?

Needless to say, similar problems arise for $\aleph_2$, etc. Since it might be natural to assume that the continuum hypothesis is not true, I would think that such problems may be sensible. By the way, I would assume the usual axioms of set theory as used by you *and* the axiom of choice
2 but *not* the continuum hypothesis, | of course.

Perhaps I will be able to visit in Princeton sometime during the Spring and would like very much to be able to see you at that time.

With my very best regards and greetings,

Sincerely,

S. M. Ulam.

Jean van Heijenoort

In 1959, W. V. Quine and Burton Dreben of the Harvard University philosophy department formulated a plan with the Harvard University Press to publish a Source Book of translations of the important founding papers in modern mathematical logic. Quine and Dreben then recruited Jean van Heijenoort to be the editor of what was originally planned to be two volumes in the Press's series of *Source books in the history of science*: the first to span the period from *Frege 1879* to Gödel's incompleteness paper, the second to begin right afterwards and continue "to the present", as van Heijenoort wrote in the early 1960s. In the event, only one volume appeared, *van Heijenoort 1967*, under the title *From Frege to Gödel*, a title to which Gödel agreed.[a] Other works of Gödel also included in the volume were the paper on the completeness of first-order logic, *1930*, and two short notes on incompleteness, *1930b* and *1932b*.

At the start of this project, Jean van Heijenoort (who used the Anglicized "John" until the mid-1960s) was on the faculty of the undergraduate mathematics department at New York University; in 1965 he became professor of philosophy at Brandeis University. He had a colorful past: born in 1912 in France, he left his *lycée* studies in Paris in 1932 to join Leon Trotsky, who was then in exile in Turkey. He served as Trotsky's personal secretary, bodyguard and translator in Turkey, France and Mexico until 1939.[b] In the early 1940s he abandoned political activity and began graduate work in mathematics, earning a Ph.D. in 1949. Although he did not publish in the area until the 1960s, van Heijenoort was keenly interested in mathematical logic and its development. Dreben thought him a good choice for editor of the Source Book, given this scholarly interest and his command of several of the original languages of the texts to be included.

Van Heijenoort first wrote Gödel on 10 May 1961, sending him a proposed table of contents for the first volume of the Source Book and a draft translation of *Gödel 1931*. Van Heijenoort had prepared the draft several years earlier, for inclusion in a volume to be published by the N.Y.U. Press, along with an account of Gödel's proof by Ernest Nagel and James Newman. This project did not come to fruition; see the introductory note to the Nagel correspondence in this volume.

[a] Letter of 26 July 1962 (not reproduced here).

[b] *van Heijenoort 1978* is a memoir of his years with Trotsky. Further biographical information can be found in *A. B. Feferman 1993*.

Jean van Heijenoort

Most of the ensuing letters concern Gödel's emendations of and additions to the translation. In his revisions, Gödel showed his habitual fastidiousness and attention to detail. On 28 August 1963, he returned the revised draft van Heijenoort had sent him on 23 March with a considerable number of changes marked in. Many of these changes were adopted without further ado, but several prompted disagreement and were discussed further. Van Heijenoort met with Gödel in Princeton in late September, and correspondence with him on various revisions continued through the fall and winter of 1963–1964. Letters 4 and 5 illustrate the type of attention Gödel devoted to the translation.

A graphic case of Gödel's intensive scrutiny of the translation concerns the last clause of the first sentence of *1931*. In the original, that clause reads, "daß das Beweisen nach einigen wenigen mechanischen Regeln vollzogen werden kann." Apparently van Heijenoort's draft translated this clause along the lines of "so that proofs can be carried out according to a few mechanical rules." Gödel wanted to change this, but van Heijenoort was resistant. Gödel persisted, and explained his objections in a letter of 12 December 1963:

> "carry out" somehow presupposes that the proof is given beforehand, and "according" does not exclude other rules. E.g., it is true that proofs can be carried out according to the rule that each sentence contain fewer than 50 symbols. But this does not mean that proofs can be formalised. Moreover, I would not use "nach" in this context but rather "in Uebereinstimmung mit", or I would say "ausgefuehrt" instead of "vollzogen".

Gödel had preferred "using a few mechanical rules", but van Heijenoort had objected to this as a dangling participle. Gödel responded by citing a grammar book, *Poutsma 1929*, as to the allowability of dangling participles in some circumstances. To break the impasse, van Heijenoort asked Quine for advice; in a letter to Gödel of 13 January 1964, he reported Quine's suggestion, first, of "so that proofs can be carried through by means of a few mechanical rules", and second, to avoid Gödel's objection to words like "carry out" or "carry through", of "so that we can prove any theorem using nothing but a few mechanical rules." Gödel agreed to a slight modification of Quine's second suggestion in a letter of 7 February (letter 6).

Gödel also provided new material to be included in the translations. Indeed, the note dated 28 August 1963 appearing at the end of *Gödel 1931* in *van Heijenoort 1967* (these *Works*, vol. I, p. 195) was on Gödel's mind as soon as he was told of the project. A text of this note was already in his first letter to van Heijenoort of 15 October 1961; and the footnote to it, footnote 70, was in his next letter, of 14 November 1961.[c] In addition, Gödel expanded or added several footnotes, and he provided an explanation as to why the scheme of italicization (or, in the eventual Source Book version, small capital letters) was not carried through the entire paper (letter 4, last paragraph). (In *van Heijenoort 1967*, this paragraph appeared in van Heijenoort's introduction to *Gödel 1931*, rather than in the paper itself. Hence it was not reprinted in volume I of these *Works*.) Gödel's final addition was the long foonote to *1932b* (these *Works*, vol. I, p. 235) which he sent to van Heijenoort on 18 May 1966.

The most interesting material in the correspondence lies in Gödel's reponses to van Heijenoort's inquiries on historical matters. Gödel commented on his own work, and works of Thoralf Skolem and of Jacques Herbrand.

In the postscript of a letter of 22 February 1964 (letter 6) Gödel remarked on his route to the incompleteness proof. In *1934*, §7, he had mentioned, as figuring in his motivation, the aim of finding a distinction between provability and truth. Apparently, in discussion with van Heijenoort he mentioned that the proof originated in his attempt to find a proof of the consistency of analysis, relative to that of arithmetic. The postscript concerns the relation of these two motivations. It parallels a similar recounting in an unsent letter to Yossef Balas, printed in these *Works*, vol. IV. In August 1964 (letters 8 and 9) Gödel gave two reasons for his having taken, as the system treated in *1931*, not the (simple) theory of types itself, but the theory of types superimposed on the non-negative integers as the individuals (objects of type 1), thus requiring the Peano axioms to be adjoined to those of type theory. His remarks here are not echoed in any other known writing of his.

In connection with the preparation of *Hilbert 1926* for publication in the Source Book, van Heijenoort asked about the relation between Gödel's proof of the consistency of the continuum hypothesis and the strategy Hilbert outlined in that paper, which was aimed at demonstrating that the continuum hypothesis is true (letter 15). Gödel answered on 8 July 1965 (letter 16); his characterization appeared in *van Heijenoort 1967*, p. 369. Gödel made some further elaborations of his remarks in 1969, in response to an inquiry from Constance Reid; see the correspondence with Reid in this volume. Shortly afterwards, Gödel wrote Paul Bernays on the topic; see letters 68b and 70 of that correspondence in these *Works*, vol. IV, as well as §3.5.4 of the introductory note.

[c]Presumably, Gödel sent the final, revised versions of the note and footnote along with letter 4; hence van Heijenoort's dating of the note. The changes he made were slight: the last sentence of the note was originally worded slightly differently, and the final version of the footnote contains "mechanical devices" where the original had "purely physical devices".

At their meeting in September 1963 van Heijenoort apparently asked Gödel about the relation of *Gödel 1930* to work of Skolem's, and in particular to *Skolem 1923a*. (Translations of that paper, *Skolem 1928*, and part of *Skolem 1920* were to be published in the Source Book.) The proof of the Skolem–Löwenheim theorem in *Skolem 1923a* is very close to Gödel's argument for completeness. (See the introductory note to *Gödel 1929* and *1930* in these *Works*, vol. I, pp. 51–55.) Gödel responded that he had not read *Skolem 1923a* before he wrote *1930* (letter 4). In response to further questioning, he noted that Skolem could have claimed to have "implicitly proved: 'Either A is provable or $\overline{A}$ satisfiable' ('provable' taken in an informal sense)." However, he added, Skolem did not make this result clear, even to himself (letter 8). Van Heijenoort continued to inquire about Skolem's use of the term "widerspruchsfrei" [consistent], which sometimes appears to be a syntactic property and sometimes a semantic one (letter 10). Gödel replied (letter 11) that Skolem "had no clear ideas in these matters. However, anyone who has can, without any further *mathematical* inferences, immediately draw the conclusion I stated in my previous letter." The same topic occupies some of Gödel's correspondence with Hao Wang a few years later, as Wang was preparing an introduction to a collection of Skolem's papers. In a letter of 7 December 1967 (letter 2 of the Wang correspondence in this volume), Gödel wrote, "The completeness theorem, mathematically, is indeed an almost trivial consequence of Skolem ⟦*1923a*⟧." The issue is taken further in the correspondence with Wang: there is consideration of *Skolem 1928*, which van Heijenoort had not alluded to; and Gödel suggested that Skolem's not becoming clear on the formulation of the theorem had to do with his wanting to restrict himself to finitary reasoning. See §1.1 of the introductory note to the correspondence with Wang.

In *1934*, Gödel had presented the first definition of the general recursive functions, and remarked that it had been suggested by Herbrand. In his initial letter to Gödel, van Heijenoort asked for details on how that suggestion had been made. In his reply (letter 2), Gödel said that it had been made in a letter of 1931, which he could not (in 1963) find, and what he had presented in *1934* as Herbrand's suggestion was exactly what was in the letter: in particular, there was no reference to computing the values of the function. Here Gödel's memory was mistaken, for the suggestion in Herbrand's letter explicitly requires proofs of calculability. For a discussion of this, and the issues raised by the discrepancy, see the introductory note to the correspondence with Herbrand in this volume.[d]

[d]See also the correspondence with Richard Büchi in these *Works*, vol. IV. Büchi had raised the same question to Gödel as van Heijenoort, four years earlier.

At their meeting in September 1963, van Heijenoort and Gödel discussed Herbrand's thesis *1930*, chapter 5 of which was to appear in the Source Book. The preceding year, Herbrand's argument for his fundamental theorem had been shown to be erroneous. At this occasion, Gödel told van Heijenoort that he had also, years before, found an error in Herbrand's argument, and devised a correction. On 12 October van Heijenoort wrote to Gödel expressing the hope that Gödel's correction could appear in the Source Book, but nothing came of this idea. For further details, see the introductory note to the correspondence with Dreben in these *Works*, vol. IV.

The issue of Herbrand's definition of recursiveness recurred a year later, in Gödel's letter of 14 August 1964 (letter 8). In a draft of his introduction to *Gödel 1931* for the Source Book, van Heijenoort had apparently said that the notion of (general) recursive function was "introduced" by Herbrand, and cited *Herbrand 1931*, which was also to appear in the Source Book. Gödel found the term too strong, and proposed "foreshadowed"; in the end, van Heijenoort used "suggested" (*van Heijenoort 1967*, p. 594). Herbrand's remarks in *1931* are rather more compressed than in his letter to Gödel; he says that functions may be introduced by any quantifier-free axioms, provided that "Considered intuitionistically, they make the actual computation of the $f_i(x_1, x_2, \ldots, x_{n_i})$ possible for every given set of numbers, and it is possible to prove intuitionistically that we obtain a well-determined result." In letter 10 van Heijenoort raises the question of what "intuitionism" meant for Herbrand, and suggests it might be closer to finitism, in Hilbert's sense, than to Brouwer's doctrine. In letter 11, Gödel disagrees, and finds support in Herbrand's use in *1931b* of the expression "intuitionism in its extreme form", which Gödel identifies as meaning finitism, thus suggesting that "intuitionism" without the modifier means Brouwerian intuitionism. Gödel is on somewhat tenuous ground here, as is suggested by Herbrand's writing "Hilbert then raised the problem of resolving the questions indicated above [consistency and the decision problem] by the use of exclusively intuitionistic arguments" (*1931b* in *Herbrand 1971*, p. 274). Moreover, it should be noted, as van Heijenoort did in letter 10, that in no writing of Herbrand is there any evidence that Herbrand had read either Brouwer or Weyl. If Herbrand did mean finitary reasoning by "intuitionistic proof", then of course the class of functions he demarcates falls far short of the class of all recursive functions.

Warren Goldfarb

A complete calendar of the correspondence with van Heijenoort appears on pp. 461–463 of this volume.

1. van Heijenoort to Gödel

100 Washington Square
New York 3, N.Y.
25 March 1963

Dear Professor Gödel,

I enclose a revised version of the English translation[a] of your 1931 paper.[b] The manuscript of the Source Book[c] is nearing completion and should go to the printer in the near future.

There is an historical question that comes up in the Source Book, on which it would perhaps be advisable to say something, namely, the origin of the (general) recursive functions. They are now generally referred to in the literature as the Herbrand–Gödel functions. In the 1934 Kleene–Rosser notes[d] footnote 19 states that the definition was "suggested" by Herbrand. It is now said and printed that there was an "oral" suggestion. There are, in general, conflicting stories on that point. My own understanding is that you took your inspiration from page 5 of Herbrand's paper, Sur la non-contradiction de l'arithmétique.[e] I am not even sure you had an opportunity to meet Herbrand. It is true that Herbrand's formulation could be a bit more explicit, and uniqueness of the value of the function is not stated, although one could consider it implied by the expressions used ("faire effectivement le calcul de...", "l'on obtient un résultat bien déterminé"). Perhaps a short note could be introduced in the Source Book, in one form or another, to clarify the record on that point.

Sincerely yours,

John van Heijenoort

---

[a] *1967.*

[b] *1931.*

[c] *van Heijenoort 1967.*

[d] *Gödel 1934.*

[e] *Herbrand 1931.*

2. Gödel to van Heijenoort

April 23, 1963

Dr. John van Heijenoort

100 Washington Square
New York 3, N.Y.

Dear Dr. Heijenoort:

Replying to your letter of March 25 I would like to say this:

I have never met Herbrand. His suggestion was made in a letter in 1931 and it was formulated *exactly* as on page 26 of my lecture notes, i.e., without any reference to computability. However, since Herbrand was an intuitionist, this definition for him evidently meant that there exists a *constructive* proof for the existence and unicity of φ. He probably believed that such a proof can only be given by exhibiting a computational procedure.* So I don't think there is any discrepancy between his two definitions as he meant them. What he failed to see (or to make clear) is that the computation, (for *all* computable functions), proceeds by *exactly the same rules*. It is this fact which makes a precise definition of general recursiveness possible.

Unfortunately I do not find Herbrand's letter among my papers.[a] It probably was lost in Vienna during World War II, as many other things were. But my recollection is very distinct and was still very fresh in 1934.

Thank you for sending me a copy of the latest version of your translation. I hope it is still possible to make a few more improvements. I find this particularly desirable in the introduction, which probably will be read by many non-specialists and where, therefore, everything should be said as clearly as possible.

I shall return the manuscript with my suggestions very soon.

Sincerely yours,

Kurt Gödel

*Note that, if Church's thesis is correct, it is *true* that, if $(\exists!\varphi)A(\varphi)$ holds
2 |intuitionistically, then $(\iota\varphi)A(\varphi)$ is general recursive, although, | in order to obtain the computational procedure for φ, it may be necessary to add some equations to those contained in $A(\varphi)$.

K. G.

---

[a]The letter was later found in Gödel's papers during the cataloging process and is published in this volume.

3. Gödel to van Heijenoort

August 28, 1963

Dr. John van Heijenoort
The Springs
East Hampton, N.Y.

Dear Dr. van Heijenoort:

I am returning herewith the manuscript of your translation. Many of the suggested changes are merely stylistical. It seems to me that you are sometimes using rather uncommon expressions and that, moreover, the style of the translation has a strong German touch. I am afraid this may bother the readers or even make it harder for them to follow the arguments.

As you doubtless know words like: also, still, actually, already, etc. are used less frequently in English than in German. Moreover sentences in English are shorter and the order of words is different. Also composite substantives are less frequent and substantives are rarely used to express possibilities such as "satisfiability". Rather these are expressed by sentences.

In many cases I think the suggested translation also corresponds more closely to the meaning of the German text. E.g., the meaning of "richtig" lies somewhere in between "correct" and "true", and "true" fits much better in the context. "After all" and "imagine" have some connotation of vagueness which the German text does not have. "Remarkable" is not the correct translation of "merkwürdig". The German "beliebig" in most cases is best translated by "any".

In a few passages I made insertions in square brackets which however only make explicit what was meant in the original.

The readers of my paper will probably be puzzled by the fact that the scheme of italisation was carried through only partially. Therefore I would request you to mention the following, perhaps in a footnote:

The idea was to use this notation only for those metamathematical concepts that had been defined in their usual sense before (namely on p⟦p⟧. 177–178). From p. 187 up to the general considerations in the end of section 2, and again in section 4 every metamathematical term referring to P is supposed to denote the corresponding arithmetical one. But, of course, because of the complete isomorphism the distinction in many cases is entirely irrelevant.

Sincerely yours,

Kurt Gödel

2 | P.S. If I were to read the manuscript again I would probably find a few more changes desirable. But I would like to let you know what I have done so far. I have kept a copy of the manuscript.

Incidentally: Isn't "formulae" the more usual plural form?

4. Gödel to van Heijenoort

October 4, 1963

Dr. John van Heijenoort
100 Washington Square
New York 3, N.Y.

Dear Dr. van Heijenoort:

I am sending you herewith another copy of p. 2. Since in several cases the final choice of words was not decided in our conversation I would like to see the final version. I assume the translation is to be designated as "approved by the author".

According to what you said I presume that in most cases you will accept my suggestions without change. Since I proceeded on the assumption that making these changes would not create any problems, the choice of words in some cases may be due to mere personal preference. But in many cases, I believe, some real improvement is accomplished.

The square brackets could perhaps be omitted in some cases, namely whenever the reason for the insertion is merely stylistic or grammatical (as, e.g., on p⟦p.⟧ 5, 15, 36). As for the other square bracket insertions it would be best to say in the footnote which I mentioned in my letter of August 28 that when I was revising the translation I suggested these insertions because I thought they would be helpful to the reader. In the same footnote the following should be added:

> The last clause of footnote 27 was not taken into account in the formulas (18) on p. 26. But an explicit formulation of the cases with fewer variables on the right hand side is actually necessary there for the formal correctness of the proof, unless the identity function, $I(x) = x$, is added to the initial functions.

As far as Skolem's paper of 1922[a] is concerned, I think I first read it

---

[a] *Skolem 1923a.*

about the time when I published my completeness paper. That I did not quote it must be due to the fact that either the quotations were taken over from my dissertation or that I did not see the paper before the publication of my work. At any rate I am practically sure that I did not know it when I wrote my dissertation. Otherwise I would have quoted it, since it is much closer to my work than the paper of 1920[b] which I did quote.

It occurred to me that "interesting" is an acceptable translation of "merkwürdig".

| "Demnaechst" could perhaps be translated by "in the near future" or "soon" if you add in square brackets that it was intended to publish the sequel in the next volume of the Monatshefte. As for "dangling participles" and similar questions it might be best to consult a philologically unbiassed educated person whose native language is English. 2

Sincerely yours,

Kurt Gödel

---

[b] *Skolem 1920.*

5. van Heijenoort to Gödel

100 Washington Square
New York 3, N.Y.
24 October 1963

Dear Professor Gödel,

I trust you have my letter of 14 October 1963.

I enclose a xerox copy of the English translation of your 1931 paper. A large number of your corrections have been incorporated in the text, and there is nothing more to say about them. For those of your corrections that raise some questions, I enclose a list of commentaries by Mr. Elder, the science editor of Harvard University Press (and a few by myself). I marked the passages to which these commentaries refer with red numbers, so that you can locate them quickly.

Mr. Elder has a long experience of editing English texts, and I have always found his judgment in these matters very sound. You can either go over these corrections and suggestions yourself or allow me to prepare, together with Mr. Elder, a final version that we will submit to you.

Skolem's papers of the 20's (Church numbers: 247*3*,[a] 247*5*,[b] 247*7*,[c] 1928, and 247*8*)[d] raise a puzzling question. He presents several proofs of Löwenheim's result or of a generalization of it, which we can briefly state: If satisfiable, then $\aleph_0$-satisfiable. The word he uses for satisfiable is, most of the time, "widerspruchslos" or "widerspruchsfrei". Hence some of his statements *read* like a completeness result (which they are not). Skolem, however, never gives a definition of "widerspruchsfrei", and the question is whether he understood the word in the semantic sense, that is, as "satisfiable", or in the syntactic sense, and then took the inference from "syntactically consistent" to "semantically consistent" (or "satisfiable") as obvious and immediate, so obvious and immediate, in fact, that it does not deserve one word of explanation. I have not been able to decide that question by | a reading of Skolem's texts. The problem is somewhat complicated by the fact that in some cases, for instance when he discusses what he calls a "Theorie", that is, a basic logical system plus some specific axiom(s) (247*7*, page 134; 247*8*), page 29), he uses "widerspruchsfrei" in the sense of "syntactially consistent". This makes his use of the same word in other contexts more puzzling.

2

Sincerely yours,

John van Heijenoort

---

[a] *Skolem 1920.*
[b] *Skolem 1923a.*
[c] *Skolem 1928.*
[d] *Skolem 1929.*

6. Gödel to van Heijenoort

February 22, 1964

Professor John van Heijenoort
35 Columbia Street
Watertown 72, Massachusetts

Dear Professor van Heijenoort:

Thank you for your letters of January 13 and February 14. Quine's second suggestion is all right with me, except that I think "one" would be better than "we" in this context. Also I forgot to mention last time that on p. 16, line 4 from below, "by saying" would be better than "thus", if you adopt the order of words I suggested.

The reason for omitting the italicizing of "beweisbar" in the end of Th. XI and in the last but one paragraph of the paper was that these remarks do not form part of the arguments, but only point out some further results. But in the mere formulation of a result (especially in a rather sketchy formulation) it is of no use to refer to the arithmetization, which is necessary only for the proof. The "beweisbar" on p. 197, line 14, should have been italicized, but then the P which I wanted for stress, would have been out of place. I think the italics should be left as they stand, since the footnote I suggested in the end of my letter of August 28 should suffice to make the situation clear. Perhaps the following could be added to it: Therefore also the italicizing of some terms defined on pp. 182–186 was omitted in some rather informal statements.

I would like, of course, to see the final draft of the translation, and also of the parts of the introduction that refer to my papers. Of course I do not want in any way to prescribe to you what to say, but there is always the possibility of errors which I might be able to prevent or ⟨of more⟩ ~~of~~ precis~~ations~~⟨e formulations⟩ which I could perhaps suggest.

Sincerely yours,

Kurt Gödel

| P.S.[a] Perhaps you were puzzled by the fact that I once said an attempted relative consistency proof for analysis le~~a~~d to the proof of the existence of undecidable propositions and another time that the heuristic principle and the first version of the proof were those given in Sect. 7 of my 1934 Princeton lectures.[b] But it was exactly the relative consistency proof which made it necessary to formalize either "truth" or "provability" and thereby forced a comparison of the two in this respect. 2

K.G.

[a]The postscript occurs only on the retained copy at the I.A.S. It has the date "February 21, 1964", but the back of the sheet carries in Gödel's hand the annotation "22./II. 64".

[b]*1934*.

7. van Heijenoort to Gödel

100 Washington Square
New York 3, N.Y.
24 June 1964

Dear Professor Gödel,

I enclose

(1) The English translation[a] of the abstract of your 1931 paper;[b]
(2) The German text of that abstract, in case you don't have it;
(3) My introductory note to your 1930 paper;[c]
(4) My introductory note to your 1931 paper.[d]

Concerning the translation, there is really very little left. At the beginning of footnote 48a, there is still the question of using "true" or "real" (see my letter of 28 April 1964). The changes indicated in your letter of 5 June have been incorporated, except that of point 4, for which I would suggest "The prompt acceptance of his results was one of the reasons that made him ... "

Concerning Skolem's claims (I wrote to you several months ago on his use of "widerspruchsfrei" and we spoke about the point), I find in one of his recent publications (*Abstract set theory*, Notre Dame, 1962, p. 47[e]):

"The theorem of Löwenheim says that if F is a well-formed formula of the first-order predicate caluclus with certain predicate variables $A, B, C, \ldots$, either $\overline{F}$ is provable or F can be satisfied in the natural numbers series by suitable determination of $A, B, C, \ldots$ in that domain of individuals. The generalization which I proved in 1919 says that the same is true for an enumerated set of such formulas, say $F_1, F_2, \ldots$, |that is, either some $\overline{F}_i$ is provable or the whole set of formulas can be satisfied by suitable determination in N of the predicates occurring in them." 2

In a paper written in Norwegian and published in 1936[f] he states that the question whether or not a given formula of the first-order predicate calculus is satisfiable in some domain is equivalent to the question whether or not the negation of the formula is formally provable (in the Hilbert–Ackermann formalism).[g]

---

[a] *1967*.

[b] *1931*.

[c] *1930b*.

[d] In the margin Gödel pencilled roman numerals: "I" beside item (1), "II" beside item (3) and "III" beside item (4).

[e] *Skolem 1962*.

[f] Presumably *Skolem 1936*.

[g] Gödel pencilled a question mark in the margin beside this paragraph.

Skolem is claiming for Löwenheim and himself the full completeness theorem. Of course, Löwenheim's proof or Skolem's own arguments do not justify that claim. He must have had some idea that he never made explicit.

Sincerely,

John van Heijenoort

There is the question where to print the abstract. Logically, it should go between my introductory note and your 1931 paper. However, the abstract is for the experts, and I feel that at that place it may somewhat confuse the ordinary reader. It is perhaps better to print it after the paper. Do you have any preference?

8. Gödel to van Heijenoort

August 14, 1964

Professor John van Heijenoort
100 Washington Square
New York 3, New York

Dear Professor Heijenoort:

I had just completed my suggestions of changes in the material you sent me on June 24, when your letter of August 12 arrived. I hope no inconvenience will result from my replying so late.

Here are my suggestions, in addition to those I marked in the translation of my paper in the "Anzeiger,"[a] which I am returning herewith. I am citing the 3 parts of the material you sent me by M(1), M(3), M(4) in accordance with your letter of June 24.

ad M(1):

1. I think "decidable" applied to "system" is not very good, nor usual either. I would suggest either "decision complete" or "decision definite" or simply "complete".

2. "imbed" on p. 1, line 1 suggests that there may be other individuals besides the integers. See also items 1 and 2 ad M(4).

ad M(3):

[a] *1930b*.

On p. 3, line 15 "1" must be replaced by "2".

ad M(4):

1. Instead of line 3 of p. 2 I would suggest: "with the logic of P.M. (...) applied to them".

2. The last sentence of (1) on p. 2 does not describe the situation correctly. I identified the individuals of PM with the integers in order to obtain a system every proposition of which has a well⟦-⟧defined meaning in 2 |classical mathematics and, therefore, viewed from the standpoint of classical mathematics, must be either true or false. The question of completeness is of philosophical interest only for systems which satisfy some requirement of this kind.

3. On p. 6, line 13 fr. bel. I would suggest ⟨you⟩ cit~~ing~~⟨e⟩ my paper in Dial. 12, 1958,[b] of course not as an expression of the views of the Hilbert School.

4. On p. 5, last line, after "Z" I would suggest to insert: "and this proof can almost literally be transferred to any system containing Z."

5. "introduced" on p. 6, line 6 is definitely an overstatement, since it was exactly by *specifying* the rules of computation that a mathematically workable and fruitful concept was obtained. Herbrand, on the other hand, *explicitly excludes* the specifying of formal rules of computation by the phrase "considerée *intuistiquement*" (and the explanation given in his footnote[5)]). Whether Herbrand's concept (as defined in J. r. angew. Math. 166[c]) is equivalent with general recursiveness is considered an undecided question by Heyting, myself, and others. In my opinion Church's Thesis is unquestionably correct for mechanical, but perhaps incorrect for intuitionistic, computability (as I clearly stated in the postscript to my 1934 lectures). Perhaps one could say "foreshadowed" instead of "introduced".

6. The second sentence of the last paragraph of p. 6 could more accurately be stated as follows: "Prof. G. approved the translation, which in many places was accommodated to his wishes."

7. On p. 4, lines 7 and 8 "many important axiom systems" should be replaced by "all known axiom systems of mathematics or any substantial part of it".

8. On p. 4, line 13 I would say "yields" instead of "is", since in Th. VII or its proof I did not state or prove anything about computability, which forms the main content of Th. V of Sect. 2.[d]

---

[b] *1958*.

[c] *Herbrand 1931*.

[d] Here van Heijenoort had checked off each item except number 3 under M(4).

I assume that my 1934 lectures will not be published in the same volume. I don't know whether I ever mentioned to you that, due to the fact that I did not read the manuscript carefully before it/s first publication, a considerable number of corrections are necessary. I have communicated those, together with several notes and a rather long postscript, with the editors of a new edition. I shall be glad to send you a copy of this material. 3

The suggestion in your letter about point 4 is all right with me. "True", I think, is much better than "real". As for Skolem, what he could justly claim, but apparently does not claim, is that, in his paper at the Congr. of Scand. Math. 1922,[e] he implicitly proved: "Either A is provable or $\overline{A}$ is satisfiable" ("provable" taken in an informal sense). However, since he did not clearly formulate this result (nor, apparently, had made it clear to himself), it seems to have remained completely unknown, as follows from the fact that Hilbert–Ackermann in 1928 do not mention it in connection with their completeness problem.

As far as your commentaries on my papers are concerned, I am in favor of placing them *in front of* the papers.

Sincerely yours,

Kurt Gödel

[e] *Skolem 1923a*.

9. Gödel to van Heijenoort

Princeton, Aug.$_\wedge$15, 1964

Dear Professor Heijenoort:

On rereading my letter of Aug.$_\wedge$14 I find that in suggestion 2. ad M(4) I have given the wrong impression that what I say there was my only reason for adjoining Peano's axioms. Another reason, of course, was the simplification of the proofs which ⟨results⟩ from it. In fact, I believe that either one of these two considerations would have been sufficient by itself. However, if the second one had been my only reason, I could have omitted the axiom of complete induction, thereby admitting other individuals besides the integers.

Sincerely yours,

Kurt Gödel

10. van Heijenoort to Gödel

100 Washington Square
New York 3, N.Y.
27 August 1964

Dear Professor Gödel,

Thank you very much for your letters of 14 and 15 August. I adopted all your suggestions. You didn't express any opinion about the question raised in the postscript to my letter of 24 June, namely, whether to print the abstract of your 1931 paper above or below the text of the paper itself.

You are very generous toward Skolem. If one reviews what he actually uses in his proofs (247*3*,[a] p. 7, lines 9–10; 247*5*,[b] p. 220, lines 5–7, and p. 221, lines 10–13—I use Church numbers for bibliographical references), one is drawn to the conclusion that he merely proves: "If A is satisfiable, A is $\aleph_0$ satisfiable." But on the other hand,

(a) Skolem frequently uses the word "widerspruchsfrei" (or "widerspruchslos"), and some of his statements read like a completeness result: "If A is consistent, A is $\aleph_0$-satisfiable." He jumps without argument from "widerspruchsfrei" to "satisfiable" (247*5*, p. 219, lines 8–7 from bottom, compared to p. 220, lines 5–6; also 247*7*,[c] p. 134, lines 10–11), and I cannot decide whether he uses "widerspruchsfrei" in the sense of "semantically consistent" (or "satisfiable") or in the sense of "syntactically consistent" and then considers the inference from "syntactically consistent" to "semantically consistent" so obvious and immediate that it does not deserve one word of explanation.

(b) In 247*7*, from p. 134, line 15 from bottom to p. 135, line 3, and in 247*8*, p. 29, line 10 to bottom, Skolem advances some syntactic considerations; but they do not establish (nor, I think, are intended to establish): "If A is not provable, A is satisfiable."

I would like to say a few words about Herbrand's "intuitionism". Judging from his expressions and references, | it seems to me that Herbrand was not acquainted with Brouwer's writings, but that his knowledge of Brouwer's intuitionism was second-hand, acquired through Hilbert's papers and perhaps conversations with von Neumann. What he understands by an intuitionistic argument (footnote 3 of 382*8*[d]) seems closer

2

[a] *Skolem 1920.*

[b] *Skolem 1923a.*

[c] *Skolem 1928.*

[d] *Herbrand 1931.*

to Hilbert's "finit" than to Brouwer's doctrine. There is a remark of von Neumann in the same sense (299*3*,[e] p. 3, lines 5–7). In a note in *Annales de l'Université de Paris* (6e année, n° 2, mars–avril 1931, p. 187[f]) Herbrand writes: "toutes les fonctions introduites devront être effectivement calculables pour toutes les valeurs de leurs arguments, *par des opérations décrites entièrement d'avance.*" (My emphasis) Although, in the lines that follow, Herbrand refers to the "brouweriens", his description of the method fits much more finitism than intuitionism. I agree with you that footnote 5 of 382*8* indicates a shift from the *formal* to the *inhaltlich*, but, by footnote 3, this *inhaltlich* is closer to Hilbert than to Brouwer.

Sincerely yours,

John van Heijenoort

---

[e] *von Neumann 1927.*
[f] *Herbrand 1931b.*

11. Gödel to van Heijenoort

September 18, 1964

Professor John van Heijenoort
100 Washington Square
New York 3, N.Y.

Dear Professor van Heijenoort:

Thank you for your letter of August 27.

I think the abstract in the "Anzeiger" should be placed above my paper.

What you say about Skolem only proves my point, namely that he had no clear ideas in these matters. However, anyone who has can, without any further *mathematical* inferences, immediately draw the conclusion I stated in my previous letter.

The passage in Herbrand which you quote is very interesting. I think, judging by Herbrand's explicit statements, what he calls "intuitionnisme" agrees pretty well with Brouwer's intuitionism. That it is not finitism follows from Ann. Un. Paris 6, p. 187,[a] where he speaks of "intuitionnisme

---

[a] *Herbrand 1931b.*

dans sa forme extreme", which is characterized by the fact that it admits as objects of mathematics only integers or what can be mapped on in- ê tegers. This, taken in the sense which Herbrand probably had in mind, is a correct demarcation of finitism within intuitionism. In footnote 3) of Crelle 166[b] he does *not* require the enumerability of the mathematical objects, and gives a definition which fits Brouwer's intuitionism very well, if one takes into account that the logical constants in intuitionism can be defined in terms of operations on proofs. The difference between "in- tuitionnisme" and "intuitionnisme dans sa forme extreme", I think, ac- ê counts for the difference in the two concepts of function in Crelle 166 and Ann. Un. Paris 6. The latter concept would seem to require that each of his functions should be computable by formal rules, but that these rules may be different for different functions. (Otherwise he would not have failed to state that they are uniform.) This concept is, in fact, co-extensive with general recursiveness (or demonstrable general recursiveness). However, since Herbrand leaves the unprecise term "operations" unanalysed, his concept is still very far from general recursiveness, because it gives no reduction to precise mathematical ideas, which is the main achievement of general recursiveness.

2 |Could you perhaps let me see the final formulation of my motives for adjoining Peano's axioms?

Sincerely yours,

Kurt Gödel

---

[b] *Herbrand 1931.*

12. Gödel to van Heijenoort

November 3, 1964

Professor John van Heijenoort
100 Washington Square
New York 3, N.Y.

Dear Professor Heijenoort:

Thank you for your letter of September 30. The new formulation of point (1) on p. 2 is all right with me. In the suggested addition to the introductory note the last sentence is unintelligible to me. What is "truth under the metamathematical coding"?

As for Herbrand's concept of intuitionism I have noticed that on p. 8 of Crelle 166[a] he in fact annuls his previous distinction between intuitionism and finitism by stating that the two very likely are coextensive. This seems to imply that he also believed the two concepts of computable functions I discussed in my preceding letter would turn out to be equivalent.

But, at any rate, in both cases the conceptual distinctions were clear to him, and he by no means considered the equivalence of the concepts in question to be obvious (see Crelle 166, p. 8, (a)).

Sincerely yours,

Kurt Gödel

---

[a] *Herbrand 1931.*

13. van Heijenoort to Gödel

100 Washington Square
New York 3, N.Y.
14 November 1964

Dear Professor Gödel,

Thank you for your letter of 3 November. Your remarks on Herbrand have helped me to follow his thought. I would like to bring one more quotation. The logician to whom Herbrand was closest at that time is probably von Neumann, who writes (Zur Hilbertschen Beweistheorie,[a] p. 3):

Dieses "inhaltliche Beweisen" [in metamathematics] muß also ganz im Sinne der Brouwer-Wey⟦l⟧schen intuitionistischen Logik verlaufen: Die Beweistheorie soll sozusagen auf intuitionistischer Basis die klassische Mathematik aufbauen und den strikten Intuitionismus so ad absurdum führen.

Perhaps Herbrand's "intuitionnisme dans sa form extrême" comes from this "strikter Intuitionismus", although I am not sure whether von Neumann means here the finitistic part of intuitionism or Brouwer's specific form of intuitionism.

---

[a] *von Neumann 1927.*

By "truth under the mathematical coding" I was simply referring to the fact that the undecidable formula was declared true because it asserts its own unprovability, not by the consideration of the complicated arithmetic statement that it expresses.

At the end of the third paragraph of page 1 of my introductory note I added the following sentences:

Gödel also brings to light the relation that his argument bears to the Richard paradox and Cantor's diagonal procedure. (Herbrand, on pages 000–000 below, and Weyl (1949,[b] pp. 219–235) particularly stress this aspect of Gödel's argument.)

Sincerely yours,

John van Heijenoort

---

[b] *Weyl 1949.*

14. Gödel to van Heijenoort

December 15, 1964

Professor John van Heijenoort
100 Washington Square
New York 3, N.Y.

Dear Professor van Heijenoort

I definitely think that in your introductory note the sentence referring to the double meaning in which the undecidable proposition is true should be omitted. If I understand your explanation correctly, what you call "truth under the ma̸th̸e̸/mathematical coding" is not a different sense of "being true", but rather truth under a different interpretation of the terms occurring in the sentence; namely if these terms are interpreted, not arithmetically, but as referring to objects of metamathematics through the latter's association with numbers. But this is *not* the meaning of the terms which is used in my proof (neither in the introduction nor in the paper itself). Hence it is not correct that <u>17 Gen *r*</u> is also true under the metamathematical coding, since what is true in this sense is *not* 17 Gen *r* (which is an arithmetical proposition) but a different (though equivalent) proposition. Since, by my explanations on p. 176, 17 Gen *r* has a quite definite meaning and leaves no room for different interpretations, what you say can only be understood in the sense that you are speaking of two meanings of "being true", which is not what you have in mind.   e/a,

What von Neumann, in the passage you quote, means by "strikter Intuitionismus" is, I believe, something entirely different from Herbrand's "int. dans sa forme extrême". It can only mean: "intuitionism including the assertion that the non-constructive inferences of classical mathematics must simply be discarded as non-sense", while a less strict intuitionism would leave open the question whether any or what meaning can be given to them.

Sincerely yours,

Kurt Gödel

P.S. Of course what I said implies that the undec. prop. asserts its own provability only if "unprovability" is italicized. But in systems containing metamathematical concepts the undec. prop. asserts its own unprovability in a strict sense.

15. van Heijenoort to Gödel

100 Washington Square
New York 3, N.Y.
25 May 1965

Dear Professor Gödel,

The Source Book includes an integral translation of Hilbert's paper, *Über das Unendliche* (1925).[a] In his review[b] of your *Consistency-proof for the generalized continuum-hypothesis*,[c] Bernays states (JSL *5*, 118): "The whole Gödel reasoning may also be considered as a way of modifying the Hilbert project for a proof of the Cantor continuum hypothesis, as described in *Über das Unendliche*, so as to make it practicable and at the same time generalizable to higher powers."

Since Hilbert's Lemmas I and II have remain⟦ed⟧vague and obscure for many readers, I wonder if you would like to comment on them and on their connections with your own notion of constructible set. I feel that any comment by you on that point would be extremely valuable and helpful.

Sincerely yours,

John van Heijenoort

---

[a] *Hilbert 1926.*

[b] *Bernays 1940a.*

[c] *1939a.*

16. Gödel to van Heijenoort

July 8, 1965

Professor John van Heijenoort
100 Washington Square
New York 3, New York

Dear Professor van Heijenoort

About the relationship of my work on the continuum problem to Hilbert's paper "Ueber das Unendliche" I can say the following:

There is a remote analogy between Hilbert's Lemma II and my Theor. 12.2 ~~(in~~⟨for $\alpha = 0$(see⟩ Ann. Math. Stud. No. 3).[a] There is, however, this great difference that Hilbert considers only strictly constructive definitions and, moreover, transfinite iterations of the defining operations only up to constructive ordinals, while I admit, not only quantifiers in the definitions, but also iterations of the defining operations up to *any* ordinal number, no matter whether or how it can be defined. The term "constructible set", in my proof, is justified only in a very weak sense and, in particular, only in the sense of "relative to ordinal numbers", where the latter are subject to no conditions of constructivity. It was exactly by viewing the situation from this highly transfinite, set-theoretical point of view that in my approach the difficulties were overcome and a *relative* finitary consistency proof was obtained. Of course there is no place in this approach for anything like Hilbert's Lemma I. Hilbert probably hoped to prove it as a special case of a general theorem to the effect that transfinite modes of inference applied to a constructively correct system of axioms lead to no inconsistency.

Sincerely yours,

Kurt Gödel

---

[a] *1940*.

17. Gödel to van Heijenoort

August 9, 1965

Professor J. van Heijenoort
100 Washington Square
New York 3, N.Y.

Dear Professor Heijenoort:

I have no objection to your quoting the passage from my letter of August 14, 1964, which appears in the enclosure of your letter of July 20. However, I don't want any guesses of mine as to what Hilbert and Ackermann could, or would, have said to be published, especially if they are as uncertain as this one. You could, instead, say something like this:

> "Professor G. remarked that at the time when these lines were written a substantial part of this problem had implicitly been solved already by Skolem in his 1922 paper."

Moreover, in order to avoid any wrong impression, it is necessary to add that I did not know this proof of Skolem's when I wrote my dissertation.

On rereading my letter of July 8, it occurred to me that "place" in line 5 from below should better be replaced by "need". Moreover, "of course", in the same sentence, should be dropped.

Sincerely yours,

Kurt Gödel

18. Gödel to van Heijenoort

December 13, 1965

Professor John van Heijenoort
Department of Philosophy
Brandeis University
Waltham, Massachusetts 02154

Dear Professor Heijenoort:

Many thanks for your letter of November 6. The latest version of the passage on the relationship of Skolem's work to the completeness theorem is all right with me, except that, amplifying what I remarked in my previous letter, I would like to say the following: Unless it is stated somewhere that I had not read the passage in question of Skolem's writings prior to completing my dissertation, the wrong impression will necessarily be created that I was acquainted with, and have cited, Skolem's paper in connection with my completeness proof.

Sincerely yours,

Kurt Gödel

John von Neumann

John von Neumann

The correspondence between von Neumann and Gödel opens with an extraordinary letter from von Neumann, written on 20 November 1930. In early September of that year von Neumann had met Gödel at a congress in Königsberg and was informed about a theorem Gödel had just discovered—(a form of) the first incompleteness theorem. Von Neumann was deeply impressed; he turned his attention to logic again and gave lectures on proof theory in the winter term of 1930–1931. As can be gathered from Herbrand's letter to Claude Chevalley,[a] von Neumann was preoccupied with Gödel's result and, as he put it in his own letter, with the methods Gödel had used "so successfully in order to exhibit undecidable properties." In reflecting on this result and Gödel's methods, von Neumann arrived at a new result that seemed remarkable to him, namely, that the consistency of a formal theory is unprovable within that theory, if it is consistent. He formulated this "new result" in the letter to Gödel, claiming—less precisely—that the consistency of mathematics is unprovable; this strong interpretation of what we know as the second incompleteness theorem was to become a point of contention between Gödel and von Neumann.

In his next letter of November 29, von Neumann acknowledges the receipt of a "Separatum" and a letter from Gödel.[b] It is most likely that the separatum was a copy of the abstract *1930b* that had been presented to the Vienna Academy of Sciences on 23 October 1930 and already contained the classical formulation of the second incompleteness theorem.[c] Von Neumann states in his response: "As you have established the theorem on the unprovability of consistency as a natural continuation and deepening of your earlier results, I clearly won't publish on this subject." Their differing views on the impact of this result for Hilbert's consistency program are discussed below. Two additional topics of scientific interest are addressed in later correspondence: (i) the

[a]Cf. the introductory note to the Gödel–Herbrand correspondence, in this volume; Herbrand wrote the letter on 3 December 1930.

[b]Unfortunately, it seems that this letter and two others in this early correspondence have not been preserved: von Neumann acknowledges in his letter of 12 January 1931 that he had received two letters from Gödel. (The von Neumann Papers in the Library of Congress do not contain these letters.)

[c]Gödel had by this time completed his *1931*; indeed, the paper had been submitted for publication on 17 November 1930. Von Neumann acknowledged receipt of the galley proofs of *Gödel 1931* in his letter of 12 January 1931. From *Mancosu 1999a* it is clear that Gödel had not sent the galleys before the end of December 1930; see the letters between Hempel and Kaufmann quoted there on pp. 35–36. Cf. also editorial note a to letter 2 below.

relative consistency of the axiom of choice and the generalized continuum hypothesis, in letters from 1937 through 1939, and (ii) the feasibility of computations (related to the now famous P vs. NP problem), in Gödel's last letter to von Neumann in 1956.

As von Neumann's life and work are well known, only the briefest biographical sketch is presented.[d] Born on 28 December 1903 in Budapest (Hungary), von Neumann grew up in a wealthy Jewish family and attended the excellent Lutheran Gymnasium in Budapest from 1914 to 1921. He then entered the University of Berlin as a student of chemistry, but switched in 1923 to the Eidgenössische Technische Hochschule in Zürich, where he earned three years later a *Diplom* degree in that subject. He obtained, also in 1926, a doctoral degree in mathematics from the University of Budapest. Von Neumann spent the academic year 1926–1927 in Göttingen supported by a Rockefeller Fellowship. He was Privatdozent in Berlin (1927–1929) and Hamburg (1929–1930). In 1930 he was appointed visiting lecturer at Princeton University with the agreement that he would be back in Berlin for the winter term of 1930–31. In 1931 he was promoted to professor of mathematics at Princeton and became two years later one of the six mathematics professors at the newly founded Institute for Advanced Study, together with J. W. Alexander, A. Einstein, M. Morse, O. Veblen and H. Weyl; he kept that position for the remainder of his life. As to the correspondence with Gödel, it obviously started while von Neumann was staying in Berlin during the winter term of 1930–1931.

In the 1920s von Neumann contributed to the foundations of mathematics not only through a series of articles on set theory (*1923*, *1925*, *1926*, *1928*, *1928a*, *1929*) but also very specifically to Hilbert's emerging finitist consistency program through his paper *Zur Hilbertschen Beweistheorie*. Though published only in 1927, the paper had already been submitted for publication in July of 1925. In it von Neumann established the consistency of a formal system of first-order arithmetic with quantifier-free induction; he also gave a detailed critique of the consistency proof in *Ackermann 1924*. What is of interest in the context of his early correspondence with Gödel is the general strategic attitude he took towards proof-theoretic research. It is expressed in the following quote from the introduction to his *1927*, where he formulates four guiding ideas of Hilbert's proof theory. (Note that "intuitionist" and "finitist" were evidently synonymous for von Neumann.) Viewing an intuitionistic consistency proof for classical formal theories as the crucial

[d]Von Neumann was born Janos, used Johann when living in Germany and Switzerland, and switched to John after having moved to the United States. For accounts see the collection of essays by Garrett Birkhoff and others in the *Bulletin of the American Mathematics Society 64* (*3*), part 2 (in particular the accessible description in *Ulam 1958*) and the book *Macrae 1992*.

aim, he articulates the final guiding idea as follows:

> Here one has always to distinguish sharply between two different ways of "proving": between the formalized ("mathematical") proving within a formal system and the contentual ("metamathematical") proving about the system. While the former is an arbitrarily defined logical game (that must be, however, to a large extent analogous with classical mathematics), the latter is a chaining of immediately evident contentual insights. This "contentual proving" has consequently to be carried out completely within the intuitionistic logic of Brouwer and Weyl: proof theory is to rebuild classical mathematics, so-to-speak, on an intuitionistic basis and in this way reduce strict intuitionism ad absurdum.[e]

The strategic goal of proof-theoretic research, as interpreted by von Neumann, also shaped his talk at the Second Conference for Epistemology of the Exact Sciences. The conference was held in Königsberg from 5 to 7 September 1930, and on the first day of the congress von Neumann talked about Hilbert's finitist standpoint in a plenary session, where Carnap and Heyting presented the logicist, respectively, intuitionist position.[f] On the next day Gödel described the results of his

[e]*von Neumann 1927*, pp. 2–3. The German text is this: "Hierbei muß stets scharf zwischen zwei verschiedenen Arten des "Beweisens" unterschieden werden: Dem formalisierten ("mathematischen") Beweisen innerhalb des formalen Systems, und dem inhaltlichen ("metamathematischen") Beweisen über das System. Während das erstere ein willkürlich definiertes logisches Spiel ist (das freilich mit der klassischen Mathematik weitgehend analog sein muß), ist das letztere eine Verkettung unmittelbar evidenter inhaltlicher Einsichten. Dieses "inhaltliche Beweisen" muß also ganz im Sinne der Brouwer-Weylschen intuitionistischen Logik verlaufen: Die Beweistheorie soll sozusagen auf intuitionistischer Basis die klassische Mathematik aufbauen und den strikten Intuitionismus so ad absurdum führen."

[f]Gödel reviewed the published versions of these presentations in *Gödel 1932e, f* and *g*. Waismann had also given a paper in the plenary session, entitled *Das Wesen der Mathematik: Der Standpunkt Wittgensteins*; his talk was not published.

From the letters between von Neumann and Carnap, quoted in *Mancosu 1999a*, we know that their Königsberg talks were published (in the form they were) only to reflect the situation before Gödel's results. Von Neumann writes, in his letter to Carnap of 7 June 1931:

> Ich halte daher den Königsberger Stand der Grundlagendiskussion für überholt, da Gödels fundamentale Entdeckungen die Frage auf eine ganz veränderte Plattform gebracht haben. (Ich weiss, Gödel ist in der Wertung seiner Resultate viel vorsichtiger, aber m. E. übersieht er die Verhältnisse an diesem Punkt nicht richtig.)
>
> Ich habe mit Reichenbach mehrfach besprochen, ob es unter diesen Umständen überhaupt Sinn hat, mein Referat zu publicieren—hätte ich es 4 Wochen später gehalten, so hätte es ja wesentlich anders gelautet. Wir kamen schliesslich überein, es als eine Beschreibung eines gewissen, wenn auch überholten Standes der Dinge doch niederzuschreiben.

In a note to the last sentence von Neumann adds: "Ich möchte betonen: *Nichts* an Hilberts Ansichten ist *falsch*. Wären sie durchführbar, so würde aus ihnen durchaus das von ihm Behauptete folgen. Aber sie sind eben undurchführbar, das weiss ich erst seit Sept. 1930."

dissertation.[g] The plenary session was complemented on 7 September by a roundtable discussion concerning the foundations of mathematics. That discussion was chaired by Hans Hahn, and its participants included Carnap, Heyting and von Neumann, but also three additional scholars, namely, Arnold Scholz, Kurt Reidemeister and Gödel. A shortened and edited transcript of this discussion was published as *Hahn et alii 1931* in *Erkenntnis*. Gödel was invited by the editors of the journal to expand on the very brief remarks about the first incompleteness theorem he had made during the discussion; the resulting note was added as a *Nachtrag* to the transcript (see *Gödel 1931a*). According to *Dawson 1997*, Gödel had already discussed the new discovery with Carnap and Waismann in Vienna, before the conference on 26 August 1930:

> The main topic of conversation was the plan for their upcoming journey to the conference in Königsberg, where Carnap and Waismann were to deliver major addresses and where Gödel was to present a summary of his dissertation results. But then, Carnap tersely noted, the discussion turned to "Gödel's discovery: incompleteness of the system of *Principia Mathematica*; difficulty of the consistency proof."[h]

This provides a sketch of the background for the meeting at which von Neumann made the acquaintance of Gödel. In his *1981*, Wang reports (Gödel's view) about the encounter with von Neumann:

> In September 1930, Gödel attended a meeting at Königsberg (reported in the second volume of *Erkenntnis*) and announced his result ⟦i.e., the first incompleteness theorem⟧. R. Carnap, A. Heyting, and J. von Neumann were at the meeting. Von Neumann was very enthusiastic about the result and had a private discussion with Gödel. In this discussion, von Neumann asked whether number-theoretical undecidable propositions could also be constructed, in view of the fact that the combinatorial objects can be mapped onto the integers, and expressed the belief that it could be done. In reply, Gödel said, "Of course undecidable propositions about integers could be so constructed, but they would contain concepts quite different from those occurring in number theory like addition and multiplication." Shortly afterward Gödel, to his own astonishment, succeeded in turning the undecidable proposition into a polynomial form preceded by quantifiers (over natural numbers). At the same time but independently of this result, Gödel also discovered his second theorem to the effect that no consistency proof of a reasonably rich system can be formalized in the system itself.[i]

[g] The abstract of Gödel's talk is *1930a*, the draft of his talk presumably **1930c*. Dawson's *1985* and *1997*, and also Mancosu's *1999a*, describe the early reception of the incompleteness theorems.

[h] *Dawson 1997*, p. 68.

[i] *Wang 1981*, pp. 654–655. The introductory note to the correspondence with Wang, in this volume, describes in section 3.2 the interaction between Gödel and Wang on which this paper is based.

As to the discovery of the second incompleteness theorem, we thus clearly know that Gödel did not have it in Königsberg and that, in contrast, the abstract *1930b* contains its classical formulation. It was Hahn who presented the abstract on 23 October 1930 to the Vienna Academy of Sciences. The full text of Gödel's *1931* was received for publication by the editors of *Monatshefte* on 17 November 1930.

There is genuine disagreement between Gödel and von Neumann on how the second incompleteness theorem affects Hilbert's finitist program. Von Neumann states his view strongly in his letters to Gödel of 29 November 1930 and 12 January 1931. (As to other views, cf. the introductory note to the correspondence with Herbrand and the exchange with Bernays, in particular, the letters of 24 December 1930, 18 January 1931, 20 April 1931 and 3 May 1931.) In his letter of 29 November to Gödel, von Neumann writes:

> I believe that every intuitionistic consideration can be formally copied, because the "arbitrarily nested" recursions of Bernays–Hilbert are equivalent to ordinary transfinite recursions up to appropriate ordinals of the second number class. This is a process that can be formally captured, unless there is an intuitionistically definable ordinal of the second number class that could not be defined formally—which is in my view unthinkable. Intuitionism clearly has no finite axiom system, but that does not prevent its being a part of classical mathematics that does have one.

From the general fact of the unprovability of a system's consistency within the system, he concludes: "There is no rigorous justification of classical mathematics." In the third letter, after having received the galleys of Gödel's *1931*, he writes even more forcefully:

> I absolutely disagree with your view on the formalizability of intuitionism. Certainly, for every formal system there is, as you proved, another formal one that is (already in arithmetic and the lower functional calculus) stronger. But intuitionism is not affected by that at all.

Denoting first-order number theory by A, analysis by M and set theory by Z, von Neumann continues:

> Clearly, I cannot prove that every intuitionistically correct *construction* of *arithmetic* is formalizable in A or M or even in Z—for intuitionism is undefined and undefinable. But is it not a fact, that not a single construction of the kind mentioned is known that cannot be formalized in A, and that no living logician is in the position of naming such ⟦a construction⟧? Or am I wrong, and you know an effective intuitionistic arithmetic construction whose formalization in A creates difficulties? If that, to my utmost surprise, should be the case, then the formalization should work in M or Z!

We know of Gödel's response to von Neumann's dicta not through a letter from Gödel, but rather through the minutes of the meeting of the

Schlick Circle that took place on 15 January 1931. These minutes report what Gödel viewed as questionable, namely, the claim that the totality of all intuitionistically correct proofs is contained in *one* formal system. That, he emphasized, is the weak spot in von Neumann's argumentation.[j] However, we also know that by December of 1933 Gödel had changed his view as follows: Finitism, considered by Gödel as the strictest form of constructive mathematics, is *narrower* than intuitionism and (its practice) can be captured in a formal system. Thus he argues, alluding to the second incompleteness theorem, the hope of succeeding along the lines proposed by Hilbert "has vanished entirely in view of some recently discovered facts". That change is made explicit in his talk **1933o* to the Mathematical Association of America.[k]

Before moving on to the further correspondence, some additional remarks on von Neumann's letter of 12 January 1931 are warranted: it contains metamathematical observations of special interest. The letter starts out by reporting:

> Incidentally, the other day I developed a method that always allows a finite decision for the effective provability question concerning propositions that are built up solely by means of the concepts "not", "or" (thus also "and", "follows", etc.), ⟦and⟧ "provable" (starting from the identical truth—consistency is for example such a proposition).

This observation seemingly anticipates (and announces a solution to) a problem Harvey Friedman formulated in *1975* as the 35th of his one hundred and two problems. *Boolos 1976* provided the first published solution.[l] A few paragraphs later von Neumann remarks on *Gödel 1931* and, in particular, on the proof of the second incompleteness theorem sketched there.

> Your paper is very nice; I am quite delighted, how briefly and elegantly you carried out the difficult and lengthy "enumeration" of formulas. However, I believe that the proof of the unprovability of consistency

[j]The minutes are found in the Carnap Archives of the University of Pittsburgh. Part of the German text is quoted in *Sieg 1988*, note 11, and more fully in *Mancosu 1999a*, pp. 36–37. Interestingly, *Bernays 1933* uses "von Neumann's conjecture" to infer that the incompleteness theorems impose fundamental limits on proof-theoretic investigations.

[k]Cf. **1933o*, these *Works*, vol. III, pp. 51–52 and also the introductory note to the correspondence with Herbrand.

[l]Around the same time, Claudio Bernardi and Franco Montagna also found a solution; their paper was submitted to *The journal of symbolic logic* shortly after Boolos' paper (thus, not accepted and indeed never published). According to Montagna in private communication, their proof was based on the algebraic semantics of provability logic due to Roberto Magari. "But apart from translating from Logic to Algebra, the proof was very similar to that of Boolos."

> can be shortened, i.e., that the general formal repetition of all considerations, as you propose, can be avoided.

Gödel had indeed suggested there a formal repetition of all considerations that lead to the unprovability in P of his sentence G, assuming P's consistency. He states on p. 197, "all notions defined (or statements proved) in Section 2, and in Section 4 up to this point are also expressible (and provable) in P." In the 1934 Princeton Notes (on p. 18) it is similarly asserted, "The fairly simple arguments of this proof can be paralleled in the formal logic...". Von Neumann's sketch of a simplified argument is obviously intended as a proof in P (though, through his appeal to Gödel's Theorem V, it is not quite). It has the flavor of the penetrating proof given by Hilbert and Bernays in *1939* via their axiomatic derivability conditions. I.e., von Neumann separates the general conditions on the provability predicate needed for the proof of the second incompleteness theorem from their verification concerning a particular formal system.

Von Neumann's admiration for Gödel's work is expressed directly in his very first letter of 20 November 1930, when he calls the first incompleteness theorem "the greatest logical discovery in a long time." That admiration is also reflected, for example, in his decision to talk about the incompleteness theorems when he lectured at Princeton in the fall of 1931. Kleene reports in his *1987b* (on p. 491) that through this lecture "Church and the rest of us first learned of Gödel's results." Von Neumann's friend Ulam states in his *Adventures of a Mathematician*:

> When it came to other scientists, the person for whom he ⟦von Neumann⟧ had a deep admiration was Kurt Gödel. This was mingled with a feeling of disappointment at not having himself thought of "undecidability." For years Gödel was not a professor at Princeton ... Johnny would say to me, "How can any of us be called professor when Gödel is not?"[m]

The letters from 13 July 1937 through 17 August 1939 are mainly focused on practical issues surrounding the publication of Gödel's work on the relative consistency of the axiom of choice and the generalized continuum hypothesis. After von Neumann finally had the opportunity to study Gödel's lectures thoroughly, he wrote on 22 April 1939:

> I would like to convey to you, most of all, my admiration: You solved this enormous problem with a truly masterful simplicity. And you reduced to a minimum the unavoidable technical complications of the proof details by a presentation of impressive persistence and drive. Reading your investigations was really a first-class aesthetic pleasure.

[m] *Ulam 1976*, p. 80.

It is quite impressive that von Neumann studied Gödel's investigations in sufficient detail to make also some "critical remarks"; perhaps even more impressive is his earlier letter of 28 February 1939 in which he directs Gödel to the paper *1938a* by Kondô that contains, in his view, "quite remarkable and surprising results on higher projective sets". He asks Gödel, "Are such matters not important for your further investigations on the continuum hypothesis...?" Gödel responds in his letter of 20 March 1939 by saying with reference to his *1938*: "The result of Kondô is of great interest to me and will definitely allow an important simplification in the consistency proof of 3. and 4. of the attached offprint." (For an explanation of the nature of these results, see Solovay's introductory note to *1938–1940*, these *Works*, vol. II, in particular pp. 14–15.)

During the following 17 years, it seems, von Neumann and Gödel did not exchange letters; after all, they were colleagues at the Institute for Advanced Study. In the spring of 1955, von Neumann took a leave from the Institute and moved from Princeton to Washington, DC, in order to work as a member of the Atomic Energy Commission to which he had been appointed by President Eisenhower. In the preface to von Neumann's posthumous *1958*, his widow Klara reports that von Neumann was diagnosed with bone cancer in August 1955. His health deteriorated quickly. By January 1956 he was confined to a wheelchair, though he still attended meetings and worked in his office. There was also some hope that X-ray treatment might be helpful. Klara von Neumann writes that by March of 1956, however, "... all false hopes were gone, and there was no longer any question of Johnny being able to travel anywhere.... In early April Johnny was admitted to Walter Reed Hospital; he never left the hospital grounds again until his death on February 8, 1957." Gödel wrote his last letter to von Neumann on 20 March 1956. He had heard, so he states in this letter, that von Neumann had undergone a radical treatment and was feeling better. "I hope and wish," Gödel continues, "that your condition will soon improve even further and that the latest achievements of medicine may, if possible, effect a complete cure." Then he formulates a striking mathematical problem and asks for von Neumann's view on it. It concerns the feasibility of computations and is closely connected to the problem that has caught, independently, the attention of mathematicians and computer scientists, the P versus NP problem.[n] For Gödel it is the question "how significantly *in general* for finitist combinatorial problems the number of steps can be reduced

[n]This is the question, whether the class P of functions computable in polynomial time is the same as the class NP of functions computable non-deterministically in

when compared to pure trial and error." The context in which he locates the general issue is noteworthy. Consider the question, whether a formula F in the language of first-order logic has a proof of length n, i.e., n is the number of symbols occurring in the proof. A suitably programmed Turing machine can answer this question. If $\psi(F, n)$ is the number of steps an "optimal" machine must take to obtain the answer and $\varphi(n) = \max_F \psi(F, n)$, then the important question is how rapidly $\varphi(n)$ grows. Gödel remarks that it is possible to prove that $\varphi(n) \geq Kn$, for some constant K. If there were a machine such that $\varphi(n)$ would grow essentially like Kn (or even Kn^2), Gödel suggests, "that would have consequences of the greatest significance. Namely, this would clearly mean that the thinking of a mathematician in case of yes-and-no questions could be completely replaced by machines, in spite of the unsolvability of the Entscheidungsproblem." In the next-to-last paragraph Gödel mentions Friedberg's recent solution of Post's problem and returns then to an issue that had been underlying much of the foundational discussion of the 1920s: In what formal framework can one develop classical analysis? Gödel reports that Paul Lorenzen has built up the theory of Lebesgue measure within ramified type theory.[o] "But," Gödel cautions, "I believe that in important parts of analysis there are impredicative inference methods that cannot be eliminated."

In the face of human mortality, Gödel thus chose to raise and discuss eternal mathematical questions.

Wilfried Sieg[p]

The translation is by Wilfried Sieg, revised using suggestions of John Dawson. A complete calendar of the correspondence with von Neumann occurs on p. 464 of this volume.

polynomial time. For a very good introduction to the rich and multifaceted problems that fall into the NP category, see *Garey and Johnson 1979*.

Part of the letter was already published in *Hartmanis 1989*; the full German letter and its English translation are found in the preface to *Clote and Krajíček 1993*. In both papers Gödel's question is related in informative ways to contemporary work in computational complexity. All the mathematical issues raised in Gödel's letter are addressed and resolved in *Buss 1995*. In particular, Buss shows that indeed $\varphi(n) \geq Kn$, for some constant K and infinitely many n, and that the n-symbol provability question raised by Gödel is NP-complete for predicate logic and, surprisingly, even for sentential logic.

[o]Gödel refers presumably to *Lorenzen 1955*.

[p]Many thanks go to Sam Buss and John Dawson for providing me with information on *Buss 1995* and *Clote and Krajíček 1993*, respectively *Hartmanis 1989*; Dawson helped in addition with a question concerning *1930b*. Solomon Feferman and Charles Parsons suggested substantive and stylistic improvements.

1. von Neumann to Gödel

Berlin, den 20. 11. 1930.
Hohenzollernstrasse 23, (Tiergarten.)

Lieber Herr Gödel!

Ich habe mich in der letzten Zeit wieder mit Logik beschäftigt, unter Verwendung der Methoden, die Sie zum ~~a~~ufweisen unentscheidbarer Eigenschaften so erfolgreich benützt haben. Dabei habe ich ein Resultat erzielt, dass mir bemerkenswert erscheint. Ich konnte nämlich zeigen, dass die Widerspruch⟦s⟧freiheit der Mathematik unbeweisbar ist. A

Dies ist genauer so:

In einem formalen System, das die Arithmetik umfasst, lässt es sich, in Anlehnung an Ihre Betrachtungen, aussprechen, dass die Formel 1 = 2 nicht Endformel eines von den Axiomen dieses Systems ausgehenden Beweises sein kann—u. zw. ist diese Formulierung eine Formel de~~f~~s genannten formalen Systems. Sie heisse 𝔚.

In einem widerspruch⟦s⟧vollen System ist jede Formel, also auch 𝔚 beweisbar. Ist die Widerspruch⟦s⟧freiheit intuitionistisch erwiesen, so lässt 2 sich durch "Übersetzung" der inhaltlich-intuitionisti|schen Betrachtungen ins Formale, 𝔚 ebenfalls beweisen. (Auf Grund Ihres Resultats könnte man an einer solchen "Übersetzbarkeit" u. U. zweifeln. Ich glaube immerhin, dass sie im genannten Falle bestehen müsste, und möchte gerne Ihre Ansicht darüber kennen.) Bei unbeweisbarem 𝔚 ist also das System widerspruch⟦s⟧frei, aber die Wi~~e~~derspruch⟦s⟧freiheit unbeweisbar.

Ich zeigte nun: in widerspruch⟦s⟧freien Systemen ist 𝔚 immer unbeweisbar, d. h. ein eventueller effektiver Beweis von 𝔚 liesse sich sicher in einen Widerspruch transformieren.—

Es würde mich *sehr* interessieren, Ihre Ansicht hierüber, besonders über die "Übersetzbarkeitsfrage" zu hören. Wenn es ~~s~~ie interessiert, teile ich Ihnen die Beweisdétails mit, sobald ich sie ~~D~~ruck- und ~~M~~itteilungsfähig aufgeschrieben habe; was bald der Fall sein wird—oder sobald wir Sie in Berlin begrüssen können. S d m

Wann erscheint Ihre Abhandlung und wann sind Korrekturen erhält- 3 |lich? Dies interessiert mich auch technisch, da ich sachlich und in den Bezeichnungen möglichst an Sie anschliessen möchte, und andererseits auch meinerseits je eher publicieren will.

Und nochmals: sehen wir Sie nicht bald in Berlin? E. Schmidt, dem ich Ihr Resultat, wie Sie es in Königsberg vortrugen, mitteilte, war davon entzückt. Er hält es, wie ich, für die grösste logische Entdeckung seit langer Zeit.

1. von Neumann to Gödel

Berlin, 20 November 1930
Hohenzollernstrasse 23 (Tiergarten)

Dear Mr. Gödel,

I have recently concerned myself again with logic, using the methods you have employed so successfully in order to exhibit undecidable properties. In doing so I achieved a result that seems to me to be remarkable. Namely, I was able to show that the consistency of mathematics is unprovable.

This is more precisely as follows:

In a formal system that contains arithmetic it is possible to express, following your considerations, that the formula $1 = 2$ cannot be the end-formula of a proof starting with the axioms of this system—in fact, this formulation is a formula of the formal system under consideration. Let it be called $\mathfrak{W}$.

In a contradictory system any formula is provable, thus also $\mathfrak{W}$. If the consistency ⟦of the system⟧ is established intuitionistically, then it is possible, through a "translation" of the contentual intuitionistic considerations into the formal ⟦system⟧, to prove $\mathfrak{W}$ also. (On account of your result one might possibly doubt such a "translatability". But I believe that in the present case it must obtain, and I would very much like to learn your view on this point.) Thus with unprovable $\mathfrak{W}$ the system is consistent, but the consistency is unprovable.

I showed now: $\mathfrak{W}$ is always unprovable in consistent systems, i.e., a putative effective proof of $\mathfrak{W}$ could certainly be transformed into a contradiction.

I would be *very* much interested to hear your views on this, in particular on the "translatability issue". If you are interested, I will send you the proof details, as soon as they are written up ready to be printed and communicated, which will soon be the case—or as soon as we can welcome you in Berlin.

When is your paper going to appear, and when are proof sheets going to be available? This is also of technical interest to me, as I would like to follow you as closely as possible both substantively and notationally, and on the other hand I would like to publish as soon as possible.

And again: aren't we going to see you soon in Berlin? E. Schmidt, to whom I communicated your result as you had presented it in Königsberg, was delighted by it. He considers it, as I do, to be the greatest logical discovery in a long time.

In der Hoffnung einer möglichst postwendenden Antwort, und des baldigsten Wiedersehens

bin ich Ihr ergebener

J. v. Neumann

2. von Neumann to Gödel

Berlin, den 29. 11. 1930

Lieber Herr Gödel!

Besten Dank für Ihren Brief und Ihr Separatum.[a] Da Sie den Satz über die Unbeweisbarkeit der Widerspruch⟦s⟧freiheit als naturgemässe Fortführung und Vertiefung Ihrer früheren Resultate bewiesen haben, werde ich natürlich über diesen Gegenstand nicht publicieren.

Auf Grund Ihrer Mitteilungen glaube ich⟦,⟧ Ihren Gedankengang reproducieren zu können, und kann Ihnen daher mitteilen, dass ich eine etwas andere Methode verwendete. Sie bewiesen $W \rightarrow A$, ich zeigte independent die Unbeweisbarkeit von W, u. zw. mit einer anderen, ~~An~~ gleichfalls Antinomien kopierenden, Schlussweise.

Hat sich seither etwas Neues ereignet? Sind Sie z. B. in der Lage entscheiden oder vermuten zu können, ob die Mathematik unvollständig ist, oder ω-~~W~~iderspruch⟦s⟧voll? w

Ich glaube, dass sich jede ~~I~~ntuitionistische Betrachtung formal kopieren lässt, denn die "beliebig hoch übereinander verschränkten" Rekursionen von Bernays–Hilbert sind ja gewöhnlichen transfiniten Rekursionen bis zu geeigneten Ordnungszahlen der zweiten Zahlklasse gleichwertig. Dies ist aber ein formal erfassbarer Prozess, es sei denn, dass eine intuitio|nistisch definierte Ordnungszahl der zweiten Zahlklasse nicht auch formal definiert werden könnte—was m. E. undenkbar ist. Wohl besitzt der Intuitionismus kein endliches Axiomensystem, aber das hindert doch nicht, das⟦s⟧ er Teil der klassischen Mathematik sei, die eins besitzt. i 2

Infolgedessen halte ich die Grundlagenfrage durch ~~i~~hr Resultat für im negativen Sinne erledigt: es gibt keine strenge Rechtfertigung für die klassische Mathematik. Welcher Sinn unserer Hoffnung, wonach sie doch I

[a]Probably *1930b*, which in view of the context of von Neumann's letter is more likely than *1930*. John Dawson has verified that Gödel did have reprints of *1930b* and distributed them to colleagues. However, since *1930b* does not contain any indication of

Hoping for an answer by return mail and to see you again very soon, I remain yours sincerely,

J. v. Neumann

2. von Neumann to Gödel

Berlin, 29 November 1930

Dear Mr. Gödel,

Many thanks for your letter and your reprint.[a] As you have established the theorem on the unprovability of consistency as a natural continuation and deepening of your earlier results, I clearly won't publish on this subject.

On the basis of your communications I think I can reproduce your line of thought, and therefore I can tell you that I employed a somewhat different method. You proved $W \rightarrow A$, I showed independently the unprovability of W, in fact with a different argument, that also mimics antinomies.

Has anything new happened since then? Are you able, for example, to decide or to conjecture whether mathematics is incomplete or ω-inconsistent?

I believe that every intuitionistic consideration can be formally copied, because the "arbitrarily nested" recursions of Bernays–Hilbert are equivalent to ordinary transfinite recursions up to appropriate ordinals of the second number class. This is a process that can be formally captured, unless there is an intuitionistically definable ordinal of the second number class that could not be defined formally—which is in my view unthinkable. Intuitionism clearly has no finite axiom system, but that does not prevent its being a part of classical mathematics that does have one.

Thus, I think that your result has solved negatively the foundational question: there is no rigorous justification for classical mathematics. What sense to attribute to our hope, according to which it is de facto

---

the proof of the second incompleteness theorem, von Neumann's remark "Auf Grund ihrer Mitteilungen..." must refer to the content of Gödel's letter.

de facto widerspruch⟦s⟧frei ist, dann zuzuschreiben ist, weis⟦s⟧ ich nicht—aber das ändert m. E. nichts an der vollendeten Tatsache.

Ich sehe mit grossem Interesse Ihren Korrekturen entgegen. Es tut mir sehr leid, dass die Aussicht, Sie hier begrüssen zu können, gering ist.

In der Hoffnung des baldigen Wi⟦e⟧derhörens

bin ich mit den besten Grüssen Ihr ergebener

Johann v. Neumann

3. von Neumann to Gödel

Berlin W. 10, den 12. 1. 1931.
Hohenzollernstrasse 23.

Lieber Herr Gödel!

Vielen Dank für Ihre zwei Briefe und die Korrektur-Fahnen. Ihre Bemerkung über die ω-Widerspruch⟦s⟧freiheit hat mich sehr interessiert. Übrigens habe ich ⟨mir⟩ neulich eine Methode überlegt, die bei Aussagen, die allein mit Hilfe der Begriffe "nein", "oder" (also auch "und", "folgt", u.s.w.), "~~B~~eweisbar" aufgebaut sind (etwa von der identischen Wahrheit ausgehend ~~)~~ —die Widerspruch⟦s⟧freiheit ist z. B. eine solche Aussage), ~~eine~~ stets eine endliche Entscheidung der effektiven Beweisbarkeitsfrage ermöglicht. (Natürlich unter Voraussetzung der ω-Widerspruch⟦s⟧freiheit.) Vielleicht, interessiert es Sie, oder Sie haben noch nicht daran gedacht, dann kann ich es aufschreiben. b

Bezüglich der ω-Widerspruch⟦s⟧freiheit bin ⟦ich⟧ eigentlich beruhigt, 2 da | sie ja aus der Widerspruch⟦s⟧freiheit des nächsten Typus folgt.

Absolut nicht einverstanden ~~mit~~ bin ich mit Ihrer Ansicht über die Formalisierbarkeit des Intuitionismus. Gewiss gibt es zu jedem formalen System, wie Sie zeigten, ein anderes formales⟦,⟧ das (schon in der Arithmetik und im engeren Funktionen-Kalkül) beweisstärker ist. Aber der Intuitionismus wird dadurch doch nicht berührt.

Um mich präziser zu fassen: Nennen wir das arithmetische Axiomensystem, in dem nur Zahlenvariablen vorkommen, aber weder Funktions- noch Mengenvariablen, dagegen die Kollektivbegriffe ((x), (Ex)) für die Zahlenvariablen frei verwendbar sind, A. Sind auch Funktionsvariable erster Stufe (etwa Funktionen mit einer einzigen Zahlenvariablen) da, 3 mit ihren Kollektivbegriffen ((f), | (Ef)), so heisse dieses System M. Schliesslich heisse etwa mein mengentheoretisches Axiomensystem Z.

consistent, I do not know—but in my view that does not change the completed fact.

I am looking forward to your proof sheets with great interest. I am very sorry that there is so little prospect of being able to welcome you here.

Hoping to hear from you again soon,

I remain with best wishes, yours sincerely

Johann v. Neumann

3. von Neumann to Gödel

Berlin W. 10, 12 January 1931
Hohenzollernstrasse 23

Dear Mr. Gödel,

Many thanks for your two letters and the proof sheets. Your remark on ω-consistency was very interesting to me. Incidentally, the other day I developed a method that always allows a finite decision for the effective provability question concerning propositions that are built up solely by means of the concepts "not", "or" (thus also "and", "follows", etc.), ⟦and⟧ "provable" (starting from the identical truth—consistency is for example such a proposition). Perhaps, if it interests you, or you have not yet thought of it, then I can write it up.

Concerning ω-consistency I am actually reassured, because it is implied by the consistency of the next type.

I absolutely disagree with your view on the formalizability of intuitionism. Certainly, for every formal system there is, as you proved, another formal one that is (already in arithmetic and the lower functional calculus) stronger. But intuitionism is not affected by that at all.

To be more precise: let us denote by A the arithmetical axiom system that contains number variables, but neither function nor set variables, and uses freely the quantifiers ((x), (Ex)) for the number variables. If also first-order function variables are available (functions of just one variable, for example) together with their quantifiers ((f), (Ef)), then this system may be called M. Finally, let for example my set theoretic axiom system be called Z.

Ich kann natürlich nicht beweisen, dass jede intuitionistisch korrekte *Konstruktion* der *Arithmetik* in A oder M oder auch nur in Z formalisierbar ist—denn der Intuitionismus ist undefiniert und undefinierbar. Aber es steht doch fest, dass keine einzige solche Konstruktion der genannten Art bekannt ist, ~~und das kein lebend~~ die nicht in A formalisierbar ist, und das⟦s⟧ kein lebender Logiker in der Lage ist, eine solche anzugeben? Oder irre ich, und kennen Sie eine ~~intuition~~ ⟨effektive⟩ intuitionistische arithmetische Konstruktion, deren Formalisierung in A Schwierigkeiten macht? Und sollte dies, zu meiner grössten Überraschung, der Fall sein, so geht doch sicher die Formalisierung in M oder in Z!

4 | Ich wäre Ihnen sehr dankbar, wenn Sie mir sagen würden, ob Sie wirklich die Existenz solcher Beispiele vermuten, oder gar welche kennen?—

Ihre Abhandlung ist sehr schön, ich bin ganz entzückt, wie kurz und elegant Sie die schwierige und langwierige "Abzählung" der Formeln ~~geb~~ durchgeführt haben. Dagegen glaube ich⟦,⟧ dass sich der Unbeweisbarkeitsbeweis für die Wi~~e~~derspruch⟦s⟧freiheit etwas kürzer führen lässt, d. h. dass die generelle formale ~~Wi(e)d~~ Wiederholung aller Betrachtungen, die ~~s~~ie vorschlagen, vermeidbar ist. S

Man kann ungefähr so schliessen:

1./ Sei a eine ~~Indu~~ ~~R~~ekursive Aussage. Dann kann man r

$$a \to B(a)$$

5 (*B* bedeute: beweisbar, in Ihrem Sinne) zeigen. (Das ist doch | ungefähr Ihr Satz 5?)[a]

2./ Ist $b = (Ex)a$, so kann man aus

$$a \to B(a)$$

auf

$$b \to B(b)$$

schliessen.

3./ Da jedes $B(a)$ diese Form hat, ist

$$B(a) \to B(B(a)),$$

für beliebige a.

4./ Sie gaben ein a mit

* $$\overline{a} \sim B(a)$$

an. Nach 3./ ist

** $$\overline{a} \to B(\overline{a}).$$

[a]Of *1931*, p. 186; in these *Works*, vol. I, p. 170.

Clearly I cannot prove that every intuitionistically correct *construction* of *arithmetic* is formalizable in A or M or even in Z—for intuitionism is undefined and undefinable. But is it not a fact, that not a single construction of the kind mentioned is known that cannot be formalized in A, and that no living logician is in the position of naming such ⟦a construction⟧? Or am I wrong, and you know an effective intuitionistic arithmetic construction whose formalization in A creates difficulties? If that, to my utmost surprise, should be the case, then the formalization should certainly work in M or Z!

I would be very grateful if you would tell me whether you are really conjecturing the existence of such examples, or whether you even know some?

Your paper is very nice; I am quite delighted, how briefly and elegantly you carried out the difficult and lengthy "enumeration" of formulas. However, I believe that the proof of the unprovability of consistency can be shortened, i.e., that the general formal repetition of all considerations, as you propose, can be avoided.

It is possible to argue roughly as follows:

1./ Let a be any recursive proposition. Then

$$a \rightarrow B(a)$$

can be shown (where B stands for provable, in your sense). (Isn't that approximately your Theorem 5?)[a]

2./ If $b = (Ex)a$, we can conclude

$$b \rightarrow B(b)$$

from

$$a \rightarrow B(a).$$

3./ As every $B(a)$ is of this form, we have

$$B(a) \rightarrow B(B(a)),$$

for arbitrary a.

4./ You constructed an a with

$*\qquad \bar{a} \sim B(a)$

According to 3./ we have

$**\qquad \bar{a} \rightarrow B(\bar{a}).$

Heisse O die Absurdität, und W die Widerspruch⟦s⟧freiheit, dann ist nach $*$ und $**$

$$\overline{a} \sim B(a) \ \& \ B(\overline{a}) \sim B(a \& \overline{a}) \sim B(O) \sim \overline{W},$$

$$W \sim a.$$

Somit ist W genau so unbeweisbar wie a.
(Sie können die Lücken meiner Darstellung sicher ausfüllen.)

Mit den besten Grüssen

Ihr Johann v. Neumann

4. von Neumann to Gödel

Budapest, den 14. 2. 1933.
V. Arany János ucca 16.

Lieber Herr Gödel!

Ich bin soeben aus Princeton nach Europa zurückgekommen, und ich würde mich sehr freuen, wenn sich während meines E̸⟨e⟩uropäischen Aufenthaltes Gelegenheit bieten würde, Sie zu treffen. Ich gehe Ende September nach Princeton zurück, wo ich ⟨mich⟩ in Zukunft 2 Semester im Jahre aufhalten werde, da ich ans neue Bamberger–Flexnersche "Institute for a̸dvanced Study" berufen wurde, und angenommen habe. A

Sie werden vielleicht über die Princetoner Verhältnisse, und die Struktur dieses neuen Instituts, dem Sie ja erfreulicherweise im nächsten Jahre angehören werden, Näheres wissen wollen, als Sie aus der bisherigen kurzen Korrespondenz erfahren haben. Ich schreibe Ihnen daher in erster 2 Linie, gleichzeitig im Auftrage Veblens, um Ihnen zu sagen, dass ich | Ihnen gerne mit jeder Information, die Sie interessieren mag, zur Verfügung stehe.

Vielleicht können wir uns auch irgendwann und irgendwo treffen—ich werde im Sommer wahrscheinlich noch (zum letzten Mal) in Berlin lesen, und wiederholt in Wien sein, bzw. durchreisen. Was ist Ihr Programm?

Mit den besten Grüssen, die ich Sie auch an Menger und Hahn weiterzugeben bitte,

Ihr

Johann von Neumann

Let O be absurdity and W consistency; then we have according to * and ** that

$$\overline{a} \sim B(a) \;\&\; B(\overline{a}) \sim B(a\&\overline{a}) \sim B(O) \sim \overline{W},$$

⟦and consequently⟧ $W \sim a$.

Thus, W is exactly as unprovable as a.
(I am sure you can fill in the gaps in my presentation.)

With best regards,

Yours sincerely,

Johann v. Neumann

4. von Neumann to Gödel

Budapest, 14 February 1933
V, Arany János ucca 16

Dear Mr. Gödel,

I just returned to Europe from Princeton, and I would be very glad if during my European stay there would be an opportunity to meet you. At the end of September I will go back to Princeton, where in the future I will spend two terms per year, as I received an offer from the new Bamberger–Flexner "Institute for Advanced Study", and I accepted.

Perhaps you would like to know more than you have gathered from the brief correspondence up to now about the conditions in Princeton and about the structure of this new Institute, of which you will fortunately be a member next year. Therefore, I write to you first of all, and also on Veblen's request, to tell you that I am happy to provide you with any information you may be interested in.

Perhaps we can also meet sometime and somewhere—this summer I will probably still lecture in Berlin (for the last time), and I will repeatedly be in or travel through Vienna. What is your program?

With best regards, which I ask you to extend also to Menger and Hahn,

Yours sincerely,

Johann von Neumann

5. Gödel to von Neumann

Wien VIII, 14./III. 1933.
Josefstädterstr. 43. Tel. A24921

Lieber Herr von Neumann!

Besten Dank für Ihren freundlichen Brief. Es würde mich sehr freuen, wenn ich im Laufe der nächsten Monate wieder einmal mit Ihnen sprechen könnte. Ich werde jedenfalls bis Ende Juni in Wien sein und voraussichtlich im Sommersemester meine erste Vorlesung halten. Bezüglich meiner Tätigkeit in Princeton hat mir Veblen geschrieben, daß er mir möglichst freie Hand lassen will. Vorlesungen möchte ich (wenigstens im 1. Semester meines Aufenthaltes) nicht halten, da sie mir unverhältnismäßig viel Mühe machen würden. Ich möchte mich lieber mit der Ausarbeitung einiger Ideen beschäftigen; z. B. habe ich eine Interpretation des Heytingschen Aussagenkalküls mit Hilfe des Begriffs "beweisbar" gefunden, konnte aber bisher die völlige Äquivalenz mit dem Heytingschen System nicht nachweisen.—Ein sehr großer Gewinn für mich wäre
2 | es, wenn ich Gelegenheit hätte, mich durch eine Ihrer Vorlesungen oder das Veblen-Seminar in die Quantenmechanik etwas einzuarbeiten. Diese Dinge interessieren mich lebhaft, ich bin aber bisher nicht dazu gekommen, mich ernstlich mit ihnen zu beschäftigen.—Näheres über die Verhältnisse in Princeton würde ich natürlich sehr gerne erfahren und danke Ihnen für Ihre Bereitwilligkeit, mich darüber zu informieren. Falls wir uns treffen sollten, kann dies wohl am besten mündlich geschehen.—Augenblicklich machen mir die amerikanischen Währungsverhältnisse einige Sorgen. Meinen Sie, daß eine wesentliche Entwertung des $ zu befürchten ist, und glauben Sie, daß ich in diesem Fall mit einer Entschädigung von Seiten des Flexner-Institutes rechnen könnte?—Ich hoffe, daß ich bald Gelegenheit haben werde, mit Ihnen persönlich zu sprechen, und verbleibe mit den besten Grüßen

Ihr Kurt Gödel

5. Gödel to von Neumann

Vienna VIII, 14 March 1933
Josefstädterstr. 43. Tel. A24921

Dear Mr. von Neumann,

Many thanks for your kind letter. I would be very glad if I could once again talk with you in the course of the next few months. I will in any case stay in Vienna until the end of June and, presumably, give my first set of lectures during the summer semester. With respect to my activities in Princeton Veblen wrote to me that he will give me as much freedom as possible. I would not like to give lectures (at least not during the first term of my stay), because they would require disproportionate efforts. I would prefer working out some ideas; for example, I have found an interpretation of Heyting's calculus for sentential logic by means of the concept "provable", but up to now I have not been able to establish the full equivalence with Heyting's system.

I would greatly gain from an opportunity to familiarize myself with quantum mechanics through one of your lectures or Veblen's seminar. I have a lively interest in these matters, but up to now I have not managed to take them up in a serious way.

Naturally, I would very much like to get to know more about the conditions in Princeton, and I am grateful to you for your willingness to inform me about them. In case we should meet, that can best be done orally.

At the moment I am worried about the state of the American currency. Do you think that there is a danger of a substantial devaluation of the dollar, and do you believe that in such a case I could count on being compensated by the Flexner-Institute?

I hope that I will very soon have the opportunity to talk with you in person, and remain with best regards,

Yours sincerely,

Kurt Gödel

6. von Neumann to Gödel[a]

Berlin, W10, den 13. 6. 1933

Lieber Herr Gödel!

In 2 Tagen werde ich in Wien sein, und würde mich sehr freuen, wenn ich Sie bei dieser Gelegenheit sehen könnte. Da ich voraussichtlich nur 24 Stunden lang (von Donnerstag früh bis Freitag früh) bleiben kann, wäre ich Ihnen sehr dankbar, wenn Sie am Donnerstag eine Zeit für mich frei halten würden. Ich werde ~~mie~~ gleich nach meiner Ankunft versuchen, Sie telephonisch zu erreichen.

In der Hoffnung des baldigen Wiedersehens

Ihr

J. v. Neumann

---

[a]On letterhead of Hausverwaltung Hohenzollernstrasse Nr. 23/24.

7. von Neumann to Gödel

Budapest, den 12. August 1934.
V, Arany János ucca 16.

Lieber Herr Gödel!

Vielen Dank für Ihren lieben Brief. Wie Sie von meinem Bruder gehört haben werden, geht es uns allen, inkl. Auto, wieder ganz gut. Wir bleiben bis zum 20-sten September hier, und reisen am 22-sten September von Genua nach New-York–Princeton.

Wie geht es Ihnen, und was gibt es Neues in Wien? Ich habe mit grossem Bedauern ⟦von⟧ Hahn/s Tod erfahren. Wie geht es Menger? Was ist Ihr Winter-Programm, und wann sehen wir Sie in Princeton wieder?

Ich werde vielleicht in einigen Wochen 1–2 Tage in Wien sein, finde ich Sie dann dort, oder gehen Sie in die "Ferien"?

Mit den besten Grüssen, auch von meiner Frau, bin ich

Ihr

J. v. Neumann

6. von Neumann to Gödel[a]

Berlin W10, 13 June 1933

Dear Mr. Gödel,

In two days I will be in Vienna, and I would be very glad if I could see you on that occasion. As I can presumably stay only for 24 hours (from Thursday morning to Friday morning), I would be very grateful to you if you would reserve some time for me on Thursday. I will try to reach you by telephone immediately after my arrival.

In the hope of seeing you again soon,

Yours sincerely,

J. v. Neumann

---

7. von Neumann to Gödel

Budapest, 12 August 1934
V, Arany Jánus ucca 16

Dear Mr. Gödel,

Thank you very much for your kind letter. As you have heard from my brother, all of us, including the car, are reasonably fine again. We are going to stay here until 20 September and will travel on 22 September from Genoa to New York–Princeton.

How are you doing, and what is new in Vienna? I have heard with great sorrow of Hahn's death. How is Menger? What is your program for the winter, and when are we going to see you again in Princeton?

Perhaps I will be in Vienna in a few weeks for one or two days; will I find you there, or are you going on vacation?

With best regards, also from my wife, I remain

yours sincerely,

J. v. Neumann

8. von Neumann to Gödel

Budapest, den 18. 9. 1934.

Lieber Herr Gödel,

vielen Dank für Ihren lieben Brief. Ich komme, zu meinem Bedauern, doch nicht nach Wien, vielmehr reisen wir heute Abend nach Genua, von wo wir übermorgen früh ~~abreis~~ mit der "Conte di Savoia" weiterreisen. Am 27-sten sind wir in N.Y.C., am 28-sten in Princeton.

Hoffentlich ist Ihre Gesundheitsstörung nichts Ernstes, und ich hoffe bestimmt, dass wir uns im Frühling in Princeton wiedersehen.

Ich wäre Ihnen recht dankbar, wenn Sie mir gelegentlich schreiben würden, meine Princetoner Adresse ist die alte.

Mit den besten Grüssen, auch von meiner Frau, Ihr ergebener

J. v. Neumann

P.S. Bitte grüssen Sie Menger herzlichst von mir.

9. von Neumann to Gödel[a]

Paris, le 29. Mai 1936.

Lieber Herr Gödel,

ich war 2 Wochen hier, da ich am obigen Institut Vorträge hielt, und bleibe noch ca. eine Woche. Dann—am 4 Juni—reise ich nach Princeton zurück, hauptsächlich um ein paar überfällige Arbeiten in Ruhe aufzuschreiben.

Leider habe ich seit Ihrer Abreise nichts mehr von Ihnen gehört. Wie geht es Ihnen, und was sind Ihre Pläne? ⟨Besonders puncto Princeton?⟩ Was haben Sie betr. Ihrer Wohlordnungssatz-Arbeit beschlossen?

---

[a] On letterhead of Faculté des Sciences, Université de Paris, Institute Henri Poincaré, 11, Rue Pierre-Curie (V^{e}).

8. von Neumann to Gödel

Budapest, 18 September 1934

Dear Mr. Gödel,

Thank you very much for your kind letter. After all, and to my regret, I won't come to Vienna; rather, we are leaving tonight for Genoa, and from there we'll continue our trip on the morning of the day after tomorrow with the "Conte di Savoia". On the 27th we'll be in N.Y.C., on the 28th in Princeton.

Hopefully your health problems are not serious, and I hope definitely that we are going to see each other again during the spring in Princeton.

I would be very grateful to you if you would write to me occasionally; my Princeton address is the old one.

With best wishes, also from my wife, I remain yours sincerely,

J. v. Neumann

P.S. Please give my warmest regards to Menger.

9. von Neumann to Gödel[a]

Paris, 29 May 1936

Dear Mr. Gödel,

I have been here for two weeks, as I am giving lectures at the above institute, and I am going to stay approximately for another week. Then —on 4 June—I'll travel back to Princeton, mainly to write up leisurely several papers that are overdue.

Unfortunately I have not heard from you since your departure. How are you, and what are your plans? In particular, with respect to Princeton? What have you decided with regard to your paper on the well-ordering theorem?

2 Ich wäre Ihnen *sehr* dankbar, wenn Sie mich | gelegentlich von Ihnen hören liessen, am liebsten noch hierher! (Adresse: 37, Rue Cambon, Hotel De Castille.)

Mit den allerbesten Grüssen, auch von meiner Frau,

Ihr

J. v. Neumann

10. von Neumann to Gödel[a]

Budapest, den 13 Juli 1937

Lieber Herr Gödel!

Ich habe mich ungemein gefreut⟦,⟧ wieder von Ihnen zu hören, und ganz besonders, zu erfahren dass es Ihnen gut geht, und dass Sie wieder Vorlesungen gehalten haben.

Ich freue mich auch sehr, dass Ihre Arbeit über das Auswahl Axiom nunmehr erscheinen wird. Wenn Sie sich entschliessen sollten, die ausführliche Abhandlung in den "Annals" zu veröffentlichen[b]—ich brauche 2 Ihne⟦n⟧ wohl nicht besonders zu versichern, | wie sehr ich da~~s~~s begrüssen würde, und auch meine Kollegen, wenn sie es wüssten—so könnten wir dieselbe innerhalb eines halben Jahres publici⟦e⟧ren. (Ich nehme an, dass die Arbeit 30–35 Seiten lang wäre? Da die "Annals" vierteljährlich erscheinen ist die Wartezeit natürlich "gequantelt", also sollte ich lieber sagen: Die Arbeit könnte bestimmt in der zweiten oder dritten Nummer —wahrscheinlich in der zweiten[1] —nach Eingang erscheinen.) Wenn Sie mir mitteilen, wann das Manuskript voraussichtlich fertig wird, so kann ich Lefschetz und Bohnenblust schreiben, um Platz zu reservi⟦e⟧ren. (Ich 3 habe | vorläufig noch niemand~~a~~m darüber etwas mitgeteilt.) e

[1]Ich könnte dies durch eine Anfrage in Princeton vermutlich in eine Gewissheit verwandeln.

[a]On letterhead of the I.A.S.

I would be *very* grateful to you if you would write to me on occasion, preferably while I am still here! (Address: 37, Rue Cambon, Hotel de Castille.)

With best regards, also from my wife,

Yours sincerely,

J. v. Neumann

10. von Neumann to Gödel[a]

Budapest, 13 July 1937

Dear Mr. Gödel,

I am extremely happy to hear from you again, and in particular, to learn that you are well and that you have been lecturing again.

I am also very glad that your paper on the axiom of choice is now going to appear. If you should decide to publish the extended presentation in the "Annals"[b]— I need not especially assure you how much I would welcome that, and also my colleagues, if they only knew about it—then we could publish it within half a year. (I assume that the paper would be 30–35 pages long? As the "Annals" appear quarterly the waiting time is naturally "quantized", and I should rather say: The paper could certainly appear in the second or third issue—probably in the second[1]—after ⟦your⟧ submission.) If you tell me when presumably the manuscript is going to be ready, I can write to Lefschetz and Bohnenblust to reserve space. (Up to now I have not told anyone about this.)

[1]I could, presumably, turn this into a certainty through an inquiry in Princeton.

[b]At this point, von Neumann is discussing the publication of a paper in the journal *Annals of mathematics*. Gödel's lecture notes were published eventually as the third volume in the new *Annals of mathematics studies*; cf. letter 19 below.

Es wäre sehr schön, wenn Sie in 1938/39 nach Princeton kommen würden. Ich habe keine Zweifel, dass das gehen wird, und würde mich sehr freuen, dies im Herbst "offiziell" zu regeln. Soll ich es tun?

Ich möchte mich über viele Dinge, insbesondere über Ihre weiteren Pläne, mit Ihnen mündlich unterhalten. Wann sind ~~s~~ie in Wien zu fin- S
4 den? Ich werde wohl in einigen Wochen | einmal hinkommen. Bitte teilen Sie mir doch Ihre Sommer-~~Plan~~-Pläne mit!

Mit den besten Grüssen auch von meiner Frau

Ihr

J. v. Neumann

P.S. Darf ich Sie um das Folgende bitten: Ich habe letztes Jahr Menger eine Abhandlung über eine mathematische Behandlung einer Nationalökonomischen Frage gegeben—für die "Mitteilungen aus dem Mathematischen Seminar, etc.". Menger hat mir wiederholt versichert, dass
5 die Abhandlung bald erscheint, aber alle diese | Termine sind lange überschritten. Wissen Sie zufällig, wie es damit steht? Oder *wer* dafür in Wien *wirklich* zuständig ist? Ich möchte die Arbeit, wenn ich nicht erfahren kann, wann ~~si~~ sie erscheinen wird, lieber zurückziehen und anderswo publici⟦e⟧ren. Ich wäre Ihnen für eine baldige Antwort sehr dankbar.

Im vornherein dankend Ihr

J. v. Neumann

(P.S.)^2 Da ich Ihrer Adresse nicht ganz sicher bin, wäre ich Ihnen
6 recht dankbar, wenn Sie mich | möglichst bald wissen liessen, ob Sie dieser Brief erreicht hat. Meine Adresse hier ist:

J. v. Neumann
Budapest (Ungarn),
V., Arany János ucca 16.

It would be very nice if you would come to Princeton in 1938/39. I have no doubt that that is possible, and I would be very glad to settle this in the fall "officially". Shall I do this?

I would like to talk with you in person about many things, in particular about your further plans. When is it possible to find you in Vienna? I assume I'll be there in a few weeks. Please let me know about your summer plans!

With best regards, also from my wife,

Yours sincerely,

J. v. Neumann

P.S. May I ask you for the following favor: Last year, I sent a paper on a mathematical treatment of an economic question to Menger—for the "Mitteilungen aus dem Mathematischen Seminar, etc.". Menger assured me repeatedly that the paper is to appear soon, but all those dates have long since passed. Do you know by any chance, what the situation is? Or *who* in Vienna is *really* responsible for it? If I cannot learn when the paper is to appear I would rather withdraw it and publish it elsewhere. I would be very grateful to you for a speedy reply.

I am thanking you in advance,

Yours sincerely,

J. v. Neumann

(P.S.)^2 As I am not completely sure of your address, I would be very grateful to you if you could let me know as soon as possible whether this letter has reached you. My address here is:

J.v. Neumann
Budapest (Ungarn),
V., Arany János ucca 16.

11. von Neumann to Gödel[a]

Sept 14., 1937.

Lieber Herr Gödel,

ich schreibe Ihnen, um zu berichten, dass ich Ihnen nach Rücksprache mit Lefschetz nunmehr offi¢iell versprechen kann, dass Ihre Arbeit über z
2 die Hypothese $\aleph_{\alpha+1} = 2^{\aleph_\alpha}$ in der | ersten Nummer der Annals in der es physisch möglich ist (d. h. deren Drucklegung $\geq$ 14 Tage nach Ankunft des Manuskriptes abgeschlossen wird) erscheint. Wünschen Sie Korrekturen nach Europa?—Ich möchte Sie aber bitten⟦,⟧ mir möglichst bald mit‿zu‿teilen, *ob* und *wann* wir auf die Arbeit rechnen können.

Am Institut sind noch Veblen und Flexner nicht zurück. Bald schrei-
3 |be[b] ich Ihnen auch über unsere Möglichkeiten betr. Ihres Besuches.

Mit den besten Grüssen

Ihr J. v. Neumann

[a]Addressed to Gödel at VIII Josephstädterstrasse 43, Wien, with annotation on envelope: With *S.S. "NORMANDIE"*, sails on *WEDNESDAY*, Sept. 15, from *NEW YORK CITY*; The Warwick, 65 West 54th St., New York.

12. von Neumann to Gödel[a]

Nov. 11., 1937.

Lieber Herr Gödel,

vielen Dank für Ihren lieben Brief. Wie Sie inzwischen von Veblen erfahren haben werden, sind wir in der Lage⟦,⟧ Ihnen ein Angebot zu machen⟦,⟧ das mit jenem von Notre Dame combini⟦e⟧rt, ganz vorteilhaft ist. Für mehr reicht's in diesem Jahr finanziell nicht. Ich hoffe aber sehr, dass wir im nächsten Jahr etwas wesentlich seriöseres—etwa im von Ih-

[a]Addressed to Gödel at "VIII Josephstädterstrasse 43, Wien" and forwarded to XIX Himmelstrasse 43/53; on letterhead of the I.A.S., with envelope bearing the annotation: With "S. S. Bremen", sails on Nov. 12, Friday, from New York.

11. von Neumann to Gödel[a]

14 September 1937

Dear Mr. Gödel,

I am writing you to tell you that I can promise you officially, after having talked to Lefschetz, that your paper on the hypothesis $\aleph_{\alpha+1} = 2^{\aleph_\alpha}$ is going to appear in the first number of the Annals in which it physically could (i.e., whose printing will be completed ≥ 14 days after the arrival of the manuscript). Would you like ⟦us to send⟧ the proof sheets to Europe?—But I would request that you tell me as soon as possible *whether* and *when* we can count on your paper.

At the Institute, Veblen and Flexner are not back yet. I'll write[b] to you soon also about our possibilties concerning your visit.

With best regards,

Yours sincerely,

J. v. Neumann

---

[b]Crossed out on the left-hand side: ⟦schrei⟧be ich Ihnen wieder über unsere Pläne. Herzl. Grüsse von Neumann. (I will write again about our plans. Cordial greetings, von Neumann.)

12. von Neumann to Gödel[a]

11 November 1937

Dear Mr. Gödel,

Thank you very much for your kind letter. As you have heard in the meantime from Veblen, we are in a position to make you an offer that, in combination with the one from Notre Dame, is reasonably advantageous. For financial reasons we can't do more this year. But I hope very much that next year we can do something more serious—in the sense you in-

nen angedeuteten Sinne—unternehmen ~~könnten~~ ⟨können⟩. Ich hoffe sehr, dass Sie sich doch noch entschliessen werden, noch in diesem akademischen Jahr hierherzu|kommen. Wenn nicht, so hoffe ich um so mehr aufs nächste Jahr.

Natürlich würde Ihr Entschluss—ob Sie nämlich schon jetzt kommen, oder nicht—die Chancen für's nächste Jahr in keiner Weise beeinträchtigen; aber ich hoffe doch sehr—aus rein egoistischen Gründen, und alle meine Kollegen sind mit mir darin einig—dass wir Sie noch dieses Jahr zu sehen bekommen.

Mit den besten Grüssen

Ihr ergebener

J. v. Neumann

13. von Neumann to Gödel

Budapest, den 13. 1. 1938.

Lieber Herr Gödel,

ich bin in verschiedenen Familien- und anderen Angelegenheiten auf kurze Zeit nach Europa gekommen, und würde mich sehr freuen Sie noch vor meiner Rückreise in Wien sehen zu können. Sind Sie im Laufe der nächsten Woche, ~~sowie~~ besonders aber am Ende der Woche (Sonntags, den 23-sten) erreichbar?

Ich werde Ihnen mündlich über die Princetoner Möglichkeiten berichten, ich glaube⟦,⟧ dass wir eine in Ihrem Sinne befriedigende Lösung herbeiführen können.

In der Hoffnung einer möglichst *postwendenden* Antwort[a]

Ihr sehr ergebener

J. v. Neumann

Z.Zt. Budapest (Ungarn),
V. Pozsonyi ut 28.

---

[a]This letter was followed by a telegram from Budapest, *Nachlaß* document #013042, dated January 22, 1938. Addressed to Gödel at Himmelgasse 43, Wien, its brief message read "Erwarte sie halle hotel sacher elf vormittags sonntag. Neumann"; ("I'll wait for you in the hall of the Hotel Sacher at 11 a.m. Sunday".)

dicated. I hope very much that you will decide after all to come here for this academic year. If not, I hope all the more for next year.

Naturally, your decision—whether or not to come now—would not in any way diminish your chances for next year; but I hope very much—for purely egotistical reasons, and my colleagues agree with me on this—that we will see you still during this year.

With best regards,

Yours sincerely,

J. v. Neumann

13. von Neumann to Gödel

Budapest, 13 January 1938

Dear Mr. Gödel,

I have come to Europe for a short time to take care of various family and other affairs, and I would be very glad to see you in Vienna before my departure. Can you be reached during the next week, in particular at the end of the week (Sunday, the 23rd)?

I will report to you orally about the possibilities in Princeton; I believe that we can achieve a solution that will be satisfactory to you.

Hoping for an answer, possibly, *by return mail*,[a]

Yours sincerely,

J. v. Neumann

Presently: Budapest (Hungary)
V. Pozsonyi ut 28.

14. Gödel to von Neumann

Wien 12. Sept. 1938.

Lieber Herr v. Neumann!

Von Veblen höre ich, dass Sie im Sommer wieder in Budapest waren u. hoffe, dass Sie mein Brief noch dort erreicht. Ich komme mit der "Hamburg" am 7. Okt. in New York an u. beabsichtige das ganze Semester in Princeton zu bleiben u. über das Kontinuumproblem zu lesen. Worüber lesen Sie? Wird insbes. Ihre Vorlesung eine Fortsetzung einer vorjährigen oder etwas Neues sein?

Meine Arbeit über das Kontinuum ist seit ca. 2 Monaten im wesentlichen fertig. Doch wollte ich noch einige Änderungen anbringen u. ausserdem mit der Publikation bis zu meiner Ankunft in U.S.A. warten, um das 2 Hin- u. Hersenden der Korrekturen zu ersparen. | Der Satz über eineindeutige stetige Bilder komplem⟦.⟧-analytischer Mengen, über den wir bei unserem letzten Beisammen sein sprachen, hat sich als falsch herausgestellt (widerlegt v. Mazurkiewicz in Fund⟦amenta Mathematicae⟧ 10).[a]

Ich habe jetzt sogar einige Resultate in entgegengesetzter Richtung, nämlich: Man kann widerspruchsfrei die beiden folgenden Sätze annehmen:

1. Es gibt nicht-messbare eineindeutige stetige Bilder kompl.-analyt. Mengen⟦.⟧
2. Es gibt unabzählbare total imperfekte kompl.-analytische Mengen⟦.⟧

Ich freue mich sehr⟦,⟧ Sie bald wiederzusehen.

Mit besten Grüssen sowie Empfehlungen an Ihre geschätzte Familie

Ihr ergebener Kurt Gödel

---

[a] *Mazurkiewicz 1927*.

14. Gödel to von Neumann

Vienna, 12 September 1938

Dear Mr. v. Neumann,

I heard from Veblen that you were again in Budapest during the summer, and I hope that my letter will still reach you there. I will arrive in New York with the "Hamburg" on 7 October and intend to spend the whole term in Princeton and lecture on the continuum problem. On what are you going to lecture? In particular, will your lectures continue ones from last year, or will they start with something new?

My paper on the continuum has been ready essentially for two months. But I wanted to make some changes and, in addition, wait for the publication until my arrival in the US in order to eliminate sending the proof sheets back and forth. The theorem on one-to-one continuous images of compl.-analytic sets, which we discussed at our last meeting, turned out to be false (refuted by Mazurkiewicz in Fund⟦amenta Mathematicae⟧ 10).[a]

I now even have some results in the opposite direction, namely: one can consistently assume the following two statements:

1. There are non-measurable one-to-one continuous images of compl.-analytic sets.
2. There are uncountable totally imperfect compl.-analytic sets.

I am very glad to see you again soon. With best greetings, as well as regards to your esteemed family,

Yours sincerely,

Kurt Gödel

15. von Neumann to Gödel[a]

Febr. 28. ⟦1939⟧

Lieber Herr Gödel,

ich möchte Ihre Aufmerksamkeit auf die folgende Abhandlung von M. Kondô[b] lenken:

> "Sur l'uniformisation des complémentaires analytiques et les ensembles projectifs de la seconde classe", Japanese Journal of Mathematics, vol XV (1938), SS. 197–230.

Die Abhandlung enthält verschiedene, m. E. sehr bemerkenswerte und überraschende Resultate über höhere projektive Mengen. Z. B.: Das letzte "Théorème" auf S. 198, wonach jede lineare projektive Menge zweiter
2 Klasse *eineindeutige* Projektion | einer (ebenen) Komplementärmenge einer (ebenen) projektiven Menge erster Klasse—d. h. eines ebenen "analytischen Komplements"—ist. Also ~~analytic~~ auch *eineindeutiges* stetiges Bild eines linearen "anal. Kompl.".
Sind solche Dinge nicht für Ihre weiteren Untersuchungen über die Kontinuumhypothese (betr. der Widerspruch⟦s⟧freiheit der Negation derselben) wichtig?

Wie geht es Ihnen, und was machen Sie? Ich wäre Ihnen sehr dankbar, wenn Sie mir die Abschrift Ihrer Princetoner Vorlesung 1938/39 irgendwie zugänglich machen könnten.

Mit den besten Grüssen stets Ihr

J. v. Neumann.

[a]On I.A.S. letterhead.

15. von Neumann to Gödel[a]

28 February ⟦1939⟧

Dear Mr. Gödel,

I would like to point your attention to the following paper by M. Kondô:[b]

> "Sur l'uniformisation des complémentaires analytiques et les ensembles projectifs de la seconde classe", Japanese Journal of Mathematics, vol XV (1938), SS. 197–230.

The paper contains various results on higher projective sets that in my view are quite remarkable and surprising. For example: The last "Théorème" on p. 198, according to which every projective set of the second class is a *one-to-one* projection of a (planar) complement of a (planar) projective set of the first class—i.e. of a planar "analytic complement". Thus also a *one-to-one* continuous image of a linear "analytic complement".

Are such matters not important for your further investigations on the continuum hypothesis (concerning the consistency of its negation)?

How are you, and what are you doing? I would be very grateful to you if you could somehow make accessible to me the copy of your Princeton lectures 1938/39.

With best regards,

Yours sincerely,

J. v. Neumann

---

[b] *Kondô 1938a.*

16. Gödel to von Neumann

Notre-Dame, 20./III. 1939.

Lieber Herr v. Neumann!

Vielen Dank für Ihren Brief vom 28./II. Das Resultat von Kondô ist von grossem Interesse für mich u. wird jedenfalls den Beweis der Widerspruchsfreiheit von 3. und 4. des beiliegenden Separatums[a] erheblich zu vereinfachen gestatten.

Was meine Princeton'er Vorlesung betrifft, so bin ich eben dabei, das Manuskript vor der Vervielfältigung noch einer gründlichen Revision zu unterziehen. Damit hoffe ich aber im Laufe dieser Woche fertig zu sein u. werde es Ihnen dann sofort zuschicken.

Mir geht es hier soweit ganz gut, bloss das Klima finde ich unangenehmer als in Princeton.

Herzliche Grüsse u. beste Empfehlungen an Ihre Frau Gemahlin.

Ihr Kurt Gödel

[a] *Gödel 1938.*

17. von Neumann to Gödel[a]

April 22., 1939.

Lieber Herr Gödel!

Veblen hat mir Ihr Manuskript des Zulässigkeitsbeweises von ⟨dem⟩ Wohlordnungssatz + ~~A~~llgemeine~~r~~ Kontinuumssatz gezeigt. Ich habe ~~Ihr~~ dasselbe eingehend studi⟦e⟧rt, und es dann Herrn Brown übergeben, der es zur Vervielfältigung vorbereitet. a m

Ich möchte Ihnen vor Allem meine Bewunderung ausdrücken: Sie haben dieses enorme Problem mit einer wirklich meisterhaften Einfachheit erledigt. Und die unvermeidlichen technischen Komplikationen der Beweisdétails haben Sie durch eine Darstellung von imponi⟦e⟧render Consequenz und Schwung | aufs Minimum reduci⟦e⟧rt. Die Lektüre Ihrer Untersuchung war wirklich ein ästhetischer Genuss erster Klasse.—Kritisch möchte ich nur dieses bemerken: 2

[a]On I.A.S. letterhead.

16. Gödel to von Neumann

Notre-Dame, 20 March 1939

Dear Mr. v. Neumann,

Thank you very much for your letter of 28 February. The result of Kondô is of great interest to me and will definitely allow an important simplification in the consistency proof of 3. and 4. of the attached offprint.[a]

As to my Princeton lectures, I am in the process of thoroughly revising the manuscript before it is duplicated. But I hope to be finished with that in the course of this week, and then I will send it to you immediately.

I am doing reasonably well here, except that I find the climate more unpleasant than in Princeton.

Warm regards and best wishes to your wife,

Yours sincerely,

Kurt Gödel

---

17. von Neumann to Gödel[a]

22 April 1939

Dear Mr. Gödel,

Veblen has shown me your manuscript of the admissibility proof for the well-ordering theorem + general continuum theorem. I studied it closely and then gave it to Mr. Brown, who is preparing it for duplication.

I would like to convey to you, most of all, my admiration: You solved this enormous problem with a truly masterful simplicity. And you reduced to a minimum the unavoidable technical complications of the proof details by a presentation of impressive persistence and drive. Reading your investigation was really a first class aesthetic pleasure.

Critically, I would like to make only the following remarks:

---

1) Warum benützen Sie das "Fundi⟦e⟧rungsaxiom"? Wenn Sie es nicht benützen, wird die Ordnungszahlen-Theorie nur um ein ganz Geringes verwickelter—ich glaube nicht, dass Ihre ganze Abhandlung um mehr als 1 bis 2 Seiten länger wird.
   Und aus $\Sigma = \Delta$ folgt das "Fundi⟦e⟧rungsaxiom" sofort.
2) Aus $\Sigma = \Delta$ folgt auch die Gültigkeit meines Axioms: "Ein Bereich ist dann und nur dann eigentlich (d. h.: keine Menge), wenn er dem Bereich aller Elemente äquivalent ist.⟦"⟧
3) D. h.: Aus Ihrem $\Sigma = \Delta$ folgt | automatisch Alles, was ich in meiner zweiten mengentheoretischen Arbeit im Crelleschen Journal[b] bewiesen hatte. Mein Vorgang dort ist ja in einem gewissen Sinne eine viel gröberes (weil auf die logische Struktur der gebildeten Mengen keine Rücksicht nehmendes) Analogon Ihrer Δ erzeugenden transfiniten Rekursion.
4) Auch die Zulässigkeit der Annahme, dass ausser ω keine regulären Anfangszahlen von Limesindex gibt, folgt jetzt (mit Hilfe des allgemeinen Kontinuumssatzes) sofort.

3

Wann kommen Sie nach Princeton, bzw. wann passi⟦e⟧ren Sie es auf der Durchreise?

Ich bin bis Mai 2. hier, von Mai 3. bis Mai 17. in Ann Arbor, Michigan, wo ich Vorlesungen halte, und von Mai 18. bis Ende Juni wieder hier. Hoffentlich sehen wir uns irgendwo! Könnten wir nicht im Anschluss an meine | Michiganer Expedition zusammenkommen—Ann Arbor ist wohl nicht sehr weit von South Bend.

4

Ich möchte gerne wieder einmal über viele mathematische Dinge mit Ihnen sprechen. Z. B. über die nicht-distributive Logik, und andere—zwar distributive, aber von der klassischen doch wesentlich verschiedene—logische Systeme, und über deren "transcendenten" Ausbau. Und auch über nicht-logische Fragen.—

In der Hoffnung des baldigen Wi⟦e⟧derhörens,

und mit den besten Grüssen,

Ihr

J. v. Neumann

[b] *von Neumann 1929*.

1) Why do you use the "foundation axiom"? If you do not use it, then the theory of ordinals is going to be only slightly more complicated —I don't believe that your whole paper would be made longer by more than 1 or 2 pages. And $\Sigma = \Delta$ immediately implies the "foundation axiom".
2) $\Sigma = \Delta$ also implies the validity of my axiom: "A domain is proper (i.e.: not a set) if and only if it is equivalent to the domain of all elements."
3) I.e.: Your $\Sigma = \Delta$ automatically implies everything I proved in my second set-theoretic paper that was published in Crelle's Journal.[b] My procedure there is in a certain sense (as it does not respect the logical structure of the built-up sets) a much cruder analogue of your transfinite recursion generating Δ.
4) Also, the admissibility of the assumption that apart from ω there is no regular initial number with limit index now follows immediately (by means of the general continuum theorem).

When are you coming to Princeton, or when are you passing through? I am going to be here until 2 May, in Ann Arbor, Michigan, from 3 May to 17 May, where I'll give lectures, and here again from 18 May to the end of June. I hope that we'll see each other somewhere! Couldn't we get together after my expedition to Michigan—Ann Arbor is presumably not very far from South Bend. I really would like to talk to you again about many mathematical topics. E.g. on non-distributive logic and other logical systems—⟦ones⟧ that are distributive, but quite different from classical logic—and on their "transcendent" extension. And also on non-logical questions.

Hoping to hear from you soon, and with best regards,

Yours sincerely,

J. v. Neumann

18. von Neumann to Gödel[a]

Sonnabend, den 13. Mai. ⟦1939⟧

Lieber Herr Gödel,

vielen Dank für Ihren lieben Brief. Ich hätte Sie ⟨so⟩ gerne in dieser Gegend noch gesehen, aber leider ist etwas dazwischen gekommen: Ich wollte dieses Wochenende zu einem Besuch in Notre Dame verwenden, aber es hat sich so ergeben, dass ich meinen Bruder, der in Buffalo ist, sprechen muss—und das wird den ganzen Sonntag in Anspruch nehmen. Und Dienstag—den 16-ten—muss ich schon nach Princeton, weil wir ein Haus umbauen, so dass Sie sich denken können, wie eilig ich es habe.

Wann kommen Sie nach dem Osten? Ich rechne darauf, dass wir uns dann in Princeton ausführlich sehen! Bitte lassen Sie mich das rechtzeitig wissen.

Beste Grüsse, auch an Mengers,

von Ihrem

J. v. Neumann.

---

[a]On stationery of the University of Michigan Union, Ann Arbor.

19. von Neumann to Gödel[a]

Mittwoch, den 19-ten Juli. ⟦1939⟧

Lieber Herr Gödel:

Ich möchte Ihnen über eine Möglichkeit berichten, die im Zusammenhang mit der Vervielfaeltigung Ihrer Princetoner ~~Vrlesungen über das~~ Vorlesungen über das Kontinuumproblem, etc., aufgekommen ist. Es besteht die Möglichkeit⟦,⟧ die Vorlesungen in einer technisch vollkommeneren Weise zu vervielfaeltigen: Ebenfalls photographisch, aber mit

[a]Typed on stationery of the Nassau Tavern, Princeton.

18. von Neumann to Gödel[a]

Saturday, 13 May ⟦1939⟧

Dear Mr. Gödel,

Many thanks for your kind letter. I really would have liked to see you around here, but unfortunately something came up: I intended to use this weekend for a visit to Notre Dame, but it turned out that I have to talk to my brother, who is in Buffalo—and that will take up all of Sunday. And on Tuesday—the 16th—I have to go to Princeton, as we are remodeling a house, so you can imagine how hurried I am.

When are you coming to the East? I count on seeing you then extensively in Princeton! Please let me know that in advance.

Best regards, also to the Mengers,

Yours sincerely,

J. v. Neumann

---

19. von Neumann to Gödel[a]

Wednesday, 19 July ⟦1939⟧

Dear Mr. Gödel,

I would like to report to you on a possibility that arose in connection with the duplication of your Princeton Lectures on the continuum problem etc. There is the possibility of duplicating the lectures in a technically more perfect way: also photographically, but with smaller, more

---

kleineren, schaerfer umrissenen, dem normalen Druck sehr aehnlichen Buchstaben, und demgemaess auf Seiten von einem kleineren, Buchaehnlicheren, Format.

Es besteht im Allgemeinen die Absicht in der Zukunft diese ~~v~~erviel-faeltigungs-Technik zu verwenden, wenigstens bei Vorlesungen⟦,⟧ die wirkliches Interesse auslösen. Diese Vervielfaeltigungen sollen dann auch allgemein kaeuflich sein, und denjenigen wissenschaftlichen Zeitschriften die Neuerscheinungen kritisch referieren (Bulletin Amer. Math. Soc., u. ae.) zum Referat eingeschickt werden. Das Komité das diese Dinge organisiert möchte mit Ihrer Vorlesung anfangen. Ich möchte Sie daher fragen: Sind Sie damit einverstanden? V

Ich persönlich glaube⟦,⟧ dass die Idee eine sehr gute ist. Falls Sie, in Hinblick auf eine spaetere Publikation in Buchform den jetzigen Absatz (etwa auf eine feste Zahl von Exemplaren) beschraenken wollen—oder falls sie vorlaeufig keine Referate ~~in der Literatur wuenschen, sol~~ in der Literatur wuenschen, so können diese besonderen Wünsche natürlich berücksichtigt werden.

Bitte lassen Sie mich so bald wie möglich wissen, wie Sie sich zu diesen Dingen stellen. Im übrigen habe ich den letzten ~~t~~eil ihres Manuskripts Ihrem Wunsche entsprechend nochmal gelesen und einige kleinere Druckfehler korrigiert. Das Manuskript ist vollkommen “Druckfertig”. T

2 Was gibt es bei Ihnen sonst Neues? Hier ist nicht viel l⟦os?⟧ | Ich schreibe ein Buch über “Continuous Geometries”[b] u. ae. für die Amer. Math. Soc. Colloquium Series, aber es geht leider sehr langsam vorwaerts. Am meisten beschaeftige ich mich mit Operatorentheorie (Operatorenaxiome, wie sie F. J. Murray und ich in den Ann. of Math., 1936[c] untersucht haben) und da habe ich verschiedene Fortschritte erzielt.

In der Hoffnung des baldigen Wiederhörens—und Wiedersehens—und mit der Bitte mich Ihrer Frau unbekannterweise zu empfehlen,

mit den besten Grüssen

stets

Ihr J. v. Neumann

Adresse: Fine Hall, Princeton, N.J., U.S.A.

---

[b] *von Neumann 1936* and *1937* were eventually reproduced in *von Neumann 1960.*

clearly outlined letters that are very similar to those used in normal printing, and thus on pages of a smaller format, similar to that for a book.

There is in general the intention to use this technique in the future for duplication, at least for lectures that stir real interest. These duplicated notes should then be available also for general purchase and should be sent to those journals (Bulletin of the Amer. Math. Soc. and similar ones) that critically review new publications. The committee organizing these matters would like to start with your lectures. Thus I would like to ask you: Is this agreeable to you?

Personally, I believe that this is a very good idea. If, in view of a future publication in book form, you want to limit the current production (say to a fixed number of copies)—or if you do not wish for the time being any reviews ⟦to appear⟧ in the literature, then these special wishes can of course be taken into account.

Please let me know as soon as possible what you think of these matters. As for the rest, I re-read according to your wish the last part of your manuscript and corrected some minor misprints. The manuscript is perfectly "ready for printing".

What is new at your end? Here not too much is going on; I am writing a book on "Continuous Geometries"[b] and similar matters for the Colloquium Series of the Amer. Math. Soc., but I am making only very slow progress. I am mostly occupied with operator theory (operator axioms in the way in which F. J. Murray and I investigated them in the Ann. of Math., 1936[c]), and in that respect I have made various advances.

I hope to hear from you soon—and to see you again—and send my regards to your wife without actually knowing her, with best wishes,

Yours sincerely,

J. v. Neumann

Address: Fine Hall, Princeton, N.J., U.S.A.

[c] *von Neumann and Murray 1936.*

20. Gödel to von Neumann

Wien 17./VIII. 1939.

Lieber Herr v. Neumann!

Vielen Dank für Ihren Brief vom 19./VII. Ich bin auch mit der neuen Art der Vervielfältigung meiner Vorlesungen vollkommen einverstanden, vorausgesetzt dass dadurch keine wesentliche Verzögerung entsteht. Natürlich nehme ich an, dass das alte Manuskript wörtlich übertragen wird u. dass Mr Brown (oder irgend jemand anderer, der die Sache inhaltlich versteht) es auf Druckfehler, besonders in den Formeln, durchsieht. Da das Manuskript für die frühere Vervielfältigungsmethode bereits fertig ist u. das Umschreiben immerhin ziemlich viel Mühe erfordern würde, bin ich persönlich eigentlich mehr dafür⟦,⟧ bei der alten Methode zu bleiben, umsomehr als ich, wie Sie wissen mit | 2 der gegenwärtigen Form des Beweises nicht restlos zufrieden bin. Bezüglich der Anzahl der Exemplare habe ich keine besondern Wünsche. Ich danke Ihnen noch vielmals, dass Sie so freundlich waren den letzten Teil des alten Manuskriptes ~~noch mals~~durchzusehen.

Bei mir gibt es nicht viel Neues; ich hatte in letzter Zeit eine Menge mit Behörden zu tun. Ende September hoffe ich wieder in Princeton zu sein.

Mit besten Grüssen, auch von meiner Frau

Ihr Kurt Gödel

Bitte richten Sie auch Empfehlungen von mir an Ihre Frau u. an Veblen aus.

21. Gödel to von Neumann

Princeton 20./III.1956.

Lieber Herr v. Neumann!

Ich habe mit grösstem Bedauern von Ihrer Erkrankung gehört. Die Nachricht kam mir ganz unerwartet. Morgenstern hatte mir zwar schon im Sommer von einem Schwächeanfall erzählt, den Sie einmal hatten, aber er meinte damals, dass dem keine grössere Bedeutung beizumessen sei. Wie ich höre, haben Sie sich in den letzten Monaten einer radikalen Behandlung unterzogen u. ich freue mich, dass diese den gewünschten Erfolg hatte u. es Ihnen jetzt besser geht. Ich hoffe u. wünsche Ihnen, dass

20. Gödel to von Neumann

Vienna, 17 August 1939

Dear Mr. v. Neumann,

Many thanks for your letter of 19 July. I am completely happy with the new way of duplicating my lectures, assuming that no delay is caused by it. Naturally I assume that the old manuscript is transcribed literally and that Mr. Brown (or someone else who understands the contents) checks for misprints, particularly in the formulas. As the manuscript is already prepared for the earlier method of duplication and as the transcription would in any event require quite a bit of effort, I personally am more in favor of using the old method, all the more because, as you know, I am not completely satisfied with the present form of the proof. With respect to the number of copies, I do not have any special wishes. I want to thank you again very much for your kindness in going through the last part of the old manuscript.

There is not too much going on here; during the last few weeks I had a lot to do with government agencies. By the end of September I hope to be in Princeton again.

Best regards, also from my wife,

Yours sincerely,

Kurt Gödel

Please give my regards also to your wife and to Veblen.

21. Gödel to von Neumann

Princeton, 20 March 1956

Dear Mr. v. Neumann,

I have heard with the greatest sympathy of your illness. The news came to me quite unexpectedly. In fact, Morgenstern had already told me in the fall about a weak spell you once had, but he believed then that no major significance should be attributed to it. As I heard, during the last few months you have undergone a radical treatment, and I am glad that it had the desired success and that you are better now. I hope and

Ihr Zustand sich bald noch weiter bessert u. dass die neuesten Errungenschaften der Medizin, wenn möglich, zu einer vollständigen Heilung führen mögen.

Da Sie sich, wie ich höre, jetzt kräftiger fühlen, möchte ich mir erlauben, Ihnen über ein mathematisches Problem zu~~s~~ schreiben, über das mich | Ihre Ansicht sehr interessieren würde: Man kann offenbar leicht eine Turingmaschine konstruieren, welche von jeder Formel F des engeren Funktionenkalküls[a] u. jeder natürl. Zahl n zu entscheiden gestattet, ob F einen Beweis der Länge n hat [Länge = Anzahl der Symbole]. Sei $\psi(F, n)$ die Anzahl der Schritte, die die Maschine dazu benötigt u. sei $\varphi(n) = \max_F \psi(F, n)$. Die Frage ist, wie rasch $\varphi(n)$ für eine optimale Maschine wächst. Man kann zeigen $\varphi(n) \geq Kn$. Wenn es wirklich eine Maschine mit ~~$\varphi(n) \sim K.(n$~~ $\varphi(n) \sim K.n$ (oder auch nur $\sim Kn^2$) gäbe, hätte das Folgerungen von der grössten Tragweite. Es würde nämlich offenbar bedeuten, dass man trotz der Unlösbarkeit des Entscheidungsproblems die Denkarbeit des Mathematikers bei ja-oder-nein Fragen vollständig[1] durch Maschinen ersetzen könnte. Man müsste ja bloss das n so gross wählen, dass, wenn die Maschine kein Resultat liefert, es auch keinen | Sinn hat⟦,⟧ über das Problem nachzudenken. Nun scheint es mir aber durchaus im Bereich der Möglichkeit zu liegen, dass $\varphi(n)$ so langsam wächst. Denn 1.) scheint $\varphi(n) \geq Kn$ die einzige Abschätzung zu sein, die man durch eine Verallgemeinerung des Beweises für die Unlösbarkeit des Entscheidungsproblems erhalten kann; 2. bedeutet ja $\varphi(n) \sim K.n$ (oder $\sim Kn^2$) bloss, dass die Anzahl der Schritte gegenüber dem blossen Probieren von N auf $\log N$ (oder $(\log N)^2$) verringert werden kann. So starke Verringerungen kommen aber bei andern finiten Problemen durchaus vor, z. B. bei der Berechnung eines quadratischen Restsymbols durch wiederholte Anwendung des Reziprozitätsgesetz~~t~~es. Es wäre interessant zu wissen, wie es damit z. B. bei der Feststellung, ob eine Zahl Primzahl ist, steht u. wie stark *im allgemeinen* bei finiten kombinatorischen Problemen die Anzahl der Schritte gegenüber dem blossen Probieren verringert werden kann. | Ich weiss nicht⟦,⟧ ob Sie gehört haben, dass "Post's ~~p~~roblem" (ob es unter den Problemen $(\exists y)\varphi(y, x)$ mit rekursivem φ Grade der Unlösbarkeit gibt) von einem ⟨ganz⟩ jungen Mann namens Richard Friedberg in positivem Sinn gelöst wurde.[b] Die Lösung ist sehr elegant. Leider will Friedberg nicht Mathematik, sondern Medizin studieren (scheinbar unter dem Einfluss seines Vaters).

[1] abgesehen von der Aufstellung der Axiome

[a] I.e., first-order predicate logic.

wish that your condition will soon improve further and that the latest achievements of medicine may, if possible, effect a complete cure.

Since, as I hear, you are feeling stronger now, I would like to take the liberty to write to you about a mathematical problem; your view on it would be of great interest to me: Obviously, it is easy to construct a Turing machine that allows us to decide, for each formula F of the restricted functional calculus[a] and every natural number n, whether F has a proof of length n [length=number of symbols]. Let $\psi(F, n)$ be the number of steps required for the machine to do that, and let $\varphi(n) = \max_F \psi(F, n)$. The question is, how rapidly does $\varphi(n)$ grow for an optimal machine? It is possible to show that $\varphi(n) \geq Kn$. If there really were a machine with $\varphi(n) \sim Kn$ (or even just $\sim Kn^2$) then that would have consequences of the greatest significance. Namely, this would clearly mean that the thinking of a mathematician in the case of yes-or-no questions could be completely[1] replaced by machines, in spite of the unsolvability of the Entscheidungsproblem. n would merely have to be chosen so large that, when the machine does not provide a result, it also does not make any sense to think about the problem. Now it seems to me to be quite within the realm of possibility that $\varphi(n)$ grows that slowly. For 1.) $\varphi(n) \geq Kn$ seems to be the only estimate obtainable by generalizing the proof of the unsolvability of the Entscheidungsproblem; 2.) $\varphi(n) \sim Kn$ (or $\sim Kn^2$) just means that the number of steps when compared to pure trial and error can be reduced from N to $\log N$ (or $\log N^2$). Such significant reductions are definitely involved in the case of other finitist problems, e.g., when computing the quadratic remainder symbol by repeated application of the law of reciprocity. It would be interesting to know what the case would be, e.g., in determining whether a number is prime, and how significantly *in general* for finitist combinatorial problems the number of steps can be reduced when compared to pure trial and error. I do not know whether you have heard that "Post's problem" (whether there are degrees of unsolvability among the problems $(\exists y)\varphi(y, x)$ with recursive φ) was solved positively by a quite young man by the name of Richard Friedberg.[b] The solution is very elegant. Unfortunately, Friedberg is not going to study mathematics, but rather medicine (seemingly under the influence of his father).

[1] Except for the formulation of axioms.

[b] *Friedberg 1957*.

Was halten Sie übrigens von den Bestrebungen, die Analysis auf die verzweigte Typentheorie zu begründen, die neuerdings wieder in Schwung gekommen sind? Es ist Ihnen wahrscheinlich bekannt, das⟦s⟧ Paul Lorenzen dabei bis zur Theorie des Lebesgueschen Masses vorgedrungen ist.[c] Aber ich glaube, dass in wichtigen Teilen der Analysis nicht eliminierbare imprädikative Schlussweisen vorkommen.

Ich würde mich sehr freuen, von Ihnen persönlich etwas zu hören, u. bitte lassen Sie es mich wissen, wenn ich irgend etwas für Sie tun kann.

Mit besten Grüssen u. Wünschen, auch an Ihre Frau Gemahlin

Ihr sehr ergebener Kurt Gödel

P.S. Ich gratuliere Ihnen bestens zu der Au⟦s⟧zeichnung⟦,⟧ die Ihnen von der amerik. Regierung verliehen wurde.

---

[c] *Lorenzen 1951b.*

What do you think of the efforts to base analysis on ramified type theory, which have recently gained some impetus? It is probably known to you that Paul Lorenzen advanced in this way up to the theory of the Lebesgue measure.[c] But I believe that in important parts of analysis there are impredicative inference methods that cannot be eliminated.

I would be very glad to hear from you personally, and please let me know if I can do anything for you.

With best regards and wishes, also to your wife,

Yours sincerely,

Kurt Gödel

P.S. My congratulations for the award that was bestowed on you by the American government.

Kurt Gödel and Hao Wang

Hao Wang

Hao Wang (1921–1995) was a logician with wide-ranging interests in logic, computer science and philosophy. He came to the United States from China in 1946 and studied at Harvard University. He held teaching appointments there, at Oxford, and from 1967 until his retirement as professor at the Rockefeller University in New York.[a]

In July 1967 Wang asked Gödel to comment on a partial draft of his introduction *Wang 1970* to the collection of Skolem's papers in logic (*Skolem 1970*). With some interruptions, exchanges between the two continued for most of the remainder of Gödel's life. His relation with Wang was probably the closest and most intense intellectual relationship into which Gödel entered in the last years of his life. The correspondence published below is only a very partial expression of their exchanges. After the two letters of 7 December 1967 and 3 March 1968, both of which were published with some omissions and revisions in *Wang 1974*,[b] Gödel's side of the exchanges was expressed almost entirely in conversation or in the very unusual arrangement by which some of the latter's views were expressed in Wang's book. It is already quite well documented, first by the relevant parts of *Wang 1974* and then by what Wang wrote about Gödel after his death, especially *1987* and *1996*, the latter of which quotes extensive remarks of Gödel from their conversations.[c] It would be beyond the scope of this note to give a full narrative of this relationship, although that is the context in which the correspondence included here must be seen.

In fact Wang first wrote to Gödel in December 1948, sending him an unpublished paper, and a month later mentioning an error in the paper. A letter of 7 March 1949, also not included here, attests that they spoke on the telephone in February. Gödel drafted but apparently did not send a reply to that letter (which dealt with the relation of simple type

[a]For a brief account of Wang's life and work, emphasizing his contributions to logic, see *Parsons 1996*. *Parsons 1998* discusses his philosophical work. *Grossi et alii 1998* is a bibliography of his writings.

Within six months of his death in May 1995, Wang's papers were given by his widow to Rockefeller University. They very likely contain documents relevant to the subject of this note. However, the Rockefeller University Archives has still not made them available to scholars.

Transcriptions of Gödel's shorthand in this note and in the editorial notes to the letters below are by Cheryl A. Dawson.

[b]They are published here (as letters 2 and 4) as Wang received them at the time.

[c]Wang warns us, however, that the remarks are reconstructed from notes.

theory and Zermelo set theory). On 7 July Wang wrote asking for an appointment with Gödel, and they did meet. In 1952, when Gödel came to Harvard to receive an honorary degree, Wang met him and his wife at a dinner given by W. V. Quine. They had a small number of meetings and telephone conversations between then and Wang's inquiry in 1967 but apparently no further correspondence.[d]

The correspondence included here has three phases, treated in the first three sections of the note below.

1. The correspondence 1967–1970

On 27 July 1967 Wang wrote (letter 1) that he was writing an introduction to the collection of Skolem's papers in logic (*Skolem 1970*) and asked if Gödel "would be willing to comment on any serious errors" in his draft of the first section, which was apparently enclosed. No reply by Gödel is known, but on 14 September Wang sent a complete draft with a brief note. Gödel replied at length on 7 December (letter 2).

1.1. *Gödel on Skolem.* Gödel's comments on Wang's manuscript nearly all concern the Skolem–Löwenheim theorem and places where Skolem comes close to his own completeness theorem. It is section 1 of the published version that discusses these matters, so that the comments were on the section mentioned and probably sent in July. In connection with the work on *van Heijenoort 1967*, there had been considerable discussion at Harvard of work on quantificational logic in the 1920s, involving principally Burton Dreben, W. V. Quine, Jean van Heijenoort and Wang.[e] Moreover, there had been correspondence on the matter between Gödel and van Heijenoort.[f] It had come to light that Skolem had obtained what was mathematically essential to the completeness theorem and had in his *1928* stated what might have been a kind of completeness for a certain refutation procedure.[g] Wang's draft apparently claimed more for Skolem than Gödel would accept. Gödel argues that although in *1923a* Skolem states the lemma needed to get from the nontermination of a procedure like that of *1928* to the existence of a model and in *1928* (as Gödel interprets him) states what amounts to

[d]*Wang 1987*, pp. 115–116, 118; *1996*, p. 133.

[e]Wang was a professor of applied mathematics and mathematical logic at Harvard from 1961 to 1967, although in 1966 he went to Rockefeller as a visitor.

[f]See their correspondence in this volume.

[g]See especially *Dreben and van Heijenoort 1967*. The procedure was a prototype of the procedures based on Herbrand's theorem in *Dreben 1952* and *Quine 1955a*.

the completeness of the procedure, he does not even state completeness in the first paper and in the second substitutes for the lemma he had proved earlier in another context "an entirely inconclusive argument". Wang accepted Gödel's points and even echoes some of his language (*1970*, pp. 22–23). However, the interpretation of the relevant passage of *Skolem 1928* (pp. 134–135) stated clearly and emphatically in Gödel's second letter (letter 4), is not incontestable.[h]

More striking are Gödel's *explanation* for Skolem's failure to obtain the completeness theorem when he had come so close and the comments to which Gödel is led on his own achievements in logic. Not only Skolem but other logicians of the time lacked "the required epistemological attitude"; rejection of non-finitary reasoning in metamathematics was widespread. In particular, the completeness theorem is a metamathematical result whose proof requires a non-constructive (thus non-finitary) step.

At this point we should separate the questions raised by Gödel's diagnosis of the thought of logicians in the 1920s from his statements about his own work. In his statement of the prevailing views, Gödel's formulation is essentially of the views of the Hilbert school, at least on an interpretation that has been common and probably prevailed at the time he was writing about. Although Skolem in *1928* does seem concerned at a crucial point to avoid a non-finitary step, and many of his views show affinity with those Gödel cites, other considerations have been advanced by interpreters to explain his not posing the completeness problem clearly or proving the completeness theorem on the basis of the ideas he had developed. Most prominent is that he came to his model-theoretic work and generalization and application of Löwenheim's theorem from the perspective of the tradition of Schröder (to which Löwenheim's work belonged), which did not have a clear concept of formal proof.[i] Apart from Skolem, Gödel may well have overestimated

[h]Goldfarb (*1971*, pp. 523–524) mentions three possibilities and argues that all have difficulties. The first is that of Gödel, adopted in *Wang 1970*; the second is that of *Dreben and van Heijenoort 1967*. The third, proposed by Goldfarb, need not concern us here.

[i]See for example the introductory note to *1929*, *1930* and *1930a*, these *Works*, vol. I, p. 45 (also by Dreben and van Heijenoort); also *Wang 1970*, p. 22. As regards the *influence* of the Hilbert school on Skolem, it might be remarked that in 1922, the time of the address that was published as *1923a*, the first publications of Hilbert and Bernays on finitism and metamathematics were just appearing. Moreover, Skolem always showed a certain skeptical reserve toward the Hilbert program.

The conceptual issues about quantificational logic, formal proofs and models in the period before his own early work that Gödel and Wang discussed have been treated in other publications besides those cited above, in particular the notes in *Herbrand 1971* as well as *Goldfarb 1979*. On semantics and completeness in the Hilbert school before *Hilbert and Ackermann 1928*, see now *Sieg 1999*, section B3.

how widespread the views he mentions really were, particularly if one seeks outside the German orbit. In Britain or America, to be sure, the problem might rather have been to engage mathematicians in metamathematical inquiry at all (as opposed to work in the direction set by *Principia mathematica*), though Emil Post was an exception. In Poland, however, a conception of metamathematics was being articulated, especially by Tarski, that involved none of the methodological restriction characteristic of the Hilbert school.[j]

At the close of the letter Gödel remarks that Skolem's epistemological attitude was "diametrically opposed" to his own. This is no doubt largely true; however, the second of the specific comments he makes on *Skolem 1929* seems to me to rest on a misreading. At the very end of the paper on p. 49, Skolem is commenting on a result of the paper, a relative consistency theorem for what he calls *Vertreteraxiome*, essentially Hilbert's ε-axioms.[k] He proves that they can be added consistently to a consistent first-order theory, as can some related additions. The final statement is:

> On the other hand all such formal extensions of mathematics give no means at all for deciding problems that are formulable without them.

It seems quite clear that Skolem is making a straightforward logical claim, that the additions in question yield conservative extensions of a first-order theory.[l]

1.2. *Gödel on objectivism in his own work.* Gödel's interesting reflections on his own work in the later paragraphs of letter 2 must of course be accorded considerable authority. Two observations should be made. As regards the incompleteness theorem, what he regards as essential to his discovering the theorem is, first, that he regarded the formulae of the formal system as meaningful, in contrast to a formalistic view associated with Hilbert, and, second, that he used "the highly transfinite

[j]See for example *Tarski 1930* and *1930a*. For this reason I would question the assertion of Davis (*1998*, p. 121) that without Gödel the completeness theorem might not have been proved for some time. Of course Tarski and others absorbed *Gödel 1930* and *1931* very quickly. But the evidence does not favor the view that they required these influences to entertain the very idea of studying the relation of formal proof and models with the usual methods of classical mathematics.

In fact even Bernays was not put off by the nonfinitary step in *Gödel 1930*; see his letter to Gödel of 24 December 1930 in volume IV of these *Works*.

[k]The theorem is essentially a different formulation of what in *Hilbert and Bernays 1939* is called the second ε-theorem, proved in §§3.1–3.2.

[l]Wang did not ask to publish this paragraph, and Gödel did not urge it on him. The first of his remarks is reiterated in letter 4. Possibly he thought better of the second.

conception of 'objective mathematical truth' as *opposed* to that of 'demonstrability'" (p. 2) as the heuristic principle of his construction. The first, though opposed to formalism, is by Gödel's own statements (for example in **1951* and **1953/9*) not opposed to intuitionism; the second, while certainly a form of realism, is well short of the version of platonism Gödel defended from *1944* on.[m] Another step in that direction, acceptance of impredicatively defined sets, is mentioned in the context of the remark on his work in set theory several years after the incompleteness theorem and is only relevant to that.

1.3. *Wang's questions.* Wang replied on 19 December (letter 3); he seems convinced by Gödel's arguments but apparently remained doubtful about the belief expressed in the P.S. to Gödel's letter that his decidability proof in *1933i* could be extended to the case with identity. In *Wang 1970*, p. 31, he states that that is Gödel's "present belief" but that others have attempted to supply the details "but have failed so far". This reflects the revision (e) that Wang refers to in the letter.[n]

Gödel's remarks about his own work prompt Wang to say that he is "eager to find out more details about your objectivistic conception outlined in your several papers". He proceeds to raise several questions. The first group refers to *Gödel 1958*, the distinction between the original and some extended finitary point of view and the relative advantages of Gödel's and Gentzen's consistency proofs for first-order arithmetic. The second asks about a remark in Gödel's letter that any non-finitary consistency proof will yield a finitary relative consistency proof. Wang seems to be looking for a more precise mathematical statement. The third asks what more Gödel had in mind in stressing the importance of his objectivistic conception for his work in logic. Wang says "there must be a lucid overall picture". The fourth asks whether the iterative conception of set is implicit in Cantor's basic explanations.

The fifth brings up a well-known contrast, between Gödel's statement that Russell's analysis of the paradoxes "brought to light the amazing fact that our logical intuitions...are self-contradictory" (*1944*, p. 131) and the statement that the set-theoretical paradoxes "are a very serious problem, not for mathematics, however, but rather for logic and epistemology" (*1964*, p. 262). (Wang also refers to *1947*, p. 518.) Wang finds it unclear in what sense they are a serious problem for logic without

[m] On the development of Gödel's platonism, see *Parsons 1995a*, §2, and for a somewhat different picture, *Davis 1998*.

[n] See the letter of Wang to Jens-Erik Fenstad, editor of *Skolem 1970*, 13 December 1967, document no. 013078 in Gödel's papers.

being one for mathematics. Although he doesn't find Gödel's position inconsistent, he asks for more elaboration.

Under (6) Wang says he is "under the impression that your Gibbs lecture ⟦*1951*⟧ ...applies the incompleteness results to argue against a certain form of mechanism or mechanical materialism" and expresses the wish to learn more about his argument. Wang very likely relied on his own memory and possibly notes from attending the lecture.

Gödel did not reply until 7 March 1968 (letter 4) and did not go into any of Wang's further questions. He did promise to do so later, and his markings on the letter (indicated in the editorial notes) show interest in many of them, and some were clearly prominent in their conversations in 1971–1972. The "objectivistic conception", the iterative conception of set and the question of mechanism about the mind are subjects of Gödel's contributions to *Wang 1974* (see below), where Gödel's explanation of the contrast in (5) is also given (pp. 187–188). Issues close to the ones Wang raises about significance of the paradoxes for logic and mathematics also arise in what is reported in *Wang 1996*.

Instead Gödel reaffirms and reinforces the statements of the first letter about the influence of finitism and formalism on logic just before his own work. The second paragraph seems to be a reply to the objection that Skolem's failure to obtain the completeness proof might be attributed to the lack of a fully articulated conception of formal proof, in particular as an object of mathematical study.[o]

Gödel does qualify the claims of the earlier letter by saying that "the formalistic point of view did not make *impossible* consistency proofs by means of transfinite models" but "only made them much harder to discover, because they are somehow not congenial to this attitude of mind". However, he finds that the need to apply the idea of the ramified hierarchy "in a totally nonconstructive way" for the concept of constructible set and the need to distinguish mathematical truth from provability were more serious obstacles.

1.4. *The Gödel class with identity.* Wang's reply of 23 April (letter 5) again expresses agreement "in all aspects where I understand you". He then turns to the problem of extending to the situation with identity Gödel's proof in *1933i* that the class of prenex quantificational formulae

[o]In a letter to Fenstad of February 7, 1968 (document no. 013079, Gödel papers) Wang had sent a revision mentioning Skolem's adherence to the tradition of "Boole, Schröder, Löwenheim, and Korselt" as an alternative to "reluctance to use nonfinitary reasoning" as an explanation for the inconclusiveness of the argument in *Skolem 1928* that Gödel had pointed out. This revision was not Wang's most immediate response to Gödel's points in letter 2; rather it reflects correspondence with Bernays.

of the form $\exists \ldots \exists\forall\forall\exists \ldots \exists$ is decidable and finitely controllable. He mentions two specific difficulties:

> As I see it, when equality is included, we have in general no longer the functions δ and η (p. 436 of your 1933 paper) and I do not know how to introduce the appropriate equivalence relation. As you undoubtedly know, Theorem II (p. 435 of your paper) can probably be refuted by a counterexample when equality is included.

The functions Wang mentions are introduced at the beginning of Gödel's proof of Theorem II. The example Wang refers to is surely one of those obtained by Stål Anderaa to show that what the introductory note to *1933i* in these *Works* calls the Gödel–Kalmár–Schütte Criterion is in the presence of equality not sufficient for satisfiability. That is contrary to Theorem II, which asserts that the Criterion implies the existence of a finite model.

Gödel did not reply directly to this letter, and Wang wrote on 9 January 1970 asking if he might see Gödel once or twice "to talk with you on general philosophical matters such as the questions I submitted to you in 1968" (actually December 1967) and proposing some possible dates. These meetings did not take place because of Gödel's poor health at the time.[p] On 3 April 1970 Dana Scott wrote on Gödel's behalf an extended reply to Wang's and Burton Dreben's inquiries about the extension to the case with equality of the decidability result of *1933i* (letter 2 of Appendix A). Wang was still not persuaded and on 15 April wrote a detailed reply (letter 6).[q] Of course this story ends after Gödel's death with the proof in *Goldfarb 1984* and *1984a* that the Gödel class with identity is undecidable.

During the next year Wang wrote only once (see note p).

2. 1971–1972: Gödel's contribution to *Wang 1974*

A new phase of their relationship, which was to include the first set of their regular conversations, began with a letter of Wang of 25 May 1971 (letter 7), in which he asked to quote a substantial part of Gödel's

[p]This is attested by Gödel's expression of regret in letter 8. Moreover, opening a letter of 13 September 1970, the first since his reply to the Scott letter, Wang writes, "I very much hope your health has greatly improved since last winter". Gödel had in fact gone through a severe crisis; see *Dawson 1997*, pp. 233–235, 237. Wang seems to have known the extent of this only later and largely at second hand (*1987*, pp. 129–130).

[q]The difficulties of Scott's proposal are pointed out in the introductory note to *1933i*, these *Works*, vol. I, pp. 229–230. Cf. also the correspondence with Dreben in volume IV of these *Works*.

letter of 7 December 1967 in the introduction to his book *Knowledge and logic* (which eventually became *Wang 1974*).

2.1. *Permission to quote Gödel's letters.* What Wang asked to quote is the substantive part of the letter except for the specific comments on Skolem. Gödel did not reply until 9 July (letter 8), although a little earlier Stanley Tennenbaum called Wang to tell him that Gödel would give the permission under certain conditions. Gödel's requirements express his characteristic meticulousness and precision. The most important are (1) and (2), asking that the letter of 7 March 1968 be included and that the selection from the earlier letter begin earlier, in the middle of the third paragraph. (3) proposes an addition to the reference in the letter to an explanation in *Gödel 1934*, and (4) makes a reference in letter 4 to a passage in letter 2 more precise. Two other small suggestions are made.

On 9 July Wang had written a letter responding to Tennenbaum's call (letter 9); it appears that Wang had already learned something of the content of Gödel's letter. Letter 9 was typed and sent on 20 July with a reply (letter 10) to Gödel's letter 8. His plan was to put the quotations from Gödel's second letter into a footnote, as the draft he enclosed indicates. But in letter 10 Wang says that he is inclined to move them into the main text, which in the event he did. In letter 9 he expresses the intention of sending Gödel some chapters of his book and in letter 10 he takes up Gödel's encouragement to come to see him, evidently thinking of a number of meetings, as Gödel at the time may not have. The next day he sent the first ten chapters of *Knowledge and logic* with a brief note.[r]

Gödel's reply of 4 August (letter 11) characteristically does not promise to read Wang's manuscript in the near future. He stresses in the form of a rather precise suggestion the importance of his second letter.[s] He then proposes an addition to the second letter, to be noted as such.[t] In his reply of 11 August (letter 12), Wang agrees to put both letters into the text and emphasizes, as more extensively concerned with Gödel's work, the chapters of his manuscript on the concept of set (chapter VI), the section on mechanical procedures (section 3 of chapter II) and a section of chapter X called "Mathematical arguments".[u] In addition to

[r] *Wang 1974* consists of twelve chapters and an appendix consisting of reprints of early philosophical essays. Wang does not seem to have submitted anything beyond chapter X to Gödel. The last two chapters are of a more general nature.

[s] The suggestion proved otiose, since it envisaged a situation where the first letter was in the main text and the second in a footnote.

[t] As published it was abbreviated from the 4 August formulation and so bears the date April 1972. See *Wang 1974*, p. 10, and letter 13 below.

[u] In *Wang 1974*, chapters II and X are entitled "Characterization of general mathematical concepts" and "Minds and machines", respectively.

the Introduction which contains the substance of the two letters, these chapters are the ones where, in the published book, previously unpublished views of Gödel are presented and which are cited in the Preface (p. x) as having been discussed between them.

2.2. *Gödel's involvement in revisions of Wang's book.* There is no known further correspondence until the following January, although by then some of their conversations had taken place. Wang was trying to get his book into final form, and he sent Gödel drafts of revisions to the book. There are no known letters of Gödel from this period; Gödel's reactions were communicated either in conversation or by telephone (in one case via Tennenbaum).

On 27 January 1972 Wang wrote sending along some revised pages and saying that he is "attempting to revise the section about minds and machines and to add a section to report on your views", expressing the hope that he will be able to bring this material to a meeting the following Wednesday (2 February). (The new section (section 7 of chapter X) was in fact added; Gödel clearly had a substantial hand in writing it (see below).) A week later Wang sent revisions of some other portions. On 6 April, he wrote concerning the acknowledgments in the Preface and in chapter VI; he had apparently gone further in attributing views to Gödel than Gödel wished. Wang was to send two more revised versions of the relevant portion of the Preface on 10 and 13 April.[v]

A detailed letter of 10 April (letter 13) states what further revisions he is making before sending the manuscript to Stuart Hampshire,[w] who is said to be negotiating the publication of the book with a publisher, undoubtedly Routledge and Kegan Paul, which published *Wang 1974* in the series International Library of Philosophy and Scientific Method.

What is most interesting in this letter is that Wang plans to send the manuscript to Hampshire with indications that revised text for certain portions is to be supplied by Gödel. These places were a passage in the Introduction, much of section 3.1 of chapter II, "Gödel on mechanical procedures and the perception of concepts," some specific discussions in chapter VI on the concept of set, and section 7 of chapter X, "Gödel on minds and machines." These include most of the places that are mentioned in the acknowledgment to Gödel in the Preface of the book

[v]The last of these (document no. 013112) still differs slightly from the published version (pp. x–xi); the latter describes in slightly less narrow terms the sections discussed with Gödel.

[w]The well-known philosopher Stuart Hampshire (1914–) was at the time Warden of Wadham College, Oxford. Wang probably knew him from the time he himself was at Oxford (1956–1961); moreover, Hampshire was from 1964 to 1970 at Princeton University.

as exceptions to Wang's general disclaimer to be representing Gödel's views. It appears from a remark about the section on mechanical procedures in Wang's letter of 26 April (letter 14) that Wang still expected Gödel to send his revisions to him, of which two in the chapter on the concept of set are also mentioned. But by 16 June, Gödel had still not sent everything. (In the meantime, Gödel had come to Rockefeller University to receive an honorary degree on 1 June, the result of Wang's initiative.) It had apparently been arranged that Gödel would send "the remaining pages attributed to you" directly to the editor of the International Library, Professor Ted Honderich of University College London.

2.3. *Wang's trip to China and Gödel's letters to Honderich.* Wang's conversations with Gödel and his work to get his book into final form took place in the context of public events that were to affect his own life considerably. These were the dramatic steps toward establishing normal relations between the United States and the People's Republic of China. Already on 15 July 1971 it was announced that Henry Kissinger, national security adviser to President Richard Nixon, had met Premier Zhou Enlai of China in Beijing and that Nixon had accepted an invitation to visit China. The visit took place in February 1972, ending with a joint communiqué on 28 February. Wang arranged to visit China for the first time since he came to the United States as a student in 1946. He left very shortly after a meeting with Gödel on 20 June and was in the People's Republic of China from 28 June until 26 July.[x] He sent Gödel a postcard from Beijing on 14 July.

In his absence, on 27 June Gödel wrote to Honderich (letter 1) enclosing the remaining revised pages on the concept of set. From Wang's letter of 26 April, these would include the motivation of the axiom of replacement attributed to Gödel on p. 186 of *Wang 1974* and the discussion of principles for setting up axioms on pp. 189–190. What remained was section 7 of chapter X, on his views on minds and machines. Here he seems to have been quite dissatisfied with Wang's last version, even though it reflected discussion with him. He promised an "abbreviated version" in a week or ten days. In a P.S. he asks to "retain the right of

[x]See *Wang 1996*, p. 145. But I am indebted to Jane Hsiaoching Wang, who accompanied her father on the trip, for supplying the precise dates.

having the pages written or revised by me reprinted if an occasion for stating my views on these matters should arise."[y]

Gödel sent the remaining section to Honderich on 19 July, with letter 2, stating that he had had to rewrite the whole section. It is not clear what were the "items Professor Wang reported in this section" that Gödel thought did not fit into the theme of the chapter. Interestingly, commenting on his Gibbs lecture **1951*, he expresses the view that the arguments and results given there can be "strengthened considerably." It is this section that contains an alternate version of Remark 3 of *Gödel 1972a* (reprinted in this volume in Appendix B).

After Wang returned their conversations resumed and continued until December. Only very slight further changes seem to have been made in the book, although Gödel read the proofs of some of it, presumably the portions of it that reported his views. From then until mid-1975 their contacts were rather limited, although Wang sent Gödel a number of items, mostly having to do with reviews of the book.[z]

3. The last phase, 1975–1977

Gödel arranged for Wang to be in residence at the Institute for Advanced Study in 1975–1976,[aa] which was to be his last year before his retirement as professor. During this time they had another series of conversations. The earlier series, though delayed by Gödel's crisis of 1970, took place during a period when Gödel's health was relatively good. This was not the case during this second series, when in addition his wife Adele was seriously ill much of the time. Gödel's withdrawal

[y]From Gödel's handwritten draft. The contract for the book contains the provision:

> It is agreed that Professor Kurt Gödel reserves the right to reprint the material contributed by him and explicitly attributed to him in the Preface, namely Subsection 3.1 of Chapter II, Section 7 of Chapter X, and those parts attributed to him in Section 2 of the Introduction and Section 1 of Chapter VI.

(Letter of Wang to Sally Spiller of Routledge and Kegan Paul, 30 November 1972, document no. 013124 in Gödel's papers.)

[z]Wang wrote later (*1987*, p. 132) that it was his preoccupation with China and the resulting interest in dialectical materialism that limited his contacts with Gödel in 1973 and 1974. He also spent 1973–1974 at IBM working on computer-related matters. One interest was computer recognition of Chinese characters, which occasioned an amusing misunderstanding with Gödel (*1996*, p. 147). He also worked on a design for a Chinese typewriter; see *Wang and Dunham 1973*. In December 1973 he visited China again and again sent Gödel a postcard.

[aa]Wang describes this residence as part-time (*1987*, p. 132), and with only one exception his letters to Gödel during this time are on Rockefeller letterhead.

from the world was rather far advanced. In particular he had largely ceased to come to his office at the Institute. Therefore the conversations took place largely by telephone. Wang also became to some degree involved with Gödel's medical and personal problems and was one of those who tried vainly to help during Gödel's final decline in the last year of his life.[ab]

3.1. *Concepts and sets.* To some extent the conversations focused on drafts written by Wang. What they discussed before November is not witnessed by any correspondence.[ac] In November Wang was evidently occupied with his paper "Large sets" (*Wang 1977*), an early version of which had been presented in a symposium at the Fifth International Congress of Logic, Methodology, and Philosophy of Science in London, Ontario, the previous summer. It seems to have been a version of that paper that Wang says in a letter of 4 November, not included here, that he is "slow in revising". He then refers to notes relating to Gödel's views, particularly as expressed in the supplement to *Gödel 1964*.

During the next month he was occupied with the attempt to formulate Gödel's views on sets and concepts. On 7 November he wrote that he was "quite confused about the place of concepts in Ackermann's system and Powell–Reinhardt's", evidently referring to *Ackermann 1956*, *Powell 1972*, and *Reinhardt 1974* and *1974a*. Ackermann had presented an axiomatization of set theory based on a reflection principle, and Powell and Reinhardt used higher-order logic to motivate large cardinal axioms by formulating them as reflection principles; Wang was evidently trying to relate this work to Gödel's reflections on concepts. With the same letter he sent a manuscript by the undersigned, in fact his contribution to the same symposium at the International Congress and an early version of *Parsons 1977*. He called Gödel's attention to a passage criticizing Wang's idea that a "multitude" is a set if it has an "intuitive range of variability" (*1974*, p. 182).[ad] Gödel wrote the shorthand comment "Often ⟦he⟧ continually confuses concepts and sets, and moreover he does not understand 'idealization' broadly enough."[ae] Another comment, apparently referring to Wang's hope to talk with him on the telephone, says in effect, "What's his hurry? I didn't have time to read

[ab]On all these matters see *Dawson 1997*, ch. XII.

[ac]Evidence on the subject could be gathered from *Wang 1996* and some other documents in Gödel's papers; however, I will not pursue the question here.

[ad]Cf. *Parsons 1977*, pp. 341–345 (*1983*, pp. 275–279).

[ae]"Oft verwechselt fortwährend concepts und sets und ausserdem versteht er 'Idealisierung' nicht genug weit." The reading 'idealistisch' might be more likely than 'Idealisierung', but it is the latter that makes sense in the context. It seems clear that the remark refers to my paper.

his manuscript."[af] On 14 November Wang sent some revisions of what he had sent before and wrote:

> I remain troubled by the relation between classes and concepts. As you have emphasized, concepts could sometimes apply to themselves. But presumably classes cannot belong to themselves.

Wang sent some more typescripts, evidently on sets and concepts from an "objectivist" point of view that would be congenial to Gödel, which he later referred to as fragments M, N and C (*1996*, p. 149). On 9 December he had an extended discussion with Gödel in person. Wang followed this up with a longer letter on 12 December (letter 15). With that letter and a brief one dated 15 December he sent a new typescript, "Objectivism of sets and concepts", which undertakes to present Gödel's views rather fully. This is what is called fragment Q (*1996*, p. 149). Wang writes in the latter place that Gödel preferred the briefer presentation in fragment M, disappointing a hope Wang had expressed in the 15 December letter.

From the fourth paragraph of the letter of 12 December it appears that the question whether Wang might publish something based on his notes of their conversations had arisen between them. He disclaimed "any intention of direct publication" but rather saw them as an aid to further discussion. Evidently after Gödel's death Wang came to think that this disclaimer no longer bound him.

Wang did not write again until 5 February 1976, sending a new version of *Wang 1977*. It may be that they continued to discuss what Wang would publish from the "fragments" mentioned above. Section 1 of *Wang 1977* does present Gödel's views, but now limited to the concept of set. It states more clearly than do Gödel's own publications a conception of sets as collections, that is as constituted by their elements (not, however, as in any literal sense resulting from an operation of collecting; cf. p. 312). But evidently Gödel was not ready to have more of his views on concepts published.

3.2. "*Some facts about Kurt Gödel.*" Early in April Gödel was briefly hospitalized in Princeton. Wang writes (*1996*, p. 150):

> From then on extended theoretical discussions virtually stopped. In June of 1976 he began to talk about his personal problems and told me a good deal about his intellectual development.

This turn of their conversations evidently gave Wang the idea of writing something on Gödel's life and the development of his work. Already on

[af] "1. Warum so eilig? (hatte keine Zeit sein Man. zu lesen)." He added, "2. Erwarte ein paar kurze präzise Fragen" ("⟦I⟧ expect a few short precise questions"). It is possible that (1) refers to the first paragraph, referring to Wang's manuscript, and (2) refers to Wang's hope for a telephone discussion.

24 June he sent Gödel a typescript, "Some facts about Kurt Gödel", the first version of *Wang 1981* (see letter 16).

Wang sent Gödel two further versions, dated 8 July and 3 November. It is thus possible to trace what Gödel did and did not do to correct Wang's paper. The first version was marked up in a number of places, and the corrections are marked in the second version, for the most part in Wang's hand. Gödel made only a few small corrections in the second version. The third embodies a few further changes, nearly all inessential. The two copies in Gödel's papers have no corrections, and the main text of *Wang 1981* agrees with it closely but not entirely. The footnotes are not present as such in any of the versions in Gödel's papers; however, a few restore items from the first version that had been deleted at Gödel's request.[ag]

In his corrections Gödel concentrated on statements concerning his work and on the one paragraph about his health (pp. 655–656 of *Wang 1981*) and not on biographical detail. Some of the dates concerning his work have been shown to be incorrect; in at least some cases it seems more likely that Gödel did not bother with these points than that his memory deceived him. But it appears that he was careful about chronology where there might be a question about priority, as in statements about von Neumann and the incompleteness theorems and the interpretation he gives of the acknowledgment in *1930* (p. 349 n.) of Hahn's suggestions (*Wang 1981*, pp. 653–654). Some insertions into the second version are marked in the first only by a sign signifying an insertion. Very likely Gödel communicated these on the telephone. That would explain probably the most egregious error in *Wang 1981*, the statement that in the spring of 1939 Gödel lectured at Rotterdam when in fact it was Notre Dame.[ah] It is likely that Gödel would have examined Wang's paper, particularly in the third version, more carefully if his physical and mental health had been better at the time.

[ag]In particular this is true of the footnote concerning Gödel's marriage (p. 655 n. 3), which misstates Adele Porkert's name as Pockert.

[ah]The first version states that from the autumn of 1938 Gödel stayed at the IAS "except for the spring term of 1939 when he was back in Vienna". Gödel wrote an insertion sign at the end of the sentence. The second version has "except for the spring term of 1939 when ⟨he lectured at Notterdam and the autumn term of 1939 when⟩ he was back in Vienna", the insertion being in Wang's hand. Of course "Notterdam" makes no sense, and in the third version it is changed to "Rotterdam". Clearly Wang did not quite get what Gödel told him on the telephone, and Gödel did not correct him later. This and a few other errors are corrected in the reprint in *Wang 1987*.

The remarks here about *Wang 1981* correct statements in *Dawson 1997*, p. 250, and *Parsons 1998*, p. 17 and n. 33.

3.3. *Final matters; the addendum to Kleene 1976.* In his letter of 3 November transmitting the third version, Wang expresses interest in learning more about Gödel's "independence proof and general consistency proof of the axiom of choice" when Gödel is willing to look into the matter.[ai] "Of course it would be even better if you were to write up some of the material yourself and consider publication." It is quite likely that Wang knew that that was now beyond Gödel's capacity.

Wang had not, however, given up engaging Gödel in the project of exposition of his ideas. On 7 December 1976 he wrote reminding Gödel that earlier an editor of *Scientific American* had expressed interest in a report by Wang of some of his discussions with Gödel. He sent some pages entitled "Mind, matter, machine, mathematics" with this end in view, although he expressed doubt whether Gödel was "at all willing to consider this matter at present" and whether a revision acceptable to him would also be acceptable to *Scientific American.* From Wang's next letter of 3 February 1977 (letter 17), enclosing revisions, we learn that they had discussed the paper on the telephone on 18 January and that Gödel was "not much interested" in the idea. Wang did not give up the idea of publishing the paper in some form, with Gödel's approval of the exposition of his views.

It is surprising that any scientific relation between them could now survive, since already the previous summer Gödel's situation had become more critical with an extended illness of his wife, and it did not really recover thereafter.[aj] However, in the spring of 1977 Wang served as Gödel's emissary in a matter concerning S. C. Kleene's paper *1976* on Gödel's work, presented on the occasion of Gödel's 70th birthday to the Association for Symbolic Logic. Gödel was dissatisfied on some matters, particularly with the treatment of his work in set theory. Evidently after talking with Gödel and on commission from him, Wang drafted a letter dated 6 April 1977. Gödel annotated this draft, and Wang wrote a clean copy dated 11 April. It is clear that the version of 6 April was not sent. It is not definitely known whether the next version was sent. However, Wang's next letter on the matter, dated 11 May (letter 3 of Appendix A), refers to telephone conversations, which it is reasonable to conjecture were prompted by the 11 April letter. The letter goes into some detail about Gödel's work on constructible sets and the consistency of the axiom of choice and the GCH, going beyond correcting what he seems to have considered Kleene's principal error,

[ai]Cf. *Wang 1981*, p. 657. These statements are unchanged from the first version. That concerning independence is more detailed and a little less cautious than what is said in the relevant letters to Alonzo Church (in volume IV of these *Works*) and Wolfgang Rautenberg (in this volume). But of course it was written by Wang.

[aj]See *Dawson 1997*, pp. 250–251.

"giving the impression that the finite axiomatization ⟦used for the proof in *Gödel 1940*⟧ is his major contribution".[ak] Kleene replied on 24 May; although Wang sent Gödel a copy of this letter, there is none in Gödel's papers.[al] Evidently it contained some self-defense but probably expressed willingness to publish an addendum to his paper. Wang discussed the matter further with Gödel and wrote to Kleene again on 22 June (letter 4 of Appendix A). The third paragraph of the letter contains a text proposed by Gödel for such an addendum. (It seems likely that it was dictated to Wang on the telephone.) In addition the letter contains two additional "not very important" objections.

The published addendum incorporates all of Gödel's proposals; except for the one mentioned in note ak, Kleene's changes are editorial. It also addresses the point concerning finite axiomatization from the previous letter. Wang sent Gödel the last letter to Kleene on the same day, with a new version of "Mind, matter, machine, mathematics", asking for Gödel's judgment of the paper "when you have the inclination." That was his last letter to Gödel, and Wang does not say in his writings whether they discussed the matter further.[am] What is attributed to Gödel in the letters to Kleene may be Gödel's last scientific writings. Wang continued to try to keep in touch with Gödel by telephone and to help to the extent that his own situation permitted. He writes that he visited him at home as late as 17 December (*1987*, p. 133). Near the end of the month, after his wife came home from a long hospitalization, Gödel entered Princeton Hospital. Wang telephoned him on 11 January. "He was polite but sounded remote." Three days later Gödel died. Wang was one of the speakers at the memorial service at the Institute on 3 March 1978.[an] He continued to write about Gödel for the

[ak]The letter gives 1937 as the date when Gödel proved that GCH holds in the constructible sets. That agrees with other evidence; see *Dawson 1997*, pp. 122–123. All versions of "Some facts about Kurt Gödel" give the date as 1938 (see *Wang 1981*, p. 656). Curiously, the 1937 date is implied by the phrase "two years later" (i.e., after 1935) in Gödel's proposed addendum (see below and letter 4 of Appendix A), but in the published addendum this is changed to "three years later", again implying the date 1938. Whether this wavering is due to Gödel, Wang or Kleene is not known.

[al]It is very likely that the original is in Wang's papers, but see note a. Although many of Kleene's scientific papers are in the possession of his family, it has not been possible to locate either a copy of this letter or the originals of Wang's letters to Kleene.

[am]No article with this title appeared in *Scientific American* or elsewhere in English. However, Wang clearly mined the manuscript for a paper he presented to a conference in Paris in 1978, eventually published in French translation as *Wang 1991*. (See *Grossi et alii 1998*, p. 34.) In the portion reporting Gödel's views (pp. 445–449) no claim of the latter's approval is made.

[an]His remarks were published as *Wang 1978*.

rest of his life, not only in his books *1987* and *1996*, but also in a number of articles.[ao]

4. Questions

The correspondence prompts some general questions. If we compare the letters of 1971–1972 with other exchanges where publication of writing of Gödel was at issue, the highly unusual arrangement by which views of Gödel were published in *Wang 1974* was worked out with relatively little fuss. One reason was surely the consistently deferential posture toward Gödel that Wang's letters show. In particular, he was willing to give Gödel final responsibility for certain texts in his book, even when they were not announced as being by Gödel (or by the two of them jointly). But it must have been the case that Gödel was eager to get some more of his philosophy published but presumably welcomed a way of doing so "indirectly"[ap] so as to minimize his own involvement in controversy. In fact both were disappointed, with reason, that the book did not attract more attention.[aq]

Another question is the following: The correspondence of 1971–1972 together with remarks of Wang in various places shows that Gödel took final responsibility for some parts of *Wang 1974* where he did not speak in his own name, as he does in the published versions of letters 2 and 4 and in *1974a*. Should these passages be regarded as part of his collected works? This might be thought, for example, about the whole of section 7 of chapter X. The present editors have taken the conservative course of so treating only the passages where Gödel speaks in his own name.

Charles Parsons[ar]

A complete calendar of the correspondence with Wang appears on pp. 465–467 of this volume.

[ao]On Wang as a source for and interpreter of Gödel's thought, see §5 of *Parsons 1998*.

[ap]As John Dawson puts it, *1997*, p. 239.

[aq]*Wang 1987*, p. 132.

[ar]I am more than usually indebted to Cheryl Dawson, for assisting with the unwieldy Wang material in the Gödel papers. I am also indebted to Ti-Grace Atkinson for assistance and to my fellow editors, especially Solomon Feferman, for comments and suggestions.

1. Wang to Gödel[a]

July 27, 1967

Dear Professor Gödel:

I am writing an introduction to a collection of Skolem's papers in logic being prepared by mathematicians in Norway.[b] So far I have completed only the draft of a first section and much fear that I may have made gross misrepresentations. I should be most grateful if you would be willing to comment on any serious errors in this draft in order that I may correct them before publicaton.

Yours respectfully,

Hao Wang

---

[a]All Wang's letters included here are on letterhead of The Rockefeller University, New York, N.Y. 10021. On letters 15–17, the letterhead adds "1230 York Avenue".

[b]Published as *Skolem 1970*, with the introduction *Wang 1970*.

2. Gödel to Wang[a]

December 7, 1967

Professor Hao Wang
Department of Mathematics
The Rockefeller University
New York, N. Y. 10021

Dear Professor Wang:

Thank you very much for sending me your manuscript about Skolem's work. I am sorry for the long delay in my reply. It seems to me that, in some points, you don't represent matters quite correctly. So I wanted to consider carefully what I have to say.

---

[a]On letterhead of The Institute for Advanced Study, Princeton, New Jersey 08540, School of Mathematics.

1. On p. 11 (line 4 from the bottom) "probably" should be replaced by "evidently" because in the passage from which you quote I am stating explicitly the assumptions necessary for the application of the infinity lemma. Moreover, in my dissertation I gave the complete proof. I omitted it in the publication, because the infinity lemma seemed to be well known at that time, and the proof is not difficult.

2. On p. 11 (lines 12–16) you say, in effect, that the completeness theorem is attributed to me only because of my attractive treatment. Perhaps it looks this way, if the situation is viewed from the present state of logic by a superficial observer. The completeness theorem, mathematically, is indeed an almost trivial consequence of Skolem 1922.[b] However, the fact is that, at that time, nobody (including Skolem himself) drew this conclusion (neither from Skolem 1922 nor, as I did, from similar considerations of his own).

As you mention yourself, Hilbert and Ackermann, in the 1928 edition of their book,[c] state the completeness question explicitly as an unsolved problem. As far as Skolem is concerned, although in 1922 he proved the required lemma, nevertheless, when in his 1928 paper[d] (at the bottom of p. 134) he stated a completeness theorem (about refutation), *he did not use his lemma of 1922 for the proof. Rather he gave an entirely inconclusive argument.* (See p. 134, line 10 from below to p. 135, line 3.)

This blindness (or prejudice, or whatever you may call it) of logicians is indeed surprising. But I think the explanation is not hard to find. It lies in a widespread lack, at that time, of the required epistemological attitude toward metamathematics and toward non-finitary reasoning.

| Non-finitary reasoning in mathematics was widely considered to be meaningful only to the extent to which it can be "interpreted" or "justified" in terms of a finitary metamathematics.[1] This view, almost unavoidably, leads to an exclusion of non-finitary reasoning from metamathematics. For, such reasoning, in order to be permissible, would require a finitary metametamathematics. But this seems to be a confusing and unnecessary duplication. Moreover, admitting "meaningless" transfinite elements into metamathematics is inconsistent with the very idea of this science prevalent at that time. For according to this idea metamathematics is *the* meaningful part of mathematics, through which the mathematical symbols (meaningless in themselves) acquire some substitute of meaning, 2

[1]Note that this, for the most part, has turned out to be impossible in consequence of my results and subsequent work.

---

[b]In our bibliography *Skolem 1923a*.

[c]See *Hilbert and Ackermann 1928*, p. 68.

[d]*Skolem 1928*.

namely rules of use. Of course, the essence of this viewpoint is a rejection of all kinds of abstract or infinite objects, of which the prima facie meanings of mathematical symbols are instances. I.e., meaning is attributed solely to propositions which speak of *concrete and finite objects*, such as combinations of symbols.

But now the aforementioned easy inference from Skolem 1922 is definitely non-finitary, and so is any other completeness proof for the predicate calculus. Therefore these things escaped notice or were disregarded.

I may add that my objectivistic conception of mathematics and metamathematics in general, and of transfinite reasoning in particular, was fundamental also to my other work in logic.

How indeed could one think of *expressing* metamathematics *in* the mathematical systems themselves, if the latter are considered to consist of meaningless symbols which acquire some substitute of meaning only *through* metamathematics?

Or how could one give a consistency proof for the continuum hypothesis by means of my transfinite model Δ if consistency proofs have to be finitary? (Not to mention that from the finitary point of view an interpretation of set theory in terms of Δ seems preposterous from the beginning, because it is an "interpretation" in terms of something which itself has no meaning.) The fact that such an interpretation (as well as any non-finitary consistency proof) yields a finitary *relative* consistency proof apparently escaped notice.

Finally it should be noted that the heuristic principle of my construction of undecidable number⟦-⟧theoretical propositions in the formal systems of mathematics is the highly transfinite concept of "objective mathematical truth", *as opposed* to that of "demonstrability",[2] with | which it was generally confused before my own and Tarski's work. Again the use of this transfinite concept eventually leads to finitarily provable results, e.g., the general theorems about the existence of undecidable propositions in consistent formal systems.

3

Skolem's epistemological views were, in some sense, diametrically opposed to my own. E.g., on p. 29 of his 1929 paper,[f] evidently because of the transfinite character of the completeness question, he tried to *eliminate* it, instead of answering it, using to this end a new definition of logical

[2]Cnf. M. Davis, The Undecidable,[e] New York 1965, p. 64.

---

[e]*Davis 1965*. The reference is to *Gödel 1934*, pp. 21–22.

[f]*Skolem 1929*.

consequence, whose idea exactly was to *avoid* the concept of mathematical truth. Moreover he was a firm believer in set theoretical relativism and in the sterility of transfinite reasoning for finitary questions (see p. 49 of his 1929 paper).

Sincerely yours,

Kurt Gödel

P.S. As far as your remark on p. 25, line 3, is concerned, I believe my proof can be extended to the case where identity occurs, as I also stated in the end of my paper.[g]

---

[g]Gödel refers to the decidability proof of *1933i*; the belief expressed here was eventually refuted in *Goldfarb 1984* and *1984a*; cf. the introductory note to *1933i* in these *Works*, vol. I, especially pp. 229–231, also letters 3 and 5 below, letter 2 of Appendix A (of Dana Scott to Burton Dreben and Wang, 3 April 1970), and the correspondence with Dreben in volume IV of these *Works*.

3. Wang to Gödel

December 19, 1967

Dear Professor Gödel:

Thank you very much indeed for your illuminating letter of December 7. After reexamining the relevant passages in the several papers in the light of your letter, I feel I have understood better the circumstances surrounding the discovery of the completeness theorem. I have accordingly written the enclosed letter to the editor of the Skolem volume to make corrections in conformity with my improved understanding of the situation. I believe the revised exposition to be more accurate, but I am not yet certain that it does represent completely faithfully the facts. If you have further comments on the manuscript as revised, I would greatly appreciate an early letter in order that corrections can be made before publication. Among other things, I am not happy over revision (e) dealing with your P.S., since I have not myself made the attempt to extend your proof to the case with identity. While I intend to undertake such a study, I feel I shall not be able to get the matter clarified before the publication of the Skolem volume.

Your letter brings out the great epistemological gap between Skolem's and your work on properties of the predicate logic. I was confused by the fact that Skolem's treatment of Löwenheim's theorem obviously involves non-finitary reasoning. I believe I see from your letter that Skolem probably thought he could not use the same sort of argument when considering the question of completeness, which is squarely in the domain of metamathematics. This may explain his persistent confusion between "consistent" and "satisfiable".

I am grateful to you both for your patient explanation of the surprising blindness of logicians at that time, and for your important information on the relation between your work in logic and your objectivistic conception of mathematics and metamathematics. Like many others, I am eager to find out more details about your objectivistic conception 2 outlined in your several | papers. And your present letter seems to me to provide a good occasion for me to ask a few questions along this line. I sincerely hope you will not find the questions too confused to deserve any replies.

1. Finitary position.

Professor Bernays's distinction between the original and the extended finitary positions is explained in your Dialectica article,[a] in which the following points are also brought out. Neither position (i.e. neither intuitive nor abstract evidence) has received a sharp formulation. For example, it is not known how much of recursive inference under ϵ_0 is valid under the original conception, or how much of Brouwer's methods is acceptable by the extended conception. Nonetheless, the two positions are sufficiently clear to yield the conclusion that the original program is impossible by your incompleteness results and that Gentzen's proof and your two interpretations do conform to the extended position.

While your relative interpretation is more attractive in giving directly a constructive meaning to each theorem of axiomatic number theory, Gentzen's proof yields a single proved proposition stating the consistency of the system Z. I do not know whether this latter fact could be viewed as a significant additional result from the extended finitary position, and whether there is a simple general theorem by which, in particular, a single statement[b] of consistency follows from your relative interpretation, as a theorem in a directly related extended system.

2. Non-finitary consistency proofs.

You state in your letter that any non-finitary consistency proof yields a finitary *relative* consistency proof. I am under the impression that you

[a] *Gödel 1958*.

[b] In the left margin Gödel wrote in shorthand "ja".

do not mean by this merely that if T has an inner model in S, then we have a finitary relative consistency proof of T to S. This of course is by now clear to many people, certainly from your result on the continuum hypothesis. Do you have a more general assertion in mind? Also, what is the position of von Neumann's consistency proof of the axiom of foundation (1929, J. f. Math. vol. 160[c]) in this connection? I believe he does not consider the question of absoluteness. But he does compare (footnote 5) his proof to the relative consistency of non-Euclidean to Euclidean geometry. Further, I do not know his relation to the finitary point of view at that time.

3. Objectivistic conception and your work in logic.

You point out in your letter that your objectivistic conception is *necessary*[d] for your most important discoveries in logic. | In addition, you mention in one case the heuristic principle supplied by the concept of objective mathematical truth. These observations are of course exceedingly important for the philosophy of mathematics. I expect that when you state that your objectivistic conception was *fundamental*[e] *to* your work in logic, you have much more in mind than is explained in your letter. I mean there must be a lucid overall picture, as well as connecting strands between your basic conception and your results. I should like to be able to formulate some more definite questions,[f] after rereading your published papers in the near future. Meanwhile, I shall confine myself to two peripheral queries.

3

4. The iterative concept of set.

This concept is explained in your 1947 paper on the continuum problem (pp. 518–519; Benacerraf and Putnam, pp. 262–263),[g] and is by now widely popularized. I have for sometime thought that this concept was already implicitly in Cantor's original definition of 1895: Unter einer "Menge" verstehen wir jede zusammenfassung *M* von bestimmten wohlunterschiedenen objekten *m* unsrer Anschauung oder unseres Denkens (welche die "Elemente" von *M* genannt werden) zu einem ganzen.[h] My feel-

[c] *von Neumann 1929*.

[d] Gödel wrote 'x' to signify a footnote and wrote "heuristically" as the note at the bottom of the page. Moreover, directly under "is *necessary*" he wrote "was fundamenta⟦l⟧".

[e] Gödel wrote in the left margin an arrow pointing to the line "your objectivistic conception was fundamental to your work in".

[f] Gödel wrote a large question mark in the left margin.

[g] I.e., *Gödel 1947*, pp. 518–519, *1964*, pp. 262–263.

[h] *Cantor 1895* (*1932*, p. 282). Obviously 'zusammenfassung', 'objekten' and 'ganzen' should be capitalized.

ing was that the "collecting together" contains a genetic aspect and thereby implies a process of iteration. Others with whom I have discussed the topic wished to say that, for example, Russell's diagonal set is not excluded by this definition. But, of course, Cantor in his letters to Dedekind also distinguishes inkonsistente Vielheiten from sets (konsistente Vielheiten). From your deep understanding of Cantor's work, do you think one can argue convincingly that the iterative concept is explicitly (or at least implicitly) Cantor's original concept?[i]

5. Set-theoretical paradoxes.

In your paper on Russell's logic,[j] you say (p. 131 and p. 215 in reprinting): "By analyzing the paradoxes to which Cantor's set theory had led, he freed them from all mathematical technicalities, thus bringing to light the amazing fact that our logical intuitions (i.e. intuitions concerning such notions as: truth, concept, being, class, etc.) are self-contradictory". Read in isolation, this statement appears superficially to disagree with your overall position. In your 1947 paper, as revised recently, you state (p. 518 and p. 262): "They (viz. the set-theoretical paradoxes) are a very serious problem, not for mathematics, however, but rather for logic and epistemology".[k] It is not clear to me in what sense they are a serious problem for logic, while not being one for mathematics.[l] Since the second quotation is only concerned with set-theoretical paradoxes, even though I can conceive of the two contrary situations together, I do not know the 4 exact way in which you regard them as a serious problem for logic.[m] | In a different direction, I can also understand if one distinguishes between foundations of set theory and foundations of mathematics. Very crudely, I am interested in knowing the different senses in which you regard set-theoretical paradoxes as misunderstandings (of the intuitive concept of set) and as antinomies.

6. Philosophical implications of the incompleteness results.

I am under the impression that your Gibbs lecture (unpublished as far as I know[n]) applies the incompleteness results to argue against certain

[i]In the left margin Gödel wrote "no" and below that "after paradoxes".

[j]*1944*; the reprinting referred to is in *Benacerraf and Putnam 1964*.

[k]*1964*, p. 262; the other page reference is to *1947*.

[l]Gödel wrote an arrow in the left margin.

[m]At the bottom of the page Gödel wrote "| intuit. (surprises) insane (mental sickness)".

[n]Subsequently published in these *Works*, vol. III, as **1951*.

form⟦s⟧ of mechanism or mechanical materialism. It would be very gratifying to me to be able to learn in some way your line of reasoning on this subject.

I hope very much you will care to comment on some of these questions.

Yours respectfully,

Hao Wang[o]

---

[o]On the reverse side of the last page Gödel wrote "Fragen Wang" (Wang's questions). The envelope of the letter contains on separate lines the notations "Wiss. Interessante" ("scientifically interesting"), "Ergänz. zu 1967/68" ("amplification of 1967–68") "und 1971/72 Disk⟦ussionen⟧" ("and 1971/72 discussions"). Only 'zu' and 'und' are in shorthand. In the *Nachlaß* the envelope apparently contained handwritten (partially shorthand) notes and drafts by Gödel, probably drafts of answers to Wang's questions. It is evidently the first of these notations that Wang refers to in *1996*, p. 134.

4. Gödel to Wang[a]

March 7, 1968

Professor Hao Wang
Department of Mathematics
Rockefeller University
New York, N.Y. 10021

Dear Professor Wang:

Thank you very much for sending me copies of your letters of December 13, February 7, and February 9 to Professor Fenstad. It seems to me the last two contradict each other in the evaluation of Skolem, 1928, p. 134–135. In my opinion the first formulation is by far the best.

I am still *perfectly convinced* that reluctance[1] to use non-finitary concepts and arguments in metamathematics was the primary reason why the completeness proof was not given by Skolem or anybody else before

[1]Herbrand also provides a very good example of this reluctance.

---

[a]On letterhead of The Institute for Advanced Study, Princeton, New Jersey 08540, School of Mathematics.

my work. It may be true that Skolem had little interest in the *formalization* of logic, but this does not in the least explain why he did not give a correct proof of *that* completeness theorem *which he explicitly stated* (loc. cit. p. 134), namely that there is a contradiction at some level n if there is an informal disproof of the formula. On the basis of his lemma of 1922 this would have been quite easy, since evidently a correct informal disproof implies the non-existence of a model. Moreover, what he tried to accomplish on page 29 of his paper of 1929 (see the last paragraph of my letter of December 7, 1967) *evidently was* to eliminate transfinite arguments from metamathematics (in a manner quite similar to Herbrand's).

That he used non-finitary reasoning for Loewenheim's Theorem proves nothing, because pure model theory, where the concept of proof does not come in, lies on the borderline between mathematics and metamathematics, and its applications to special systems with a finite number of axioms actually belong to mathematics, at least for the most part. This also explains von Neumann's use of a transfinite model of set theory, which, by the way, is rather trivial.

On rereading my letter of December 7 I find that the phrasing of the last but one paragraph on page 2 is perhaps a little too drastic. It must 2 be understood cum grano salis. Of course, the formalistic point of view | did not make *impossible* consistency proofs by means of transfinite models. It only made them much harder to discover, because they are somehow not congenial to this attitude of mind. However, as far as, in particular, the continuum hypothesis is concerned, there was a special obstacle which *really* made it *practically impossible* for constructivists to discover my consistency proof. It is the fact that the ramified hierarchy, which had been invented *expressly for constructivistic purposes*, has to be used in an *entirely non-constructive way.* A similar remark applies to the concept of mathematical truth, where formalists considered formal demonstrability to be an *analysis* of the concept of mathematical truth and, therefore, were of course not in a position to *distinguish* the two.

I would like to add that there was another reason which hampered logicians in the application to metamathematics, not only of transfinite reasoning, but of mathematical reasoning in general. It consists in the fact that, largely, metamathematics was not considered as a science describing objective mathematical states of affairs, but rather as a theory of the human activity of handling symbols.

Unfortunately I was very busy the past few weeks with rewriting one of my former papers.[b] But I hope to be able soon to answer the other

[b] *1958*; Gödel evidently refers to the work on *1972*; see p. 217 of the introductory note to *1958* and *1972* in these *Works*, vol. II.

questions raised in your letter of December 19.

Sincerely yours,

Kurt Gödel

5. Wang to Gödel

April 23, 1968

Dear Professor Gödel:

Thank you for your letter of March 7. I have been so angry with myself that I delayed looking at the matter again. I have now uncovered many of my unforgivable confusions and come to agree with you in all aspects where I understand you. I have just written up another correction in which I attempt to organize all the salient points made by you and Professor Bernays. I hope you will find this version satisfactory.

I am looking forward to your answer to the other questions raised in my letter of December 19.

I have spent a good deal of time on $\exists^m \forall^2 \exists^n$ case with equality but remain unclear about the true state of affairs. I feel I do not know any coherent treatment of the decision problem of logic especially when equality is involved. As I see it, when equality is included, we have in general no longer the functions δ and η (p. 436 of your 1933 paper[a]) and I do not know how to introduce the appropriate equivalence relation. As you undoubtedly know, Theorem II (p. 435 of your paper) can probably be refuted by a counterexample when equality is included. I intend to look into the problem more. Meanwhile, whatever information and comments which you care to give me will be most welcome.

With best wishes,

Yours respectfully,

Hao Wang

[a] *1933i*.

6. Wang to Gödel

15 April 1970

Dear Professor Gödel,

I am very sorry indeed to hear that you are presently not feeling well and sincerely hope that with the arrival of spring your health will greatly improve.

I am grateful to you for having Dana Scott send me a report on your procedure for extending your solvable case to include equality.[a] The vague recollection is that similar ideas have occurred independently to other people but nobody has been able to carry them through. The reduction to a "sharp" form presents no problem, but the problem of solving formulas in sharp form does not seem to fall naturally under the scope of the original method for formulas without equality.

Consider for simplicity the *satisfiability* problem of a formula in the simple form:

(A) $\forall x\, \forall y\, \exists z(x \neq y \supset (x \neq z \wedge y \neq z \wedge Mxyz))$, where M contains no $=$.

It is not clear how we should construe this sharp formula as one without equality. One natural way would seem to be replacing $=$ by a dyadic predicate letter, say I. Another natural way would be to drop the bounds and consider just $\forall x\, \forall y\, \exists z\; Mxyz$. In either way, applying the original method to the result need not give the same answer as (A), even if we add expressible natural properties of $=$ such as Ixx, $Ixy \equiv Iyx$, and require that there are at least, say, three objects. (Of course we have no obvious way of adding transitivity.)

Stål Aanderaa who has spent a good deal of effort on the general problem has examples to support the last assertion.

2 | Let M in (A) be $(\neg Gx \vee \neg Gy) \wedge (Gz \equiv (\neg Gx \wedge \neg Gy))$. Then (A) has a simple model with three objects, while $\forall x\, \forall y\, \exists z\; Mxyz$ has no model since the first clause requires every object not to have G but the second clause requires some object to have G.

A more complex example shows that an unsatisfiable (A) becomes satisfiable when $=$ is replaced by a neutral predicate I. Let M be the conjunction of the following four clauses:

(1) $Fzz \equiv (\neg Fxx \wedge \neg Fyy)$.
(2) $\neg Fxx \vee \neg Fyy$.
(3) $Fxx \vee Fyy \vee (Fzx \equiv \neg Fzy)$.
(4) $Fxx \supset (Fyz \wedge \neg Fzy)$.

---

[a]Letter 2 of Appendix A, dated 3 April 1970.

Using this conjunction as M, (A) has no model because (1), (2), (3) require that there be no more than three objects but (1) and (4) require at least four objects. On the other hand, if we replace $=$ by I in (A) , the result is satisfiable.

In order to check that the original method for formulas without equality applies to sharp formulas, it would, therefore, seem necessary to specify more exactly how the sharp formulas are to be taken so that special properties of $=$ can be avoided.

I have discussed this letter with Dreben and he is writing a more detailed letter with special reference to your original paper.

I am sending a copy of this letter to Scott.

With very best wishes,

Yours respectfully,

Hao Wang

7. Wang to Gödel

25 May 1971

Dear Professor Gödel:

I am writing to ask your permission to make a lengthy quotation from your letter to me dated December 7, 1967. I would like to quote from the last paragraph of p. 1 to the fourth line of p. 3 and include these paragraphs in the introduction of my book *Knowledge and logic* which is near completion.[a] I enclose herewith a table of contents. As soon as the revised introduction is typed, I will send you a copy. In fact, I hope to send you some of the other chapters too and would of course welcome any comments and criticisms you would care to make.

I sincerely hope you will be willing to grant me the permission, since I believe many people would find the paragraphs highy illuminating. If you prefer not to be quoted in the particular context, I would appreciate your letting me know early so that I can make the appropriate deletions or corrections.

Yours respectfully,

Hao Wang

---

[a]Eventually published as *Wang 1974*.

8. Gödel to Wang[a]

July 9, 1971

Professor Hao Wang
The Rockefeller University
New York, N.Y. 10021

Dear Professor Wang:

As you probably have heard from Professor Tennenbaum already, I have no objection whatsoever to your publishing my letter of December 7, 1967. In fact I am very much in favor of these things becoming generally known.

I only have to require:

1. that you also publish my letter of March 7, 1968 from the second paragraph to the last but one paragraph, both included. Between the two letters something like this should be said: "To some objections Gödel, in a letter of March 7, 1968, responded the following;"
2. that the quotation from my letter of December 7, 1967 start in the middle of the third paragraph, i. e., with: "The completeness theorem, mathematically, is...:"
3. that to the footnote ** of my letter of December 7[b] the following be added: "[where I explain the heuristic argument by which I arrived at the incompleteness results (added in this edition)];"
4. that in the fourth paragraph of my letter of March 7 after "paragraph on page 2" the following be added: "[i. e. the paragraph beginning with: "Or how could one give"]."

Finally, I would suggest that, immediately before the quotation from my letter of December 7 you insert the following: "about the role which his philosophical views played in his work in mathematical logic."

I believe that these additions (in particular 1.–4.) are really important for making everything as clear as possible.

I am sorry that, in consequence of my illness, our meeting, proposed for January 1970, never materialized. I shall be very glad to see you sometime this year at your convenience.

2 | Thank you very much for sending me your manuscript on Metalogic.[c]

Sincerely yours,

Kurt Gödel

---

[a]On letterhead of The Institute for Advanced Study, Princeton, New Jersey 08540, School of Mathematics, carbon paper containing the word COPY in large letters roughly where the text of the letter falls.

[b]Here footnote 2.

[c]A version of *Wang 1974a*.

P.S. Perhaps in the third paragraph of my letter of Dec. 7, after "in the 1928 edition of their book" one might ~~may~~ add: "on p. 68".

9. Wang to Gödel

20 July 1971
(Written on July 9, 1971)

Dear Professor Gödel:

Stanley Tennenbaum telephoned me around the end of June to say that you are willing to grant me permission to quote your letters. He did not remember all the details of his conversation with you on this question but did mention your second letter.

I have, therefore, decided to expand footnote 3 to incluce your additional observations in the form as enclosed. I would be very grateful to receive your comments if you find anything particularly inaccurate or inadequate.

I do have a set of all the chapters but, on second thought, I feel that it would be an imposition on you to send you these bulky manuscripts. The book, when it appears, will be more compact.

Yours respectfully,

Hao Wang

10. Wang to Gödel[a]

20 July 1971

Dear Professor Gödel:

Thank you very much for your letter dated July 9. Before receiving your letter, I had drafted a letter and some additions, as enclosed. These were

[a] Above and to the right of the letterhead Gödel wrote, "p. 45 oben & in part of expressing metamath. in math. ⟨itself⟩ which does not contain any symbols denoting the handling of symb." Gödel may be alluding to the following remark in letter 4: "It ⟦a reason which hampered logicians in applying mathematical reasoning in metamathematics⟧ consists in the fact that, largely, metamathematics was not considered as a science describing objective mathematical states of affairs, but rather as a theory of the human activity of handling symbols." The addition to the letter of 7 March proposed in the third paragraph of the following letter seems to be a revised version of the handwritten comment. The remark just quoted appears in document no. 013099, a revision of the footnotes of the Introduction to Wang's book that is probably what was sent with the present letter and thus would be his first response to Gödel's request that he include material from the 7 March letter.

meant to take care of your points 1 and 4 regarding the letter of March 7, 1967.[b] The manner in which I have dealt with the two points is slightly different from your suggestions. I would like to know whether you find my way of handling the March 7 letter acceptable. I am perfectly willing to adhere to your instructions if you prefer. There is a minor point on which I cannot yet decide, viz. whether the material should be moved into the main text. Unless you express a preference on this, I shall decide (probably to put it into the main text) when I organize other revisions into the introduction.

I will obey your three instructions regarding your letter of December 7, 1967 (viz. points 2, 3, and "finally").

Of course I want very much to see you and discuss philosophy with you. I feel there is much I can learn from you. I am mostly away in the Boston area during the summer. I intend to come to visit you after the beginning of September at times convenient to you. Sometimes I have the vague idea of asking Carl Kaysen for a simple lodging place at the Institute so that I may come to see you frequently.[c]

Yours respectfully,

Hao Wang

---

[b] "1967" is evidently a slip and should be "1968".

[c] The economist Carl Kaysen (1920–) was at the time Director of the Institute for Advanced Study.

11. Gödel to Wang

August 4, 1971

Professor Hao Wang
The Rockefeller University
New York, New York 10021

Dear Professor Wang:

Thank you for your letters of July 20 and the manuscript of your book on factualism. Since I shall not have the time to read it in the near future, I would request, in case you want any comments from me, to mention to me the passages where my name occurs.

As far as my letters of 1967 and 1968 are concerned it seems to me that the second one is even more important than the first. Therefore, if you don't want to include it in the text, *there ought to be* some explicit reference to it in the text, e.g.: "See moreover Gödel's letter published in footnote 3 below." The numeral referring to this footnote could then appear after the addition suggested under "Finally" in my letter of July 9, 1971.

Finally, I would like to mention that in my letter of March 7 I omitted one important point. This defect could easily be corrected by adding, in the last but one paragraph, after "reasoning in general," the following: "[and, most of all, in expressing metamathematics in mathematics itself, since the latter contains no symbols denoting symbols (added in this edition)]."[a]

I shall not suggest any further additions and should appreciate it, if you would let me know the final version of these passages.

Sincerely yours,

Kurt Gödel

P.S. I would like to repeat here the suggestion made in the postscript of my letter of July 9, 1971.

---

[a]The location would be *Wang 1974*, p. 10. However, it was partly removed before publication; see letter 13 below.

12. Wang to Gödel

11 August 1971

Dear Professor Gödel,

Thank you very much for your letter of August 4. I will try to organize your instructions into the introduction of my book and send you the final version for approval and comments. In particular, I intend to respect your wish and put both letters into the text.

In connection with the chapters I sent to you, I believe your name occurs more frequently than any other single name. In particular, I would specially appreciate getting your comments on the following parts which are more extensively concerned with your published works:

(1) Chapter VI. The concept of set.
(2) Chapter II, section 3. Mechanical procedures.
(3) Chapter X, section 6. Mathematical arguments.

I enclose herewith a tentative version of the revised pages.

Yours respectfully,

Hao Wang

13. Wang to Gödel

April 10, 1972

Dear Professor Gödel:

Following your telephone call on Friday, I have made the following revisions to carry out your instructions. If you are dissatisfied with any of them, please let me know so that I can make further corrections.

* * *

⟦The omitted portion concerns changes in the Preface.⟧

(2) Introduction. On p. 16, line⟦s⟧ 2 to 4, delete the half sentence beginning with "and, most of all,". [I am unhappy about the deletion, especially the explicit statement of symbols denoting symbols. I recall somebody (perhaps Tennenbaum) specifically commending this half sentence.][a]

Delete the five paragraphs beginning from the fourth paragraph on p. 17, and state that corrected text will be supplied by Gödel.[b]

(3) Mechanical procedures.

* * *

⟦Wang mentions some changes for the most part in section 3.1 of chapter II, including two instances where revised text is to be supplied by Gödel.⟧

---

[a]See note a to letter 11 above.

[b]This probably refers to the remarks on Hilbert on pp. 11–12 of *Wang 1974*.

(4) The concept of set.

* * *

⟦Wang mentions some small corrections in the earlier pages of chapter VI on this subject.⟧

pp. 11–13. Delete the two paragraphs beginning from line 3 of p. 11 and state that corrected text will be supplied by Gödel.

pp. 17–18. Delete long paragraph beginning about middle of p. 17 (to end of p. 18) and state that corrected text will be supplied by Gödel.[c]

(5) Minds and machines.

Delete section 7 (pp. 51–59) and state that corrected text will be supplied by Gödel.

- - -

With these revisions made, I am sending the resulting texts to Stuart Hampshire who is negotiating the publication of my book with a publisher. In any case, the book will not be published until the passages related to your approval are handled to agree with your wishes.

Yours respectfully,

Hao Wang

[c]Although it is hard to reconcile with the page numbers Wang gives, these passages must be the paragraph on the axiom of replacement (*Wang 1974*, p. 186) and remarks on principles for setting up axioms of set theory (pp. 189–190). Cf. the following letter and *Wang 1974*, pp. x–xi.

14. Wang to Gödel

April 26, 1972

Dear Professor Gödel:

I have made an attempt to revise the earlier parts of the pages on the iterative concept of set. Pp. 1–4 and 11–13 are to be replaced by the enclosed pages.[a] While you are revising the paragraphs on the jump operation and the summary of five principles for setting up axioms,[b] you may wish to look at these pages (in particular, the new p. 2, p. 3, p. 4, p. 11) and determine whether there are serious mistakes. There is no mention of your name in the four new pages.

My home phone number is now 212-734-1309. This time I expect the number to remain the same for the for⟦e⟧seeable future.

With regard to the section on mechanical procedure, I think, ⟨before I make up the corrected version⟩, I should wait till you let me have the revised pages, in order to avoid too many different versions.

Looking forward to seeing you on Wednesday, May 3rd.

Respectfully yours,

Hao Wang

---

[a]The first pages probably correspond roughly to pp. 181–183 of *Wang 1974*. It is hard to be certain in identifying the second, but they probably correspond to somewhere between p. 186 and p. 189 of the book.

[b]Cf. *Wang 1974*, pp. 186 and 189–190, and note c to the preceding letter.

15. Wang to Gödel

December 12, 1975

Dear Professor Gödel:

Thank you very much for coming out on such a rainy and chilly day to the Institute on Tuesday. I have enjoyed and learnt from the long talk together. I am especially happy to see that you appear to be in very good condition.

For quite some time I had been worried about finding a proper way of organizing the material on objectivism. It is a relief that a way is found. I now plan to write the paper in three parts. (1) Quotations from Gödel. (2) Exposition and commentary (3) Interpretation and broader issues. In this way, I shall feel free to express my own opinions in (2) and (3), and your theory will be reported in (1) untainted.

I have written a draft of the first six sections of (1), and am enclosing a copy of this part.[a] The remainder of (1) will be devoted to stating the various arguments for objectivism you have told me or published. Most of the arguments are directed to mathematical objects rather than concepts. Since the whole of (1) is attributed to you, you will undoubtedly wish to examine it thoroughly.

On Tuesday you asked me several questions ⟦to⟧ which I did not then have articulate answers. I should like to answer them more completely now. Now that the question of organization is basically solved, I do hope, with your approval in due time, to publish the paper containing (1), (2), and (3) as sketched above, perhaps in the proceedings of the International Congress in Canada last summer.[b] My wish to write out more completely the notes I have taken from conversations with you is not connected with any intention of direct publication. Rather I have in mind the prospect of using such notes as a basis for further discussion with you so that, in particular, I shall better be able to ask you questions with a view to achieve a clearer understanding of your views. There are so many aspects of your thoughts which I would like to learn that it is necessary to work out an order of priority. This will of course depend on your inclinations, as well as on the relative importance | of different topics and my capabilities for appreciation of a topic. An obvious candidate is for me to learn your overview of the whole field of mathematical logic.[1] An understanding of your overview will help me to form the backbone of my plan to survey mathematical logic today. There are also more philosophical topics on which I have ~~mind~~ ⟨much⟩ to learn from you, but I do not at present have any clear conception as to what would be an appropriate topic. In any case, there will be some time before the paper on objectivism can be completed.

Yours respectfully,

Hao Wang

[1] I mean the principles you use to see the field.

---

[a] This would be document no. 013167 in Gödel's papers, a nine-page typescript entitled "Objectivism of sets and concepts". Wang sent a continuation, document no. 013169, on 19 December, with his letter dated 15 December. Together these are evidently what Wang refers to as fragment Q (*1996*, p. 149). Both typescripts contain some annotations by Gödel.

[b] That is, to incorporate the material into *Wang 1977*.

16. Wang to Gödel

June 24, 1976

Dear Professor Gödel,

I hope the conditions of your health and Mrs. Gödel's have had some improvement.

Just now I have written up some of the facts you told me about your life and especially your intellectual development. These notes are not well written at all. But more than anything else, I would like them to be free from misrepresentations.

When you are in the mood, you might wish to correct the mistakes in these notes. I am enclosing two sets in the hope that you will at some time send me a corrected set.[a]

With all best wishes,

Yours respectfully,

Hao Wang

---

[a]The enclosure was the first version of *Wang 1981*. See §3.2 of the introductory note.

17. Wang to Gödel

3 February 1977

Dear Professor Gödel,

I very much hope that your health is better. I have got the impression that you prefer not to be disturbed by telephone calls.

After your comments on Jan. 18 over the telephone, I have revised the manuscript 'Mind, matter, machine, mathematics' from page 5 on.[a] The enclosed pages are to replace the old pages 5 through 10 which I sent to you before.

While I still vaguely entertain the idea that this short paper will, after revisions and approval by you, be submitted to *Scientific American* for

---

[a]A revision of a typescript first sent by Wang on 7 November 1976; see §3.3 of the introductory note.

publication, I realize that you are not much interested in such an action. Hence, the main purpose of sending these pages to you is the hope that some time you may wish to look at them and tell me your reactions at least up to the part of page 7 where I conclude my report on your views.

Yours respectfully,

Hao Wang

Ernst Zermelo

Ernst Zermelo

The three letters reproduced here are the only ones exchanged between Gödel and Ernst Zermelo. All have been published before, the second and third in *Grattan-Guinness 1979* and the first in *Dawson 1985a*. The circumstances surrounding the correspondence are described in the former source, as well as in *Dawson 1985*, *Dawson 1997* and *Moore 1980*.[a]

Briefly, Gödel and Zermelo met for the first and only time in September 1931 at the annual meeting of the German Mathematical Union (*Deutsche Mathematiker-Vereinigung*), held that year in the town of Bad Elster. Both men spoke there on the afternoon of 15 September, Gödel on the incompleteness theorems[b] (his first lecture on those results outside Vienna) and Zermelo on his own singular conception of logic, in which universal and existential quantifiers were construed as infinitary conjunctions and disjunctions of any (ordinal) number of propositions and the notion of proof was identified with that of logical consequence.

In his address Zermelo ridiculed "the doctrine of Skolem-ism": the theorem, first proved in *Skolem 1920*, that a set of sentences in a denumerable first-order language, if satisfiable at all, must be satisfiable in a denumerable model. Zermelo regarded Skolem's result as a resurrection of Richard's paradox, which, he asserted, sprang from the unwarranted assumption ("the finitistic prejudice") that "all mathematical concepts and theorems must be representable by a fixed, finite system of signs".

Zermelo's semantic conception of proof entailed that every true sentence of arithmetic must be provable. Consequently he could not accept Gödel's claim to have demonstrated the existence of undecidable arithmetical statements.

Less than a week after the meeting, Zermelo wrote Gödel to announce that he had found a gap in Gödel's proof. He questioned whether Gödel's example of an undecidable proposition was actually a statement within Gödel's system and endeavored to show that it was not by asserting that the same method would yield "a contradiction analogous to Russell's antinomy" if Gödel's formal provability predicate Bew were replaced by the corresponding truth predicate.

[a]The papers of Grattan-Guinness and Moore were written prior to the discovery of Zermelo's first letter to Gödel, which was found during the cataloging of Gödel's *Nachlaß*.

[b]His paper *1931* had appeared in print a few months before the meeting.

In fact, of course, as Gödel pointed out in his reply, it was the truth predicate that was not formally representable in the system. Gödel went on to give a masterfully clear and precise explication of the significance of his result and of the distinctions involved in its proof. But Zermelo still failed to grasp the essence of the argument. In his second and last letter he said that he now understood better what Gödel had meant; yet he continued to maintain, both there and in the published version

1. Zermelo to Gödel

Freiburg i. Br. 21. IX 31
Karlstraße 60.

Sehr geehrter Herr Gödel,

beifolgend sende ich Ihnen einen Abzug meiner Fundamenta-Arbeit[a] und würde mich freuen, wenn ich Sie zu den wenigen zählen dürfte, die es wenigstens versucht haben, die dort entwickelten Gedanken und Methoden aufzunehmen und für die eigene Forschung fruchtbar zu machen. Während ich nun damit beschäftigt war, von meinem in Elster gehalte-
2 nen Vortrag ein kurzes Referat[b] zu machen, und dahin auch | auf den Ihrigen Bezug nehmen mußte, kam es mir nachträglich zum deutlichen Bewußtsein, daß Ihr Beweis für die Existenz unentscheidbarer Sätze eine wesentliche *Lücke* aufweist. Um einen "unentscheidbaren" Satz aufzustellen, definieren Sie auf S. 175 ein "Klassenzeichen" (eine Satzfunktion mit *einer* freien Variablen) $S = R(q)$ und zeigen dann, daß weder $[R(q); q] = A$ noch seine Negation $\overline{A}$ "beweisbar" sei. Aber gehört dann

$$S = \overline{\text{Bew}}\,[R(n); n]$$

wirklich Ihrem "System" an und sind Sie berechtigt, diese Funktion mit
3 $R(q)$ zu iden|tifizieren? Nur, weil es ein "Klassenzeichen" ist? Ich weiß, es folgt nachher eine ausführliche Theorie der "Klassenzeichen". Aber zur Kritik genügt hier folgende Überlegung. Lassen Sie in Ihrer Formel (1) die Zeichenverbindung "Bew" fort und schreiben dafür

$$(1)^* \qquad n \in K^* \cdot \equiv \overline{[R(n); n]} = S^*.$$

[a] *Zermelo 1929* or *1930*; *1930* is the more likely.

of his talk (*Zermelo 1932*), that Gödel's result was due to a "finitistic restriction" which was applied to proofs but not to statements in general. Consequently, he claimed, the former but not the latter were denumerable.

The exchange ended there, no doubt because Gödel saw no point in continuing it.

John W. Dawson, Jr.

1. Zermelo to Gödel

Freiburg im Breisgau, 21 September 1931
Karlstraße 60

Dear Mr. Gödel,

I am sending you, enclosed, a proof-sheet of my *Fundamenta* paper,[a] and I would be pleased if I might count you among the few who have at least tried to take up the ideas and methods developed there and make them fruitful for their own research. While I was engaged in preparing a short abstract of my Elster lecture,[b] in the course of which I had also to refer to yours, I came subsequently to the clear realization that your proof of the existence of undecidable propositions exhibits an essential *gap*. In order to produce an "undecidable" proposition, you define on page 175 a "class sign" (a propositional function of *one* free variable) $S = R(q)$, and then you show that neither $[R(q); q] = A$ nor its negation $\overline{A}$ would be "provable". But does

$$S = \overline{\text{Bew}}\,[R(n); n]$$

really belong to your "system", and are you justified in identifying this function with $R(q)$, just because it is a "class sign"? I know that later on there follows a detailed theory of "class signs", but for a critique the following consideration suffices here: in your formula (1), let the sign combination "Bew" be omitted and write instead

$$(1)^* \qquad n \in K^* \cdot \equiv \overline{[R(n); n]} = S^*.$$

[b] *Zermelo 1932*

Setzen Sie dann wieder $S^* = R(q^*)$, so folgt, daß der Satz

$$A^* = R(q^*; q^*)$$

weder "wahr" *noch* "falsch" sein kann, d. h. Ihre Annahme führt auf einen *Widerspruch*, analog der Russel⟦l⟧'schen Antinomie. Der Fehler beruht —ebenso wie in der Richard'schen und der Skolem'schen Paradoxie—auf der (irrigen) | Voraussetzung, daß jeder mathematisch definierbare Begriff durch eine "endliche Zeichenverbindung" (nach einem *festen* System!) ausdrückbar sei, also das, was ich das "finitistische Vorurteil" nenne. In Wirklichkeit steht es ganz anders, und erst nach Überwindung dieses Vorurteils, die ich mir zur besonderen Aufgabe gemacht habe, wird eine vernünftige "Meta-Mathematik" möglich sein. Gerade Ihre Beweisführung würde, richtig gedeutet, sehr viel dazu beitragen und damit der Wahrheit einen wesentlichen Dienst leisten können. Aber so wie Ihr "Beweis" jetzt steht, kann ich ihn als bindend nicht anerkennen. Das wollte ich Ihnen frühzeitig mitteilen, um Ihnen Zeit zur Nachprüfung zu geben.

4

Mit bestem Gruß

E. Zermelo

2. Gödel to Zermelo

Wien VIII Josefstädterstrasse 43
12./X. 1931

Sehr geehrter Herr Professor!

Besten Dank für Ihren Brief vom 21./IX. Ich konnte leider nicht sofort antworten, weil ich für einige Tage verreisen musste, will dies aber jetzt umso ausführlicher tun.

Zunächst möchte ich feststellen, daß die ersten 3 Seiten meiner Arbeit natürlich kein bindender Beweis sein sollen. Auf S. 174 habe ich ja ausdrücklich gesagt, daß es sich vorerst bloß um die Skizzierung des Hauptgedankens eines Beweises handelt. Insbesondere wird natürlich die Lücke, auf die Sie hinweisen⟦,⟧ später ausgefüllt (vgl. S 182 ff.)—ja man kann sagen, daß gerade dies der Hauptzweck der folgenden Überlegungen ist —u. zw. geschieht der Beweis in der Weise, daß die Definition der Klasse K auf einfache arithmetische Definitionsweisen (rekursive Def. etc.), welche in PM sicher formal ausdrückbar sind, zurückgeführt wird.—Ich glaube aber, daß man auch schon auf Grund der ersten 3 Seiten meiner

If you then once more set $S^* = R(q^*)$, it follows that the proposition

$$A^* = R(q^*; q^*)$$

can be *neither* "true" *nor* "false"; that is, your assumption leads to a *contradiction* analogous to Russel⟦l⟧'s antinomy. Just as in the Richard and Skolem paradoxes, the mistake rests on the (erroneous) assumption that every mathematically definable notion is expressible by a "finite combination of signs" (according to a *fixed* system!)—what I call the "finitistic prejudice". In reality, the situation is quite different, and only after this prejudice has been overcome (a task I have made my particular duty) will a reasonable "metamathematics" be possible. Correctly interpreted, precisely your line of proof would contribute a great deal to this and could thereby render a substantial service to the cause of truth. But as your "proof" now stands, I cannot acknowledge it as binding. I wanted to impart this to you early on, to give you time to check it over.

With best regards,

E. Zermelo

2. Gödel to Zermelo

Vienna VIII, 12 October 1931
Josefstädterstraße 43

Dear Professor,

Thank you for your letter of 21 September. Unfortunately I could not answer it right away, because I had to go away for a few days. Now, however, I want to do so all the more thoroughly.

First of all I would like to point out that the first three pages of my paper should of course not be taken as a binding proof. Indeed, on p. 174 I have expressly said that at the outset it is a question just of sketching the main ideas of a proof. In particular, of course, the gap to which you refer is later filled in (see pp. 182ff)—indeed one can say that that is exactly the main goal of the deliberations that follow—and in fact, the way the proof is done the definition of the class K is traced back to simple arithmetical means of definition (recursive definitions, etc.) which are certainly formally expressible in PM—but I believe that even just on the

2 | Arbeit zu der Überzeugeung von der Richtigkeit des Beweises kommen kann, wenn man die Sache genau durchdenkt.

Um dies auseinanderzusetzen⟦,⟧ knüpfe ich an Ihren Einwand an. Sie definieren eine Klasse K^* durch die Festsetzung: "n gehört zu K^*, wenn $[R(n); n]$ nicht *richtig* ist", während ich eine Klasse K definiere durch: "n gehört zu K, wenn $[R(n); n]$ nicht *beweisbar* ist". Die Annahme, daß K^* durch ein Klassenzeichen des gegebenen Systems ausdrückbar ist, führt dann auf einen Widerspruch (dies ist aber nicht meine, sondern Ihre Annahme). D. h. also man kann zeigen, daß die Klasse K^* in dem gegebenen System nicht vorkommt.[1] Von der Klasse K dagegen kann man dies *nicht* zeigen, sondern man kann im Gegenteil beweisen, daß sie mit einem Klassenzeichen des gegebenen Systems umfangsgleich ist, vorausgesetzt
3 daß in dem gegebenen Syst. gewisse einfache | arithmetische Begriffe (Addition u. Multiplikation) enthalten sind.

Um den Grund für dieses verschiedene Verhalten von K und K^* einzusehen, muß man zunächst die Definition von K^* in korrekter Form schreiben. Man kann nämlich *nicht* setzen

$$n \in K^* = \overline{[R(n); n]},$$

weil die Zeichenverbindung $\overline{[R(n); n]}$ keinen Sinn hat. Ein Negationsstrich hat ja nur Sinn über einer Zeichenverbindung, die eine Behauptung ausdrückt (über der Ziffer 5 etwa ist ein Neg.-Strich sinnlos). Die Zeichenverbindung "$[R(n); n]$" drückt aber *keine Behauptung* aus. "$[R(n); n]$" ist ja gleichbedeutend etwa mit folgenden deutschen Worten: "diejenige Formel der Princ. Math., welche aus dem n-ten Klassenzeichen bei Einsetzung der Zahl n für die Variable entsteht". "$[R(n); n]$" ist nicht etwa selbst diese Formel, "$[R(n); n]$" is ja überhaupt keine Formel der
4 Princ. Math., (denn das Zeichen [;] kommt ja gar | nicht in den Princ. Math. vor), sondern "$[R(n); n]$" ist lediglich eine Abkürzung der unter Anführungszeichen stehenden deutschen Worte. Diese Worte drücken aber offenbar keine Behauptung aus, sondern sind die eindeutige Charakterisierung einer Formel (d. h. einer räumlichen Figur), ganz ebenso wie etwa die Worte "die erste Formel jenes Buches" keine Behauptung ausdrücken, wenn auch vielleicht die Formel, welche durch diese Worte charakterisiert wird, eine Behauptung ausdrückt. "$[R(n); n]$" ist also für jede bestimmte Zahl n *ein Name* (eine eindeutige Beschreibung) für eine bestimmte Formel (d. h. eine räumliche Figur) und ein Negationsstrich darüber hat daher ebensowenig Sinn, wie etwa über der Formel "$5 + n$", welche für jede Zahl n ein Name für eine ~~für eine~~ bestimmte nat. Zahl

[1]daß es solche Klassen geben muß, folgt natürlich einfacher aus dem Diagonalverfahren oder Mächtigkeitsbetrachtungen.

basis of the first three pages of my paper one can become convinced of the correctness of the proof, if one thinks through the matter precisely.

In order to elucidate this I refer to your objection. You define a class K^* by the specification: "n belongs to K^* if $[R(n); n]$ is not correct", whereas I define a class K by: "n belongs to K if $[R(n); n]$ is not *provable*". The assumption that K^* is expressible by a class sign of the given system then leads to a contradiction (that, however, is not my assumption, but yours). That is, one can thereby show that the class K^* does not occur in the given system.[1] On the other hand, one can *not* show that for the class K, but, on the contrary, one can prove that it is coextensive with a class sign of the given system, assuming that certain simple arithmetical concepts (addition and multiplication) are contained in the given system.

In order to grasp the reason for this different behavior of K and K^* one must first write the definition of K^* in correct form. Namely, one can *not* set

$$n \in K^* = \overline{[R(n); n]},$$

because the symbol complex $\overline{[R(n); n]}$ has no meaning. A negation stroke, after all, only has meaning with reference to a symbol complex that expresses an assertion (with reference to the number 5, say, a negation stroke is meaningless). But the symbol complex "$[R(n); n]$" does *not* express an *assertion.* "$[R(n); n]$" means about the same as the following ⟦English⟧ words: "that formula of Principia Mathematica which results from the nth class sign by substitution of the number n for the variable". "$[R(n); n]$" is not itself that formula. Indeed, "$[R(n); n]$" is not a formula of Principia Mathematica at all (for the sign [;] doesn't occur at all in Principia Mathematica); rather, "$[R(n); n]$" is merely an abbreviation of the ⟦English⟧ words in quotation marks. Those words, however, obviously express no assertion, but are rather the unique characterization of a formula (that is, of a spatial figure), just as, say, the words "the first formula of that book" express no assertion, even if perhaps the formula that is characterized by those words does express an assertion. For each particular number n, "$[R(n); n]$" thus is a *name* (a unique description) for a particular formula (i.e., a spatial figure), and a negation stroke over it therefore has just as little meaning as ⟦it would⟧, say, over the formula "$5 + n$", which, for every number n, is a name for a particular natural

[1] That such classes must exist follows of course simply from the diagonal procedure or cardinality considerations.

ist. Die ganze Schwierigkeit rührt offenbar daher, daß es in der Metamathematik außer den Zeichen für Zahlen, Funktionen, etc. auch *Zeichen für Formeln* gibt und daß man ein | Symbol, welches eine Formel bezeichnet, deutlich unterscheiden muß von dieser Formel selbst.

Die Definition für K^* muß man daher korrekt so schreiben

$$n \epsilon K^* = \overline{W}[R(n); n], \tag{1*}$$

wobei $W(x)$ bedeuten soll: "x ist eine richtige Formel" oder genauer: "x ist eine Formel, die eine wahre Behauptung ausdrückt."[2] Jetzt zeigt sich ganz deutlich, daß in der Definition für K^* ein neuer Begriff, nämlich der Begriff "richtige Formel" bzw. die Klasse der richtigen Formeln vorkommt. Dieser Begriff läßt sich aber *nicht* ohne weiteres auf eine kombinatorische Eigenschaft der Formeln zurückführen (sondern stützt sich auf die Bedeutung der Zeichen) u. läßt sich daher in der arithmetisierten Metamathematik nicht auf einfache arithmetische Begriffe zurückführen; oder anders ausgedrückt: Die Klasse der richtigen Formeln | ist *nicht* durch ein Klassenzeichen des gegebenen Systems ausdrückbar[3] (daher auch nicht die daraus definierte Klasse K^*). Ganz anders steht es mit den Begriff "beweisbare Formel" (bzw⟦.⟧ der Klasse der beweisbaren Formeln, welche in der Definition von K vorkommt). Die Eigenschaft einer Formel, beweisbar zu sein, ist eine rein kombinatorische (formale), bei der es auf die Bedeutung der Zeichen *nicht* ankommt. Daß eine Formel A *in einem bestimmten System* beweisbar ist, heißt ja einfach, daß es eine endliche Reihe von Formeln gibt, welche mit irgendwelchen Axiomen des Systems beginnt u. mit A endet, und welche außerdem die Eigenschaft hat, daß jede Formel der Reihe aus irgendwelchen der vorhergehenden durch Anwendung einer der Schlußregeln hervorgeht; wobei als Schlußregeln, im wesentlichen nur die Einsetzungs- | und die Implikationsregel in Betracht kommen, welche lediglich auf einfache kombinatorische Eigenschaften der Formeln Bezug nehmen. Die Klasse der beweisbaren Formeln[4] läßt sich daher auf einfache arithmetische Begriffe zurückführen, d. h. sie kommt unter den Klassenzeichen des gegebenen Systems vor u. ebenso die daraus abgeleitete Klasse K. Der ausführliche Beweis dafür findet sich auf S. 182 ff meiner Arbeit.

Im Anschluß an das Gesagte kann man übrigens meinen Beweis auch so führen: Die Klasse W der richtigen Formeln *ist niemals* mit einem

[2]Jetzt ist die rechte Seite von (1*) eine Behauptung geworden u. daher negierbar, ebenso wie etwa die Worte "die erste Formel jenes Buches ist richtig" eine Behauptung ausdrücken.

[3]genauer gesprochen, handelt es sich natürlich um die Klasse derjenigen *Zahlen*, welche richtigen Formeln zugeordnet sind.

[4]genauer: die entsprechende Klasse natürlicher Zahlen.

number. The whole difficulty obviously is due to the fact that in metamathematics there are, besides the symbols for numbers, functions, etc., also *symbols* for *formulas*, and that one must clearly distinguish a symbol that denotes a formula from that formula itself. The definition for K^* must therefore properly be written as

$$n \epsilon K^* = \overline{W}[R(n); n], \tag{1*}$$

where $W(x)$ is supposed to mean: "x is a correct formula", or, more precisely, "x is a formula that expresses a true assertion".[2] Now it becomes quite clearly manifest that in the definition for K^* a new concept, namely the concept "correct formula", or, respectively, the class of correct formulas, occurs. This concept, however, may not, without further ado, be traced back to a combinatorial property of formulas (but rather rests upon the meaning of the symbols), and therefore may not be traced back in arithmetized metamathematics to simple arithmetical concepts; or, in other words: The class of correct formulas is *not* expressible by means of a class sign of the given system[3] (hence neither is the class K^* defined from it). The situation is quite otherwise for the concept "provable formula" (respectively, the class of provable formulas, which occurs in the definition of K). The property of a formula, that it is provable, is a purely combinatorial (formal) one, in that it does *not* depend on the meaning of the symbols. That a formula A is provable *within a specified system* simply means that there is a finite sequence of formulas that begins with some axioms of the system and ends with A, and which, in addition, has the property that each formula of the sequence arises from some of the preceding ones by application of a rule of inference (where as rules of inference, in essence, only the substitution rule and the rule of implication come into play, which refer merely to simple combinatorial properties of formulas). The class of provable formulas[4] may therefore be traced back to simple arithmetical concepts—that is, it occurs among the class signs of the given system—and so likewise may the class K derived from it. The detailed proof of that is to be found on pp. 182ff of my paper.

In connection with what has been said, one can moreover also carry out my proof as follows: The class W of correct formulas *is never* coex-

[2]The right side of (1*) has now become an assertion and is therefore negatable, just as, say, the words "the first formula of that book is correct" expresses an assertion.

[3]More precisely stated, it is of course a question of the class of those *numbers* that are assigned to correct formulas.

[4]More precisely: the corresponding class of natural numbers.

Klassenzeichen desselben Systems umfangsgleich (denn die Annahme, daß dies der Fall sei, führt auf einen Widerspruch). Die Klasse B der beweisbaren Formeln *ist* mit einem Klassenzeichen desselben Systems umfangsgleich (wie man ausführlich zeigen kann); folglich können B und W nicht
8 miteinander umfangsgleich sein. Weil aber $B \subseteq W$, so | gilt $B \subset W$, d. h. es gibt eine richtige Formel A, die nicht beweisbar ist. Weil A richtig ist, so ist auch non-A nicht beweisbar, d. h. A ist unentscheidbar. Dieser Beweis hat aber den Nachteil, daß er keine Konstruktion des unentscheidbaren Satzes liefert u. intuitionistisch nicht einwandfrei ist.

Ich möchte noch bemerken, daß ich den wesentlichen Punkt meines Resultates nicht darin sehe, daß man über jedes formale System irgendwie hinausgehen kann (das folgt schon nach dem Diagonalverfahren) sondern darin, daß es für jedes formale System der Mathem. Sätze gibt, die sich innerhalb dieses Systems *ausdrücken*, aber aus den Axiomen dieses Systems *nicht entscheiden* lassen, und daß diese Sätze sogar von rel. einfacher Art sind, nämlich der Theorie der pos. ganzen Zahlen angehören. Daß man die ganze Mathematik nicht in ein formales System einfangen kann, folgt schon nach dem Cantorschen Diagonalverfahren, aber es blieb
9 trotzden denk|bar, daß man wenigstens gewisse Teilsysteme der Mathematik vollständig (d. h. entscheidungsdefinit) formalisieren könnte. Mein Beweis zeigt, daß auch dies unmöglich ist, wenn das Teilsystem wenigstens die Begriffe der Addition u. Multiplikation ganzer Zahlen enthält. (Dabei ist unter Formalisierung zu verstehen: Zurückführung auf endlich viele Axiome und Schlußregeln.) Gewiß sind die rel. unentscheidbaren Sätze in höheren Systemen immer entscheidbar, worauf ich in meiner Arbeit auch ausdrücklich hingewiesen habe (vgl. S. 191 Fußnote 48[a]), aber auch in diesen höheren Systemen bleiben unentscheidbare Sätze derselben Art übrig u.s.w. in inf.

Es würde mich freuen, wenn meine Ausführungen Sie überzeugt hätten. Natürlich sollen auch diese kein "Beweis" sein. Den Beweis finden
10 Sie vielmehr nur an den genannten Stellen meiner Arbeit. | Ich danke Ihnen bestens für die Zusendung Ihrer Fundamenta-Arbeit, doch ist dieselbe bisher leider nicht in meine Hände gelangt (vielleicht bei der Post verloren gegangen). Ich hatte Ihre Arbeit übrigens schon bald nach ihrem Erscheinen gelesen und es waren mir damals verschiedene Bedenken aufgetaucht, die ich Ihnen, falls sie Sie interessieren, nächstens gerne mitteilen will.

Mit den besten Grüßen

Ihr ergebener Kurt Gödel

P.S. Gleichzeitig übersende ich Ihnen ein Separatum meiner Arbeit über den Funktionenkalkül, die ich in Bad Elster nicht bei mir hatte.

tensive with a class sign of that same system (for the assumption that that is the case leads to a contradiction). The class B of provable formulas *is* coextensive with a class sign of that same system (as one can show in detail); consequently B and W can not be coextensive with each other. But because $B \subseteq W$, $B \subset W$ holds, i.e., there is a correct formula A that is not provable. Because A is correct, not-A is also not provable, i.e., A is undecidable. This proof has, however, the disadvantage that it furnishes no construction of the undecidable statement and is not intuitionistically unobjectionable.

I would still like to remark that I see the essential point of my result not in that one can somehow go outside any formal system (that follows already according to the diagonal procedure), but that for every formal system of metamathematics there are statements which are *expressible* within the system but which may *not be decided* from the axioms of that system, and that those statements are even of a relatively simple kind, namely, belonging to the theory of the positive whole numbers. That one can *not* capture all of mathematics in one formal system already follows according to Cantor's diagonal procedure, but nevertheless it remains conceivable that one could at least formalize certain subsystems of mathematics completely (in the syntactic sense). My proof shows that that is also impossible if the subsystem contains at least the concepts of addition and multiplication of whole numbers. (By formalization is to be understood: traceability back to finitely many axioms and rules of inference.) To be sure, the relatively undecidable statements are always decidable in higher systems, to which I have also expressly alluded in my paper (cf. p. 191, footnote 48[a]); but even in those higher systems undecidable statements of the same kind remain, and so on ad infinitem.

I would be pleased if my explanations had convinced you. Of course, they too should not be taken as "proof". The proof is rather to be found at the places of my paper I've cited. I thank you very much for sending your Fundamenta paper, yet up to now it has unfortunately not come into my hands (perhaps having been lost in the mails). Anyway, I had already read your paper soon after its appearance and at that time various thoughts occurred to me which, if they interest you, I will be glad to impart to you next time.

With best wishes,

Yours, Kurt Gödel

P.S. At the same time I am sending you an offprint of my paper about the functional calculus, which I did not have with me in Bad Elster.

3. Zermelo to Gödel

Freiburg i. Br. 29. Oktober 1931. Karlstraße 60

Sehr geehrter Herr Gödel!

Ich danke Ihnen für Ihren freundlichen Brief, aus dem ich nun besser als aus Ihrer Abhandlung und Ihrem Vortrag entnehmen kann, wie Sie es eigentlich meinen. Also nur für die *beweisbaren* Sätze, Ihres "PM-Systems", nicht für dessen *Sätze überhaupt* soll Ihre "finitistische Einschränkung" (wie ich es nenne) der Typenbildung zur Geltung kommen, während Sie für die *Sätze* des Systems *freie* Neubildungen nach Art des Cantorschen Diagonal-verfahrens zulassen. Dann erhalten Sie natürlich ein *nicht abzählbares* System möglicher Sätze, unter denen nur eine *abzählbare* Teilmenge "beweisbar" wäre, und es muß sicherlich "unentscheidbare" Sätze geben. Daß diese "unentscheidbaren" Sätze in einem "höheren System" doch wieder "entscheidbar" werden, geben Sie ja zu. Aber dieses "höhere" System unterscheidet sich von dem ursprünglichen keineswegs durch Aufnahme neuer *Sätze*, wie man nach Ihren Formulierungen denken könnte, sondern lediglich durch neue *Beweismittel*, und alles, was Sie in der Abhdl. beweisen, kommt darauf hinaus, was auch *ich* immer betone, daß ein "finitistisch beschränktes" Beweis-Schema *nicht ausreicht*, um die Sätze eines nicht-abzählbaren mathematischen Systems zu "entscheiden". Oder können Sie etwa "beweisen", das *Ihr* Schema das "einzig mögliche" ist? Das geht doch wohl nicht; denn was ein "Beweis" eigentlich ist, ist nicht selbst wieder "beweisbar" sondern muß in irgend einer Form *angenommen*, *vorausgesetzt* werden. Und darum handelt es sich hier eben: was versteht man unter einem Beweis? Ganz allgemein versteht man darunter ein System von Sätzen derart, daß unter Annahme der Prämissen die Gültigkeit der Behauptung *einsichtig* gemacht werden kann. Und es ist nur noch die Frage, was alles "einsichtig" ist? Jedenfalls *nicht bloß*, und das zeigen Sie gerade selbst, die Sätze irgend eines finitistischen Schemas, das ja auch in Ihrem Falle immer wieder *erweitert* werden kann. Aber damit wären wir eigentlich einig: nur daß ich eben von vorn herein ein *allgemeineres* Schema, das nicht erst erweitert zu werden *braucht* zugrunde lege. Und in *diesem* System sind auch wirklich *alle* Sätze entscheidbar.

Mit bestem Gruß ergebenst E. Zermelo

3. Zermelo to Gödel

Freiburg im Breisgau
Karlstraße 60
29 October 1931

Dear Mr. Gödel,

I thank you for your friendly letter, from which I can now infer, better than ⟦I could⟧ from your paper and your lecture, what you really mean. Thus only for the *provable* statements of your "PM-system", not for its *statements in general*, does your "finitistic restriction" on the formation of types (as I call it) apply, whereas for the *statements* of the system you do permit *free* new formation in the manner of the Cantor diagonal procedure. Then of course you obtain an *uncountable* system of possible statements, among which only a *countable* subset are "provable", and there must certainly be "undecidable" statements. You admit that these "undecidable" statements nevertheless again become "decidable" in a "higher system". But this "higher" system is in no way distinguished from the original by the admission of new *statements*, as one might think from your formulation, but merely by new *means of proof*, and everything that you prove in the paper comes down to what *I* also always stress, that a "finitistically restricted" proof schema *does not suffice* to "decide" the statements of an uncountable mathematical system. Or can you perhaps "prove" that *your* schema is the "only possible one"? But that can hardly be; for what a "proof" really is is itself not in turn provable, but must in some form be *taken for granted, presupposed.* And here it is just a question: What does one understand by a proof? Quite generally, one understands by that a system of statements such that under the assumption of the premises, the validity of the assertion can be made *discernible.* And there remains only the question, what all is "discernible"? In any case, *not merely*—and that you yourself show precisely—the statements of some finitistic system, which in your case too can always be *extended.* But on that we would really agree: only that at the outset I take as a basis a *more* general schema that does not first *need* to be extended. And in *this* system all statements really are decidable as well!

With best regards, yours most sincerely, E. Zermelo

Calendars of correspondence

Where an item was undated, an educated guess has been made on the approximate date if possible. Where it has proved impossible to make a reasonable surmise beyond the year, the date is indicated by question marks, and the letter follows all others for that year. "Form" describes the general character of the letter; incidental characteristics are footnoted. When two sources are given, the second entry in the form column or the identification column corresponds to the second source.

In the calendars, the following abbreviations are used.

For the form of the letter:

AL: Autograph letter, unsigned
ALR: Autograph letter, retained copy
ALS: Autograph letter, signed
APS: Autograph postcard, signed
D: Draft
SH: Shorthand
TG: Telegram
TL: Typed letter, unsigned
TLR: Typed letter, retained copy
TLS: Typed letter, signed

For the source:

APS: American Philosophical Society
BgN: *Briefe grosser Naturforscher und Ärzte in Handscriften*[a]
BN: Behmann *Nachlaß*
CAH: Center for American History
ETH: Eidgenössische Technische Hochschule Bibliothek
HA: Heyting Archief
IAS: Institute for Advanced Study
IIT: Menger *Nachlaß*, Illinois Institute of Technology
LC: Library of Congress, von Neumann papers
NN: Nagel *Nachlaß*
PS: Private source
SIU: SIU Carbondale, Schilpp papers
SBPK: Staatsbibliothek zu Berlin—Preussischer Kulturbesitz, Nachl. 196,[b] Box 46, Gödel
U Pitt: Carnap collection, University of Pittsburgh
UFB: Universitätsbibliothek Freiburg im Breisgau
WSB: Wiener Stadt- und Landesbibliothek, Handscriftenabteilung
YU: Yale University
ZBPW: Zentralbibliothek für Physik in Wien

---

[a] *Wiedemann 1989.*

[b] Gotthard Günther

Correspondence included in these volumes

Date	*Correspondent*	*Letter number*	*To/From Gödel*	*Form*	*Source*	*Identifier*
09/14/1928	Feigl	1	From	ALS	BgN	
11/20/1930	von Neumann	1	To	ALS	IAS	013029
11/29/1930	von Neumann	2	To	ALS	IAS	013029.5
12/24/1930	Bernays	1	To	ALS	IAS	010015.44
01/12/1931	von Neumann	3	To	ALS	IAS	013030
01/18/1931	Bernays	2	To	ALS	IAS	010015.45
01/20/1931	Tarski	1	From	ALS	IAS	012760
02/10/1931	Behmann	1	To	TLS	IAS	010015.42
02/22/1931	Behmann	2	From	ALS	BN	
02/25/1931	Behmann	3	To	TLS	IAS	010015.421
03/18/1931	Behmann	4	From	ALS	BN	
03/25/1931	Behmann	5	To	TLS	IAS	010015.422
03/30/1931	Menger	2	From	ALS	IIT	
04/02/1931	Bernays	3	From[a]	ALS	IAS	
04/07/1931	Herbrand	1	To	TLS	IAS	10837.5
04/20/1931	Bernays	4	To	ALS	IAS	010015.46
04/22/1931	Behmann	6	From	ALS	BN	
05/03/1931	Bernays	5	To	ALS	IAS	10015.47
05/18/1931	Behmann	7	To	TLS	IAS	010015.425
07/25/1931	Herbrand	2	From	ALS	IAS	010837.6
08/22/1931	Heyting	1	To	TLS	HA; IAS	No number; 010839
08/??/1931	Heyting	2	From	AL	IAS	010839.5
09/03/1931	Heyting	3	From	ALS	HA	
09/21/1931	Zermelo	1	To	ALS	IAS	013270
09/24/1931	Heyting	4	To	ALS	IAS	010840
10/12/1931	Zermelo	2	From	ALS	UFB	
10/29/1931	Zermelo	3	To	TL	UFB; IAS	013271
11/01/1931	Skolem	1	From	ALS	PS	
??/??/1931	Menger	1	To	TLS	IAS	011491
02/23/1932	Carnap	1	To	TLS	IAS	010280.86
04/11/1932	Carnap	2	From	ALS	U Pitt	

[a] Acquired after cataloging.

Date	*Correspondent*	*Letter number*	*To/From Gödel*	*Form*	*Source*	*Identifier*
06/02/1932	Menger	3	To	ALS	IAS	011492
06/11/1932	Heyting	5	To	ALS	IAS	010842
06/17/1932	Church	1	From	ALS	IAS	010329.09
07/01/1932	Heyting	6	From	ALS	HA	
07/17/1932	Heyting	7	To	ALS	IAS	010843
07/20/1932	Heyting	8	From	ALS	HA	
07/26/1932	Heyting	9	To	ALS	IAS	010844
07/27/1932	Church	2	To	ALS	IAS	010329.1
08/04/1932	Heyting	10	From	ALS	HA	
08/04/1932	Menger	5	From	ALS	IIT	
08/15/1932	Heyting	11	To	TLS	HA; IAS	No number; 010845
08/27/1932	Heyting	12	To	ALS	IAS	010847
09/11/1932	Carnap	3	From	ALS	U Pitt	
09/25/1932	Carnap	4	To	TLS	IAS	010280.87
09/27/1932	Carnap	5	To	TLS	IAS	010280.88
10/29/1932	Heyting	13	To	ALS	IAS	010848
11/15/1932	Heyting	14	From	ALS	HA	
11/24/1932	Heyting	15	To	AL	IAS	010850
11/26/1932	Heyting	15	To	ALS	IAS	010851
11/28/1932	Carnap	6	From	ALS	U Pitt	
??/??/1932	Menger	4	To	ALS	IAS	011493
??/??/1932?	Menger	6	To	ALS	IAS	011495
02/14/1933	von Neumann	4	To	ALS	IAS	013031
03/11/1933	Finsler	1	To	TLS	IAS; ETH	010632; Hs648:17
03/14/1933	von Neumann	5	From	ALS	IAS	013032
03/25/1933	Finsler	2	From	AL	IAS	010632.5
03/25/1933	Finsler	3	From	AL	IAS	010632.51
04/03/1933	Menger	7	From	ALS	IIT	
04/06/1933	Menger	8	From	ALS	IIT	
04/15/1933	Heyting	16	To	ALS	IAS	010852
05/07/1933	Heyting	17	To	APS	IAS	010852.5
05/16/1933	Heyting	18	From	ALS	HA; IAS	No number; 010853
06/13/1933	von Neumann	6	To	ALS	IAS	013033
06/19/1933	Finsler	4	To	TLS	IAS; ETH	010633; Hs648:20
08/24/1933	Heyting	19	To	ALS	IAS	010854

Date	*Correspondent*	*Letter number*	*To/From Gödel*	*Form*	*Source*	*Identifier*
09/30/1933	Heyting	20	To	ALS	IAS	010855
08/12/1934	von Neumann	7	To	ALS	IAS	013035
09/18/1934	von Neumann	8	To	ALS	IAS	013036
11/07/1934	Carnap	7	To	TLS	IAS	010280.90
05/29/1936	von Neumann	9	To	ALS	IAS	013037
05/22/1937	Menger	9	To	TLS	IAS	011497
07/03/1937	Menger	10	From	ALS	IIT	
07/13/1937	von Neumann	10	To	ALS	IAS	013038
09/12/1937	Menger	11	To	TG	IAS	011498
09/14/1937	von Neumann	11	To	ALS	IAS	013039
11/03/1937	Menger	12	To	TLS	IAS	011499
11/11/1937	von Neumann	12	To	ALS	IAS	013043
12/15/1937	Menger	13	From	ALS	IIT	
01/13/1938	von Neumann	13	To	ALS	IAS	013041
05/20/1938	Menger	14	To	TLS	IAS	011503
06/25/1938	Menger	15	From	ALS	IIT	
09/12/1938	von Neumann	14	From	ALS	LC	Container 8
10/19/1938	Menger	16	To	ALS	IAS	011505
10/??/1938	Menger	17	From	ALS	IIT	
10/29/1938	Post	1	To	APS	IAS	011717.3
10/30/1938	Post	2	To	ALS	IAS	011717.4
11/11/1938	Menger	18	From	ALS	IIT	
12/??/1938	Menger	19	To	ALS	IAS	011506
01/16/1939	Ulam	1	To	ALS	IAS	012877
02/28/1939?	von Neumann	15	To	ALS	IAS	013043
03/12/1939	Post	3	To	ALS	IAS	011717.5
03/20/1939	Post	4	From	ALS	Phyllis P. Goodman	
03/20/1939	von Neumann	16	From	ALS	LC	Container 4
04/22/1939	von Neumann	17	To	ALS	IAS	013044
05/13/1939?	von Neumann	18	To	ALS	IAS	013045
06/19/1939	Bernays	6	From	ALS	ETH	Hs.975:1692
06/21/1939	Bernays	7	To	ALS	IAS	010015.49
07/19/1939?	von Neumann	19	To	TLS	IAS	013046
07/20/1939	Bernays	8	From	ALS	ETH	Hs.975.1673
08/17/1939	von Neumann	20	From	ALS	IAS	013047
08/30/1939	Menger	20	From	ALS	IAS	011510
09/28/1939	Bernays	9	To	TLS	IAS; ETH	010015.6; Hs.975:1674

Date	*Correspondent*	*Letter number*	*To/From Gödel*	*Form*	*Source*	*Identifier*
12/12/1939	Bernays	10	To	APS	IAS	010015.7
12/29/1939	Bernays	11	From	ALS	ETH	Hs.975:1695
01/16/1942	Bernays	12	From	ALS	ETH	Hs.975:1696
06/22/1942	Menger	21	From	ALS	IIT	
08/14/1942	Ulam	2	To	TLS	IAS	012879
09/07/1942	Bernays	13	To	TLS	IAS; ETH	010016; Hs.975:1697
11/18/1942	Schilpp	1	To	TLS; TLR	IAS; SIU	012109; 20/7/4
11/30/1942	Schilpp	2	From	ALS	SIU; IAS	20/7/4; 012111
??/??/1942	Bernays	14	From	SH D	IAS	010017
??/??/1942?	Ulam	3	From	ALR	IAS	012881
03/27/1943	Schilpp	3	From	ALS	SIU; IAS	20/7/4; 012114
04/18/1943	Schilpp	4	From	ALS	SIU	20/7/4
05/18/1943	Schilpp	5	From	ALS	SIU; IAS[b]	20/7/4; 012116
05/22/1943	Schilpp	6	To	TLS; TLR	IAS; SIU	012117; 20/7/4
05/26/1943	Schilpp	7	From	ALS	SIU	20/7/4
05/31/1943	Schilpp	8	To	TLS; TLR	IAS; SIU	012118; 20/7/4
06/07/1943	Schilpp	9	From	ALS	SIU; IAS	20/7/4; 012119
06/27/1943	Schilpp	10	To	TLS; TLR	IAS; SIU	012120; 20/7/4
07/14/1943	Schilpp	11	To	TLS; TLR	IAS; SIU	012123; 20/7/4
07/28/1943	Schilpp	12	To	TLS; TLR	IAS; SIU	012126; 20/7/4
08/25/1943	Schilpp	13	To	TLS; TLR	IAS; SIU	012131; 20/7/4
09/13/1943	Schilpp	14	From	ALS; ALS[c]	SIU; IAS	20/7/4; 012132
09/20/1943	Schilpp	15	From	ALS	SIU	20/7/4
09/22/1943	Schilpp	16	To	TLS; TLR	IAS; SIU	012136; 20/7/4
09/27/1943	Schilpp	17	From	ALS	SIU	20/7/4

[b]5/17/43.

[c]Draft.

Date	*Correspondent*	*Letter number*	*To/From Gödel*	*Form*	*Source*	*Identifier*
09/28/1943	Russell	1	From	ALR	IAS	012137
10/07/1943	Schilpp	18	To	TLS; TLR	IAS; SIU	012138; 20/7/4
04/27/1944	Tarski	2	To	APS	IAS	012774
05/12/1944	Tarski	3	From	ALS	IAS	012775
07/10/1946	Schilpp	19	To	TLS; TLR	IAS; SIU	012170; 20/15/8
07/25/1946	Schilpp	20	From	ALS	SIU	20/15/8
07/30/1946	Schilpp	21	To	TLS; TLR	IAS; SIU	012171; 20/15/8
08/12/1946	Schilpp	22	From	ALS	SIU	20/15/8
08/17/1946	Schilpp	23	To	TLS; TLR	IAS; SIU	012172; 20/15/8
12/10/1946	Tarski	4	To	ALS	IAS	012780
07/15/1947	Schilpp	24	From	ALS	SIU	20/15/8
12/06/1947	Ulam	4	To	ALS	IAS	012882
03/09/1949	Schilpp	25	To	TLS; TLR	IAS; SIU	012178; 20/15/8
12/21/1949	Schilpp	26	From	ALS	SIU	20/15/8
02/27/1950	Gödel, Marianne	1	From	ALS	WSB	
05/15/1953	Schilpp	27	To	TLS; TLR	IAS; SIU	012186; 20/21/4
07/02/1953	Schilpp	28	From	ALS	SIU	20/21/4
07/06/1953	Schilpp	29	To	TLS; TLR	IAS; SIU	012188; 20/21/4
09/14/1953	Menger	22	From	ALS	IIT	
03/28/1954	Schilpp	30	From	ALS	SIU	20/21/4
04/29/1954	Günther	1	To	ALS	IAS	010754
05/15/1954	Günther	2	From	ALS	SBPK	Sheets 1–2
05/23/1954	Günther	3	To	ALS	IAS	010756
06/30/1954	Günther	4	From	ALS	SBPK	Sheets 5–7
10/02/1954	Günther	5	To	ALS	IAS	010757
06/19/1955	Günther	6	To	ALS	IAS	010758
08/10/1955	Günther	7	From	ALS	SBPK	Sheets 8–10
09/07/1955	Seelig	1	From	TLS	ETH	Hs.304:648
09/18/1955	Günther	8	To	ALS	IAS	010759
11/02/1955	Schilpp	31	To	TPS	IAS	012199
11/14/1955	Schilpp	32	From	ALS	SIU	20/21/4
11/18/1955	Seelig	2	From	TLS	ETH	Hs. 304:649
03/20/1956	von Neumann	21	From	ALS	LC	Container 5

Date	*Correspondent*	*Letter number*	*To/From Gödel*	*Form*	*Source*	*Identifier*
09/20/1956	Günther	9	To	TLS	IAS	010764
12/04/1956	Boone	1	From	TLR	IAS	010166
12/28/1956	Bernays	15	To	ALS	IAS	010018
12/31/1956	Bernays	16	To	ALS	IAS	010019
02/06/1957	Bernays	17	From	ALS; SH D	ETH; IAS	Hs.975:1698; 010020
02/25/1957	Nagel	1	From	TLS[d]	IAS	011590.05
03/09/1957	Nagel	2	To	TLS	IAS	011590.1
03/14/1957	Nagel	3	From	TLS[e]	IAS	011590.15
03/18/1957	Nagel	4	From	TLR	IAS	011590.25
03/21/1957	Nagel	5	To	TLS	IAS	011590.3
03/25/1957	Nagel	6	From	TLR	IAS	011590.35
04/04/1957	Günther	10	From	ALS	SBPK	Sheets 11–12
04/07/1957	Günther	11	To	TLS	IAS	010776
04/09/1957	Angoff	1	From	TLR	IAS	020388
04/09/1957	Nagel	7	From	TLS	NN	
04/22/1957	Angoff	2	To	TLS	IAS	020390
05/06/1957	Angoff	3	From	TLR	IAS	020391
05/28/1957	Angoff	4	To	TLS	IAS	020393
06/03/1957	Angoff	5	From	TLR	IAS	020394
06/06/1957	Angoff	6	To	TLS	IAS	020395
06/25/1957	Angoff	7	From	TLR	IAS	020396
07/11/1957	Schilpp	33	From	ALS	SIU	20/21/4
08/16/1957	Nagel	8	From	TLS	NN	
08/22/1957	Nagel	9	To	TLR	NN	
08/29/1957	Follett	1	From	TLR	IAS	020397
08/29/1957	Nagel	10	From	TLR	IAS	011590.6
09/09/1957	Schilpp	34	From	ALS[f]	IAS	012209
09/26/1957	Follett	2	To	TLS	IAS	020398
10/25/1957	Ulam	5	To	TLR	APS	
11/05/1957	Büchi	1	To	TLS	IAS	010280.62
11/08/1957	Ulam	6	From	TLS	APS	
11/20/1957	Bernays	18	From	TLS; TLR	ETH; IAS	Hs.975:1699; 010021
11/22/1957	Günther	12	To	TLS	IAS	010778

[d] Apparently unsent.

[e] Apparently unsent.

[f] Not sent.

Date	*Correspondent*	*Letter number*	*To/From Gödel*	*Form*	*Source*	*Identifier*
11/26/1957	Büchi	2	From	TL	IAS	010280.621
12/04/1957	Bernays	19	To	TLS	IAS; ETH	10022; Hs.975:1700
12/23/1957	Günther	13	From	ALS	SBPK	Sheet 13
01/28/1958	Ulam	7	From	TLR	IAS	012882.7
03/05/1958	Bernays	20	To	ALS	IAS	010026
03/14/1958	Bernays	21	From	ALS	ETH	Hs.975:1702
06/02/1958	Bernays	22	To	TLS; TL[g]	IAS; ETH	010028; Hs.975:1703
09/20/1958	Pitts	1	From	TLR	IAS	011709.5
09/30/1958	Bernays	23	From	TLS; TL	ETH; IAS	Hs.975:1704; 010029
10/12/1958	Bernays	24	To	ALS	IAS	010030
10/30/1958	Bernays	25	From	ALS	ETH	Hs.975:1705
11/24/1958	Bernays	26	To	ALS; ALS[h]	IAS; ETH	010031; Hs.975:1706
01/07/1959	Bernays	27	From	ALS	ETH	Hs.975:1707
01/07/1959	Günther	14	From	TLS	SBPK; IAS	Sheet 14; 010786
01/22/1959	Bernays	28	To	TLS	IAS; ETH	010033; Hs.975:1708
01/24/1959	Schilpp	35	To	TLR	SIU	20/21/4
02/03/1959	Schilpp	36	From	ALS; TLR	SIU; IAS	20/21/4; 012212
02/06/1959	Schilpp	37	To	TLR	SIU	20/21/4
10/08/1959	Bernays	29	From	ALS	ETH	Hs.975:1709
10/09/1959	Bernays	30	To	ALS	IAS	010034
05/11/1960	Bernays	31	To	ALS; SH D	IAS; ETH	010037; Hs.975:1711
12/20/1960	Bernays	32	To	ALS	IAS	010041
12/21/1960	Bernays	33	From	ALS	ETH	Hs.975:1712
03/16/1961	Bernays	34	From	ALS	ETH	Hs.975:1713
03/23/1961	Bernays	35	To	ALS; SH D	IAS; ETH	010042; Hs.975:1714
04/20/1961	Bernays	36	To	TL; TL[i]	IAS; ETH	010042.5; Hs.975:1716

[g]Initialed.

[h]Draft.

[i]Continued 05/05/1961.

Date	*Correspondent*	*Letter number*	*To/From Gödel*	*Form*	*Source*	*Identifier*
05/05/1961	Bernays	36	To	ALS; ALS[j]	IAS; ETH	010042.5; Hs.975:1716
05/11/1961	Bernays	37	From	ALS	ETH	Hs.975:1718
05/19/1961	Bernays	38	To	ALS	IAS	010043
07/10/1961	Bernays	39	To	TLS; TL[k]	IAS; ETH	010045; Hs.975:1719
07/23/1961	Gödel, Marianne	2	From	ALS	WSB	
08/11/1961	Bernays	40	From	ALS	ETH	Hs.975:1720
08/14/1961	Gödel, Marianne	3	From	ALS	WSB	
09/12/1961	Gödel, Marianne	4	From	ALS	WSB	
10/06/1961	Gödel, Marianne	5	From	ALS	WSB	
10/12/1961	Bernays	41	To	TLS; TL	IAS; ETH	010048; Hs.975:1721
11/08/1961	Boone	2	From	TLS	IAS	010216
12/15/1961	Bernays	42	From	ALS	ETH	Hs.975:1722
12/31/1961	Bernays	43	To	ALS	IAS	010051
??/??/1961?	Tarski	5	From	AL[l]	IAS	012237
07/30/1962	Bernays	44	From	ALS	ETH	Hs.975:1723
08/02/1962	Rappaport	1	From	TLR	IAS	011831
10/12/1962	Bernays	45	To	TLS; TL	IAS; ETH	010054; Hs.975:1724
12/31/1962	Bernays	46	To	ALS	IAS	010059
01/09/1963	Bernays	47	From	ALS	ETH	Hs.975:1725
02/23/1963	Bernays	48	To	TLS	IAS	010056
03/06/1963	Dreben	1	To	TLS	IAS	010510
03/14/1963	Bernays	49	To	ALS	IAS	010056.5
03/25/1963	van Heijenoort	1	To	TLS	IAS; CAH	012905; no number
04/23/1963	van Heijenoort	2	From	TLS	CAH; IAS	No number; 012907
06/20/1963	Cohen	1	From	TLR	IAS	010352
08/28/1963	van Heijenoort	3	From	TLS	CAH	
10/04/1963	van Heijenoort	4	From	TLS	CAH; IAS	No number; 012925
10/24/1963	van Heijenoort	5	To	ALS	IAS; CAH	012927; no number
12/18/1963	Bernays	50	From	ALS	ETH; IAS	Hs.975:1728; 010060

[j] Continuation of 04/20/1961.

[k] Initialed.

[l] Probably unsent.

Date	*Correspondent*	*Letter number*	*To/From Gödel*	*Form*	*Source*	*Identifier*
01/06/1963[m]	Ulam	8	To	TLS	IAS	012883
01/01/1964	Bernays	51	To	ALS	IAS	010061
01/22/1964	Cohen	2	From	TLR	IAS	010390.5
02/10/1964	Ulam	9	From	TLR	IAS	012885
02/16/1964	Schilpp	38	To	TLS	IAS	012213.1
02/22/1964	van Heijenoort	6	From	TLS	CAH; IAS[n]	No number; 012943
04/10/1964	Popper	1	From	TLR	IAS	011717.25, 011717.26
06/01/1964	Carnap	8	From	ALS	U Pitt	
06/24/1964	van Heijenoort	7	To	ALS	IAS; CAH	012959; no number
07/23/1964	Schilpp	39	From	TLR	IAS	012213.2
08/14/1964	van Heijenoort	8	From	TLS	CAH; IAS	No number; 012962
08/15/1964	van Heijenoort	9	From	TLS	CAH; IAS	No number; 012963
08/27/1964	van Heijenoort	10	To	TLS	IAS; CAH	012964; no number
09/18/1964	van Heijenoort	11	From	TLS	CAH; IAS	No number; 012966
11/03/1964	van Heijenoort	12	From	TLS	CAH; IAS	No number; 012973
11/14/1964	van Heijenoort	13	To	TLS	IAS; CAH	012974; no number
12/15/1964	van Heijenoort	14	From	TLS	CAH; IAS	No number; 012975
01/11/1965	Cohen	3	From	TLR	IAS	010394
02/05/1965	Bernays	52	From	ALS	ETH	Hs.975:1729
02/16/1965	Goddard	1	To	ALS	IAS	010709.3
03/02/1965	Church	3	From	TLR	IAS	010332
05/25/1965	van Heijenoort	15	To	TLS	IAS; CAH	012982; no number
07/08/1965	van Heijenoort	16	From	TLS	CAH; IAS	No number; 012987
08/09/1965	van Heijenoort	17	From	TLS	CAH	
08/13/1965	Cohen	4	From	TLR	IAS	010409

[m] Although the letter is dated 1963, internal evidence indicates that the year should be 1964.

[n] The retained copy at the I.A.S. has a postscript dated 02/21/1964, but a note by Gödel on the back says "22.II.$\overline{64}$".

Date	*Correspondent*	*Letter number*	*To/From Gödel*	*Form*	*Source*	*Identifier*
09/01/1965	Reid	1	To	TLS	IAS	011840
09/17/1965	Bernays	53	To	ALS; SH D	IAS; ETH	010071; Hs.975:1732
09/27/1965	Bernays	54	From	TLS; TL	ETH; IAS	Hs.975:1733; 010073
11/08/1965	Bernays	55	To	ALS	IAS	010074
12/02/1965	Bernays	56	From	TLS; TL	ETH; IAS	Hs.975:1734; 010075
12/10/1965	Bernays	57	To	TLS	IAS; ETH	010076; Hs.975:1735
12/13/1965	van Heijenoort	18	From	TLS	CAH	
??/??/1965?	Goddard	2	From	AL°	IAS	010709.5
01/25/1966	Bernays	58	From	TLS; TL	ETH; IAS	Hs.975:1736; 010079
02/17/1966	Ulam	10	To	TLS	IAS	012885.5
03/22/1966	Reid	2	From	TLR; TLS	IAS; ETH	011845; Hs.1001:1
04/24/1966	Bernays	59	To	ALS	IAS	010080
05/22/1966	Bernays	60	From	ALS	ETH	Hs.975:1737
05/24/1966	Dreben	2	To	TLS	IAS	010513
06/27/1966	Church	4	To	ALS	IAS	010334.2
07/19/1966	Dreben	3	From	TLR	IAS	010516
08/10/1966	Church	5	From	TLR	IAS	010334.25
09/29/1966	Church	6	From	TLR	IAS	010334.86
01/24/1967	Bernays	61	From	ALS	ETH	Hs.975:1738
04/27/1967	Cohen	5	From	TLR	IAS	010417.6
06/30/1967	Rautenberg	2	From	TLR	IAS	011834
07/07/1967	Robinson, A.	1	From	TLR	IAS	011944
07/27/1967	Wang	1	To	ALS	IAS	013069
07/31/1967	Plummer	1	From	TLR	IAS	011714, 011715
10/27/1967	Bernays	62	To	ALS	IAS	010082
12/07/1967	Wang	2	From	TLS	IAS	013073–5
12/19/1967	Wang	3	To	TLS	IAS	013076
12/20/1967	Bernays	63	From	ALS	ETH	Hs.975:1739
??/??/1967?	Rautenberg	1	To	ALS	IAS	011833
01/02/1968	Menger	23	To	TLS	IAS	011513

°Unsent.

Date	*Correspondent*	*Letter number*	*To/From Gödel*	*Form*	*Source*	*Identifier*
01/22/1968	Menger	24	From	TL	IAS	011515
03/07/1968	Wang	4	From	TLS	IAS	013085–6
04/23/1968	Wang	5	To	TLS	IAS	013087
05/16/1968	Bernays	64	From	ALS	ETH	Hs.975:1740
07/20/1968	Bernays	65	To	TLS	IAS; ETH	010084; Hs.975:1741
12/17/1968	Bernays	66	From	ALS	ETH	Hs.975:1742
12/20/1968	Perlis	1	From	TLR	IAS	011709.42
01/02/1969	Heyting	21	To	TLS	IAS	010863
01/06/1969	Bernays	67	To	ALS	IAS	010086
02/10/1969	Reid	3	To	TLS	IAS	011846
03/12/1969	Heyting	22	From	TLR[p]	IAS	010864
06/25/1969	Reid	4	From	TLS	Constance Reid	
07/??/1969	Bernays	68a	From	AL, D	IAS	010090
07/25/1969	Bernays	68b	From	ALS	ETH; IAS	Hs.975:1743; 010089
12/10/1969	Brutian	1	From	TLR	IAS	010280.5
12/30/1969	Dreben	4	To	TLS	IAS	010517
01/07/1970	Bernays	69	To	TLS; TL	IAS; ETH	010097; Hs.975:1745
04/15/1970	Dreben	5	To	TLS	IAS	010517.5
04/15/1970	Wang	6	To	TLS	IAS	013091
07/12/1970	Bernays	70	To	ALS	IAS; ETH	010097.5; Hs.975:1746
07/14/1970	Bernays	71	From	ALS	ETH	Hs.975:1747
09/12/1970	Bernays	72	To	ALS	IAS; ETH	010099; Hs.975:1748
10/02/1970	Bernays	73	From	TG; AL[q]	ETH; IAS	Hs.975:1749; 010100
12/22/1970	Bernays	74	From	ALS	ETH	Hs.975:1750
12/31/1970	Bernays	75	To	ALS	IAS	010102
??/??/1970	Balas	1	From	AL, D	IAS	010015.37
03/17/1971	Hwastecki	1	To	TLS	IAS	010897
03/??/1971	Robinson, A.	2	From	AL[r]	IAS	011956
04/14/1971	Robinson, A.	3	To	TLS; TLR	IAS; YU	011957; no number
05/25/1971	Wang	7	To	TLS	IAS	013094

[p]Unsent.

[q]Draft.

[r]Unsent draft.

Date	*Correspondent*	*Letter number*	*To/From Gödel*	*Form*	*Source*	*Identifier*
07/09/1971	Wang	8	From	TL	IAS	013097
07/20/1971	Wang	9	To	TLS	IAS	013098
07/20/1971	Wang	10	To	TLS	IAS	013100
08/04/1971	Wang	11	From	TL	IAS	013104
08/11/1971	Wang	12	To	TLS	IAS	013105
09/22/1971	Blackwell	1	To	TLS	IAS	010120.1
11/22/1971	Robinson, A.	4	From	ALS; AL, D	YU; IAS	No number; 011959
??/??/1971?	Blackwell	2	From	AL, D	IAS	010120.3
??/??/1971?	Hwastecki	2	From	D[s]	IAS	010898
01/15/1972	Menger	25	To	ALS	IAS	011517
03/16/1972	Bernays	76	To	TLS; TL[t]	IAS; ETH	010109; Hs.975:1751
04/10/1972	Wang	13	To	TLS	IAS	013110
04/18/1972	Henkin	1	From	TLR	IAS	010834
04/20/1972	Menger	26	From	ALS	IIT	
04/26/1972	Wang	14	To	ALS	IAS	013113
05/31/1972	Thirring	1	To	TLS	IAS	012867
06/27/1972	Honderich	1	From	TLR	IAS	013116
06/27/1972	Thirring	2	From	ALS	ZBPW	
07/19/1972	Honderich	2	From	TLR	IAS	013121
07/21/1972	Bernays	77	To	APS	IAS	010110
10/28/1972	Bernays	78	To	ALS; SH D	IAS; ETH	010111; Hs.975:1751
12/26/1972	Bernays	79	From	ALS; SH D	ETH; IAS	Hs.975:1753; 010106
12/29/1972	Robinson, A.	5	From	AL, D	IAS	011962
01/04/1973	Robinson, A.	6	To	ALS	IAS	011963
02/21/1973	Bernays	80	To	TLS	ETH	Hs.975:1754
04/23/1973	Robinson, A.	7	To	TLS; TLR	IAS; YU	011968; no number
07/02/1973	Robinson, A.	8	From	TLS; TLR	YU; IAS	No number; 011972
07/18/1973	Robinson, A.	9	From	TLR	IAS	011977
10/02/1973	Greenberg	1	From	TLS	IAS	010729.65
10/15/1973	Greenberg	2	To	TLS	IAS	010729.7

[s] Unsent.

[t] Initialed.

Date	*Correspondent*	*Letter number*	*To/From Gödel*	*Form*	*Source*	*Identifier*
10/16/1973	Robinson, A.	10	To	TLR	YU	
10?/??/1973?	Greenberg	3	From	AL, D	IAS	010729.75
11/22/1973	Robinson, A.	11	From	ALS	YU	
12/18/1973	Bernays	81	From	ALS	ETH; IAS	Hs.975:1755; 010107
02/01/1974	Sawyer	1	To	TLS	IAS	012108.97
03/20/1974	Robinson, A.	12	From	AL, D	IAS	011998
05/11/1974	Robinson, Renee[u]		From	AL, D	IAS	012001
05/20/1974	Mostow[v]		From	TLR	IAS	012003
07/03/1974	Bohnert	1	To	TLS	IAS	010134
07/16/1974	Grandjean	1	To	TLS	IAS	010720
07/22/1974	Suppes	1	From	TLR	IAS	012462
09/17/1974	Bohnert	2	From	TLR	IAS	010135
12/16/1974	Bernays	82	To	TLS	ETH	Hs.975:1756
12/17/1974	Bernays	83	From	ALS	ETH	Hs.975:1757
??/??/1974?	Sawyer	2	From	AL[w]	IAS	012108.98
01/12/1975	Bernays	84	From	ALS	ETH	Hs.975:1758
01/24/1975	Bernays	85	To	TLS	IAS; ETH	010114; Hs.975:1759
08/19/1975[x]	Grandjean	2	From	TLR	IAS	010728
12/02/1975	Grandjean	3	To	TLS	IAS	010729.25
12/12/1975	Wang	15	To	TLS	IAS	013166
??/??/1975?	Grandjean	4	Neither[y]		IAS	010729
??/??/1975?	Grandjean	5	From	AL[z]	IAS	010727
06/24/1976	Wang	16	To	ALS	IAS	013187
02/03/1977	Wang	17	To	TLS	IAS	013196

[u]See the end of the correspondence with Abraham Robinson.

[v]See the end of the correspondence with Abraham Robinson.

[w]Unsent.

[x]Gödel had originally dated the letter August 1975, but later changed the month to January. Presumably the year should also have been changed to 1976.

[y]Questionnaire.

[z]Unsent.

Individual calendars of correspondence

Arend Heyting

Item	*Date*	*To/From Gödel*	*Form*	*Source*	*Identifier*
1	08/22/1931	To	TLS	HA; IAS	No number; 010839
2	08/??/1931	From	AL	IAS	010839.5
3	09/03/1931	From	ALS	HA	
4	09/24/1931	To	ALS	IAS	010840
	09/28/1931	To	ALS	IAS	010841
5	06/11/1932	To	ALS	IAS	010842
6	07/01/1932	From	ALS	HA	
7	07/17/1932	To	ALS	IAS	010843
8	07/20/1932	From	ALS	HA	
9	07/26/1932	To	ALS	IAS	010844
10	08/04/1932	From	ALS	HA	
11	08/15/1932	To	TLS	HA; IAS	No number; 010845
12	08/27/1932	To	ALS	IAS	010847
13	10/29/1932	To	ALS	IAS	010848
	11/06/1932	To	ALS	IAS	010849
14	11/15/1932	From	ALS	HA	
15	11/24/1932	To	AL[a]	IAS	010850
15	11/26/1932	To	ALS	IAS	010851
16	04/15/1933	To	ALS	IAS	010852
	03/??/1933	To	Card	Lost	
17	05/07/1933	To	APS	IAS	010852.5
18	05/16/1933	From	ALS	HA; IAS	No number; 010853
19	08/24/1933	To	ALS	IAS	010854
	09?/??/1933	From	ALS	Lost	

[a]Continued 11/26.

Item	*Date*	*To/From Gödel*	*Form*	*Source*	*Identifier*
20	09/30/1933	To	ALS	IAS	010855
	09/09/1957	To	TLS	IAS	010856
	10/07/1957	From	TL	IAS	010858
	10/26/1957	To	ALS	IAS	010859
	11/05/1957	From	TLR[b]	IAS	010861
	11/19/1957	To	TLS	IAS	010862
21	01/02/1969	To	TLS	IAS	010863
22	03/12/1969	From	TLR[c]	IAS	010864

[b]Unsent.

[c]Unsent.

Karl Menger

Item	*Date*	*To/From Gödel*	*Form*	*Source*	*Identifier*
1	03/30/1931	From	ALS	IIT	
2	??/??/1931	To	TLS	IAS	011491
3	06/02/1932	To	ALS	IAS	011492
	07/25/1932	From	ALS	IAS	011494
4	08/04/1932	From	ALS	IIT	
5	??/??/1932	To	ALS	IAS	011493
6	??/??/1932?	To	ALS	IAS	011495
	??/??/1932/3?	To	TLS	IAS	011495.5
7	04/03/1933	From	ALS	IIT	
8	04/06/1933	From	ALS	IIT	
	12/10/1933	From	ALS	Menger Papers, Duke University	
	09/15/1935	From	ALS	IIT	
9	05/22/1937	To	TLS	IAS	011497
10	07/03/1937	From	ALS	IIT	
11	09/12/1937	To	TG	IAS	011498
12	11/03/1937	To	TLS	IAS	011499
13	12/15/1937	From	ALS	IIT	
	03/23/1938	To	TLS	IAS	011502
14	05/20/1938	To	TLS	IAS	011503
15	06/25/1938	From	ALS	IIT	
	10/15/1938?	From	SH[a]	IAS	011504
16	10/15/1938	To	ALS	IAS	011505
17	10/19/1938	From	ALS	IIT	
18	11/11/1938	From	ALS	IIT	
19	12/??/1938	To	ALS	IAS	011506
	??/??/1938	To	ALS	IAS	011507

[a]Draft of 10/19/38.

Item	*Date*	*To/From Gödel*	*Form*	*Source*	*Identifier*
	01/06/1939	To	TG	IAS	011508
20	08/30/1939	From	ALS, D	IAS	011510
	01/16/1941	From	ALS	IIT	
21	06/22/1942	From	ALS	IIT	
	11/20/1943	To	TLS	IAS	011511
22	09/14/1953	From	ALS	IIT	
23	01/02/1968	To	TLS	IAS	011513
24	01/22/1968	From	TL	IAS	011515
25	01/15/1972	To	ALS	IAS	011517
26	04/20/1972	From	ALS	IIT	
	04/21/1972	From	D	IAS	011518
	??/??/????	To	ALS	IAS	011516

Ernest Nagel

Item	*Date*	*To/From Gödel*	*Form*	*Source*	*Identifier*
1	02/25/1957	From	TLR	IAS	011590.05
2	03/09/1957	To	TLS	IAS	011590.1
3	03/14/1957	From	TLS[a]	IAS	011590.15
4	03/18/1957	From	TLR	IAS	011590.25
5	03/21/1957	To	TLS	IAS	011590.3
6	03/25/1957	From	TLR	IAS	011590.35
	03/27/1957	To	TLS	IAS	011590.4
	03/30/1957	To	TLS	IAS	011590.45
	04/08/1957	From	TLR	IAS	011590.5
7	04/09/1957	From	TLS	NN	
	08/14/1957	From	TLR[b]	IAS	011590.55
8	08/16/1957	From	TLS	NN	
9	08/22/1957	To	TLR	NN	
	08/29/1957	From	TLR	IAS	011590.6

[a]Unsent?

[b]Carbon copy of the letter of 08/16/1957.

Abraham Robinson

Item	*Date*	*To/From Gödel*	*Form*	*Source*	*Identifier*
	09/29/1960	To	ALS	IAS	011936
	08/28/1961	To	ALS	IAS	011937
	01/16/1967	To	TLS	IAS	011938
	04?/??/1967	From	AL, D	IAS	011939.1
	04/27/1967	From	TLR	IAS	011939.2
	05/04/1967	To	TLS	IAS	011941
	07?/??/1967	From	AL, D	IAS	011942
	07?/??/1967	From	AL, D	IAS	011943
1	07/07/1967	From	TLR	IAS	011944
	07/11/1967	To	TLS	IAS	011945
	01/21/1971	From	ALS; AL, D	YU; IAS	No number; 011952
	01/29/1971	To	TLS; TLR	IAS; YU	011953; no number
	02/08/1971	To	TLS	IAS	011954
2	03/??/1971	From	AL, D[a]	IAS	011956
3	04/14/1971	To	TLS;TLR	IAS; YU	011957; no number
4	11/22/1971	From	ALS; AL, D	YU; IAS	no number; 011959
	01/03/1972	To	TLS	IAS	011960
5	12/29/1972	From	AL, D	IAS	011962
6	01/04/1973	To	ALS	IAS	011963
	03/16/1973	To	TLR	YU	
	04/02/1973	To	AL[b]	IAS	011965
	04/06/1973	To	TL[c]; TLR	IAS; YU	011966; no number
7	04/23/1973	To	TLS; TLR	IAS; YU	011968; no number

[a] Unsent.

[b] Initialed.

[c] Initialed.

Item	*Date*	*To/From Gödel*	*Form*	*Source*	*Identifier*
	07?/??/1973	From	AL, D	IAS	011970
8	07/02/1973	From	TLS; TLR	YU; IAS	No number; 011972
	07/06/1973	To	ALS; ALS	IAS; YU	011975; no number
	07?/??/1973	From	AL, D	IAS	011976
9	07/18/1973	From	TLR	IAS	011977
	08/23/1973	To	TLS	IAS	011979
10	10/16/1973	To	TLR	YU	
11	11/22/1973	From	ALS	YU	
	03?/??/1974	From	AL, D	IAS	011997
12	03 /20/1974	From	AL, D	IAS	011998

Paul Arthur Schilpp

Item	*Date*	*To/From Gödel*	*Form*	*Source*	*Identifier*
1	11/18/1942	To	TLS; TLR	IAS; SIU	012109; 20/7/4
2	11/30/1942	From	ALS	SIU; IAS	20/7/4; 012111
	12/06/1942	To	TLS; TLR	IAS; SIU	012112; 20/7/4
	03/22/1943	To	APS	IAS; SIU	012113; 20/7/4
3	03/27/1943	From	ALS	SIU; IAS	20/7/4; 012114
4	04/18/1943	From	ALS	SIU	20/7/4
	04/20/1943	To	TLS; TLR	IAS; SIU	012115; 20/7/4
5	05/18/1943	From	ALS	SIU; IAS[a]	20/7/4; 012116
6	05/22/1943	To	TLS; TLR	IAS; SIU	012117; 20/7/4
7	05/26/1943	From	ALS	SIU	20/7/4
8	05/31/1943	To	TLS; TLR	IAS; SIU	012118; 20/7/4
9	06/07/1943	From	ALS	SIU; IAS	20/7/4; 012119
10	06/27/1943	To	TLS; TLR	IAS; SIU	012120; 20/7/4
	07/05/1943	To	TLS; TLR	IAS; SIU	012121; 20/7/4
	07/12/1943	From	ALS	SIU; IAS	20/7/4; 012122
11	07/14/1943	To	TLS; TLR	IAS; SIU	012123; 20/7/4
	07/??/1943	From	TG[b]	SIU	20/7/4
	07/16/1943	To	TLS; TLR	IAS; SIU	012124; 20/7/4
	07/17/1943	To	TPS[c]	IAS	012125
12	07/28/1943	To	TLS; TLR	IAS; SIU	012126; 20/7/4
	08/02/1943	From	TG	SIU	20/7/4
	08/02/1943	To	TG, D	SIU	20/7/4
	08/04/1943	From	ALS	SIU	20/7/4

[a]The I.A.S. copy is dated 05/17/43.

[b]Retained transcript only.

[c]Signed by Irma Corbett, Assistant to the Editor.

Item	*Date*	*To/From Gödel*	*Form*	*Source*	*Identifier*
	08/05/1943	To	TLS; TLR	IAS; SIU	012128; 20/7/4
	08/10/1943	From	ALS	SIU; IAS	20/7/4; 012129
	08/12/1943	To	TLS[d]	IAS; SIU	012130; 20/7/4
13	08/25/1943	To	TLS; TLR	IAS; SIU	012131; 20/7/4
14	09/13/1943	From	ALS	SIU	20/7/4
	09/16/1943	To	TLS; TLR	IAS; SIU	012134; 20/7/4
15	09/20/1943	From	ALS	SIU	20/7/4
16	09/22/1943	To	TLS; TLR	IAS; SIU	012136; 20/7/4
17	09/27/1943	From	ALS	SIU	20/7/4
18	10/07/1943	To	TLS; TLR	IAS; SIU	012138; 20/7/4
	10/14/1943	To	TLS; TLR	IAS; SIU	012139; 20/7/4
	10/26/1943	From	ALS; AL, D	SIU; IAS	20/7/4; 012133
	11/18/1943	To	TLS; TLR	IAS; SIU	012143; 20/7/4
	11/20/1943	From	ALS	SIU	20/7/4
	11/23/1943	To	APS	IAS	012145
	12/02/1943	To	TLS; TLR	IAS; SIU	012147; 20/7/4
	12/07/1943	To	TLS; TLR	IAS; SIU	012148; 20/7/4
	12/18/1943	To	TLS; TLR	IAS; SIU	012150; 20/7/4
	12/19/1943	From	ALS	SIU	20/7/4
	12/21/1943	From	ALS	SIU	20/7/4
	12/22/1943	From	ALS	SIU	20/7/4
	12/24/1943	To	TPS	IAS	012155
	02/14/1944	To	TLS; TLR	IAS; SIU	012157; 20/7/4
	02/24/1944	From	ALS	SIU	20/7/4
	02/25/1944	From	TG	SIU	20/7/4
	08/21/1945	To	TLS; TLR	IAS; SIU	012163; 20/7/4
	09/02/1945	From	ALS	SIU	20/7/4
	09/12/1945	To	TLS; TLR	IAS; SIU	012165; 20/7/4

[d]Signed by Mary Hart Noelck, Secretary to Professor Schilpp.

Item	*Date*	*To/From Gödel*	*Form*	*Source*	*Identifier*
	09/14/1945	From	ALS	SIU	20/7/4
	05/27/1946	To	TG	SIU	20/7/4
19	07/10/1946	To	TLS; TLR	IAS; SIU	012170; 20/15/8
20	07/25/1946	From	ALS	SIU	20/15/8
21	07/30/1946	To	TLS; TLR	IAS; SIU	012171; 20/15/8
22	08/12/1946	From	ALS	SIU	20/15/8
23	08/17/1946	To	TLS; TLR	IAS; SIU	012172; 20/15/8
	02/18/1947	From	ALS	SIU	20/15/8
	02/26/1947	To	TLS; TLR	IAS; SIU	012173; 20/15/8
	03/19/1947	To	TLS; TLR	IAS; SIU	012174; 20/15/8
	06/28/1947	To	TLS; TLR	IAS; SIU	012175; 20/15/8
24	07/15/1947	From	ALS	SIU	20/15/8
	12/17/1947	To	TLS	IAS	012176
	04/17/1948	To	TPS	IAS	012177
	05/11/1948	From	ALS	SIU	20/15/8
25	03/09/1949	To	TLS; TLR	IAS; SIU	012178; 20/15/8
	03/??/1949	From	TL, D	IAS	012179
	03/18/1949	To	TPS[e]	IAS	012180
	04/21/1949	To	TLS	IAS	012181
	04/25/1949	From	ALS	SIU	20/15/8
	04/26/1949	From	AL; D	IAS	012182; 012183
	06/10/1949	To	TPS	IAS	012184
	07/20/1949	From	ALS[f]	SIU	20/15/8
	12/05/1949	From	ALS	SIU	20/15/8
26	12/21/1949	From	ALS	SIU	20/15/8
27	05/15/1953	To	TLS; TLR	IAS; SIU	012186; 20/21/4
	06/29/1953	To	TPS	IAS	012187

[e]Signed by Schilpp's secretary, Judith N. Ash.

[f]To Mrs. Ash, inquiring about proof sheets.

Item	*Date*	*To/From Gödel*	*Form*	*Source*	*Identifier*
28	07/02/1953	From	ALS	SIU	20/21/4
29	07/06/1953	To	TLS; TLR	IAS; SIU	012188; 20/21/4
	10/14/1953	To	TPS	IAS	012189
30	03/28/1954	From	ALS	SIU	20/21/4
	03/31/1954	To	TPS	IAS	012190
	05/14/1954	From	ALS	SIU	20/21/4
	05/15/1954	To	TLS; TLR	IAS; SIU	012192; 20/21/4
	12/20/1954	To	TLS; TLR	IAS; SIU	012193; 20/15/8
	12/28/1954	From	ALS	SIU	20/21/4
	01/31/1955	From	TG	SIU	20/21/4
	02/26/1955	To	TPS	IAS	012195
	03/08/1955	From	ALS	SIU	20/15/8
	03/10/1955	From	ALS	SIU	20/21/4
	03/10/1955	To	TPS	IAS	012196
	03/11/1955	To	TPS	IAS	012197
	10/25/1955	To	TPS	IAS	012198
31	11/02/1955	To	TPS	IAS	012199
32	11/14/1955	From	ALS	SIU	20/21/4
	12/12/1955	To	TLS; TLR	IAS; SIU	012200; 20/21/4
	04/23/1956	To	TPS	IAS	012202
	06/13/1956	To	TLS	IAS	012203
	09/01/1956	To	TPS	IAS	012204
	09/19/1956	To	TLS	IAS	012205
	07/08/1957	To	TLS; TLR	IAS; SIU	012206; 20/21/4
33	07/11/1957	From	ALS	SIU	20/21/4
34	09/09/1957	From	ALS[g]	IAS	012209
35	01/24/1959	To	TLR	SIU	20/21/4
36	02/03/1959	From	ALS; TLR	SIU; IAS	20/21/4; 012212

[g]Unsent.

Item	*Date*	*To/From Gödel*	*Form*	*Source*	*Identifier*
37	02/06/1959	To	TLR	SIU	20/21/4
38	02/16/1964	To	TLS	IAS	012213.1
39	07/23/1964	From	TLR	IAS	012213.2
	04/02/1971	To	TLS	IAS	012213.25
	04/19/1971	From	TLR	IAS	012213.35
	04/20/1971	To	TLS	IAS	012213.4
	05/14/1971	From	TLR	IAS	012213.5
	05/19/1971	To	TLS	IAS	012213.6

Alfred Tarski

Item	*Date*	*To/From Gödel*	*Form*	*Source*	*Identifier*
1	01/20/1931	From	ALS	IAS	012760
	06/11/1942	To	APS	IAS	012761
	09/30/1942	To	APS	IAS	012762
	12/12/1942	To	ALS	IAS	012763
	12/09/1943	To	ALS	IAS	012765
	02/12/1944	To	APS	IAS	012767
	03/09/1944	To	APS	IAS	012768
	03/29/1944	From	SH, D	IAS	012771
	04/04/1944	To	APS	IAS	012772
	04/05/1944	To	ALS	IAS	012773
2	04/27/1944	To	APS	IAS	012774
3	05/12/1944	From	ALS	IAS	012775
	03/14/1945	To	ALS	IAS	012776
	09/24/1945	To	APS	IAS	012777
	11/02/1945	To	ALS	IAS	012778
4	12/10/1946	To	ALS	IAS	012780
	01/28/1947	To	ALS	IAS	012781
5	??/??/1961?	From	AL[a]	IAS	012237
	05/19/1970	To	TLS	IAS	012782
	??/??/1970?	From	AL[b]	IAS	012783
	01/14/1972	From	TG, D	IAS	012784
	??/??/????	To	ALS	IAS	012779

[a]Probably unsent.

[b]Unsent.

Stanisław Ulam

Item	*Date*	*To/From Gödel*	*Form*	*Source*	*Identifier*
1	01/16/1939	To	ALS	IAS	012877
2	08/14/1942	To	TLS	IAS	012879
3	??/??/1942?	From	ALR	IAS	012881
4	12/06/1947	To	ALS	IAS	012882
	05/17/1956	To	TLR	APS	
	05/23/1956	From	ALS	APS	
	01/25/1957	From	ALS	APS	
	10/21/1957	From	ALS	APS	
5	10/25/1957	To	TLR	APS	
6	11/08/1957	From	TLS	APS	
	12/26/1957	To	TLS	IAS	012882.5
	01/23/1958	From	TLS	APS	
	01/27/1958	To	TLR	APS	
7	01/28/1958	From	TLR	IAS	012882.7
8	01/06/1963[a]	To	TLS	IAS	012883
9	02/10/1964	From	TLR	IAS	012885
10	02/17/1966	To	TLS	IAS	012885.5
	04/26/1973	To	TLS	IAS	012886
	05/08/1973	From	TLR	IAS	012888

[a]Internal evidence indicates this date should be 01/06/1964.

Jean van Heijenoort

Item	*Date*	*To/From Gödel*	*Form*	*Source*	*Identifier*
	05/10/1961	To	TLR	CAH	
	07/10/1961	To	TLR	CAH	
	10/15/1961	From	TLS	CAH	
	10/17/1961	To	TLR	CAH	
	11/14/1961	From	TLS	CAH	
	11/21/1961	To	TLR	CAH	
	05/29/1962	To	TLR	CAH	
	07/26/1962	From	TLS	CAH	
	08/07/1962	To	TLR	CAH	
1	03/25/1963	To	TLS	IAS; CAH	012905; no number
2	04/23/1963	From	TLS	CAH; IAS	No number; 012907
	05/02/1963	To	TLR	CAH	
	07/01/1963	From	TLS	CAH	
	07/10/1963	From	TLS	CAH	
	08/05/1963	To	TLR	CAH	
	08/21/1963	From	TLS	CAH	
3	08/28/1963	From	TLS	CAH	
	08/31/1963	To	TLR	CAH	
	09/10/1963	From	TLS	CAH	
	09/30/1963	To	TLR	CAH	
4	10/04/1963	From	TLS	CAH; IAS	No number; 012925
	10/14/1963	To	ALS	IAS; CAH	012926; no number
5	10/24/1963	To	ALS	IAS; CAH	012927; no number
	11/07/1963	To	TLR	CAH	
	11/21/1963	From	TLS	CAH	

Item	*Date*	*To/From KG*	*Form*	*Source*	*Identification*
	12/10/1963	From	TLS	CAH	
	12/16/1963	From	TLS	CAH	
	01/13/1964	To	TLR	CAH	
	02/14/1964	To	TLR	CAH	
6	02/22/1964	From	TLS; TL[a]	CAH; IAS	No number; 012943
	03/07/1964	To	TLR	CAH	
	03/19/1964	To	TLR	CAH	
	04/03/1964	To	TLR	CAH	
	04/10/1964	From	TLS	CAH	
	04/13/1964	To	TLR	CAH	
	04/22/1964	From	TLS	CAH	
	04/26/1964	To	TLR	CAH	
	04/28/1964	To	TLR	CAH	
	05/05/1964	To	TLR	CAH	
	06/05/1964	From	TLS	CAH	
7	06/24/1964	To	ALS	IAS; CAH	012959; no number
	08/12/1964	To	TLR	CAH	
8	08/14/1964	From	TLS	CAH; IAS	No number; 012962
9	08/15/1964	From	TLS	CAH; IAS	No number; 012963
10	08/27/1964	To	TLS	IAS; CAH	012964; no number
11	09/18/1964	From	TLS	CAH; IAS	No number; 012966
	09/30/1964	To	TLR	CAH	
12	11/03/1964	From	TLS	CAH; IAS	No number; 012973
13	11/14/1964	To	TLS	IAS; CAH	012974; no number
14	12/15/1964	From	TLS	CAH; IAS	No number; 012975
	03/23/1965	To	TLR	CAH	
	04/02/1965	From	TLS	CAH	
	04/07/1965	To	TLR	CAH	
15	05/25/1965	To	TLS	IAS; CAH	012982; no number
16	07/08/1965	From	TLS	CAH; IAS	No number; 012987

[a]The retained copy at the I.A.S. has a postscript dated 02/21/1964.

Item	*Date*	*To/From KG*	*Form*	*Source*	*Identification*
	07/20/1965	To	TLR	CAH	
17	08/09/1965	From	TLS	CAH	
	09/25/1965	To	TLR	CAH	
	10/14/1965	From	TLS	CAH	
	11/06/1965	To	TLR	CAH	
18	12/13/1965	From	TLS	CAH	
	12/20/1965	To	TLR	CAH	
	01/18/1966	To	TLR	CAH	
	01/26/1966	From	TLS	CAH	
	02/07/1966	To	TLR	CAH	
	02/26/1966	To	TLR	CAH	
	05/18/1966	From	TLS	CAH	
	06/05/1966	To	TLR	CAH	
	09/??/1975	To	TLR	CAH	

John von Neumann

Item	*Date*	*To/From Gödel*	*Form*	*Source*	*Identifier*
1	11/20/1930	To	ALS	IAS	013029
2	11/29/1930[a]	To	ALS	IAS	013029.5
3	01/12/1931[b]	To	ALS	IAS	013030
4	02/14/1933	To	ALS	IAS	013031
5	03/14/1933	From	ALS	IAS	013032
6	06/13/1933	To	ALS	IAS	013033
7	08/12/1934	To	ALS	IAS	013035
8	09/18/1934	To	ALS	IAS	013036
9	05/29/1936	To	ALS	IAS	013037
10	07/13/1937	To	ALS	IAS	013038
11	09/14/1937	To	ALS	IAS	013039
12	11/11/1937	To	ALS	IAS	013043
13	01/13/1938	To	ALS	IAS	013041
14	09/12/1938	From	ALS	LC	Container 8
15	02/28/1939?	To	ALS	IAS	013043
16	03/20/1939	From	ALS	LC	Container 4
17	04/22/1939	To	ALS	IAS	013044
18	05/13/1939?	To	ALS	IAS	013045
19	07/19/1939?	To	TLS	IAS	013046
20	08/17/1939	From	ALS	IAS	013047
21	03/20/1956	From	ALS	LC	Container 5

[a]Reply to a letter no longer extant.

[b]Acknowledges two letters that have not been found.

Hao Wang

Item	*Date*	*To/From Gödel*	*Form*	*Source*	*Identifier*
	12/21/1948	To	TLS	IAS	013064
	01/10/1949	To	ALS	IAS	013065
	03/07/1949	To	ALS	IAS	013066
	05/11/1949	From	ALS	IAS	013067
	07/07/1949	To	ALS	IAS	010368
1	07/27/1967	To	ALS	IAS	013069
	09/14/1967	To	TLS	IAS	013070
2	12/07/1967[a]	From	TLS	IAS	013073–5
3	12/19/1967	To	TLS	IAS	013076
4	03/07/1968[b]	From	TLS	IAS	013085–6
5	04/23/1968	To	TLS	IAS	013087
	01/09/1970	To	TLS	IAS	013088
	01?/??/1970	From	AL[c]	IAS	013089
	04/03/1970	Neither[d]	TL	IAS	013090
6	04/15/1970	To	TLS	IAS	013091
	09/13/1970	To	ALS	IAS	013093
7	05/25/1971	To	TLS	IAS	013094
	06/14/1971	To	ALS	IAS	013095
8	07/09/1971[e]	From	TL	IAS	013097
9	07/20/1971	To	TLS	IAS	013098
10	07/20/1971	To	TLS	IAS	013100
	07/21/1971[f]	To	ALS	IAS	013102

[a]Document 013070 is an unsigned, undated pencil draft of letter 2.

[b]Document 013084 is an unsigned, undated pencil draft of letter 4.

[c]Unsent reply to 013088.

[d]From Dana Scott on behalf of Gödel to Wang and Burton Dreben. See Appendix A to this volume.

[e]Document 013096 is an undated, unsigned pencil draft of letter 8.

[f]Document 013103 is an undated, unsigned draft of 013102.

Item	*Date*	*To/From Gödel*	*Form*	*Source*	*Identifier*
11	08/04/1971	From	TL	IAS	013104
12	08/11/1971	To	TLS	IAS	013105
	01/27/1972	To	TLS	IAS	013107
	02/09/1972	To	ALS	IAS	013108
	04/06/1972	To	ALS	IAS	013109
13	04/10/1972	To	TLS	IAS	013110
	04/13/1972	To	ALS	IAS	013111
14	04/26/1972	To	ALS	IAS	013113
	06/16/1972	To	ALS	IAS	013114
	07/14/1972	To	APS	IAS	013119
	06/28/1973	To	APS	IAS	013127
	08/23/1973	To	ALS	IAS	013128
	12?/??/1973	To	APS	IAS	013129
	12/26/1973	To	APS	IAS	013130
	04/11/1974	To	ALS	IAS	013132
	07/01/1974	To	ALS	IAS	013135
	02/24/1975	To	ALS	IAS	013142
	04/23/1975	To	ALS	IAS	013145
	05/15/1975	To	ALS	IAS	013150
	11/04/1975	To	ALS	IAS	013156
	11/07/1975	To	ALS	IAS	013157
	11/14/1975	To	ALS	IAS	013158
	11/18/1975	To	ALS	IAS	013160
	12/02/1975	To	ALS	IAS	013163
15	12/12/1975	To	TLS	IAS	013166
	12/15/1975	To	ALS	IAS	013168
	02/05/1976	To	ALS	IAS	013179
	02/09/1976	To	ALS	IAS	013180
	03/11/1976	To	TLS	IAS	013182

[g]Christmas card.

Item	*Date*	*To/From Gödel*	*Form*	*Source*	*Identifier*
	03/16/1976	To	TLS	IAS	013183
	04/04/1976	To	ALS	IAS	013185
16	06/24/1976	To	ALS	IAS	013187
	11/03/1976	To	TLS	IAS	013191
	12/07/1976	To	TLS	IAS	013194
17	02/03/1977	To	TLS	IAS	013196
	02/09/1977	To	TLS	IAS	013198
	06/22/1977	To	ALS	IAS	013204

Finding aid

Kurt Gödel Papers (C0282):

1905–1980, bulk 1930–1970

A
Finding Aid
Prepared
by
John W. Dawson, Jr.
1984

Revised by
Rebecca Schoff
1997

and
Barbara Volz
1998*

With microfilm information
added by Cheryl Dawson
for these *Works*

Manuscripts Division
Department of Rare Books and Special Collections
Princeton University Library

*With further corrections and editing for these *Works*.

Kurt Gödel Papers (C0282)
Table of Contents

[a]Omitted here. See Reference section of this volume.
[b]Added by Cheryl Dawson. Note also that columns with italic headings were added for this volume.

Kurt Gödel Papers (C0282) Introduction

The papers of Kurt Gödel (1906–1978), foremost mathematical logician of the twentieth century, were bequeathed by him to his wife Adele, who donated them to the Institute for Advanced Study in his memory prior to her death in 1981. Under terms of her will, literary rights to the papers are also vested in the Institute for Advanced Study. In 1985 the papers were placed on deposit in the Manuscripts Division, Department of Rare Books and Special Collections, at Princeton University Library. The Institute for Advanced Study reserves all copyrights and other literary rights to the materials, which may not be reproduced in any form, or published without the prior written permission of the Institute.

The papers comprise documents relating to all periods of Gödel's life, including scientific correspondence, notebooks, drafts, unpublished manuscripts, academic, legal, and financial records, and all manner of loose notes and memoranda. Family correspondence is notably absent, as are financial records after Gödel's emigration in 1940. Of the manuscript material, a substantial part is in Gabelsberger shorthand, a German system originally devised by Franz Xaver Gabelsberger (1789–1849) and published in his textbook, *Anleitung zur deutschen Redezeichenkunst oder Stenographie* (1834).[c] For reference a photocopy of Karl Ludwig Weizmann's *Lehr- und Übungsbuch der Gabelsbergerschen Stenographie*, 2d edition (1915)[d] is in the Manuscripts Division's Vertical File. A bibliography of Gödel's 700-book library is available upon request from the librarian of the Institute's Historical Studies Library. Preprints and offprints sent to Gödel by others are briefly described in the finding aid. They are now stored in seven cartons in the Department of Rare Books and Special Collections, and in filing cabinets in the Institute's Historical Studies Library. Extent: The Gödel papers occupy 41 archival boxes, 1 flat box, and 1 extra large box. In addition, ancillary materials occupy 7 cartons (8 cubic feet). There are approximately 9000 items of primary documentary material (14.5 cubic feet).

[c] *Gabelsberger 1834*.

[d] *Weizmann 1915*.

Kurt Gödel Papers (C0282)
Biographical Sketch

Kurt Friedrich Gödel was born 28 April 1906, in Brünn, Moravia, and died 14 January 1978 in Princeton, New Jersey. His life may be divided into three periods, corresponding both to his place of residence and to the nature of his intellectual endeavors.

Gödel's childhood and youth were spent in Brünn where his father worked as a manager of a textile factory. He attended German language primary and secondary schools, graduating with honors in 1924. After graduation he enrolled at the University of Vienna, where he joined his brother Rudolf (born 1902). Gödel remained at the University of Vienna, first as student and later as *Privatdozent* (an unpaid lecturer), until his emigration to America in 1940. He became an Austrian citizen in 1929. Later that year, in his doctoral dissertation, Gödel established the completeness of the first-order predicate calculus, a work that marked the beginning of a decade of fundamental contributions to mathematical logic, including especially his proofs of the incompleteness of formal number theory (1930, published 1931) and of the relative consistency of the axiom of choice and the generalized continuum hypothesis (1935 and 1937, published 1938–1940). His residence in Vienna was interrupted by three trips to the United States, where he visited the Institute for Advanced Study (1933–1934, Autumn 1935, and Autumn 1938) and the University of Notre Dame (Spring 1939). He married Adele Nimbursky (née Porkert) in Vienna, 20 September 1938.

In January 1940, fearing conscription into the Nazi army, Gödel left Europe with his wife via the trans-Siberian railway. Arriving in San Francisco on 4 March 1940, the Gödels settled in Princeton, where he resumed his membership in the Institute for Advanced Study. He became a permanent member in 1946, a U.S. citizen in 1948, professor at the Institute in 1953, and professor emeritus in 1976. At the Institute, Gödel interests turned to philosophy and physics. He studied the works of Gottfried Wilhelm von Leibniz in detail and, to a lesser extent, those of Immanuel Kant and Edmund Husserl. In the late 1940s he demonstrated the existence of paradoxical solutions to Albert Einstein's field equations in general relativity. His last published paper appeared in 1958. He shared the first Einstein Award in 1951 and was awarded the National Medal of Science in 1975. He also received honorary doctorates from Yale, Harvard, and Rockefeller universities, and from Amherst College.

John W. Dawson, Jr.

⟦For references cited, see *Dawson 1983*, *Dawson 1984a*, *Dawson 1997*, and *Kreisel 1980* in addition to these *Works.*⟧

Kurt Gödel Papers (C0282)
Acknowledgments

The original organization and description of Gödel's papers was done under stipendiary support from the Institute for Advanced Study in 1982–1984, together with sabbatical support from the Pennsylvania State University during the academic year 1983–1984. Appreciation is also due to the late Hermann Landshoff of New York City for assistance in deciphering Gödel's shorthand, and especially to Cheryl Dawson, without whose help the original project could not have been completed in the span of two years. In addition to helping with the decipherment, she cataloged Gödel's books and journals and helped to sort many of the preprints and offprints.

The following people shared their expertise unstintingly: Helen Samuels, Massachusetts Institute of Technology; Professor John Stachel, Boston University and the *Collected Papers of Albert Einstein*; Professor Reese Jenkins, Tony Appel, and Thomas E. Jeffrey of the *Papers of Thomas A. Edison*, Rutgers University (New Brunswick) and the Edison Historic Site (West Orange, N.J.); Spencer Weart and Joan Warnow, American Institute of Physics; Professor Richard Nollan, University of Pittsburgh; and Leon Stout, Pennsylvania State University. And lastly, the secretarial assistance of Irene Gaskill, Carolyn Underwood, and Dorothy Phares, of the Institute for Advanced Study, in preparing the original folder list should not be forgotten.

Revisions to the finding aid and the preparation of the papers for preservation microfilming by Rebecca Schoff in 1997–1998 were funded by the Alfred P. Sloan Foundation as part of Sloan Foundation Grant 95-10-14 to the Gödel Editorial Committee.

Kurt Gödel Papers (C0282)
Scope and Content Note

The Kurt Gödel Papers include documents spanning the years 1905–1980, with the bulk of the material falling between 1930 and 1970. Of greatest extent and significance are Gödel's scientific correspondence (Series I), his notebooks (Series III), and numerous drafts, manuscripts, and galleys of his articles and lectures, published and unpublished (Series IV).

Prior to its first arrangement in 1983–1984, the collection was stored in filing cabinets and moving cartons in a cage in the basement of the Institute for Advanced Study's Historical Studies Library. In 1985 the papers were placed on deposit in Firestone Library by the Institute for Advanced Study. At that time, the 22 Paige boxes and oversized container used in the first organization of the collection were split into 41 archival boxes, one flat box, one extra large box and seven cartons. In 1997, the finding aid was revised and amplified. Most of the oversized correspondence was merged with that in the main series, as were items from Series III and IV.

Although gathered together in some haste and disarray following Gödel's death, most of the items were found in envelopes labelled by Gödel himself; on that basis, an attempt was made to retain or, where necessary, restore Gödel's original order. An exception is the division of the correspondence into two series (I: Personal and Scientific Correspondence, II: Institutional, Commercial, and Incidental Correspondence) for convenience of scholarly access; the former is arranged alphabetically by correspondent, the latter by subject. Folders are numbered sequentially within each series. Documents were originally assigned six-digit item numbers, also sequentially within each series, with the first two digits corresponding to series designation. Thus, item 11013 is the thirteenth item in Series XI. With the introduction after the original cataloging of two new series for the oversized items, this correspondence will not hold there. Envelopes used by Gödel to organize his papers have been assigned the same number as the item they contained (or the first item among the items they contained). On folder labels, Gödel's own designations are enclosed in quotation marks, while dates in square brackets refer to citations of Gödel's own works in Dawson's bibliography.[e] With the publication of a new standard bibliography in the Oxford edition of Gödel's *Collected Works*, pencil annotations were added to folder labels in accordance with the citations of the new bibliography (Volume III, pp. 487–491). The finding aid, revised in 1997, follows the new bibliography with publication dates in italics.

---

[e] *Dawson 1983.*

The correspondence in Series I bulks betweeen 1950 and 1975, but includes earlier items from such correspondents as Paul Bernays, Rudolf Carnap, Jacques Herbrand, Arend Heyting, Karl Menger, Emil Post, Oswald Veblen, John von Neumann, and Ernst Zermelo. Other major correspondents include William Boone, Paul J. Cohen, Georg Kreisel, Oskar Morgenstern, Abraham Robinson, Paul A. Schilpp, Dana Scott, Gaisi Takeuti, and Hao Wang. Approximately two-thirds of the correspondence is incoming. Family correspondence is virtually absent, but about 1000 pages of Gödel's letters to his mother are preserved in the Wiener Stadt- und Landesbibliothek, Vienna.

Early records in the collection include patent correspondence of Gödel's father, birth and baptismal certificates, and Gödel's notebooks and report cards from elementary and secondary schools. Some university course notebooks are also preserved, but there are no enrollment or grade records from the University of Vienna.

Financial records (Series IV) are quite detailed for the period 1930–1939 but are totally absent after Gödel's emigration in 1940. They include account books, cancelled checks and deposit slips, ledgers, and various bills and receipts.

Gödel's personal notes and notebooks (Series III, V, and VI) span most of his life. They are largely in Gabelsberger shorthand, as are also some drafts of letters and lectures (Series I and IV). However, there are often longhand headings in German and English, and where mathematics and logic are involved, computational or symbolic notes make some material accessible to readers who cannot read the shorthand. Manuscript items are almost entirely in pencil.

An assortment of loose manuscript notes and memoranda (also largely in Gabelsberger shorthand) comprise series V and VI, including reading notes, library request slips, bibliographic memoranda, computation sheets (especially concerning Gödel's work in relativity theory), and personal notes on diverse subjects, including American history, languages, philosophy, and theology. Especially prominent are voluminous notes on the works of Gottfried Wilhelm von Leibniz.

Smaller categories include medical records (Series X), photographs (Series XI), and ephemera (Series XII). A few folders of correspondence from other sources have also been incorporated as addenda.

Ancillary materials include the books from Gödel's library, now shelved at the Institute for Advanced Study. A list of the books is available at the Historical Studies Library of the Institute, and includes a record of journals received by Gödel (unannotated and not slated for retention). Assorted preprints and offprints sent to him by others have been segregated into three groups for retention with the papers: presentation copies, items accompanied by correspondence, and items bearing annotations or with accompanying notes; others with none of these features are in filing cabinets at the Institute's Historical Studies Library. Items within each of these groups are filed alphabetically by author but are not numbered or otherwise indexed.

A few non-documentary items, donated with the papers, have been transferred for safekeeping to the Director's office at the Institute for Advanced Study. They include Gödel's briefcase, door plate, and National Medal of Science (medallion and lapel pin).

Kurt Gödel Papers (C0282)
Series Descriptions

Series I: Personal and Scientific Correspondence, 1929–1978

Boxes 1a–3c. Incoming and outgoing correspondence filed together, alphabetically by correspondent, and therein chronologically. Unidentified correspondents at end of series. Approximately 3,500 items. Bulk dates: 1930–1975.

Incoming letters, and copies of outgoing letters, with friends and scientific colleagues; also letters of recommendation. Virtually no family correspondence. Major correspondents include Paul Bernays, William Boone, Paul J. Cohen, Gotthard Günther, Arend Heyting, Georg Kreisel, Karl Menger, Oskar Morgenstern, Abraham Robinson, Paul A. Schilpp, Dana Scott, Gaisi Takeuti, Jean van Heijenoort, John von Neumann, and Hao Wang.

Filed along with correspondence are notes taken by Gödel during oral discussion with some correspondents. They are indicated separately as "discussion notes" under the name of the correspondent. Some short preprints related to accompanying correspondence are also filed here.

Series II: Institutional, Commercial and Incidental Correspondence

Boxes 4a–5a. Incoming and outgoing correspondence filed together, alphabetically by subject or type of correspondence. Approximately 1,600 items.

Includes requests for biographical information, charitable solicitations, lecture invitations, editorial correspondence, honors bestowed, internal correspondence of the Institute for Advanced Study, literary solicitations, offprint and permissions requests, and correspondence with professional societies. Also includes unsolicited correspondence from cranks and autograph seekers.

Series III: Topical notebooks

Boxes 5b–7a. Filed alphabetically by subject or by Gödel's title (as indicated in the finding aid). Approximately 150 items. Mostly in Gabelsberger shorthand.

Includes school exercise books, university course notes, vocabulary notebooks, notes for Gödel's lectures at Vienna and Notre Dame, and several series of notebooks on mathematical logic, philosophy, and current events. Among the notebooks are sixteen "Arbeitshefte" (mathematical workbooks), fourteen labelled "Allgemeine Bildung", nine history notebooks, six designated as "Logic and foundations" and four as "Results on foundations", and fifteen philosophical notebooks including material from before May 1941 until the end of Gödel's life. The philosophical notebooks are designated as "Max 0–XV", of which volume XIII is missing; the second of three theological notebooks is also lost.

Many of the notebooks were filled from both directions, with one subject or sequence beginning on the first page and proceeding toward the back of the notebook and another beginning on the last page and proceeding toward the front of the notebook. These notebooks are designated in the finding aid as written both directions. Gödel also habitually filed separate pages with notebooks in three different manners described in the finding aid as follows:

Intercalated: Placed within the notebook, apparently with some attention to a meaningful sequence, often numbered in sequence with the pagination of the notebook.

Inserted: Placed inside the notebook (usually inside the front or back cover) without obvious attention to sequence.

Loose: Simply filed in a folder with the notebook.

Series IV: Drafts and Offprints

Boxes 7a–9b. Filed chronologically under titles or descriptive headings. Approximately 500 items.

Drafts, manuscripts and typescript "Reinschriften," galleys and offprints of Gödel's articles, lectures, and reviews, published and unpublished, in English, German, and Gabelsberger shorthand. Important unpublished items include Gödel's 1951 Gibbs lecture (**1951*), a longer version of Gödel's essay *1949* on relativity theory and idealistic philosophy, "Is mathematics syntax of language?" (**1953/9*), (intended for *Schilpp 1963*), and a revised English version (*1972*) of the *Dialectica* paper (*1958*).

Published works are designated by italic years of publication as established in the Gödel bibliography published in Volume III of the Oxford edition of Gödel's *Collected Works*, pp. 487–491.

Series V: Bibliographic Notes and Memoranda

Boxes 9b–11a. Grouped by subject or by Gödel's titles, as indicated by quotation marks in the finding aid, and therein by date (where known). About 250 numbered groups (slips not all separately numbered).

Diverse notes and memoranda slips, including reading notes, library request slips, bibliographic excerpts and memoranda, and memoranda books. Extensive notes on history, philosophy and theology, especially the works of Gottfried Wilhelm von Leibniz (primary and secondary sources). Largely in Gabelsberger shorthand.

Series VI: Other Loose Manuscript Notes

Boxes 11b–12. Filed alphabetically by subject. Approximately 800 items.

Includes computation sheets, reference lists of formulas, miscellaneous mathematical notes and fragmentary drafts, and notes which once accompanied books and papers of others.

Series VII: Academic records and notices

Box 13a. Filed alphabetically by document type. Approximately 250 items.

Includes elementary and secondary school report cards, course announcements and enrollment slips for courses taught by Gödel, homework graded by Gödel, and administrative correspondence and announcements from the University of Vienna, Notre Dame University, and the Institute for Advanced Study.

Series VIII: Legal and Political Documents

Box 13a. Filed alphabetically by document type. Approximately 200 items.

Includes apartment rental agreements, birth, baptismal, marriage, and citizenship certificates, copyright and publishing agreements, patent documents and correspondence of Gödel's father, passports, and powers of attorney.

Series IX: Financial Records, 1930–1939

Box 13b–13c. Filed alphabetically by document type, unclassified items at end of series. Approximately 1,100 items.

Account books, bank statements, currency exchange vouchers, cancelled checks, deposit slips, ledgers, securities transactions, and various bills and receipts.

Series X: Medical Records

Box 14a. Filed alphabetically by type of document. Approximately 150 items.

Dosage records, medical and dietary memoranda, lists of doctors, prescriptions, and temperature records.

Series XI: Photographs

Boxes 14a, 14b, 15, 16. Filed by print size, and grouped therein by subject. Approximately 200 photographic prints.

Snapshots of Gödel alone and with family, friends, and colleagues. Some formal portaits. Photographs of the Gödel home at 145 Linden Lane. Some photographs of Gödel's ancestors and his wife's parents. N.A.S.A. Mars and lunarscape photos.

Series XII: Ephemera

Box 14c. Filed alphabetically by type of document. Approximately 250 items.

Material not falling under any of the other series; includes advertisements, newspaper articles and clippings, annotated envelopes, concert programs, publishers' catalogs, and Nazi proclamations issued to University of Vienna faculty.

Series XIII: Oversize Items and Addenda

Box 15. Unclassified oversize items, followed by items grouped into Series VII–XII. Addenda to the collection filed at end of series.

Contains items from Series VII, IX, XI, and XII that are too large to be filed in logical sequence. Each of these items is cross-referenced at its logical place in the finding aid. The logical physical placement of each item is marked with a separation notice in the regular series box. In addition, there are unclassified oversize items not belonging under any of the other series, including unidentified notes and an unidentified draft. Addenda to the collection from three sources are filed at the end of the series. These include photocopied correspondence from the Heyting Archief and the Eidgenössische Technische Hochschule, and original material contributed by Dana Scott regarding the publication of *1972a* and *1980*. Document numbers in this series are no longer in the original sequence set by John W. Dawson.

Series XIV: Extra Large Items

Box 16. Items filed in order of the series in which they logically belong.

Contains items from Series I–II, VI–VII, IX, XI, and XII that are too large to be filed in logical sequence. Each of these items is cross-referenced at its logical place in the finding aid. The logical physical placement of each item is marked with a separation notice in the regular series box. Includes Gödel's doctoral diploma and honorary certificates. Document numbers in this series are no longer in the original sequence set by John W. Dawson.

Series XV: Preprints and Offprints

Boxes 17–23. Items sorted into three groups: Presentation Copies; Preprints and offprints annotated or with accompanying notes; Preprints and offprints with accompanying correspondence. Filed alphabetically by author.

Gödel's collection of preprints and offprints (also includes some TMss and AMss) of his colleagues' work. Preprints and offprints which were sent to Gödel with accompanying correspondence are labelled to indicate the item number of the letter with which they were sent. Correspondence is filed separately in the correspondence Series I and II. Likewise, manuscript notes which once accompanied preprints or offprints have been filed separately, mostly in Series VI, as indicated by the item number recorded on each preprint or offprint. There are additional offprints and preprints at the Institute for Advanced Study Library.

Kurt Gödel Papers (C0282)
Index to Item Numbers

In addition to the organization of items into series, during the earliest organization of the collection, items were stamped with an item number. The first two digits of the number indicate the series in which the item belongs. The last four digits indicate the item's logical place in the sequence of each series, with decimal intercalations for items discovered later in the cataloging process (e.g., item 040250 is the 250th item in Series IV). The following list cross-references item numbers with box, series, and folder numbers.

Box	**Range of Series/Folders**	**Range of Item Numbers**	*Film Reel(s)*
1a	I/00–I/21	010001–010280.8	2, 3
1b	I/22–I/42	010280.81–010558.3	3, 4
1c	I/43–I/69	010558.5–010897	4, 5
2a	I/70–I/97	010898.5–011327	5, 6
2b	I/98–I/111	011327.5–011590	7
2c	I/112–I/141	011590.05–012052	8, 9
3a	I/142–I/167	012053–012462	9, 10
3b	I/168–I/189.5	012463–012888.37	10, 11
3c	I/190–I/219	012888.5–013304	11, 12
4a	II/01–II/22	020001–020383	13
4b	II/23–II/42	020384–020806	14
4c	II/43–II/68	020807–021333	15
5a	II/69–11/84	021334–021550	16
5b	111/01–III/11.5	030001–030015.5	17
5c	III/12–III/30	030016–030039	17, 18
5d	III/31–III/49	030040–030072	18, 19
6a	III/50–111/62.5	030073–030085.5	19, 20
6b	III/63–III/78	030086–030107	20, 21
6c	III/79–III/106	030108–030128.98	22, 23
7a	III/107–IV/11	030129–040019.5	23, 24
7b	IV/12–IV/36	040020–040143	24, 25
7c	IV/37–IV/55	040144–040197	26
8a	IV/56–IV/76	040198–040262	27, 28
8b	IV/77–IV/98	040263–040306	28
8c	IV/99–IV/124.6	040307–040411.5	29, 30
9a	IV/125–IV/140	040412–040448	30, 31

Box	**Range of Series/Folders**	**Range of Item Numbers**	*Film Reel(s)*
9b	IV/141–V/09	040449–050045	31, 32
9c	V/10–V/25	050046–050120	33, 34
10a	V/26–V/42	050120.1–050144	35, 36
10b	V/43–V/54	050144.1–050173	37, 38, 39
10c	V/55–V/68	050174–050209	39, 40, 41
11a	V/69–V/78	050210–050253	41, 42
11b	VI/01–VI/16	060001–060178	42, 43, 44
11c	VI/17–VI/35	060179–060524	44, 45, 46
12	VI/36–VI/52	060525–060777	46, 47
13a	VII/00–VIII/15	070001–080146	47, 48
13b	IX/01–IX/32	090001–090621	48, 49
13c	IX/33–IX/59	090622–091134	48, 50
14a	X/01–XI/17	100001–110221	50, 51, 52
14b	Loose items; unnumbered photographs in folder XI/0	110000–110145	52
14c	XII/O1–XII/11	120001–120261	53

Boxes 15 and 16 ⟦microfilmed on reels 53 and 54⟧ contain oversize and extra large items that are numbered according to their logical places in the sequences above. In boxes 1a–14c these items are replaced with separation notices indicating the physical location of the items in the oversize and extra large series.

In addition, box 15 contains unclassified oversize items (Folder 0), photocopied correspondence from the Heyting Archief and the ETH-Bibliothek (Folders 6 and 8), and material contributed by Dana Scott re the publication of *1980* (Folder 7), all without item numbers. Box 16 contains four unnumbered items: a framed certificate for the National Medal of Science, two certificates of academy membership, and a portfolio of N.A.S.A. lunarscape photographs.

Kurt Gödel Papers (C0282)
Detailed Collection Outline

Box	Folder		Date(s)	*Initial document*	*Reel*	*Initial frame*
		I. Personal and Scientific Correspondence, 1929–1978				
1a	0	Addison, John	1956, 1958	010001	2	3
	1	Miscellaneous "A"	1952–1976	010000		36
	2	Behman⟦n⟧, Heinrich	1930–1935	010015.415		88
		Bernays, Paul:				
	3		1930–1931	010015.44		122
	4		1939, 1942, 1956–1959	010015.48		180
	5		1960–1961	010037		253
	6		1962–1965	010054		304
	7		1966–1970	010079		395
	8		1971–1975	010105		472
		[See also Series XIII: Folders 6 and 7]				
		Boone, William:				
	9		1954	010136		505
	10		1955	010143		526
	11		1956–1957	010153		550
	12		January–May, 1958	010177		597
	13		June–August, 1958	010185		617
	14		1959–1961	010204		665
	15		1962–1963	010218		709
	16		1964–1966	010233		762
	17		1967–1969, 1974-1976, 1978-1979	010259		805
	18		n.d.	010266		836
	19	Brutian, George A.	1969–1970	010280.05		856
	19.5	Burks, Arthur	1961, 1964	010280.676		880

Box	Folder		Date(s)	*Initial document*	*Reel*	*Initial frame*
		I. Personal and Scientific Correspondence, 1929–1978 (cont.)				
		Miscellaneous "B":				
1a	20	T. R. Bachiller to Errett Bishop	1933–1973	010015.325	2	901
	21	Max Black to Terrell Ward Bynum		010140.45	3	1
1b	22	Carnap, Rudolf [See also Series XIV: Folder 1]	1929–1939	010280.81		124
	23	Chang, C. C.	1963, 1965	010282.5		152
	24	Chomsky, Noam	1957–1958	010291		187
	25	Chuaqui Kettlun, Rolando B.	1969–1972, n.d.	010308		217
	26	Church, Alonzo	1932, 1946, 1965–1966	010329.09		280
		Cohen, Paul J.:				
	27		April 24–July 17, 1963	010347		345
	28		July 20–September 27, 1963	010360		401
	29		October 4–December 13, 1963	010373		458
	30		1964	010387		512
	31		1965	010393		536
	32		1966–1969, 1975	010415		583
	33		n.d. (Fragmentary letter drafts by Gödel)	010439		659
		Miscellaneous "C":				
	34	Ronald Calinger to D. V. Choodnovsky	1949–1976	010280.81		679
	35	Jeffrey Cohen to Haskell B. Curry	1932–1977	010334.5		738
		Davis, Martin:				
	36		1950, 1963	010443.9		806
	37		1964	010463		837
	38		1965, n.d.	010494		881

Box	Folder		Date(s)	*Initial document*	*Reel*	*Initial frame*
		I. Personal and Scientific Correspondence, 1929–1978 (cont.)				
1b	39	Dreben, Burton S.	1959–1971	010503	3	940
	40	Miscellaneous "D"	1931–1975	010443.8	4	3
		[Einstein, Albert: see Miscellaneous "E"]				
		Ellentuck, Erik:				
	40.5		1963–1966	010524.85		47
	41		1967–1968, 1971–1973	010524.801		128
	42	Miscellaneous "E"	1943–1977	010524.3		218
1c		Feferman, Solomon:				
	43		1957–1961	010559		279
	44		1963–1964, 1967, 1969, 1971, 1974	010581		322
		Feigl, Herbert:				
	45		1927, 1929–1931, 1933–1934	010607		383
	46		1950, 1956–1957, 1961–1962	010616		410
	47	Fisher, Edward R., Jr.	1968–1970	010636		429
	48	Flexner, Abraham	1933–1939	010648.651		470
	49	Ford, Lester R. (re *1947*)	1945–1946	010648.7		490
	50	Friedburg, Robert	1956, 1963	010661		502
		Friedman, Harvey:				
	51		1966–1971	010672.1		535
	52		1974–1975	010673		552
	53	Miscellaneous "F"	1930–1975	010558.5		602
	54	Gandy, R. O.	1959-1960, 1973	010697		653
	54.5	Gödel Family: Miscellaneous	1939–1966	010709.7		677
	54.6	Gödel, Rudolf (brother)	1966–1972	010711.2		694
	55	Grandjean, Burke	1974-1975	010720		786

Box	Folder		Date(s)	*Initial document*	*Reel*	*Initial frame*

I. Personal and Scientific Correspondence, 1929–1978 (cont.)

Box	Folder		Date(s)	*Initial document*	*Reel*	*Initial frame*
		Günther, Gotthard:				
1c	56		1953	010732	5	3
	57		1954–1955	010755		47
	58		1956–1957	010762		101
	59		1957–1961, includes undated notes	010782		163
	60	Miscellaneous “G”	1932–1975	010696.4		198
	61	Halpern, James, includes discussion notes	1965–1966	010732		264
	62	Hasenjaeger, G.	1963–1965	010776		313
	63	Henkin, Leon	1960-1972	010817		363
	64	Herbrand, Jacques	1931	010837.5		424
		Heyting, Arend:				
	65		1931–1933	010839		443
	66	[See also Series XIII: Folder 8]	1957, 1969	010856		479
	67	Howard, William A. (includes discussion notes)	1964–1973	010874.5		496
		Miscellaneous “H”:				
	68	John Haag to Stephen Hechler	1933–1975	010731.1		562
	69	Radcliffe Heermance to Ralph Hwastecki	1931–1976	010812.5		623
2a	70	Miscellaneous “I” and “J”	1937–1972	010898.5		695
		Jech, Tomás:				
	71		1969, 1972	010900		712
	72		1973–1974	010924		781
	73	Jeffrey, Richard	1963–1974, n.d.	010941		833
	74	Jørgensen, Jørge	1932–1933	010960		869
	75	Kleene, Stephen C.	1936, 1956, 1965, 1975	010964.2		872

Box	Folder		Date(s)	*Initial document*	*Reel*	*Initial frame*

I. Personal and Scientific Correspondence, 1929–1978 (cont.)

Box	Folder		Date(s)	Initial document	Reel	Initial frame
2a	76	Kochen, Simon (includes discussion notes)	1965–1975	010968	5	889
	77	Kondô, Motokiti	1961–1963	010998	6	3
		Kreisel, Georg:				
	78		1955–1958	011013		31
	79	Accompanying notes by Gödel	1959–1960	011061		128
	80		1959–1960	011078		154
	81	Accompanying notes by Gödel	1959–1960	011100		220
	82		1961–February 1962	011106		235
	83		March–December 1962	011134		315
	84		1963–April 1964	011045		387
	85	Discussion notes	March 1964	011157		430
	86		June 1964–December 1964	011158		458
	87		1965	011180		512
	88		1966	011193		552
	89	Discussion notes	April 1963–August 1966	011208		599
	90		1967	011226		656
	91		1968	011244		698
	92		1969	011261		747
	93		1970	011271		782
	94		1971	011282		817
	95	Discussion notes	October 1966–1971	011301		881
	96		1972, n.d.	011313		927
	97	Undated notes [See also Takeuti, Gaisi]		011319		968

Box	Folder		Date(s)	*Initial document*	*Reel*	*Initial frame*
		I. Personal and Scientific Correspondence, 1929–1978 (cont.)				
		Kuroda, Sigekatu:				
2b	98		1954–1958	011341.5	7	3
	99		1959–1960, 1962, 1972, n.d.	011367		65
	100	Miscellaneous "K"	1937–1975	010961.3		119
	101	Leonard, Henry	1959–1967	011419		209
	102	Levy, Azriel	1958–1966	011440		254
	103	Lorenzen, Paul	1958, n.d.	011457		288
		Miscellaneous "L":				
	104	Daniel Lacombe to S. Lefschetz	1943–1977	011398		334
	104.50	Walter Leighton to W. A. J. Luxemburg	1934–1980	011418.5		386
	105	Menger, Karl	1931–1968, n.d.	011491		416
	106	Morel, Anne C.	1959–1960	011535		481
	107	Morgenstern, Oskar	1946–1948, 1956	011550		503
		Copies of Morgenstern's correspondence with others	1965, 1972–1974			
		Myhill, John:				
	108		1957–1958	011591		549
	109		1962–1968, includes undated notes	011570		592
		Miscellaneous "M":				
	110	Angus Macintyre to Terrance Millar	1932–1977	011474		673
	111	Charles F. Miller to Jan Mycielski	1931–1977	011526		748
2c	112	Nagel, Ernest (re *Nagel and Newman 1958*)	1957	011590.05	8	3
	113	Nakamura, Akira	1965	011591		29

Box	Folder		Date(s)	*Initial document*	*Reel*	*Initial frame*

I. Personal and Scientific Correspondence, 1929–1978 (cont.)

Box	Folder		Date(s)	*Initial document*	*Reel*	*Initial frame*
		Natkin, Marcel:				
2c	114		1927–1929, 1931, 1936	011605	8	51
	115		1957, 1961–1962	011616		80
	116	Newsom, C. V. (re *1947*)	1947–1948	011632		102
		Novikov, P. S.:				
	117		1964	011642		115
	118		1965–1967	011676		167
	119	Miscellaneous "N" and "O"	1931–1975	011590.9		213
	119.50	Popper, Karl R.	1934–1964	011717.2		240
	120	Post, Emil	1938–1939	011717.3		254
		Pour-El, Marian Boykan:				
	121		1958	011718		264
	122		1962–1965, 1970–1971, n.d.	011728		280
	123	Powell, William C., with discussion notes	1972–1975	011754		333
		Prikry, Karel:				
	124		1973–1974	011778		390
	125		1975	011792		443
	126	Miscellaneous "P" and "Q"	⟦1935⟧–1976	011706.9		478
	127	Rabin, Micael Oser	1956–1974	011806		561
	128	Rappaport, Leon	1962	011826		603
	129	Reid, Constance	1965–1966, 1969	011840		628
	130	Reid, Constance: fragmentary letter drafts by Gödel	1966	011851		670
		Reinhardt, William N.:				
	131		1966–1967, 1971–1971	011863		720

Box	Folder		Date(s)	*Initial document*	*Reel*	*Initial frame*

I. Personal and Scientific Correspondence, 1929–1978 (cont.)

Box	Folder		Date(s)	*Initial document*	*Reel*	*Initial frame*
		Reinhardt, William N. (cont.):				
2c	132		1973–1974, n.d.	011885	8	784
	133	Robbin, Joel W.	1965–1967	011920.5	9	3
		Robinson, Abraham:				
	134		1960–1961, 1967–1968	011936		33
	135		1971–1972	011952		63
	136		1973	011963		86
	137	Discussion notes:	1971, 1973	011988		125
	138	With texts from memorial service	1974–1975	011997		171
	139	Robinson, Robert W.	1966–1967	012011		214
	140	Rucker, Rudy	1971–1976	012038		246
	141	Miscellaneous "R"	1931–1973	011824		337
3a	142	Sacerdote, George S.	1973–1976	012053		403
		Sacks, Gerald:				
	143		1963–1968	012081.5		454
	143.50		1971–1975	012103		484
		Schilpp, Paul A. (re *1944*):				
	144		November 1942–July 1943	012109		524
	145		August–November 1943	012128		562
	146		December 1943–March 1944	012147		598
	147	(includes undated notes)	August–September 1945	012163		634
		Schilpp, Paul A. (re *1949*):				
	148		1946–1949	012170		648

Box	Folder		Date(s)	*Initial document*	*Reel*	*Initial frame*
		I. Personal and Scientific Correspondence, 1929–1978 (cont.)				
		Schilpp, Paul A. (re *1953*):				
3a	149	(includes correspondence re **1953/9-III* and **1953/9-V*	1953–1957, 1959	012186	9	676
	150		1964–1971	012213.1		721
	151	Schütte, Kurt	1958–1959	012223		738
		Scott, Dana:				
	152		1955, 1961–1962	012232		749
	153		1965–1966	012243		794
	154		1967–1968	012262		840
	155		1969	012274		873
	156	[See also Series XIII: Folder 7]	1970, 1972	012280		920
	157	Discussion notes	1956, 1966, 1968–1970, n.d.	012284		937
	158	Shepherdson, John C. (re 1934 lectures)	1962–1965	012305.5	10	3
	159	Shepherdson, John C. et al. (re integration in finite terms)	1964–1967	012334		69
	160	Shoenfield, Joseph	1956–1960, n.d.	012342		80
		Solovay, Robert:				
	161		1955, 1965–1967	012360		103
	162	Discussion notes:	1964–1967, 1972	012385		141
	163		1968, 1970–1971, 1974, n.d.	012394		173
	164	Spector, Clifford	1957–1961	012413		207
	165	Stahl, Gerold	1968–1970	012435		256
		Miscellaneous "S":				
	166	E. Sarton to Laurence Shepley	1931–1974	012108.95		288
	167	Margaret Shields to Patrick Suppes	1947–1975	012341.5		339

Box	Folder		Date(s)	*Initial document*	*Reel*	*Initial frame*

I. Personal and Scientific Correspondence, 1929–1978 (cont.)

Box	Folder		Date(s)	*Initial document*	*Reel*	*Initial frame*
		Tait, William W.:				
3b	168		1960–1962	012463	10	383
	169	(includes undated notes)	1964–1965, 1973	012485		421
		Takeuti, Gaisi:				
	170		1958–1960	012514		473
	171		1962	012540		514
	172		1963–1965	012560		548
	173		1966	012585		586
	174	with Georg Kreisel	June–August 1966	012621		653
	175		1967	012634.5		686
	176		1970–1972	012657		762
	177		1973	012666		793
	178		1974	012675		845
	179	(includes undated notes)	1975–1976	012689.5		878
		Tamari, Dov:				
	180		1959–1961, 1966–1967	012696	11	3
	181	re Taman vs. Technicon Israel:	1968, 1971	012727		64
	182		1973, n.d.	012745		104
	183	Tarski, Alfred contains letter that became **1970c*	1931, 1942–1947, 1970, 1972	012760		131
		Tenenbaum, Stanley:				
	184		1964–1968, 1970–1975, n.d.	012788		189
	185	Discussion notes:	1966–1968, 1970–1971	012824		254
	186	Discussion notes:	1972–1974, n.d.	012844		319
	187	Tharp, Leslie	1966–1973, n.d.	012861		378

Box	Folder		Date(s)	*Initial document*	*Reel*	*Initial frame*
		I. Personal and Scientific Correspondence, 1929–1978 (cont.)				
	188	Miscellaneous "T"	1932–1977	012507	11	402
	189	Ulam, S. M.	1939–1973	012877		458
3b	189.50	Unger, Georg	1955–1966, n.d.	012888.3		496
		van Heijenoort, Jean:				
3c	190		1958, 1961–1962	012892		522
	191		March–October 1963	012905		539
	192		November–December 1963	012929		580
	193		January–June 1964	012937		602
	194		August–December 1964	012960		639
	195		1965	012977		677
	196		1966, 1975, n.d.	013002		710
	197	Veblen, Oswald	1933–1949	013024.5		769
	198	von Neumann, John [See also Series VI: Folder 33 under "M" for AMs draft of memorial letter]	1930–1939	013029		824
	199	Miscellaneous "V"	1939–1979	012888.5		896
	200	Wajsberg, N.	1929–1932	013058	12	3
		Wang, Hao:				
	201		1948–1949	013064		24
	202		1967	013069		35
	203		1968–1970	013078		71
	204	(re *Wang 1974*)	1971–1973	013094		105
	205	(re *Wang 1974*)	1974–April 1975	013132		178
	206	(re *Wang 1974*)	May–November 1975	013148		217
	207	Quotations from Gödel	December 1975	013161		244
	208	Discussion notes:	1971–1975	013170		304

Box	Folder		Date(s)	*Initial document*	*Reel*	*Initial frame*
		I. Personal and Scientific Correspondence, 1929–1978 (cont.)				
	209	(includes drafts of *Wang 1981*)	February–July 1976	013181	12	367
3c		Wang, Hao (cont.):				
	210		November 1976–February 1978	013191		415
	211	Miscellaneous "W"	1935–1978	013207.1		470
	212	Yasuhara, Mitsuru	1970–1972	013217		520
	213	Yourgrau, Wolfgang	1962–1976	013246		561
	214	Miscellaneous "Y"	1957–1966	013242.4		593
	215	Zermelo, Ernst	1931	013270		607
	216	Zuckerman, Martin	1967	013272		612
	217	Miscellaneous "Z"	1935–1973	013262		632
	218	Multiple Correspondents	1955–1970	010855.5		661
	219	Unidentified Correspondents [See also Series XIV: Folder 1]	1939–1971	013288		675
		II. Institutional, Commercial, and Incidental Correspondence				
4a	1	Announcements of academic and cultural events	1939–1974	020001	13	3
	2	Appreciation, Letters of	1958–1973	020018		32
		Autograph Requests:				
	3	A–M	1949–1976	020023		45
	4	N–W	1953–1976	020036		75
	5	Bibliographic Requests, Miscellaneous	1956–1975	020051		109
	6	Biographical Requests from Individuals	1969–1976	020066		138
		Biographical Requests from Institutions and Biographical Registers:				
	7		1946–1954	020075		166

Box	Folder		Date(s)	*Initial document*	Reel	*Initial frame*
		II. Institutional, Commercial, and Incidental Correspondence (cont.)				
		Biographical Requests from Institutions and Biographical Registers (cont.):				
4a	8		1960–1965	020100	13	230
	9		1966–1980, n.d.	020125		274
	10	Booksellers, Miscellaneous	1929–1960	020145		316
	11	Charitable Solicitations	1935–1975	020152.1		332
		Conference and Colloquium Announcements:				
	12		1929–1935, 1938–1939	020168		364
	13		1960–1964	020199		419
	14		1965–1967	020224		528
	15		1968	020242		565
	16		1969–1972	020257		607
	17		1973–1975, n.d.	020285		653
		Crank Correspondence:				
	18	A–G	1963–1977	020301		717
	19	H–O	1937–1977	020322		778
	20	R–Sh	1959–1977	020339		850
	21	Sk–Z, except W	1967–1976	020359		914
	22	W	1960–1976	020363		931
4b		Editorial Correspondence:				
	23	New York University Press (re *Gödel's Proof* by Nagel and Newman, 1958[f])	1957	020384	14	3
	24	New York University Press, undated notes		020402		30
	25	La Nuova Italia Editrice (re *1968*)	1968	020411		64

[f] *Nagel and Newman 1958.*

Box	Folder		Date(s)	*Initial document*	*Reel*	*Initial frame*
		II. Institutional, Commercial, and Incidental Correspondence (cont.)				
4b	26	Oliver and Boyd/Basic Books (re Meltzer's translation of *1931*)[g]	1963	020426	14	88
	27	Oliver and Boyd/Basic Books (re Meltzer's translation of *1931*)	1964–1966	020458		137
	28	Oliver and Boyd/Basic Books (re Meltzer's translation of *1931*)	1968–1969, n.d.	020493.1		207
	29	Princeton University Press (re *1940*)	1961–1970	020507		242
	30	Time, Inc. (re *Bergamini et alii 1963*)	1963	020514		276
	31	*Zentralblatt für Mathematik* (re reviews)	1930–1937	020523		310
	32	*Zentralblatt für Mathematik* (re Gödel/Arend Heyting Collaboration)	1931–1935	020582		412
	33	Miscellaneous, including collected works solicitations	1930–1978	020610		456
		Honors:				
	33.50	Einstein Award	1951	020638.5		562
	34	Honorary Degrees: Amherst College (1967), Cambridge University (1972–1973), Harvard University (1952)	1952, 1967, 1972–1973	020639		566
	35	Honorary Degrees: Princeton University	1975	020646		577
	36	Honorary Degrees: Rockefeller University	1972	020656		644
	37	Membership: American Academy of Arts and Sciences, American Philosophical Society, British Academy	1961–1972	020683		708
	38	Membership: Institut de France, Institut International des Sciences Théoriques	1947–1972	020707		748

[g]Note that the targets on the microfilm for folders 26–28 misstate the topic as *1962*.

Box	Folder		Date(s)	*Initial document*	*Reel*	*Initial frame*
		II. Institutional, Commercial, and Incidental Correspondence (cont.)				
4b	39	Membership: London Mathematical Society, National Academy of Sciences, Royal Society, Trinity Mathematical Society	1955–1968	020734	14	798
		National Medal of Science:				
	40		May–June 1975	020748		819
	41		September–October, 1975	020768		847
	42	Letters of Congratulation, 1975	1975	020795		895
4c		Institute for Advanced Study:				
	43	Financial memoranda	1933–1967	020807	15	3
		Internal memoranda (re applicants)				
	44		1956–1962	020835		46
	45		1970–1971	020856		80
	46		1972–1975	020881.5		117
	47	Miscellaneous Internal Correspondence	1934–1973	020802		157
	48	Inquiries from Students and Amateurs	1956–1976	020938		224
	49	Inquiries, Miscellaneous	1946–1977	020958		267
	50	Interview Requests	1964–1969	020962		275
		Invitations to Lecture:				
	51		1933–1939	020977		301
	52		1949,1957–1958	021002		344
	53		1960–1965	021018		383
	54		1966–1969	021035		413
	55		1970–1974, 1976	021053		445
	56	Invitations, Other Scholarly	1957–1966	021092		505
	57	Invitations, Social	1935–1975	021103		530

Box	Folder		Date(s)	*Initial document*	*Reel*	*Initial frame*
		II. Institutional, Commercial, and Incidental Correspondence (cont.)				
4c	58	Job Search Inquiries	1958–1967	021118	15	551
	59	Legal Correspondence, Miscellaneous	1932–1938	021129		566
	60	Leibniz Microfilming Project	1949–1953	021141		583
	61	Library Correspondence, Miscellaneous	1935–1973	021186		648
		Literary Solicitations:				
	62		1933–1953	021207		695
	63		1960, 1962-1963	021228		733
	64		1964–1967, 1969	021243		767
	65		1971–1972, 1975–1976	021272		826
		[Miscellaneous, see Unclassified]				
		[Moving Company, see Rental Agency and Moving Company]				
		Offprints, Requests for:				
	66		1938–1952	021281		844
	67		1954–1962	021301		881
	68		1963–1965, 1970, 1975–1976	021320		913
5a		Permission Requests:				
	69	Quotation	1975	021334	16	3
	70	Reprintings and Photocopies	1959–1973	021471.5		9
		Translations:				
	71		1953–1969	021358		79
	71.50		1973–1977	021369		139
	72	(re *1980*)	1978–1980	021383		191
	73	Photo Requests	1952–1981	021401		214

Box	Folder		Date(s)	*Initial document*	*Reel*	*Initial frame*
		II. Institutional, Commercial, and Incidental Correspondence (cont.)				
		Professional Associations:				
		National Academy of Sciences:				
5a	74		1956–1964	021424	16	259
	75		1966–1968	021441		325
	76		1970, 1972–1973, 1975	021451		346
	77	Miscellaneous	1937–1975	021466		397
		Publisher's Announcements:				
	78		1934–1959	021475		410
	79		1960–1976	021490		449
	80	Recommendation Requests, Miscellaneous	1952–1971	021508		511
	81	Rental Agency and Moving Company, Vienna	1937–1939	021530.05		555
	82	Souvenir Cards	1967	021531		580
	83	Subscriptions	1936–1972	021531.5		630
	84	Unclassified	1937–1976	021546		658

III. Topical Notebooks

"Allgemeine Bildung": A set of notebooks labelled "Allg. Bild.," numbered, and dated by Gödel as given below. The notebooks contain notes (apparently intended to contribute to Gödel's "general education") on political and cultural issues. There are entries on major "newsmakers" of the period, ranging from politicians to icons of popular culture. Headings are in German or English, notes are in Gabelsberger shorthand.

Box	Folder		Initial document	Reel	Initial frame
5b	1	No. I: 1953–May 1957, written both directions	030001	17	4

Box	**Folder**		*Initial document*	*Reel*	*Initial frame*
		III. Topical Notebooks (cont.)			
		"Allgemeine Bildung" (cont.)			
5b	1., cont.	No. II: July 1957–February 1958, written both directions. Backward direction is labelled "Diff. Geom. Engl. Vok., Org. ⟦v. Neum. System der Mengenl.⟧ Chemie II, 1937" and contains mathematical computations and work using symbolic chemistry notation.	030002	17	76
	2	No. 3: March 1958–June 1958	030003		144
		No. 4: July 1958–September 1958	030004		188
	3	No. 5: September 1958–December 1958	030005		231
		No. 6: December 1958–May 1959	030006		275
	4	No. 7: May 1959–August 1959, plus one loose sheet	030007		319
	5	No. 8: August 1959–December 1959	030008		364
	6	No. 9: December 1959–March 1960	030009		388
	7	No. 10: March 1960–July 1962	030010		412
	8	No. 11: July 1962–December 1965	030011		440
	9	No. 12: December 1965–?	030012		465
	10	No. 13: June 1967–September 1967	030013		501
	11	No. 14: September 1967–April 1972	030014		512
		No. 15: April 1972–May 1974	030015		588
	11.5	**American History**: One notebook, headings in English, notes in English and Gabelsberger shorthand.	030015.5		659

Box	Folder		Initial document	Reel	Initial frame
		III. Topical Notebooks (cont.)			
		"**Arbeitshefte**": Set of notebooks labelled "Arb. H." and numbered 1–16 by Gödel. Any additional labelling is given below in quotes. These "mathematical workbooks" contain theorems, definitions, proofs, and computations with headings in German.			
5c	12	Index to the "Arbeitshefte": AMs notes cross-referencing subject headings with notebooks and page numbers	030016	17	679
	13	No. 1: Written both directions, plus 8 loose pages	030017		689
	14	No. 2: "Kontinuum": Plus 8 loose pages	030018		770
	15	No. 3	030020		821
	16	No. 4: Written both directions, plus 7 loose pages	030021		833
	17	No. 5: Plus 2 loose pages	030023	18	4
	18	No. 6: Plus 5 loose pages	030024		55
	19	No. 7: Written both directions, plus 24 intercalated pages and 1 loose page	030025		89
	20	No. 8	030026		154
	21	No. 9: "Int. Math": Plus 9 intercalated pages	030027		173
	22	No. 10	030028		223
	23	No. 11	030029		248
	24	No. 12: Plus 8 intercalated pages	030030		281
	25	No. 13: Plus 2 intercalated pages	030031		333
	26	No. 14: with insert dated 1966	030032		389
	27	No. 15	030033		445
	28	No. 16: Plus 11 intercalated pages and 1 page inserted in back	030034		499

Box	Folder		Initial document	Reel	Initial frame
		III. Topical Notebooks (cont.)			
		Astronomy: (re: angular orientation of nebulae, etc.) Two notebooks without external labelling, contents as follows:			
5c	29	Front page headed: "Liste d. grossen Nebel" ("List of the Large Nebulae"), headings in German with tables of numeric figures, plus 8 loose pages	030035	18	554
	30	Unlabelled: Tables of numeric figures, plus 3 loose pages	030038?		596
		[See also Physics: "Physik 1935"]			
		[**Bibliographic Notes**: See Loose-leaf Notebook, and Physics: "Lit Physik"]			
		[**Chemistry**: See "Allgemeine Bildung": II, and School: Geography, Geology, Chemistry]			
5d		**History**: Notebooks numbered 1–9 by Gödel. Additional labelling appears below in quotes. Headings of historic dates, figures, and works in German, Latin, and English with notes in Gabelsberger shorthand.			
	31	Partial index to the History Notebooks: alphabetized references to entries in the notebooks, primarily names and dates	030040		620
	32	No. 1: "Lexicon, etc"	030041		668
	33	No. 2: "Lexicon, etc": all loose pages	030042		711
	34	Notes on Leopold von Ranke, *History of Popes* (filed by Gödel with History Notebook No. 2)	030044		787
	35	No. 3	030053		848
	36	No. 4	030054	19	4
	37	No. 5	030055		86
	38	No. 6	030056		148
	39	No. 7			
		Loose sheets	030058		210
		Notebook 7	030057		212
	40	No. 8	030059		275

Box	Folder		Initial document	Reel	Initial frame
		III. Topical Notebooks (cont.)			
		History (cont.)			
5d	41	No. 9: The title "Höhere Math." is stricken; the first four pages contain mathematical notes.	030060	19	319
		[See also American History]			
	42	**IAS Committee Memoranda, 1953–1967**: One notebook containing notes in Gabelsberger shorthand with English and German headings, and budget tables, plus 3 loose pages.	030061?		328
	43	**Language Practice: Italian**: One notebook containing transcriptions in Italian, plus three loose pages inserted in front. [For other language practice, see School: Latin Practice; and Vocabulary: English, Latin, Dutch.]	030065		355
		Logic and Foundations: Set of notebooks, each labelled "Logik und Grundlagen" and numbered by Gödel as given below. Gödel's pagination is continuous throughout the set as listed below. The notebooks contain work written in symbolic logic notation and entries in Gabelsberger shorthand.			
	44	Index to the Logic and Foundations Notebooks, 3 loose sheets	030066		406
	45	No. 1: pp. 1–54	030067		411
	46	No. 2: pp. 55–134	030069		460
	47	No. 3: pp. 135–206	030070		504
	48	No. 4: pp. 207–266	030071		545
	49	No. 5: pp. 267–343	030072		579
6a	50	No. 6: pp. 344–440	030073		626
	51	**Loose-leaf Notebook**: Mostly bibliographic notes in English, German, and Gabelsberger shorthand.	030074		683

Box	Folder		Initial document	Reel	Initial frame
		III. Topical Notebooks (cont.)			
		Mathematics: Various notebooks containing mathematical work, labelled by Gödel as quoted below. In the absence of designation by Gödel, some physical description of the notebook is offered.			
6a	51.5	"Amerika 1933/34 (Grundl[agen])": Mathematical computation with notes in Gabelsberger shorthand, written both directions	030074.5	19	755
	52	"Auswahlaxiom, Mengenlehre II, Amerika 1935": Mathematical computation (re the axiom of choice) with notes in German and Gabelsberger shorthand, written both directions	030075		800
	53	"Amerika II, 1935": Mathematical computation, notes in Gabelsberger shorthand, written both directions, plus two loose pages	030076	20	4
	54	"Beweis d. Gültigkeit d. int. Ax.": Notes filed by Gödel with "Amerika 1935" notebooks, notes in English and Gabelsberger shorthand, with mathematical computation, total 21 loose sheets	030077		59
	55	"Lekt. Math. I": Notes on Institute for Advanced Study lectures, in Gabelsberger shorthand with mathematical computation, plus 3 intercalated pages, and 3 pages inserted in back	030078		84
	56	"Nur Weyl, Geometrie der Zahlen": Lecture notes on Hermann Weyl, Institute for Advanced Study lectures [Autumn 1940], theorems and definitions, in mathematical notation, notes in Gabelsberger shorthand, written both directions	030080		136
	57	"Siegel, Fubini, and div. bekannte Theoreme (Sommer 1941?)": Notes on Institute for Advanced Study lectures, in mathematical notation; lectures given by Carl Ludwig Siegel and Guido Fubini	030081		197

Box	Folder		Initial document	Reel	Initial frame
		III. Topical Notebooks (cont.)			
		Mathematics (cont.)			
6a	58	"Altes Excerpten Heft I (1931–. . .)": Notes on mathematical literature, headings in German and English, notes in Gabelsberger shorthand, plus 3 loose pages and 2 inserted cards	030079	20	236
	59	"Aflenz 1936 Analysis, Physik": Mathematical computation with notes in Gabelsberger shorthand (taken while in Aflenz, Austria?), written both directions, plus 4 loose pages	030082		315
	60	Undesignated: Mathematical computation with notes in Gabelsberger shorthand, written both directions	030083		382
	61	Undesignated: Back cover missing: Mathematical computation	030084		436
	62	Undesignated: Both covers missing, binding disintegrated: Mathematical computation with notes in Gabelsberger shorthand	030085		499
	62.5	Undesignated: Both covers missing, binding intact: Mathematical computation, written both directions	030085.5		539

[See also Arbeitshefte; History: No. 9; Philosophy: "[Heinrich] Gomperz;" and School: Geography, Latin Practice, Mathematics]

Box	Folder		Initial document	Reel	Initial frame
	III. Topical Notebooks (cont.)				
		Philosophy: Set of notebooks labelled and dated by Gödel as quoted below (names of months have been translated from abbreviations), containing notes on philosophical "maxims" in Gabelsberger shorthand with headings in German or English.			
6b	63	"Philosophie I soll heissen Max 0"	030086	20	563
	64	"Max I Zeiteinteilung"	030087		607
	65	"Max II Zeiteinteilung": Plus 21 loose pages	030088		651
	66	"Max III": Plus 5 inserted pages	030089		729
	67	"Max IV" (May 1941–April 1942)	030090		812
		"Max V" (May 1942–?)	030091		883
	68	"Max VI" (?–July 1942)	030092	21	4
		"Max VII" (15 July 1942– 10 September 1942)	030093		54
	69	"Max VIII" (15 September 1942– 18 November 1942)	030094		105
		"Max IX" (18 November 1942– 11 March 1943)	030095		167
	70	"Max X" (12 March 1943– 27 January 1944)	030096		218
		"Max XI" (28 January 1944– 14 November 1944)	030097		268
	71	"Max XII" (15 November 1944– 5 June 1945)	030098		349
		[Vol. XIII missing, see note in "Phil. XIV"]			
	72	"Phil. XIV" (July 1946–May 1955)	030099		413
		"Max XV, Letztes" (May 1955–)	030100		481
	72.5	"Gomperz, Geschichte der europ. Phil., Winter 1925": Philosophy course notes (in Gabelsberger shorthand with German headings) on the history of European philosophy, shorthand with German headings) on the history of European philosophy, written both directions	030100.4		503
		Backward direction: Mathematical notes			536

Box	Folder		*Initial document*	*Reel*	*Initial frame*
		III. Topical Notebooks (cont.)			
		Philosophy (cont.)			
6b	72.6	"Gomperz, 1926": Philosophy course notes, written in both directions	030100.5	21	577
		Backward direction: Mathematical notes			604
		Physics: Various notebooks containing notes on physics, labelled by Gödel as quoted below. Where left undesignated, notebooks are distinguished by content.			
	73	"Kottler, Sommer 1926": Kinetic theory of matter, course notes in mathematical notation (From the lectures of Friedrich Kottler?)	030101		617
	74	"Physik 1935 Stat. Mech., Optik": Notes on statistical mechanics and optics in mathematical notation with diagrams and headings in German	030102		659
	75	"Physik 1935": Mathematical computation with diagrams and headings in German, written both directions, plus 3 loose pages	030103		727
	76	"Hobart Physik 1935": Mathematical computation with headings in German, written both directions	030104		811
	77	"Physik 1935": Includes section titled "Astronomische Zahlen," mathematical computation with headings in German, written both directions, plus one page intercalated	030105		888
	78	"Physik Quantenmech. I": Mathematical computation with notes in Gabelsberger shorthand and headings in German and English, written both directions	030106		967
		"Physik Quantenmech. II" [1935]: Mathematical computation with notes in Gabelsberger shorthand	010307		1024

Box	Folder		Initial document	Reel	Initial frame
		III. Topical Notebooks (cont.)			
		Physics (cont.)			
6c	79	"Physik Excerpte f. d. Prufüng": Loose notes filed with physics notebooks by Gödel (examination excerpts), mathematical computation	030108	22	3
	79.5	Undesignated (re rotating universes): Mathematical computation and extensive notes in English, plus one loose sheet	030109.1		36
	79.6	Undesignated (re rotating universes): Mathematical computation, plus one loose page	030109.6		108
	79.7	"Physik (rein)" (re relativity): Mathematical computation with headings in German, written in both directions, plus one loose page	030109.7		154
	80	"Lit Physik": Notebook on literature in Physics and accompanying loose bibliographic notes, in both German and Gabelsberger shorthand	030110		207
	81	"**Protokoll**": One notebook labelled "Prot." by Gödel. Contains notes on conversations and lectures, etc., in Gabelsberger shorthand with German and English headings.	030114		266
		Results on Foundations: A set of notebooks, each labelled "Resultate Grundlagen" and numbered by Gödel as listed below. Gödel's pagination throughout the set is continuous, as shown. Notebooks contain work written in symbolic logic notation and notes in Gabelsberger shorthand with German headings.			
	82	Index to the Results on Foundations Notebooks: AMs notes cross-referencing subject headings to notebooks and page numbers	030115		308

Box	Folder		Initial document	Reel	Initial frame
		III. Topical Notebooks (cont.)			
		Results on Foundations (cont.)			
6c	83	I: pp. 1–52	030116	22	312
	84	II: pp. 53–153	030117		344
	85	III: pp. 154–267	030118		398
	86	IV: pp. 268–368	030119		459
		School: Set of notebooks from Gödel's schooling, ranging from grade school to high school. The notebooks are arranged below with the earliest three listed first, and thereafter alphabetized by subject.			
	87	Arithmetic (Evangelische Bürgerschule in Brünn, 1912/1913): Practice writing numerals and sums	030120		526
	88	Penmanship (Evangelische Bürgerschule in Brünn, 1912/1913): Practice writing alphabet	030121		538
	89	Composition (Evangelische Bürgerschule in Brünn, 1914/1915): Essays in German, plus one loose page	030123		574
	90	Composition (Staatsrealgymnasium in Brünn, 1920/1921): Essays in German	030124		588
		Drafting: Three notebooks, undesignated, distinguishable by physical characteristics			
	91	Plain white, covers missing: contains practice constructing geometrical figures	030124.1		615
	92	Plain white, folded open: contains practice constructing geometrical figures	030124.2		670
	93	Plain white, staples removed: contains practice constructing geometrical figures, plus three loose sheets	030125		743

Box	Folder		Initial document	Reel	Initial frame
		III. Topical Notebooks (cont.)			
		School (cont.)			
6c	94	Geography, Geology, Chemistry: One notebook labelled "Mittelschule Geog." and then "org. Chemie (1937)," contains notes in German and Gabelsberger shorthand on geography and geology, and notes in German on chemistry. Written both directions, backward direction includes some mathematical calculations	030126	22	783
	95	Latin Practice (Staatsrealgymnasium in Brünn, 1920/1921): Practice essays in Latin, Algebra practice pencilled in back	030126.5		810
	96	Logic: Labelled inside cover "Übungsheft Log." Contains exercises in symbolic logic notation	030127		830
		Mathematics: Set of notebooks covering Gödel's early education in mathematics, grouped chronologically, with undated material at end.			
	97	Workbook(Staatsrealgymnasium in Brünn,1918/1919): Labelled "Mathematische Schularbeiten," contains mathematical schoolwork with teacher's comments in German	030128		878
	98	Geometry (Staatsrealgymnasium in Brünn, 1920/1921): Labelled "Geom. Hausübungen," contains geometry "homework," including the construction of figures	030128.1		892
	99	Workbook (Staatsrealgymnasium in Brünn, 1923/1924): Labelled "Mathematische Schularbeiten," contains schoolwork with algebraic and trigonometric functions, and calculus, with teacher's comments in German	030128.2?	23	4

Box	Folder		*Initial document*	*Reel*	*Initial frame*
		III. Topical Notebooks (cont.)			
		School (cont.)			
6c	100	Calculus Workbook: Unlabelled, contains pencilled practice in calculus	030128.3	23	44
	101	Miscellaneous: Labelled "Gödel Kurt," contains mathematical computation with notes in Gabelsberger shorthand, written both directions	030128.4?		82
	102	Miscellaneous: Inside cover labelled "Algebra," contains mathematical computation with notes in Gabelsberger shorthand, written both directions	030128.5		102
	103	Mathematics/German Culture: Unlabelled, contains pencilled practice in mathematical computation and notes on German language, history, and culture in German and Gabelsberger shorthand	030128.6?		153
	104	Natural History: One notebook labelled "Naturgeschichts Heft.," contains notes in German with hand-drawn diagrams	030128.7?		211
		Physics: Two notebooks from Gödel's early schooling in physics, labelled by Gödel as quoted below			
	105	"Früh Phy. 1924": From Staatsrealgymnasium in Brünn, contains diagrams, notes in German, and mathematical computation, written both directions	030128.8		237
	106	"Physik Heft": From Staatsrealgymnasium in Brünn [1921/1922], contains diagrams, notes in German, and mathematical computation, plus 10 loose pages	0301028.9?		296

Box	Folder		*Initial document*	*Reel*	*Initial frame*
		III. Topical Notebooks (cont.)			
		Theology: A set of two notebooks, labelled and numbered by Gödel as below. Numbering suggests a third notebook, #2, now missing.			
7a	107	"Theologie 1 Nur Vorlesungen": Headings in German, French and Latin, with notes in Gabelsberger shorthand, plus 4 intercalated pages	030129	23	343
	108	"Theol. 3": Headings in German, notes in Gabelsberger shorthand, including tables of content for the Bible	030130		389
		Vocabulary: Set of 16 notebooks containing vocabulary lists and language practice, four of which belonged to Gödel's brother, Rudolf. Arranged alphabetically by language, with Rudolf's notebooks at the end. Labelled as quoted below.			
		[Dutch: See Latin: "Dutch"]			
		English: Seven notebooks total			
	109	"English": English translated into Gabelsberger shorthand	030132		421
		Unlabelled: English translated into German or Gabelsberger shorthand, written both directions	030131		471
	110	"English": English translated into Gabelsberger shorthand	030133		536
		"English": English translated into Gabelsberger shorthand or French, transcription of Charles Dickens in English at back	030135		588
		"Englische V, alt": English translated into Gabelsberger shorthand, includes table of vowel pronunciation, written both directions	030134		612
	111	Unlabelled: English translated into Gabelsberger shorthand, some French as well, plus 7 loose pages.	030136		640

Box	Folder		*Initial document*	*Reel*	*Initial frame*
		III. Topical Notebooks (cont.)			
		Vocabulary (cont.)			
7a	112	"English": English translated into Gabelsberger shorthand or German, includes practice sentences in English, written in both directions, plus 4 loose pages	030138	23	692
		Latin: Five undated notebooks of Latin vocabulary and language practice, designated as quoted below:			
	113	Unlabelled: Latin translated into Gabelsberger shorthand	030140		726
		"8": Latin translated into Gabelsberger shorthand	030141		800
	114	"Lat. Vok.": Latin translated into Gabelsberger shorthand, Greek translated into Latin or German	030142		852
		Unlabelled notepad: Latin translated into Gabelsberger shorthand, plus 9 loose pages	030143		855
	115	"Dutch": Written both directions	030146		
		Forward direction: Dutch translated into Gabelsberger shorthand			883
		Backward direction: Latin translated into Gabelsberger shorthand			899
		Rudolf Gödel: Four school notebooks belonging to Kurt Gödel's brother, Rudolf, from Staatsrealgymnasium in Brünn			
	116	"Deutsch Heft I" (1918/1919): Headings in German, notes in shorthand	030147		967
		"Deutsch Heft II" (1918/1919): Headings in German, notes in shorthand	030148		988

Box	Folder		Initial document	Reel	Initial frame
		III. Topical Notebooks (cont.)			
		Vocabulary (cont.)			
7a	117	"Deutsch Heft III" (1919/1920): Headings in German, notes in shorthand	030149	23	1012
		"Deutsch Heft IV" (1919/1920): Headings in German, notes in shorthand, plus two intercalated pages	030150		1035

IV. Drafts and Offprints

Box	Folder		Date(s)	Initial document	Reel	Initial frame
		***1929*, Über die Vollständigkeit des Logikkalküls" (Doctoral Dissertation):**				
7a	1	AMs Notebook (in Gabelsberger shorthand) labelled "Diss. unrein," written both directions	[1929?]	040001	24	4
	2	TMs [carbon] (in German) labelled "Dissertation," with autograph corrections, 34 pp.	[1929?]	040002		43
	3	TMs [photocopy] of final typescript deposited at Universität Wien	[1929]	not filmed		
		***1930*, "Die Vollständigkeit der Axiome des logischen Funktionenkalküls":**				
	4	TMs (in German), labelled "Vollständigkeit d. Axiome" with autograph corrections, 20 pp.	[1930?]	040004		81
		Printed page proof with autograph corrections	[1930?]	040005		101
		Offprint	1930	040006		113
		Galley with autograph annotations	[1930]	040007		126
		****1930c*, Lecture, "Über die Vollständigkeit des Funktionenkalküls" (delivered on dissertation, September 6, 1930?):**				
	5	TMs (in German) with autograph corrections, pp. 10	[1930]	040009		144
		AC describing contents of original file	n.d.			143

Box	Folder		Date(s)	Initial document	Reel	Initial frame
		IV. Drafts and Offprints (cont.)				
		***1930c*, (cont.)**				
7a	5 (cont.)	TMs (in German) with autograph corrections, p. 6, back labelled "Vortrag über Vollständigkeit Fnkt. Kalk"	[1930?]	040009	24	154
		***1931a*, "Diskussion zur Grundlegung der Mathematik" (a prepared script of the proceedings on September 7, at the Königsberg conference of 1930, published in *Erkenntnis 2* with a "Nachtrag" by Gödel):**				
	6	Printed Matter: from publisher, "Beispiele für Umfang and Art der Autoreferate"	ca. 1930	040009.5		235
		TMs [carbon] of discussion (in German) with autograph corrections, p. 23	ca. 1930	040010		236
		TMs of discussion, with autograph corrections, pp. 21 and 23, back labelled "Tagungs Königsberg Diskuss."	ca. 1930	040011		237
		TMs of Nachtrag ("Supplement"), with autograph corrections, 3 pp., back labelled "Erkenntnis"	[1931?]	040012		240
		TMs of Nachtrag with autograph corrections, p. 3		040012		244
		Offprint with autograph annotations	[1931]	040013		158
		Copy of Erkenntnis 2 with autograph annotations	1931	040013.5		178
		Undecidability Results (early drafts of *1931*):				
	7	AMs (in Gabelsberger shorthand) in 2 Notebooks, one inserted in the other, labelled "Unentsch. unrein," written both directions	[1930?]	040014		247
	8	AMs Notebook (in Gabelsberger shorthand), labelled "Unentsch. unrein," written both directions	[1930?]	040015		292

Box	Folder		Date(s)	Initial document	Reel	Initial frame
		IV. Drafts and Offprints (cont.)				
		***1930b*, "Einige metamathematische Resultate über Entscheidungsdefinitheit"**				
7a	9	TMs (in German) with autograph corrections in two hands, 3 pp.	1930	040016	24	369
		Offprint	1930	040017		372
		Galley		040018		373
		Project for "Ergebnisse", A proposed joint book with Arend Heyting (see Correspondence Series I: Heyting, Arend):				
	10	Drafts by Gödel:				
		AMs Notebook (in Gabelsberger shorthand)	[1930–1934]	040019		378
	11	Drafts by Arend Heyting:				
		TMs (in German) with autograph corrections in two hands. Sections labelled as follows: III: 1–7, V: 1–26, VII: 1–4, Unlabelled: 1–4, VI: 1–2, Unlabelled: 1–19, VII: 1–5, References: 1–4, 1–4, 1–4	[1930–1934]	040019.5		482
		Autograph Material: Envelope	n.d.	040019.5		481
7b		***1931*, "Über formal unentscheidbare Sätze der Principia Mathematica and verwandter Systeme I":**				
	12	TMsS (in German) with autograph corrections, front page labelled "... formal unentsch. Sätze," 44 pp.	[1930?]	040020		579
	13	Printed page proof with autograph corrections	[1931]	040021		627
		Offprint with autograph annotations	1931	040024		653
		Galley with autograph annotations		040021		682
		Untitled Lecture: Hans Hahn's logic seminar, 1931/1932:				
	14	Autograph Material: Envelope	[1932]	040025		738
		AMs (in Gabelsberger shorthand and German) of notes (for lecture draft?)	n.d.	040025		739
		Autograph Material: Envelope		040026		846

Box	Folder		Date(s)	*Initial document*	*Reel*	*Initial frame*

IV. Drafts and Offprints (cont.)

Untitled Lecture: Hans Hahn's logic seminar (cont.)

Box	Folder		Date(s)	*Initial document*	*Reel*	*Initial frame*
7b	14 (cont.)	AMs (in Gabelsberger shorthand and German), first page headed "Vollständigkeit des Funkt. Kalküls"	[1931–1932]	040026	24	847
	15	AMs of computational notes on Arend Heyting's propositional calculus for Hans Hahn's seminar, 14 pp., with envelope	1931–1932	040028		884
		AMs of computational notes on Arend Heyting's propositional calculus, miscellaneous pages	n.d.	040029		913
	16	Autograph Material: Folder	[1932]			
		Prepared notes for Hahn's seminar, in two versions:				
		TMs [carbon] (in German) with autograph corrections, 36 pp.	1931–1932	040030	25	4
		TMs [carbon] (in German) with autograph corrections, 57 pp.	1932	040031		63
		TMs [carbon] (in German) of notes from the proceedings of Verein Ernst Mach	n.d.	040031.5		129

Reviews for Zentralblatt für Mathematik: Autograph Drafts (in German):

Box	Folder		Date(s)	*Initial document*	*Reel*	*Initial frame*
	17	*1931b*: Review of *Ludwig Neder 1931*, AMsS, 2 pp.	[1931]	040055		192
		1931c: Review of *David Hilbert 1931*, AMsS, 2 pp.	1931	040032		153
		1931f: Review of *Helmut Hasse and Heinrich Scholz 1928*, AMs, fragmentary draft, 1 p.	[1931]	040032.5		155
		1932d: Review of *Thoralf Skolem 1931*, AMsS, 2 pp.	[1932]	040033		157
		1932e: Review of *Rudolf Carnap 1931*, AMsS, 3 pp.	[1932]	040056		195
		1932f: Review of *Arend Heyting 1931*, AMsS, 2 pp.	[1932]	040057		198

Box	Folder		Date(s)	*Initial document*	Reel	*Initial frame*
		IV. Drafts and Offprints (cont.)				
7b	17	**Reviews for Zentralblatt für Mathematik: (cont.) Autograph Drafts (in German):**				
		1932g: Review of *John von Neumann 1931*, AMsS, 2 pp.	[1932]	040058	25	200
		1932h: Review of *Fritz Klein 1931*, AMsS, 2 pp.	[1932]	040034		159
		1932i: Review of *Franz G. Hoensbroech 1931*, AMsS, 1 p.	1932	040035		161
		1932j: Review of *Fritz Klein 1932*, AMsS, 1 p.	1932	040036		162
		1932k: Review of *Alonzo Church 1932*, AMsS, 2 pp.	1932	040037		163
		1932l: Review of *Lászlo Kalmár 1932*, AMsS, 1 p.	1932	040038		165
		1932m: Review of *Edward V. Huntington 1932*, AMsS, 1 p.	1932	040039		166
		1932n: Review of *Thoralf Skolem 1932*, AMsS, 2 pp.	1932	040059		202
		1933j: Review of *Stefan Kaczmarz 1932*, AMsS, 1 p.	[1933]	040040		167
		1933k: Review of *Clarence I. Lewis 1932*, AMsS, 1 p.	[1933]	040041		168
		1933l: Review of *Lászlo Kalmár 1933*, AMsS, 2 pp.	1933	040042		170
		⟦Not in printed list but on film:⟧ *1934a*: Review of *Skolem 1933*, AMsS, 2 pp.	1934	040043		172
		1934b: Review of *Willard V. Quine 1933*, AMsS, 1 p.	[1934]	040044		174
		1934c: Review of *Thoralf Skolem 1933a*, AMsS, 2 pp.	[1934]	040045		175
		1934d: Review of *Kien-Kwong Chen 1933*, AMsS, 1 p.	[1934]	040046		177
		1934e: Review of *Alonzo Church 1933*, AMsS, 2 pp.	[1934]	040047		178

Box	Folder		Date(s)	*Initial document*	*Reel*	*Initial frame*
		IV. Drafts and Offprints (cont.)				
7b	17	**Reviews for Zentralblatt für Mathematik: (cont.) Autograph Drafts (in German):**				
		1934f: Review of *Bernard Notcutt 1934*, AMsS 1 pp.	1934	040048	25	180
		1935: Review of *Thoralf Skolem 1934*, AMsS, 1 pp.	[1935]	040049		181
		1935a: Review of *Edward V. Huntington 1934*, AMsS, 2 pp.	[1935]	040050		182
		1935b: Review of *Rudolf Carnap 1934*, AMsS, 2 pp.	[1935]	040051		184
		1935c: Review of *Lászlo Kalmár 1934*, AMsS, 1 p.	[1935]	040052		186
		1936b: Review of *Alonzo Church 1935*, AMs (in Gabelsberger shorthand), labelled "Resension II Church System", 2 pp., and AMsS (in German), 2 pp.	[1936]	040053		187
	18	**Reviews for Zentralblatt für Mathematik: Printed Matter, 1931–1936:**				
		1931b: Review of *Ludwig Neder 1931*, Galley	1931	040060		205
		1931a: Review of *David Hilbert 1931*, Galley	1931	040061		206
		1932d: Review of *Thoralf Skolem 1931*, Galley	1932	040062		207
		1932e: Review of *Rudolf Carnap 1931*, Galley	1932	040063		208
		1932f: Review of *Arend Heyting 1931*, Galley with autograph corrections	1932	040064		209
		1932g: Review of *John von Neumann 1931*, Galley	1932	040064		209
		1932h: Review of *Fritz Klein 1931*, Galley	1932	040065		210
		1932l: Review of *Franz G. Hoensbroech 1931*, Galley	1932	040066		211
		1932j: Review of *Fritz Klein 1932*, Galley ⟦with autograph annotations⟧	1932	040067		212

Box	Folder		Date(s)	*Initial document*	Reel	*Initial frame*
		IV. Drafts and Offprints (cont.)				
7b	18	**Reviews for Zentralblatt für Mathematik: (cont.) Printed Matter, 1931–1936:**				
		1932k: Review of *Alonzo Church 1932*, Galley	1932	040068	25	213
		1932l: Review of *Lászlo Kalmár 1932*, Galley	1932	040068		213
		1932m: Review of *Edward V. Huntington 1932*, Galley	1932	040069		214
		1932n: Review of *Thoralf Skolem 1932*, Galley	1932	040070		215
		1933j: Review of *Stefan Kaczmarz 1932*, Galley	1933	040072		217
		1933k: Review of *Clarence I. Lewis 1932*, Galley	1933	040073		218
		1933l: Review of *Lászlo Kalmár 1933*, Galley	1933	040074		219
		1934a: Review of *Thoralf Skolem 1933*, Galley	1934	040075		220
		1934b: Review of *Willard V. Quine 1933*, Galley	1934	040075		220
		1934c: Review of *Thoralf Skolem 1933a*, Galley	1934	040076		221
		1934f: Review of *Bernard Notcutt 1934*, Galley	1934	040077		222
		1935: Review of *Thoralf Skolem 1934*, 2 Galleys	1935	040079		224
		1935a: Review of *Edward V. Huntington 1934*, 2 Galleys	1935	040079		224
		1935b: Review of *Rudolf Carnap 1934*, Galley	1935	040081		227
		1935c: Review of *Lászlo Kalmár 1934*, Galley	1935	040081		227
		1936b: Review of *Alonzo Church 1935*, Galley	1936	040082		228

Box	Folder		Date(s)	*Initial document*	*Reel*	*Initial frame*
		IV. Drafts and Offprints (cont.)				
		***1932*, "Zum intuitionistischen Aussagenkalküls":**				
7b	19	TMsS (in German) with autograph corrections, 2 pp.	1932	040083	25	230
		Printed Matter: Offprint	1932	040084		232
		Galley		040085		233
		***1932a*, "Über einen Spezialfall des Entscheidungsproblem der theoretischen Logik":**				
	20	TMs (in German), 3 pp.	[1932]	040086		235
		Printed Matter: Offprint	1932	040087		238
		***1932b*, "Über Vollständigkeit and Widerspruchsfreiheit":**				
	21	TMs (in German) with autograph label: "Menger Kolloquium" (prepared for Karl Menger's colloquium), 3 pp.	[1932]	040088		241
		***1932c*, "Eine Eigenschaft der Realisierungen des Aussagenkalküls":**				
	22	TMs (in German) with autograph insertions and label: "Realisierung Aussagenkalk d. (Menger Koll.)" (prepared for Karl Menger's colloquium), 1 p.	[1932]	040089		246
		***1933e*, "Zur intuitionistischen Arithmetik and Zahlentheorie":**				
	23	TMs (in German) with autograph label "Zur int. Arithm... Zahlentheorie", pp. 10	[1933]	040090		249
		***1933i*, "Zum Entscheidungsproblem des logischen Funktionenkalküls":**				
	24	TMsS (in German) with autograph corrections, 16 pp.	[1933]	040092		263
		AMs, single page with diagram	1933	040091		261
		Printed page proof with autograph corrections	[1933]	040093		281
		Offprint	1933	040096		287
		Galley with autograph annotations	[1933]	040094		295

Box	Folder		Date(s)	*Initial document*	Reel	*Initial frame*
		IV. Drafts and Offprints (cont.)				
		Untitled Lecture Notes: Vienna, Summer 1933				
7b	25	Autograph Material: Envelope	1933	040097	25	319
		AMs (in German), loose pages numbered 1–20, 1–6, plus 6 unnumbered pages and one bound notebook	[1933]	040104		371
		***1933o*, "The Present Situation in the Foundations of Mathematics" (Lecture given to the American Mathematical Society, 30 December 1933):**				
	26	AMs draft fragments (in Gabelsberger shorthand and English), 6 pp., each numbered 2, 3, or 4	n.d.	040105		387
		AMs draft (in English), pp. 11–24	n.d.	040109		393
		AMs draft fragments (in English), p. 3, 7, 10, 19.1	n.d.	040110		407
		AMs (in English and German), labelled "Vortrag US" and "Suppl. I, II, III, IV", includes loose pages for insertion	n.d.[h]	040111		411
		AMs draft fragments (in English), pp. 1–3, 5–6, 8–9	n.d.	040112		419
		AMs draft (in English), pp. 1–22 plus one unnumbered	n.d.	040113		426
	27	AMs of final text (in English), labelled "Vort. Cambridge", 31 pp.	[1933]	040114		453
		***1934*, "On Undecidable Propositions of Formal Mathematical Systems" (Lectures in Princeton, February–May, 1934):**				
	28	AMs draft fragments (in English), pp. 1-9, 9-10, 10-12, 18-25	[1934]	040115		487
		AMs insertions, pp. 26.1, 31, 32, 34, 35	[1934]	040116		511
		AMs, labelled "Vorles. Spring 34 Princeton", pp. 1–3, 3–7, 6–16	1934	040117		516

[h] Shorthand in label says "Höchst wahrscheinlich 1934 oder 33".

Box	Folder		Date(s)	*Initial document*	Reel	*Initial frame*
		IV. Drafts and Offprints (cont.)				
		***1934*, "On Undecidable Propositions of Formal Mathematical Systems" (Lectures in Princeton, February–May, 1934) (cont.):**				
7b		AMs draft fragments, pp. 1–5, 9–10, N1, N2, plus 5 unnumbered	[1934]	040118	25	537
		AMs, labelled "Erganzung Vorl. 34," 5 pp.	1934	040119		555
		AMs draft fragments, pp. 1–3, 5–8, 9.1, 9.2, 10–20	[1934]	040120		562
		AMs insertions, 5 pages, first 2 labelled "I", next 3, "II"	[1934]	040121		582
	29	Notes to Gödel's lectures, prepared by S. C. Kleene and J. B. Rosser: TMs with Gödel's and others' autograph corrections, 30 pp.	1934	040122		590
		"The Existence of Undecidable Propositions in Any Formal System Containing Arithmetic" (Lecture given to the Philosophical Society of New York University, April 18, 1934):				
	30	AMs (in English), labelled "Vortr. New York", 23 pp.	[1934]	040123		623
		TMs, Abstract of 1931 as handed out to attendees of the lecture	n.d.	040123		649
		Untitled Lecture of Summer 1935:				
	31	Autograph Material: Envelope, labelled "Vorlesung, Sommer 1935"	1935	040125		656
		AMs (in German), 27 pp.	[1935]	040126		657
		Consistency Proof for the Axiom of Choice:				
	32	AMs Notebook (in Gabelsberger shorthand and English), written both directions: front labelled "Amerika 1935" (English outline of consistency proof), back labelled "Auswahl-axiom/Mengenlehre I (Aflenz)" (shorthand draft)	1935	040131		700

Box	Folder		Date(s)	*Initial document*	*Reel*	*Initial frame*
		IV. Drafts and Offprints (cont.)				
		***1936a*, "Über die Länge von Beweisen":**				
7b	33	AMs (in Gabelsberger shorthand) labelled "Note für Menger Koll." (notes for Karl Menger's collo-, quium), 1 p.	[1936]	040132	25	759
		Galley	[1936]	040133		761
		Offprint	1936	040134		763
		Untitled Lectures on Set Theory:				
	34	AMs Notebook (in German and Gabelsberger shorthand), labelled "Vorl. Max & Vorlesung (elem) 1 (Ax. Meng. Lehre)"	n.d.	040135		768
	35	AMs Notebook (in Gabelsberger shorthand), labelled "Vorlesung (elem) 2 (Ax. Meng. L.")"	n.d.	040137		812
		Untitled Lecture in Vienna, Summer 1937 (and Menger Colloquium talks?):				
	36	Autograph Material: Envelope	1937	040139		837
		AMs Notebook (in Gabelsberger shorthand and German), labelled "Eigene Vorles. So. 1937", and 14 pp. loose	1937	040140		840
		****1938a* Untitled Lecture at Zilsel's, 29 January 1938:**				
7c	37	Autograph Material: Envelope, labelled "Vortrag bei Zilsel"	1938	040144	26	3
		AMs (in Gabelsberger shorthand), labelled "Reinschrift" 17 pp. plus 23 fragmentary notes	n.d.	040145		4
		Lectures on the Consistency of the Continuum Hypothesis (Princeton, Autumn 1938):				
	38	Talk given to Princeton Mathematics Club: AMs Notebook (in English), labelled "B Club Talk, 1938, Herbst, Princeton (Continuum)"	1938	040149		69
	39	Lectures 1–3: AMs Notebook (in English), labelled "Vorlesung 1–3 Kontin. Princeton 1938"	1938	040150		123

Box	Folder		Date(s)	*Initial document*	Reel	*Initial frame*
		IV. Drafts and Offprints (cont.)				
		Lectures on the Consistency of the Continuum Hypothesis (Princeton, Autumn 1938): (cont.)				
7c	40	Lecture 4: AMs Notebook, labelled "Princeton 1938 Vorlesung 4"	1938	040151	26	206
	41	Lecture 5: AMs Notebook, labelled "Princeton 1938 Vorlesung 5"	1938	040152		269
	42	Lecture 6: AMs Notebook, labelled "Princeton 1938 Vorlesung 6"	1938	040153		313
	43	Lecture 7: AMs Notebook, labelled "Princeton 1938 Vorlesung 7"	1938	040154		393
	44	Fragments: AMs, miscellaneous insert sheets	[1938]	040155		448
	45	Miscellaneous: AMs loose sheets, and 1 sheet TMs entitled "The Consistency of Cantor's Continuum Hypothesis and of the Axiom of Choice," with autograph corrections	n.d.	040165		533
		***1938*, "The Consistency of the Axiom of Choice and of the Generalized Continuum-hypothesis":**				
	46	AMsS, 4 pp.	n.d.	040174		583
		2 TMss [carbon], 3 pp.	n.d.	040175		588
		2 Offprints with autograph corrections	1938	040177		594
		"The Consistency of the Generalized Continuum-hypothesis" (Lecture to the American Mathematical Society, 28 December 1938):				
	47	AMs Notebook (in English), labelled "Vortrag Math Soc. Dec. 1938 Continuum"	1938	040179		600
	48	AMs Notebook, labelled "Entw. vortr. Math Soc. Dec. 1938	1938	040180		629
	49	Autograph Material: Envelope	1938	040183		718
		2 AMss, loose drafts, pp. 1–19 and 1–15, plus 9 unnumbered sheets	[1938]	040181		666

Box	Folder		Date(s)	*Initial document*	*Reel*	*Initial frame*
		IV. Drafts and Offprints (cont.)				
		***1939*, "The consistency of the generalized continuum hypothesis" (Abstract of the AMS lecture on 28 December 1938):**				
7c	50	AMsS, 2 pp.	n.d.	040186	26	735
		TMs [carbon], 1 p.	n.d.	040187		737
		***1939a*, "Consistency proof for the generalized continuum hypothesis"**				
	51	AMsS, pp. 1–13	n.d.	040188		739
		2 TMsS [carbon], pp. 1-9	n.d.	040189		752
		2 Offprints, with autograph annotations	1939	040192		774
		Galley with autograph annotations	[1939]	040191		768
		Notre Dame Continuum Lectures, Spring 1939:				
	52	Lecture I: AMs Notebook (in English), labelled "Cont. Vorl. Notre Dame I," plus one insert: "Prüfungsfragen"	[1939]	040194		784
	53	Lecture II: AMs Notebook (in English and Gabelsberger shorthand), labelled "Cont. N.D. Il (Vorl.)"	[1939]	040195		838
	54	Lecture III: AMs Notebook (in English), labelled "Cont. Vorl. N.D. III"	[1939]	040196		898
	55	Lecture IV: AMs Notebook, labelled "Cont. Vorl. N.D. IV"	[1939]	040197		951
8a	56	Lecture V: AMs Notebook, labelled "Cont. Vorl. N. D. V"	[1939]	040198	27	4
	57	Fragments: AMs miscellaneous loose pages	n.d.	040199		39
		Notre Dame Lectures on Logic, Spring 1939:				
	58	Preliminary work: AMs Notebook (in English), labelled "Vorl. Log. N.D. 0"	[1939]	040209		107
	59	Lecture I: AMs Notebook (in English and Gabelsberger shorthand), labelled: "Log. Vorl. Notre Dame I"	[1939]	040210		152

Box	Folder		Date(s)	*Initial document*	*Reel*	*Initial frame*
		IV. Drafts and Offprints (cont.)				
		Notre Dame Lectures on Logic, Spring 1939 (cont.):				
8a	60	Lecture II: AMs Notebook (in English), labelled "Log. Vorl. Notre Dame II"	[1939]	040211	27	195
	61	Lecture III: AMs Notebook, labelled "Log. Vorl. N. D. III"	[1939]	040212		242
	62	Lecture IV: AMs Notebook, labelled "Log. Vorl. N.D. IV"	[1939]	040214		318
	63	Lecture V: AMs Notebook, labelled "Log. Vorl. N.D. V"	[1939]	040215		368
	64	Lecture VI: AMs Notebook, labelled "Log. Vorl. N.D. VI"	[1939]	0401217		421
	65	Lecture VII: AMs Notebook, labelled "Logik Vorl. N.D. VII"	[1939]	040218		478
	66	Fragments: AMs Miscellaneous loose pages	n.d.	040220		521
	67	Notes: TMs [mimeograph] of seminar notes, 23 pp.	n.d.	040235		635
		***1939b*, Untitled Lecture (Given at Göttingen 15 December 1939):**				
	68	Autograph Material: Envelope labelled "Vortrag Göttingen," unrein	[1939]	040236		659
		AMs draft (in Gabelsberger shorthand), 12 pp.	[1939]	040236		660
	69	Autograph Material: Envelope labelled "Vortrag Göttingen rein"	[1939]	040237		716
		AMs draft (in Gabelsberger shorthand and German), 24 pp., plus 1 loose note	[1939]	040237		685
		Lectures on Constructible Sets (Given at the Institute for Advanced Study, April 1940):				
	70	AMs Notebook (in English and Gabelsberger shorthand) labelled "Vorl. April 1940 (Princeton I)"	1940	040239		719
	71	AMs Notebook, labelled "'Vorl. April 1940 II"	1940	040242		779

Box	Folder		Date(s)	*Initial document*	Reel	*Initial frame*
		IV. Drafts and Offprints (cont.)				
		***1940*, "The consistency of the axiom of choice and of the generalized continuum hypothesis":**				
8a	72	AMs of miscellaneous draft pages	n.d.	040246	27	853
		AL, G. W. Brown to Gwen Blake (Institute for Advanced Study secretary), re Addenda to Ms	1939	040244		851
	73	AMs draft, 31 pp.	n.d.	040259		923
	74	TMs with autograph corrections, 35 pp.	n.d.	040260	28	3
	75	⟦Copies of reviews⟧		040261.5		78
		Printed Material: First edition of monograph with autograph corrections for second printing	1940	040261		80
		Printed Material: Photo-ready copy of Chapter VIII, with autograph annotations	n.d.	040260.5		119
		***1940b*, "Consistency of Cantor's continuum hypothesis" (Lecture given at Brown University, 15 November 1940):**				
	76	AMs Notebook, labelled "Vortrag widfr. Continuum (Harvard & Brown)"	[1940]	040262		138
		"In what sense is intuitionistic logic constructive?" (Lecture given at Yale University, 15 April 1941):				
8b	77	AMs Notebook (in English and Gabelsberger shorthand), labelled "Vortrag Yale"	[1941]	040263		202
		***1944*, "Russell's mathematical logic" (Essay on Bertrand Russell for *Schilpp 1944*):**				
	78	TMs with autograph corrections, 39 pp. and AMs of addenda, 4 pages	[1944]	040264		246
	79	4 Offprints with autograph annotations	1944	040265		293
		AMs of notes and corrections, 2 pp.	n.d.	040268		359

Box	Folder		Date(s)	*Initial document*	*Reel*	*Initial frame*
		IV. Drafts and Offprints (cont.)				
		***1946*, "Remarks at the Princeton Bicentennial Conference on Problems of Mathematics" (Given 17 December 1946):**				
8b	80	AMs Notebook (in English and Gabelsberger shorthand), labelled "Bicentennial Dec. 1946"	1946	040270	28	363
	81	TMs with autograph corrections, 5 pp.	[1946]	040271		412
		Offprint, late edition	1968	040272		417
	82	Printed Material: Copy of *Problems of Mathematics* (Princeton Bicentennial Conference Report)	1946	040273		423
		***1947*, "What is Cantor's Continuum Problem?":**				
	83	TMs with autograph corrections, of miscellaneous unordered pages	n.d.	040274		442
	84	TMs of first draft, pp. 1–11 and 1–3	[1947]	040275		455
	85	TMs of second draft, pp. 1–13, i–iv, 1–2	[1947]	040276		471
	86	TMs [carbon] of final(?) draft, with autograph corrections, pp. 1–16 and 1–6	[1947]	040277		492
		AMs of addenda and definitions, 3 pp.	n.d.	040277		515
		Offprint with autograph corrections	1947	040278		520
		***1949b*, Lecture on Rotating Universes (Given at the Institute for Advanced Study, May 1949):**				
	87	Autograph Material: File folder	1949	040279		593
		AMs (in English), 43 pp., plus 6 pages of loose notes	[1949]	040279		528
		***1949*, "Example for a new type of cosmological solutions of Einstein's field equations of gravitation":**				
	88	TMs [carbon] with autograph corrections, 16 pp., and duplicate page 12	[1949]	040280		596
		Offprint	1949	040282		633

Box	Folder		Date(s)	*Initial document*	Reel	*Initial frame*
		IV. Drafts and Offprints (cont.)				
		***1949a*, "A remark about the relationship between relativity theory and idealistic philosophy"**				
8b	89	AMs Notebook (in English and Gabelsberger shorthand)	n.d.	040282.5	28	639
		TMs with autograph corrections, 11 pp.	[1949]	040284		679
		AL draft to Paul A. Schilpp, re changes to Ms	n.d.	040283		697
		Offprint	1949	040285		691
		"Rotating universes in general relativity theory" (Lecture given to International Congress of Mathematicians, Cambridge, Massachusetts, 31 August 1950):				
	90	AMs, 40 pp., plus 3 loose pages containing calculations and addenda	[1950]	040286		702
		"The consistency of the continuum hypothesis" (Second printing of 1940):				
	91	2 Offprints with autograph annotations	1951	040287		755
		****1951*, "Some basic theorems on the foundations of mathematics and their philosophical implications" (Gibbs Lecture given to the American Mathematical Society in Providence, Rhode Island, December 1951):**				
	92	AMs of first draft (in Gabelsberger shorthand), 19 pp., plus 6 pages of loose notes (in English and Gabelsberger shorthand). NB: Back side of a half page contains TL draft to "Mama" dated "1056" [sic].	[1951]	040290		840
		Autograph Material: Envelope containing library request slips	1951	040289		864
	93	AMs (in English), 39 pp.	[1951]	040293		886
	94	AMs of addenda to the text, 13 pp.	[1951]	040294		934
	95	AMs of footnotes and various loose sheets, 18 pp.	[1951]	040295		954
		[See also Folder 97 for correspondence re Gibbs lecture]				

Box	Folder		Date(s)	*Initial document*	*Reel*	*Initial frame*
		IV. Drafts and Offprints (cont.)				
		***1952*, "Rotating universes in general relativity theory":**				
8b	96	AMsS, pp. 1–24, 1–3, and 2 pp. of addenda	[1952]	040297	28	987
	97	TMs [carbon] with autograph corrections, 12 pp., plus AMs of insertions, 2 pp.	[1952]	040299		1021
		TMs and TMs [carbon] of footnotes with autograph corrections, 8 pp.	[1952]	040301		1035
		AL draft to "Prof. Graves," re corrections to the printing of the "Cambridge lecture" [See "Rotating universes in general relativity theory" (Lecture given to International Congress of Mathematicians, Cambridge, Massachusetts, 31 August 1950)]	⟦1950⟧	040298		1020
	98	TMs with autograph corrections and TMs [carbon], pp. 1–11, plus 1 p. of footnotes	[1952]	040303		1045
		Offprint	[1952]	040306		1079
		Galley with autograph annotations	1952	040305		1074
		***1958*, "Über eine bisher noch nicht benützte Erweiterung des finiten Standpunktes":**				
8c	99	TMs of abstract (in English), 1 p.	n.d.	040307	29	3
		TMs [carbon] of abstract (in German), 1 p.	n.d.	040307		4
		TMs [carbon] with autograph corrections, pp. 1–8 plus 3 pp. of footnotes	[1958]	040307		5
		Offprint with autograph corrections	[1958]	040308		17
		Offprint with autograph dedication to Clifford Spector	[1958]	040309		23

Box	Folder		Date(s)	*Initial document*	Reel	*Initial frame*
		IV. Drafts and Offprints (cont.)				
		***1964*, "What is Cantor's continuum problem?" (published revision of *1947*):**				
8c	100	Offprint of *1947* with autograph corrections for *1964*	n.d.	040310	29	31
	101	AMs of "Supplement to second edition," pp. 1–12, plus 2 pp. headed "New Footnotes"	[1964]	040311		82
		AMs of notes, 8 pp., and insertions I–V, 5 pp.	[1964]	040313 040314		69 96
	102	TMs with autograph corrections, labelled "Copy 2," pp. 1–15, footnotes pp. i–v, supplement pp. 1–6, new footnotes p. I, plus 2 pp. of addenda	[1964]	040315		102
	103	2 TMss [carbon] with autograph corrections, pp. 1–15, footnotes pp. i–v, supplement pp. 1–5, new footnotes p. I, plus 2 pp. of addenda, and 12 miscellaneous duplicate pages	[1964]	040316		134
		***1965*, "On undecidable propositions of formal mathematical systems" (Revision of *1934* appearing in *Davis 1965*):**				
	104	Autograph Material: 2 Envelopes and one File Folder	n.d.	040321		223
		TMs [photocopy] of 1934 with autograph and typescript revisions, 30 pp.	n.d.	040326		265
		AMs of corrections, insertions, and postscriptum, 8 pp.	n.d.	040325		257
		TMs [photocopy] of corrected passages, 6 pp.	n.d.	040322		224
		TMs and 2 TMss [carbon] of postscriptum with autograph corrections, 2 pp.	n.d.	040320		215
		TMs [carbon] of revised paper with autograph corrections, 23 pp., plus TMs and TMs [carbon] of postscriptum, 1 p., and additions, 3 pp.	[1965]	040323		231

Box	Folder		Date(s)	*Initial document*	*Reel*	*Initial frame*
		IV. Drafts and Offprints (cont.)				
		***1965*, "On undecidable propositions of formal mathematical systems" (Revision of *1934* appearing in *Davis 1965*) (cont.):**				
8c	104 (cont.)	Printed copy with autograph corrections [photocopy], of unordered pages	n.d.	040318.5	29	206
		English translations of *1933e* ("Zur intuitionistischen Arithmetik and Zahlentheorie") and *1936a* ("Über die Länge von Beweisen"), and an edition of *1946* ("Remarks before the Princeton Bicentenntial Conference on problems of mathematics"), all published in *Davis 1965*:				
	105	TMs with autograph corrections [photocopy], of "On intuitionistic arithmetic and number theory," 6 pp.	[1965]	040329		306
		TMs with autograph corrections [photocopy] of "On the length of proofs," 2 pp.	[1965]	040330		312
		TMs with autograph corrections [photocopy] of *1946*, 5 pp.	[1965]	040331		314
	106	AMs (in English and Gabelsberger shorthand) of loose notes on the revisions for *Davis 1965*	n.d.	040332		320
		"The consistency of the continuum hypothesis" (7th edition of *1940*, published in 1966) with proposed biographical sketch:				
	107	AMs revision of *1940*, including bibliography, 13 pp.	n.d.	040343		399
		TMs [carbon] and TMs [carbon] with autograph corrections of additions and bibliography, 13 pp.	1965	040345		414
	108	AMs (in English, German, and Gabelsberger shorthand) of biographical sketch, including summary lists of published and unpublished work, 6 pp.	n.d.	040347		421
		TMs, TMs with autograph corrections, and TMs [photocopy] of biographical sketch	n.d.	040348		423

Box	Folder		Date(s)	*Initial document*	Reel	*Initial frame*
		IV. Drafts and Offprints (cont.)				
		"The consistency of the continuum hypothesis" (7th edition of *1940*, published in 1966) with proposed biographical sketch: (cont.)				
8c	108 (cont.)	7 TMss [photocopy] of Gödel's bibliography, each with different autograph corrections	n.d.	040352[i]	29	430
	109	Offprint of 1940 with autograph corrections for seventh printing	n.d.	040364		463
		Offprint of the seventh printing	1966			
		"What is Cantor's continuum problem?" (Revisions for proposed 3rd edition of *1947*):				
	110	AMsS (in English and Gabelsberger shorthand) of revisions, 9 pp.	1966	040366		626
		TMs [carbon] of revisions, 3 pp.	[1966]	040367		623
		***1967*, English translations of *1930* and *1930b*: Jean van Heijenoort's editions of the English translations of *1930* ("Die Vollständigkeit der Axiome des logischen Funktionenkalküls"), *1930b* ("Einige metamathematische Resultate über Entscheidungsdefinitheit und Widerspruchsfreiheit") and *1931* ("Über formal unentscheidbare Sätze der Principia mathematica and verwandter Systeme I"):**				
	111	Autograph Material: File folder	1964	040369		638
		TMs [photocopy] of Jean van Heijenoort's introductory note to "The completeness of the axioms of the functional calculus of logic," (English translation of *1930*) with autograph corrections, 3 pp.	n.d.	040369		639
		Printed Material with autograph annotation [photocopy]: First edition (in German) of *1930b*, 1 p.	n.d.	040370		642
		"Some metamathematical results on decidability and consistency," (English translation of *1930b*) TMs with autograph corrections [Photocopy], 3 pp.	n.d.	040371		644

[i]Since the documents were not filmed in order of document number, the initial frame number does not correspond to the initial document number for this segment.

Box	Folder		Date(s)	*Initial document*	Reel	*Initial frame*
		IV. Drafts and Offprints (cont.)				
		***1967*, English translations of *1930* and *1930b* (cont.):**				
8c	111 (cont.)	TMs [photocopy] of Jean van Heijenoort's introductory note to "On formally undecidable propositions of Principia mathematica and related systems 1" (English translation of *1931*) with autograph corrections, 7 pp.	n.d.	040372	29	647
	112	Autograph Material: File folder	n.d.	040374		655
		1967: TMs with autograph corrections [mimeograph], 43 pp.	n.d.	040374		656
	113	Autograph Material: File folder		040375		701
		1967: TMs with autograph corrections [photocopy], labelled "Revised Version 7 March 1964," 43 pp.	1964	040375		702
	114	*1967*: AMs of notes on corrections (in English and Gabelsberger shorthand), 5 pp.	n.d.	040377		748
		1967: TMs of commentary on 1964 version, with autograph corrections, 3 pp.	1964	040378		758
		1967: TMs [mimeograph] and TMs [carbon] of further corrections, pp. 1–13, and 1–2	n.d.	040379		761
		Italian translations of *1944* ("Russell's mathematical logic"), *1947* ("What is Cantor's continuum problem?"), and *1946* ("Remarks before the Princeton Bicentennial Conference on Problems of Mathematics"):				
	115	Italian translation of *1944*: Printed page proof with autograph corrections, pp. 48–73	n.d.	040382		781
		Italian translation of *1947*: Printed page proof with autograph corrections, pp. 74–91	n.d.	040383		807
		Italian translation of *1946*: Printed page proof with autograph corrections, pp. 92–95	n.d.	040384		825

Box	Folder		Date(s)	*Initial document*	*Reel*	*Initial frame*
		IV. Drafts and Offprints (cont.)				
		Italian translations of *1944*, *1947* and *1946* (cont.)				
8c	115 (cont.)	Printed page proof of notes to the translations with autograph corrections (in Italian), p. 18–26	n.d.	040385	29	829
		***1974*, Remarks on Non-Standard Analysis (Given in lecture March 26, 1973, and published in *Robinson 1974*, Non-Standard Analysis):**				
	116	AMsS (in English), 3 pp.	1973	040386		839
		TMs and 2 TMss [carbon], 2 pp.	1973	040387		842
		Printed Material [photocopy]: of title page and p. x of *Robinson 1974*	1974	040390		849
		Contributions to *Wang 1974*:				
	117	AMs and TMs with autograph corrections of miscellaneous pages	n.d.	040392		854
		"Consistency proof for the generalized continuum hypothesis" (1979 reprint of *1939a*):				
	117.5	Offprint	1979	no number filmed	30	3
		***1980*, "On a hitherto unexploited extension of the finitary standpoint"(English translation of *1958* by Wilfrid Hodges and Bruce Watson):**				
	118	Offprint	1980	040404		11
		On undecidable sentences (Vienna lecture texts?):				
	119	TMs with autograph corrections, back labelled "Über form. unentsch. Sätze frühere Fassung," 9 pp.	n.d.	040405		22
	120	TMs with autograph corrections, back labelled "unentsch. Sätze (Vortrag?)," 5 pp.	n.d.	040406		33
		Lectures on intuitionism (Princeton 1941?):				
	121	Lecture 1: AMs Notebook (in English), labelled "Vorl. 1"	n.d.	040407		40
	122	Lecture 2: AMs Notebook labelled "Vorl. 2"	n.d.	040408		97
	123	Notes: AMs, 13 loose pages originally filed with notebooks on intuitionism	n.d.	040409		166

Box	Folder		Date(s)	*Initial document*	Reel	*Initial frame*
		IV. Drafts and Offprints (cont.)				
		****1933?*, "Vereinfachter Beweis eines Steinitzschen Satzes" (unpublished manuscript intended for Karl Menger's colloquium):**				
8c	124	TMs [carbon] labelled "Steinit⟦zschen⟧... Menger Koll."	n.d.	040410	30	182
		****193?*, Lecture on undecidable Diophantine sentences (Never delivered):**				
	124.5	AMs Notebook (in English) labelled "Vortrag unentsch. Sätze & Polynome (nicht gehalten)"	n.d.	040411		187
		****1961/?*, "The modern development of the foundations of mathematics in the light of philosophy" (Preliminary lecture draft):**				
	124.6	Autograph Material: Envelope		040411.5		220
		AMs (in Gabelsberger shorthand and English), 5 pp.	n.d.	040411.5		221
		Vienna Workbooks on the Consistency of the Continuum Hypothesis:				
9a	125	AMs Notebook (in Gabelsberger shorthand), labelled "Arb. Kontinuum I (Reinschr)"	n.d.	040412		234
	126	AMs Notebook (in English and Gabelsberger shorthand), labelled "Arb. Kontinuum II (Reinschr.)," also includes 7 loose pages	n.d.	040413		279
	127	AMs Notebook (in Gabelsberger shorthand), labelled "Kont. Arb. Reinschrift"	n.d.	040414		326
		Theorem on real functions (filed by Gödel with Notre Dame logic lectures):				
	128	TMs [carbon] (in German) with autograph corrections	n.d.	040415		350
		****1946/9*, "Some observations about the relationship between theory of relativity and Kantian philosophy" (long version of 1949a):**				
	129	AMs Notebook, labelled "I," pp. 1–51a	n.d.	040416		359
	130	AMs Notebook, labelled "II," pp. 52–71 and footnote pages I–XLI	n.d.	040417		414

Box	Folder		Date(s)	*Initial document*	Reel	*Initial frame*
		IV. Drafts and Offprints (cont.)				
		***1946/9*, (cont.)**				
9a	131	AMs Notebook, labelled "Sec," containing second drafts of miscellaneous pages	n.d.	040418	30	474
	132	Autograph Material: File folder		040419		598
		TMs [carbon] with autograph corrections, labelled "Man. A," pp. 1–23 plus 5 pp. footnotes	n.d.	040419		505
		**1946/9-B2*:TMs and TMs [carbon] with autograph corrections, labelled "Man. B,"pp. 1–21, I–VIII	n.d.	040422		536
	133	Autograph Material: File folder and envelope containing notes in Gabelsberger shorthand	n.d.	040424		601
		**1946/9-C1*: AMs with TMs inserts, labelled "Man. C," pp. 1–30, I–XIV	n.d.	040428		616
	134	TMs with autograph corrections and TMs [photocopy] of "Man. C.," pp. 1–14	n.d.	040429		680
		***1953/9*, "Is mathematics syntax of language?":**				
	135	First version: AMs (in English) labelled "I Fassung," pp. 1–29, 1–7, footnotes I–XIV	n.d.	040433		710
	136	Second version: TMs with autograph corrections, labelled "II Fassung," pp. 1–19, footnotes I–XIX	n.d.	040434		830
	137	**1953/9-III*: Third version: TMs with autograph corrections, unlabelled, pp. 1–42 (24, 37, 40 are missing)	n.d.	040436	31	3
	138	Autograph Material: Envelope		040437		71
		**1953/9-III*: Notes: AMs loose notes in English with miscellaneous pages of text	n.d.	040437		72

Box	Folder		Date(s)	*Initial document*	*Reel*	*Initial frame*
		IV. Drafts and Offprints (cont.)				
		***1953/9*, "Is mathematics syntax of language?" (cont.):**				
9a	139	Fourth version: AMs with TMs insertions, labelled "IV Fassung," pp. 1–35, 1'–6', 7', 37	n.d.	040438	31	122
	140	**1953/9-V*: Fifth version: ⟦AMs and⟧ TMs with autograph corrections, labelled "V Fassung," pp. 1–12	n.d.	040441		193
		Sixth version: ⟦AMs and⟧ TMs [carbon] labelled "VI Fassung," pp. 1–8	n.d.	040444		216
		***1972*, English Translation of *1958*, ("Über eine bisher noch nicht benützte Erweiterung des finiten Standpunktes") by Kurt Gödel and Leo Boron:**				
9b	141	AMs labelled "I Copy," pp. 1–26, 1F–5'F	n.d.	040450		263
		TMs [mimeograph] with autograph corrections in two hands, labelled "I Copy," pp. 1–11, plus TMss [mimeograph] of abstract (in English and German)	n.d.	040449		251
	142	TMs [mimeograph] with autograph corrections, labelled "II," pp. 1–11, also AMs of unordered pages headed "Notes"	n.d.	040451		321
	143	TMs with autograph corrections [photocopy of copy "II"], labelled "For Prof. Gödel"	n.d.	040453		365
	144	TMs with autograph corrections, note here attributes work as Gödel's revision of Boron's translation, pp. 1–19 (9 and 14 are missing)	n.d.	040454		385
	145	TMs and AMs loose notes, insertions, and addenda to the translation	n.d.	040457		405
	146	Autograph Material: Envelope which once contained the loose notes now filed in this and the next two folders	n.d.	040464		467

Box	Folder		Date(s)	*Initial document*	Reel	*Initial frame*
		IV. Drafts and Offprints (cont.)				
		***1972*, English Translation of *1958*, (cont.)**				
9b	146 (cont.)	AMs (in Gabelsberger shorthand and English), miscellaneous notes on half pages and scraps	n.d.	040465	31	469
	147	AMs (in Gabelsberger shorthand and English), miscellaneous notes on half pages and scraps	n.d.	040483		545
	148	Autograph Material: Envelope	n.d.	040489		644
		AMs (in Gabelsberger shorthand and English), miscellaneous notes on half pages and scraps	n.d.	040489		645
	148.5	AMs and TMs of translation with autograph corrections, unordered loose pages	n.d.	040498.01		704
		[See *1972a* for galley with autograph corrections]				
		***1972a*, Three notes on undecidability (Appended to English translation of *1958* by Kurt Gödel and Leo Boron):**				
	149	AMs (in Gabelsberger shorthand), labelled "Endgültige Form," 18 pp.	n.d.	040507		907
		TMs (in English) with autograph corrections, 5 pp.	n.d.	040501		876
		Galleys of *1972* and *1972a*, which were to be published together in *Dialectica*, pp. 1–9, 2 printed copies with autograph corrections	n.d.	040456		854
		On schemes of ramification (unpublished manuscript once filed with notes dated 1973-1975):				
	150	AMs (in English), 4 pp.	n.d.	040509		929
		AMs of references (in English and Gabelsberger shorthand) p. 1–5	n.d.	040509		933

Box	Folder		Date(s)	*Initial document*	*Reel*	*Initial frame*
		IV. Drafts and Offprints (cont.)				
		***1970a, *1974b*, Drafts on scales of functions (unpublished manuscript, first version circulated May 1970):**				
9b	151	**1970a*, AMsS, entitled "Some considerations leading to the probable conclusion that the true power of the continuum is ..." labelled "I Fassung," 5 pp.	[1970]	040511	31	940
		**1970b*, AMsS, entitled "A proof of Cantor's continuum hypothesis from a highly plausible axiom about orders of growth," labelled "II Fassung," 5 pp.	n.d.	040512		947
		[See also Tarski correspondence, Series I, folder 183]				
		Summary of John von Neumann's work in set theory:				
	152	AMs (in English), labelled "v. Neumann's Arbeiten über die Grundlagen," 3 pp.	n.d.	040513		965
		TMs [carbon] with autograph corrections, 2 pp.	n.d.	040514		969

V. Bibliographic Notes and Memoranda

Box	Folder		*Initial document*	*Reel*	*Initial frame*
9b	1	"Alte Lit[eratur] Grund[lagen]"	050001	32	3
	2	"Literatur alt"	050012		68
	3	Law and Nationality	050018		156
	4	"Lit[eratur] Grundl[agen] Philo[sophie] d[er] Math[ematik], [19]52–[19]55)"	050021		256
	5	Philosophy, 1936–1940	050024		316
	6	Psychology, Neuropsychology, Psychiatry	050025		343
	7	"Prog[ramm] 1959–1967"	050028		379
	8	"U. & Prog[ramm]," 1972–1975	050041		643
	9	"Phil[osophische] Lit[eratur], [19)52–[19]58": includes library request slips	050045		681

Box	Folder		Initial document	Reel	Initial frame
		V. Bibliographic Notes and Memoranda (cont.)			
9b	10	"Alte Literat[ur] Math[ematik]"	050046	33	3
	11	Continuum Problem [See also Series VI: Folder 38]	050055		82
	12	Reference work for *1949a* ("A remark about the relationship between relativity theory and idealistic philosophy"), 1947–1949 [See also Bibliographic "Zettel"]	050059		164
	13	Foundations	050066		350
	14	Classical Latin Authors	050067		471
	15	David Hawkins, 1968-1969	050070		555
	16	Idealism, 1961–1962: Friedrich Bouterwek, H. Heinsoeth, Johann Jakob Wagner	050071		574
	17	Bierens de Haan, Karl Ludwig Michelet, Georg Feder, Christian Wolff	050075		662
	18	Georg Wilhelm Friedrich Hegel and Friedrich Wilhelm Joseph Schelling	050079		764
		History:			
9c	19	Ancient Roman	050082		872
	20	Greek, Near Eastern	050085	34	3
	21	Various: 1200–1900s	050089		87
	22	Edmund Husserl	050099		310
	23	"idealist[i⟦s⟧che] Ph[ilosophie]"	050113		487
		Gottfried Wilhelm Leibniz:			
	24	Biographies	050114		633
	25	Bibliographies	050115		844
10a	26	"Bibliographic curiosa"	050120.1	35	3
	27	Literature on Leibniz	050121		46
	28	Works: Academy Edition and Dutens	050123		201
	29	Jean Baruzi, etc.	050124		232
	30	Diverse sources	050125		327
	31	Joannes Eduardus Erdmann	050126		379
	32	Louis Alexandre Foucher de Careil	050127		430
	33	Reports on unpublished manuscripts	050128		521
	34	Theological writings and Louis Alexandre Foucher de Careil	050129		583
	35	C. I. Gerhardt	050130		763

Box	Folder		Initial document	Reel	Initial frame
		V. Bibliographic Notes and Memoranda (cont.)			
		Gottfried Wilhelm Leibniz (cont.):			
10a	36	C. I. Gerhardt: vol. 5, New essays on human understanding; vol. 7, correspondence with Samuel Clarke	050131	36	3
	37	Secondary Sources: On life, letters, philosophy	050132		148
	38	Secondary sources: including Bertrand Russell	050134		281
	39	Logic and Foundations (before 1952) [See also Series VI: Folder 9]	050135		313
	40	Logic and Mathematics: "laufend"	050136		487
	41	Phenomenology and Existentialism	050143		660
		Philosophy:			
	42	1952–1958	050144		714
10b	43	1959–?	050144	37	4
	44	After 1960	050146		243
	45	Philosophy of Science	050147		443
	46	Philosophy, History of Science, Psychology, Theology	050148		470
		Theology:			
	47	(1952–?)	050149	38	3
	48	(1952–?)	050153		164
	49	(1952–?)	050155		324
	50	(1952–?)	050158		476
	51	Women	050165		811
		Jozef Maria Wronski:			
	52	Miscellaneous	050167		842
	53	Miscellaneous	050168	39	3
		Library request slips ("Bestellscheine"):			
	54	1924–1929	050173		278
10c	55	1930–1931	050174		345
	56	1932	050176		443
	57	1933–1936	050177		545
	58	1937–1938	050180		691
	59	1939, after 1940, and n.d.	050181		944
	60	1959–1963: Philosophy (New York Public Library)	050184	40	4

Box	Folder		*Initial document*	Reel	*Initial frame*
		V. Bibliographic Notes and Memoranda (cont.)			
		Bibliographic "Zettel" (Fragmentary notes):			
10c	61	Book lists and library request slips	050187	40	173
	62	Book lists and library request slips: Philosophy, 1959–?	050188		407
	63	Notes once filed with library request slips	050200	41	4
	64	Notes re: *1949a*	050201		90
	65	Fiction, etc.	050203		185
	66	Greek	050204		227
	67	History, History of science, Theology	050205		269
	68	Philosophy, Philology	050209		427
11a	69	Miscellaneous	050210		603
	70	Notes taken in the U.S., before 1952	050215		717
		Memoranda Books: These small notebooks contain fragmentary notes, including bibliographical lists, computations, calendar schedules, sketches, and (in the early books) lists of homework assignments:			
	71	General, before 1940	050217		809
	72	General, before 1940	050231	42	4
	73	General, n.d.	050235		142
	74	Mathematics, before 1940	050236		187
	75	Miscellaneous memoranda slips: including Philosophy and Theology	050241		344
	76	Excerpts of newspaper articles on contemporary politics	050245		545
	77	Linguistics	050247		584
	78	Mathematical Reviews, 1940–1958	050253		711
		VI. Other Loose Manuscript Notes; includes notes on offprints, books or lectures by various author⟦s⟧. For notes on works by an individual, see section (in this Series) title⟦d⟧ "Notes..."			
11b	1	American constitution and government	060001	42	812
	2	American history	060012		901
		Computations:			
	3	"Fehlerrechn[ung] für] Winkel"	060016		936
	4	$R_{\alpha\beta}$	060018	43	3
	5	Various[j]	060033		14

[j] Papers in this folder were filmed in reverse order of document number.

Box	Folder		Initial document	Reel	Initial frame
		VI. Other Loose Manuscript Notes (cont.):			
11b	6	Dialectica interpretation	030034	43	75
		[See also correspondence with Bernays, Paul]			
	7	"Diskussion einzelner Gruppen" (Rotation groups, etc.)	060049		192
	8	"Div. oft benötigte Formeln"	060061		370
	9	"Entscheidungsverfahren für positiven Aussagenkalk[ül]"	060074		455
	10	Erroneous/superfluous results and computations	060075		468
	11	"Finitismus;" "Cohensche Methode 1942"	060103		674
		Formulas and computations:			
	12	General	060106		701
	13	Rotating universes, "Div. alt"	060130		832
	14	Rotating Universes, cont.	060153	44	3
	15	General remarks on mathematical philosophy, foundations, and undecidability	060163		126
		[Grammar, see Notes on grammar]			
	16	Greek: Pronunciation, etc.	060169		182
11c	17	Isotropic universes and non-rotating anisotropic universes	060179		261
	18	Latin Vocabulary	060198		410
		[Mathematical philosophy, see General remarks on mathematical...]			
		[Mathematics and foundations, see Questions and remarks on...]			
	19	Miscellaneous mathematical memoranda, "wertlos[e] Notizen" 1935–1945	060205		447
		Miscellaneous mathematical notes:			
	20	Four-color problem, etc.	060235		531
	21	Includes notes on reprints by Davis and van Heijenoort	060248		576
	22	Includes computation sheets, 1937–1940	060273		703
	23	Includes computation sheets and fragmentary drafts	060274		866
		Miscellaneous notes:			
	24	Languages, etc.	060313	45	3
	25	Various	060336		47
	26	Fragmentary scraps ("Zettel")	060357		114

VI. Other Loose Manuscript Notes (cont.):

<u>Notes filed with offprints</u>: These brief autograph manuscripts, often computational notes, were found filed with offprints; they have been gathered here and filed by author of offprint. They are in a mixture of Gabelsberger shorthand, German, and English. The author of the offprint is noted in brackets on each manuscript. Only major authors have been listed:

Box	Folder			*Initial document*	*Reel*	*Initial frame*
11c	27	B-G;	includes Paul Bernays, William Boone, Rudolf Carnap, Alonzo Church, Paul J. Cohen, Haskell Curry, Burton S. Dreben, Albert Einstein, Erik Ellentuck, Solomon Feferman, Harvey Friedman, and Gotthard Günther	060361	45	137
	28	H-Z;	includes James Halpern, Arend Heyting, Richard Jeffrey, Stephen C. Kleene, Georg Kreisel, Akira Nakamura, J. Robert Oppenheimer, Robert W. Robinson, Dana Scott, William W. Tait, Gaisi Takeuti, and Ernst Zermelo	060379		242

<u>Notes or items inserted in books</u>: These items (including both autograph notes and printed material) were found inserted in the books of Gödel's library; they are here filed by author of book. Each folder contains a packet of source slips which may be used to identify the books and pages marked by each insert. Only major authors eliciting autograph notes from Gödel have been listed:

Box	Folder			*Initial document*	*Reel*	*Initial frame*
	29	A-C;	includes Rudolf Carnap	060417		410
	30	D-Hue		060435		502
	31	Husserl, Edmund		060455		581
	32	J-M;	includes Oskar Morgenstern	060459		629
	33	N-R;	includes Abraham Robinson and Bertrand Russell	060477		726
	34	S-T;	includes Alfred Tarski and Gaisi Takeuti	060491		777
	35	V-Z;	includes Jean van Heijenoort, John von Neumann, Hao Wang, Ernest B. Zeisler (including a TLS [photocopy] from Albert Einstein to Zeisler)	060508		859
12	36	Notes on Emil Artin's lectures on foundations of geometry (Notre Dame, 1939)		060525	46	3
	37	Notes on Cohen's independence proofs [See also Series VI: Folder 11 "Finistismus"]		060526		11

Box	Folder		Initial document	Reel	Initial frame
		VI. Other Loose Manuscript Notes (cont.):			
12	38	Notes on continuum problem and scales of functions, 1969–1974 [See also Series V: Folder 11 and Tarski correspondence, Series 1: Folder 183]	060541	46	72
	39	Notes on grammar	060551		274
	40	Notes on symposium "Philosophic implications of the Gödel incompleteness theorem" [See also Series XV: Box 21, Preprints and Offprints with accompanying correspondence, "Benacerraf, Paul" for audio tape of proceedings]	060562		314
	41	"Ontologischer Beweis"	060565		331
	42	Philosophical remarks, 1965–1967	060567		340
	43	"Phil[osophische] Varia" (mostly after 1961)	060570		412
	44	Questions and remarks on mathematics and foundations ("alte")	060574		451
		Rotating universes:			
	45	Basic formulas	060575		497
	46	"Beginning"	060600		683
	47	Diverse negative and minor results	060601		706
	48	Miscellaneous computations	060620	47	3
	49	"I. u[nd] II. Näherung"	060638		82
	50	Stationary solutions [See also Formulas and computations]	060647		166
		[Rotation groups, see "Diskussion einzelner Gruppen"]			
	51	"Wirkungsfunktionen und Erhaltungssätze"	060660		328
	52	Unclassified loose notes [See also Series XIV: Folder 2]	060672		432
		[Undecidability, see General remarks on]			
		VII. Academic Records and Notices			
13a	0	Annual reports (with report cards): Deutsches Staats-Realgymnasium in Brünn 1919–1924, with bound newsletters for 1921–1922 and 1922–1923; Evangelische Privat- Volks-und Bürgerschule in Brünn, 1913-1916; K. K. Staatsrealgymnasium mit deutscher Unterrichtssprache in Brünn, 1916–1919	070001	47	665

Box	Folder		*Initial document*	Reel	*Initial frame*
		VII. Academic Records and Notices (cont.)			
13a	1	Announcements of lectures and meetings, Vienna, 1930–1932 and n.d.	070021	47	737
	2	Course material for courses taught by Gödel: Enrollment slips and announcements, Universität Wien, 1933–1937, University of Notre Dame, 1939; Homework graded by Gödel (as Hans Hahn's assistant) Universität Wien, 1932–1933 [See also Miscellaneous administrative announcements]	070057.5		758
	3	Dean's correspondence, Universität Wien: Official documents regarding appointment, leaves of absence, rescinding of "Lehrbefugnis" (teaching license)	070080.5		865
		[Doctoral Diploma, Universität Wien see Series XIV: Folder 3]			
		[Honorary degree diplomas: Harvard University, see Series XIV: Box 16; Yale University, see Series XIII: Folder 1]			
		[Membership certificates: American Academy of Arts and Sciences, National Academy of Science see Series XIV: Folder 4]			
		Miscellaneous administrative announcements:			
	4	Universität Wien: 1933–1936	070083		872
	5	Universität Wien: 1935–1940, n.d. University of Notre Dame: Logic homework of F.P. Jenks, Student	070121		919
	6	University of Notre Dame: March–May 1939	070177	48	3
	7	Promotion to Professor, Institute for Advanced Study (TLS from J. Robert Oppenheimer to Gödel)	070199.5		43
	8	Publication lists of Institute for Advanced Study members and applicants	070200		45
		Programs from events at Universität Wien:			
	9	600th anniversary celebration, 1965	070229		138
	10	Gödel memorial colloquium, 1980	070232		154
	11	Reference information and admission cards for various libraries, 1930–1956	070232.5		157
		VIII. Legal and Political Documents			
	1	Apartment rental agreements	080170.5		186
	2	Birth and baptismal certificates, Kurt Gödel and Adele Porkert	no number filmed		192

Box	Folder		*Initial document*	*Reel*	*Initial frame*
		VIII. Legal and Political Documents (cont.)			
13a	3	Copyright agreements and publishing contracts	080171	48	201
	4	Employment permit	080007.5		217
	5	Marriage certificates, Kurt and Adele Gödel (1938)	no number filmed		219
	6	Naturalization and citizenship documents: Austrian and American	080172.4		223
	7	Passports, Adele Gödel (1953–1978)	080025		261
	8	Passport, Kurt Gödel (1940): Includes identity card and records of immigration to U.S. via trans-Siberian railroad	no number filmed		299
		Patent documents of Rudolf Gödel (father):			
	9	Correspondence: 1905	080030?		314
	10	Correspondence: 1906–1907, 1909, and n.d.	080081		388
	11	Miscellaneous, n.d.	080125		498
	12	British patent specifications	080132		535
	13	Powers of attorney (1938–1939)	080139		593
	14	Social Security and voter registration cards	080141		598
	15	Vaterländische Front documents (1935–1936)	080146		609
		IX. Financial Records			
13b	1	Account books: Princeton Bank and Trust Company, First Bank and Trust Company of South Bend	no number filmed		617
	2	Assets, Statements of: 1939	090005		632
		Bank Miscellany:			
	3	Anglo-Čechoslovakischen and Prager Creditbank, Brünn	090007		635
	4	Princeton Bank and Trust Company, First Bank and Trust Company of South Bend	090019		654
		Bank Statements:			
	5	Anglo-Čechoslovakischen and Prager Creditbank, Brünn (1933–1935)	090032		680
	6	Anglo-Čechoslovakischen and Prager Creditbank, Brünn (1936–1939)	090041		703
	7	First Bank and Trust Company, South Bend, Indiana (1939)	090050		729

Box	Folder		*Initial document*	Reel	*Initial frame*
		IX. Financial Records (cont.)			
		Bank Statements (cont.):			
13b	8	Princeton Bank and Trust Company (1935–1936)	090057	48	739
	9	Princeton Bank and Trust Company (1937–1938)	090071		756
	10	Princeton Bank and Trust Company (1939)	090088		774
	11	Booksellers' invoices and receipts	090097		784
	12	Cancelled checks, stubs, and debits	Not filmed		
	13	Clothing bills	090231		854
	14	Contractors' bills and receipts	090277.01		917
	15	Currency exchange/Funds transfer	090277.5	49	3
	16	Customs receipts	090348		95
	17	Deposit slips	090356		105
	18	Dues notices and receipts [See also Legal and political fee/contribution receipts]	090365		116
	19	Electric bills and receipts	090387		156
	20	Freight company receipts	090414		192
	21	Fuel bills: coal, coke, wood	no number filmed		234
	22	Gas bills and receipts	090453		254
	23	Honeymoon receipts	090481		287
		Hotel and sanatorium receipts:			
	24	192?–1935	090492		303
	25	1936	090523		348
	26	1938–1939, n.d. (except honeymoon receipts)	090552		383
	27	Housecleaning bills	090588		430
	28	Household furnishings, bills and receipts	090592		436
	29	I.A.S.: salary receipts, 1938	090614		467
		Income/expense ledgers:			
	30	before 1932	090618		473
	31	January 1932, July 1933–December 1934 "Ausgleichsrechnungen"	090619		548
	32	August 1933–March 1934	090621		565
	33	1933–1936 (loose balance sheets)	090622		572

Box	Folder		*Initial document*	*Reel*	*Initial frame*
		IX. Financial Records (cont.)			
		Income/expense ledgers (cont.):			
13c	34	1931–1935, Brünn (in Marianne Gödel's(?) hand)	090645	49	614
	35	1930–1937, Vienna (in Marianne Gödel's(?) hand)	090646?		624
	36	January–August, 1933 (in Marianne Gödel's(?) hand)	090654		724
	37	September 1933–August 1934 (in Marianne Gödel's(?) hand)	090655		798
	38	1935–1937, Brünn (in Marianne Gödel's(?) hand)	090658		905
	39	1937 (in Marianne Gödel's(?) hand)	090664	50	3
	40	1933–1934, "Villa Gödel" (in Pauline Handschuh's(?) hand)	no number filmed		71
	41	1933–1934, Brünn (in Pauline Handschuh's(?) hand)	no number filmed		88
		[See also Series XIII: "Brno Villa: Records of receipts and expenditures"]			
	42	Insurance documents	090676		119
	43	Legal and political fee/contribution receipts	090713		211
	44	Lecture fee receipts, Universität Wien	090720		222
	45	Medical/dental bills and receipts, 1927–1948, n.d. [See also Hotel and sanatorium receipts]	090725		228
	46	Miscellaneous bills and receipts, 1929–1939, n.d.	090776		301
	47	Princeton University Store, 1933–1934	090807		376
	48	Redlich, Friedrich (Textile Company): Financial statements	090813		383
	49	Rent bills, 1938–1939	090822		394
	50	Rent receipts, 1937–1939	090828		401
		Securities transactions:			
	51	Anglo-Čechoslovakischen and Prager Creditbank; Österreichische Creditanstalt [See also Series XIV: Folder 5]	090876		457
	52	Ost-Österreichische Landeshypothekenanstalt and other	090915		486
	53	Tax documents and governmental fee receipts, 1934–1948	090946		521

Box	Folder		Initial document	Reel	Initial frame
		IX. Financial Records (cont.)			
13c	54	Telephone/telegraph bills and receipts, 1928-1934	090958	50	537
	55	Transportation receipts	090976		560
		Unclassified financial memoranda:			
	56	"Alte Abrechnungen and Zettel"	091006		600
	57	"Aufstellungen"	091024		625
	58	Miscellaneous	091050		676
	59	Miscellaneous [See also Series XIV: Folder 5]	091072		712
		X. Medical Records			
14a	1	Correspondence related to health care, 1932–1937	100001		770
		[Doctors, see Names and addresses of]			
		Dosage records:			
	2	May-June, 1942	no number		776
	3	1946-1963	100007		779
	4	1964-1968	100009	51	3
	5	1970-May 1974	100011		52
	6	June 1974-February 1976	100012		89
		Medical memoranda:			
	7	before 1960	100017		117
	8	1962–1963, 1967–1969	100022.6		127
	9	Dr. Hulbeck: 1963–1969, n.d.	100044.1		188
	10	1970–May 1974	100045		288
	11	May 1974–November 1975	100048		363
	12	n.d., includes dietary notes	100049.2		402
	13	Names and addresses of doctors	100111.01		548
	14	Prescriptions	100111.5		596
	15	Temperature records ("Fiebertabellen"), 1949–1953, n.d.	100112		656
	16	Treatment reports, drug and dietary recommendations	100125		682

Box	**Folder**		**Photo ID #**	*Reel*	*Initial frame*
		XI. Photographs			
		A. Photographs between 4″ × 5″ and 6″ × 9″			
14a	0	Kurt Gödel: 2 head and shoulder portraits, n.d.	110150–110151	52	5
		Adele Gödel: Posed by piano ca.1912, 3 head and shoulder portraits, n.d.	110152–110155		7
		Josef Porkert (Adele's father): Head and shoulder portrait, n.d.	110156		12
		Hildegarde Porkert (Adele's mother): Head and shoulder portrait, n.d.	110157		14
		Eduard Wette: on the H.M.S. Queen Elizabeth II	110158		16
		Rudolf (brother) and Kurt Gödel: as children, n.d.	110159		18
		Marianne (mother) and Kurt Gödel: at the Institute for Advanced Study	110160		20
		Rudolf, Adele, and Marianne Gödel: at a picnic	110161		22
		Hildegarde Porkert, Adele and Kurt: at table	110162		24
		Adele Gödel, Stephen C. Kleene, and his parents: on the Kleene farm in Maine	110163		26
		Stephen Kleene and his parents: on the Kleene farm	110164		27
		B. Photographs between 6″ × 9″ and 8″ × 10″			
	1	Kurt and Adele Gödel, wedding portrait, 23 September 1938	110200		29
	2	Kurt and Adele Gödel, 10th wedding anniversary portrait, 1948	110201		31
	3	Participants in Albert Einstein's 70th birthday celebration: including Kurt Gödel, Einstein, Hermann Weyl, and J. Robert Oppenheimer, 19 March 1949	110202		34
	4	Einstein award presentation: Albert Einstein, Gödel, Julian Schwinger, and Lewis Strauss, 14 March 1951	110203		36

Box	Folder		Photo ID #	Reel	Initial frame
		XI. Photographs (cont.)			
		B. Photographs between 6″ × 9″ and 8″ × 10″ (cont.)			
14a	5	Harvard honorary degree recipients: including Gödel and John Foster Dulles, 19 June 1952	110204	52	38
	6	Kurt Gödel: Portrait by Arnold Newman, 1956	110205		40
	7	Rockefeller University honorary degree recipients: 4 portraits, 1 June 1972	110206–110209		42
	8	Kurt Gödel: Portrait, 7 October 1973	110210		47
	9	National Medal of Science presentation: 3 portraits of Saunders Mac Lane on behalf of Gödel, 18 September 1975[k]	110211–110213		50
	10	Kurt Gödel: Early portrait (crudely retouched), n.d.	110214		53
	11	Kurt Gödel: Studio portrait by Marcel Natkin, n.d.	110215		55
	12	Adele Gödel: Studio portrait, n.d.	110216		58
	13	Adele Gödel: Profile, n.d.	110217		60
	14	Rudolf Gödel (father): 2 portraits, 1924 and 1927	110218–110219		62
	15	Albert Einstein: heading home, by Alan Richards	110220		65
	16	John von Neumann: Studio portrait, n.d.	110221		67
	17	Unidentified photographs: Unknown man, dated 1950; Einstein with unidentified man, n.d.; Josef Porkert(?), n.d.; Unknown man, n.d.			69
		C. Photographs less than 4″ × 5″			
14b		Kurt Gödel: 42 pictures, posed and candid, ca.1925–1969, n.d.	110000–110041		76

[k]The first of the three portraits has been filed under "Oversized Photographs" in Folder 3 of what was later redesignated as Series XIII.

Box	Folder		Photo ID #	*Reel*	*Initial frame*
		XI. Photographs (cont.)			
		C. Photographs less than 4″ × 5″ (cont.)			
14b	17 (cont.)	Adele Gödel: 5 portraits, n.d.	110042–110046	52	119
		Marianne Gödel: 7 pictures, posed and candid, 1959–1961, n.d.	110047–110053		125
		Josef Porkert: 3 portraits, 1940, n.d.	110054–110056		133
		Hildegarde Porkert: 4 portraits, n.d.	110057–110060		137
		Rudolf Gödel (brother): 1 portrait, n.d.	110061		142
		Hao Wang: candid picture, 1972	110062		144
		Oswald Veblen: portrait, n.d.	110063		146
		Kurt and Adele Gödel: 21 pictures, posed and candid, 1935–1964, n.d.	110064–110084		148
		Josef and Hildegarde Porkert: 3 pictures, candid and posed, 1938–1951	110085–110087		170
		Marianne Gödel with son Rudolf or Kurt: 9 pictures, 1959–1966	110088–110096		174
		Hildegarde Porkert with daughter Adele: Posed together, n.d.	110097		184
		Kurt Gödel and brother Rudolf: Posed together, n.d.	110098		186
		Kurt or Rudolf Gödel with colleagues (including Stephen C. Kleene, Oskar Morgenstern, Alfred Tarski, Hao Wang): 8 pictures, 1960–1972, n.d.	110099–110106		188
		Unidentified couple: silhouette, n.d.	110107		197
		Gödel family members: (18) 1960–1964, n.d.	110108–110124		199
		Gödel with family and friends: (5) 1941–1968	110125–110129		218
		Gödel home at 145 Linden Lane, Princeton: (18) (as of 9/30/85, photos 110137-110145 are missing), 1962, n.d.	110130–110145		224

Box	Folder		Photo ID #	*Reel*	*Initial frame*
		XI. Photographs (cont.)			
14c		[See Series XII. Ephemera]			
		D. Photographs between 8″ × 10″ and 11″ × 14″ (Filed in Series XIII: Oversize Items and Addenda)			
15	3	National Medal of Science Presentation: (Saunders Mac Lane on behalf of Gödel) 1 portrait, signed by President Ford	110211	53	973
		Kurt Gödel: Mounted studio portrait (2 copies) by Marcel Natkin	110215		974
		Kurt Gödel: 3 portraits by Alfred Eisenstadt	110300–110302		976
		Adele Porkert and schoolmates: School portrait	110303		982
	5	N.A.S.A. surveys of Mars: Prints from the Mariner 9 mission, also includes relief map of Mars, 1971-1972	no numbers	54	117
		E. Photographs larger than 11″ × 14″ (Filed in Series XIV: Extra Large Items)			
16	6	N.A.S.A. Surveys of the moon: Lunarscapes from the Apollo 15 mission, 1971	no numbers		389
		Participants in the Princeton Bicentennial Conference on Problems of Mathematics: Group portrait, 1946	110304		388

XII. Ephemera

Box	Folder		*Initial document*	*Reel*	*Initial frame*
14c	1	Advertisements and other commercial items	120001	53	3
	2	Apartment-hunting memoranda	120041		211
		Articles and clippings:			
	3	About Gödel (including memorial tributes): [See also Series XIV: Folder7]	120098		299
	4	About Einstein [See also Series XIV: Folder8] .	120122		341
	5	Sent to Gödel by OskarMorgenstern [See also Series XIV: Folder91	120140		355

Box	Folder		*Initial document*	*Reel*	*Initial frame*
		XII. Ephemera (cont.)			
14c	6	Miscellaneous annotated envelopes [See also Series XIII: Folder 4]	120161	53	562
	7	Miscellaneous enclosures from Stanley Tenenbaum correspondence	120166		577
	8	Nazi broadsides and advertisements [See also Series XIII: Folder 4]	120173		620
	9	Programs: concerts, conventions, church services, etc.	120217		643
	10	Catalogs: Publishers', etc. [See also Series XIII: Folder 4]	120228.5		733
	11	Unclassified Ephemera: includes TMsS [carbon] of anti-Semitic tract by "Dr. Austriacus" against Gödel's mentor, Moritz Schlick, ca. 1936; and TMs [carbon] of open letter to Einstein from Soviet scientists, 1947	120248		811
		XIII. Oversize Items and Addenda			
15	0	Unclassified oversize items: includes AMs notes on printed material (in German), and AMs lecture draft(?) (in Gabelsberger shorthand)	000002?		915
	1	Yale honorary degree certificate	070228.5		929
	2	Oversize financial items from Series IX: Brno villa records of receipts and expenditures	091110		933
	3	Oversize photographs from Series XI: Photographs between 8"x 10" and 11 "x 14"	numbers not filmed		973
	4	Oversize ephemera from Series XII: Miscellaneous	120247	54	2
	5	Oversize photographs from Series XI: N.A.S.A. photographic surveys of Mars	numbers not filmed		117
		Addenda to the collection:			
	6	LsS [photocopies] from the Gödel/Paul Bernays correspondence stored at the Eidgenössische Technische Hochschule, Zürich	not filmed		

Box	Folder		Initial document	Reel	Initial frame
		XIII. Oversize Items and Addenda (cont.)			
		Addenda to the collection: (cont.)			
15	7	Gödel material contributed by Dana Scott: Includes *1972a* TMs "Some remarks on the undecidability results"; Wilfrid Hodges/Dana Scott correspondence re *1980*; Paul Bernays/ Dana Scott correspondence re Gödel and Gödel's *Nachlaß*	no numbers	54	182
	8	LsS [photocopies] from the Gödel/Arend Heyting correspondence stored at the Heyting Archief	not filmed		
		XIV. Extra Large Items			
16	1	Correspondence: Carnap(?), Noether(?), and Unidentified	013287.7		297
	2	Unclassified loose notes	060695		309
	3	Doctoral Diploma: Universität Wien	no number		335
	4	Academy certificates: American Academy of Arts and Sciences, National Academy of Science	no numbers		337
	5	Financial items: Miscellaneous ledger sheets	091104		340
16	6	Photographs: Portfolio of Apollo 15 lunarscapes,	no number		389
		Group portrait of participants in Princeton Bicentennial Conference, 1946	110304		388
		Articles and clippings:			
	7	About Gödel	120118		403
	8	About Albert Einstein	120131		413
	9	Sent to Gödel by Oskar Morgenstern	120157		430
	10	Miscellaneous	120216		447
	(loose)	National Medal of Science, framed certificate	no number		292
	(loose)	Harvard University, honorary degree certificate	no number		294

XV. Preprints and Offprints ⟦Not filmed⟧

Box		Dates
	Presentation Copies:	
17	Adjan to Yourgrau	1927–ca.1976

Box		Dates
	XV. Preprints and Offprints (cont.)⟦Not filmed⟧	
	Preprints and offprints annotated or with accompanying notes:	
18	Almgren to Kondo	1928–1970s
19	Kreisel to Tait	1930–1970s
20	Takahashi to Zurich	1929–1970s
	Preprints and offprints with accompanying correspondence:	
21	Abian to Knoll	1928–1970s
22	Kreisel to Rota	1940s–1970s
23	Rucker to Yilmaz	1940s–1970s

Kurt Gödel Papers (C0282)
Note on the 1998 Preservation Microfilm

In 1998, the Kurt Gödel Papers were microfilmed for the purpose of preservation. Funded by the Sloan Foundation Grant (95-10-14) in support of the *Collected Works* of Kurt Gödel, Oxford University Press, the preservation microfilm includes the entire collection, including a section of the annotations in the books from Gödel's library[1] (which are housed separately at the Institute for Advanced Study). There are a few exceptions which are listed below. These items are largely photocopies of original material from other collections, preserved by other institutions. Also omitted are Gödel's collection of the preprints and offprints of his colleagues' work, and a small collection of cancelled checks, stubs and debits, most blank and physically difficult to film. The microfilm reels of Kurt Gödel's papers are available at the Historical Studies Library of the Institute for Advanced Study, Princeton, and in the Rare Books and Special Collections Department of Princeton University's Firestone Library. The preservation microfilm does not constitute a microfilm edition, and is subject to the same copyright restrictions as the original or published material.

Portion not Filmed

Only the following items have been omitted from the 1998 filming of the Kurt Gödel Papers:

Series	Box(es)	Folder	
IV	7a	3	TMs [photocopy] of the final typescript of Gödel's doctoral dissertation, [1929]. Original is deposited at Universität Wien.
IX	13b	12	Cancelled checks, stubs, and debits.
XIII	15	6	LsS [photocopies] from the Gödel/Paul Bernays correspondence. Originals are stored at the Eidgenössische Technische Hochschule, Zürich.
XIII	15	8	LsS [photocopies] from the Gödel/Arend Heyting correspondence. Originals are stored at the Heyting Archief.
XV	17–23		Gödel's collection of the preprints and offprints of his colleagues' work (the entirety of Series XV).

[1]For more information on the reels containing the annotations from the books, see the addendum to this finding aid.

Kurt Gödel Papers (C0282)
Items Not Stored at Princeton University Library

Description	Location
Gödel's personal library: 700 books, including his annotated books; the textbook for Gabelsberger shorthand, available copies of the bibliography and additional offprints and preprints	Historical Studies Library, Institute for Advanced Study
Microform reel of *1934* (Gödel's I.A.S. lectures, "On undecidable propositions of formal mathematical systems" as transcribed by S. C. Kleene and J. B. Rosser)	Historical Studies Library, Institute for Advanced Study
Gödel's briefcase, doorplate, and National Medal of Science (medallion, lapel pin, and tape of memorial service)	Office of the Director, Institute for Advanced Study
Bound galley of *1931* ("Über formal unentscheidbare Sätze der *Principia Mathematica* and verwandter Systeme I"), with autograph annotations	Rosenwald Rare Book Collection, Institute for Advanced Study
Original TMs Dissertation (the collection at Firestone includes a TMs [photocopy])	Universität Wien
Letters by Gödel to his mother	Wiener Stadt- und Landesbibliothek, Vienna
Correspondence between Gödel and Arend Heyting (the collection at Firestone includes some photocopies)	Heyting Archief, Haarlem
Correspondence between Gödel and Bernays, and Gödel and Seelig (the collection at Firestone includes some photocopies)	Eidgenössische Technische Hochschule, Zurich

Addendum

Selection criteria for annotations in books

Reels 55 and 56 of the preservation microfilm contain the annotations from Gödel's books. Among his 700 books, there are 154 that have been identified as having marks of some sort. These marks consist either of handwriting—longhand or Gabelsberger shorthand—or of mere underlining, tick marks and "bullets". Of those 154 books, 61 were selected for filming on the basis of having something to reveal about Gödel's thought and interests. In addition ten books previously identified during the original sorting as having annotations are now missing, along with his entire collection of books on and by Leibniz. One two-volume set of the latter was filmed from a photocopy. A list of these books is provided on reel 55 of the preservation microfilm.

The user should remember that (1) these books were inherited by the Institute for Advanced Study from Gödel's wife, Adele, and therefore some of the 700 books are very likely to have been hers; (2) there are also various books with the name of Gödel's brother on them and so it is unclear who did the underlining or other marks, as even the shorthand in some of these is of doubtful authorship; and (3) some of the books were purchased used and therefore underlining and marks that are not specifically identifiable as Gödel's handwriting may not be his. Where there was doubt about a book that was indeed filmed, an explanation or admonition was provided in the list of books filmed.

The remaining 93 books were eliminated from filming for one of the following reasons: (1) the handwriting was identifiably *not* Gödel's; (2) the only annotation was a remark or remarks so short that it was more efficient to provide a transcription (see below); (3) the book was clearly purchased used and the only marks could not be identified as Gödel's; (4) in language books, the annotations dealt purely with vocabulary or were merely artifacts of language study; (5) the only annotation was a personal index that revealed nothing special; (6) the marks were a trivial designation of bibliographical entries, either in text or in a list of references.

A detailed list of books not filmed is also provided on reel 55. This list includes transcriptions of the short annotations that might be of interest in books not filmed, together with reasons for each for its omission from the film.

Cheryl Dawson

Appendix A

Letters by others written on Gödel's behalf

1. Felix Kaufmann to Heinrich Behmann[a]

Wien, am 19. Oktober 1930

Sehr geehrter Herr Behmann!

Eben ist Herr Dr. Gödel bei mir, der gegenüber meiner Konstruktivitätsbehauptung bzw. Ihres⟨m⟩ Beweises derselben folgende Einwände erhebt:

Er behauptet sehr wohl Beispiele konstruieren zu können, wo offenkundig eine Existentialbehauptung bewiesen ist, ohne daß doch eine Konstruktion angegeben werden könne. Das einfachste von Herrn Gödel genannte Beispiel, welches aber auch prinzipiell als Repräsentant für die übrigen Beispiele gelten darf, ist folgendes:

Gegeben sei eine eineindeutige Zuordnung zwischen den natürlichen Zahlen und gewissen Rationalzahlen des Intervalles $(0, 1)$. Dann läßt sich auf die bekannte Art zeigen, daß die angegebene Folge von Rationalzahlen einen Häufungspunkt hat, ohne daß es jedoch allgemein möglich wäre, einen solche⟨n⟩ anzugeben. Betrachten wir ein Beispiel, wo man einen Häufungspunkt sicher nicht angeben kann! Eine Zahl heiße Goldbach-Zahl, wenn alle kleineren geraden Zahlen Summe zweier Primzahlen sind. Die Folge rationaler Zahlen werde nun so definiert: Der natürlichen Zahl n sei die Zahl $1/n$ zugeordnet, wenn n eine Goldbach-Zahl ist und die Zahl $1 - 1/n$, wenn n keine Goldbach-Zahl ist. Dann kann man beweisen, daß diese Folge einen Häufungspunkt hat (und zwar einen rationalen) nämlich entweder 0 oder 1. Es läßt sich aber ohne Lösung des Goldbach'schen Problems kein Häufungspunkt angeben.

Herr Gödel hat auch mit Herrn Carnap darüber gesprochen und diesem erscheint der gemachte Einwand sehr einleuchtend.

Ich wäre Ihnen sehr dankbar, wenn Sie mir mitteilen wollten, wie Sie sich zu diesem Einwand stellen, und will mir inzwischen auch selbst darüber den Kopf zerbrechen.

---

[a]This letter is published in *Mancosu 2002*. On its background see the introductory note to the correspondence with Heinrich Behmann in volume IV of these *Works*. The translation is by Paolo Mancosu, revised by Charles Parsons, using suggestions of Wilfried Sieg.

The letter has in the upper left-hand corner the return address stamp "Dr. FELIX KAUFMAN, Wien XIX, Döblinger Hauptstr. 9".

Herr Gödel fügt noch hinzu, daß es sich hier um eine Disjunktion zwischen *zwei* Möglichkeiten handle, daß dies aber nicht prinzipiell sei; es könne auch die Disjunktion zwischen unendlich vielen Fällen unentschieden bleiben.

Mit den herzlichsten Grüßen bin ich Ihr,

Felix Kaufmann[b]

2 | Wien, am 19. Oktober 1930

Nachtrag

Ich zweifle trotz der vorgebrachten Einwände durchaus nicht an der Richtigkeit des Konstruktivitätsatzes. Mir scheint der Weg zur Entkräftung der gemachten Einwände in der Richtung gelegen, daß man bei jenen Fällen, wo die angebliche Existen~~t~~tialbehauptung in einer Disjunktion von Behauptungen ~~darstellt~~ ⟨zerfällt⟩, zeigt, daß es sich hier nach Ausschaltung der terminologischen Verkürzungen gar nicht mehr um eine Existentialbehauptung handelt; während man bei jenen Behau⟨p⟩tungen, wo eine sogenannte unendliche Kette von Disjunktionen vorliegt, mit Begriffen höherer Stufe operiert wird,[c] nach deren Ausschaltung ebenfalls ~~die~~ ⟨jede⟩ nicht-konstru~~ierbare~~⟨ktive⟩ Existenzbehauptung verschwindet. Aber diese Ausführungen sind, wie ich mir wohl bewußt bin, noch recht vage und ich wäre sehr erfreut, von Ihnen recht bald präzise Formulierungen in Hinblick auf diesen Punkt zu erfahren.

Nochmals viele herzliche Grüße!

Ihr Felix Kaufmann[d]

⟦Translation follows⟧

---

[b] The signature is handwritten. In the left margin of the page is written in Gödel's hand, "Die besten Grüße, Ihr erg. Kurt Gödel".

[c] It is very likely that the presence in this clause either of 'man' or of 'wird' is a slip on Kaufmann's part, since "während man...mit Begriffen höherer Stufe operiert" and "während...mit Begriffen höherer Stufe operiert wird" make perfectly good sense in the context, while the text as it stands does not. It makes no real difference to the meaning which is omitted. The translation assumes this correction.

[d] Handwritten signature.

Vienna, 19 October 1930

Dear Mr. Behmann,

Dr. Gödel is right now with me and he raises the following objections against my constructivity claim and your proof of it:

He claims to be able to construct examples where clearly an existential claim is proved although one cannot give a construction. The simplest example mentioned by Gödel—which also serves in principle as representative for all the remaining examples—is the following:

Let there be given a one-to-one mapping between the natural numbers and certain rational numbers from the interval $(0, 1)$. Then one can prove in the usual fashion that the given sequence of rational numbers has an accumulation point, although it would not in general be possible to give it explicitly. Let us consider an example where one certainly cannot give an accumulation point explicitly. A number is called Goldbachian if all smaller even numbers are sums of two prime numbers. The sequence of rational numbers is now defined as follows: if n is Goldbachian the number $1/n$ is associated to it; if n is not Goldbachian the number $1 - 1/n$ is associated to it. Then one can prove that this sequence has an accumulation point (and indeed a rational one), that is, either 0 or 1. However, without a solution of Goldbach's problem no accumulation point can be given explicitly.

Gödel has also discussed the issue with Carnap who finds the objection made by Gödel very plausible.

I would be very grateful if you could let me know your reaction to this objection, and in the meantime I will also rack my brains over it.

Gödel also adds that although in this case it is a question of a disjunction between two possibilities, this is not essential; the disjunction between infinitely many cases could also remain undecided.

With most cordial greetings,

Yours,

Felix Kaufmann

Vienna, 19 October 1930

Postscript

Despite the objections presented, I do not doubt at all the correctness of the constructivity theorem. It seems to me that the way to invalidate the objections made lies in the direction that one shows that, in the case where the existential claim resolves into a disjunction of claims, there is no longer an existential claim at all after terminological abbreviations are eliminated; and that in those claims which involve a so-called infinite chain of disjunctions use is made of concepts of higher order, after whose elimination every non-constructive existence claim also disappears.

But these remarks are still quite vague, as I am well aware, and I would be very pleased to hear from you very soon precise formulations with respect to this issue.

Once again most cordial greetings!

Yours, Felix Kaufmann

2. Dana S. Scott to Burton Dreben and Hao Wang[a]

April 3, 1970

Dear Dreben and Wang:

Professor Gödel is presently feeling in rather poor health and consequently has been trying to put his papers in order before it is too late. There are various loose ends here and there in his published works, and he has entrusted me with the task of helping him tie a few of them up. In particular in a recent telephone conversation he mentioned that he was able to recall the method by which he dealt with the presence of *equality* in the matrix of the formula in his solvable prefix class.[b] Since both of you have thought about this question, he asked me to write you the idea.

Suppose one has a formula of the form:

$$(*) \qquad \forall x_0\, \forall x_1\, \ldots\, \forall x_{n-1}\, \exists y\, \exists z\, \forall w_0\, \forall w_1\, \ldots\, \forall w_{m-1}\, M,$$

where M is quantifier free, that is to be tested for *validity*. Gödel says it is sufficient to have the decision procedure for such formulas with what I call "sharp" quantifiers; that is, quantifiers which demand that the value

[a]This letter is evidently a response to Wang's letter to Gödel of 23 April 1968 (in this volume) and a fuller response to Dreben's letter of 24 May 1966 (in volume IV of these *Works*). Note Dreben's reply of 15 April 1970 in volume IV and Wang's reply, also of 15 April 1970, in this volume. For discussion see pp. 229–230 of the introductory note to *1933i* in volume I of these *Works*, as well as the introductory note to the correspondence with Dreben in volume IV.

The letter is on letterhead of Princeton University, Department of Philosophy, 1879 Hall, Princeton, New Jersey 08540.

[b]Cf. *1933i*, p. 443.

of the quantified variable be *distinct* from all the values of all the previously quantified variables.

I shall try to illustrate Gödel's method on the ordinary (i.e. non-sharp) formula:

$$\forall x\ \exists y\ \exists z\ \forall w\ M(x, y, z, w).$$

In the first place this is equivalent to:

$$\forall x\ [\exists z\ \forall w\ M(x, x, z, w) \vee \exists y \neq x\ \exists z\ \forall w\ M(x, y, z, w)],$$

where $\exists y \neq x[\ldots$ means $\exists y[y \neq x \ldots$. In turn this is equivalent to:

2

$$\forall x[\forall w\ M(x, x, x, w) \vee \exists z \neq x\ \forall w\ M(x, x, z, w) \vee$$
$$\exists y \neq x(\forall w\ M(x, y, x, w) \vee \forall w\ M(x, y, y, w) \vee$$
$$\exists z \neq x, y\ \forall w\ M(x, y, z, w)]],$$

where $z \neq x, y$ means $[z \neq x \wedge z \neq y]$. In the second disjunct we can rewrite z by y and distribute the quantifiers so:

$$\forall x[\forall w\ M(x, x, x, w) \vee$$
$$\exists y \neq x[\forall w\ M(x, x, y, w) \vee \forall w\ M(x, y, x, w) \vee$$
$$\forall w\ M(x, x, y, w) \vee \exists z \neq x, y\ \forall w\ M(x, y, z, w)]]$$

Now *if* we assume that the domain has at least *two* elements (I think Gödel overlooked this point—or I misunderstood what he told me), then we have the "sharp" distributive law:

$$[P \wedge \exists y \neq x\ Q(y)] \leftrightarrow \exists y \neq x\ [P \wedge Q(y)],$$

where y is not free in P. Similarly, if we assume at least *three* elements, we can distribute $\exists z \neq x, y$. Putting the whole mess into prenex form we next obtain:

$$\forall x\ \exists y \neq x\ \exists z \neq x, y\ \forall w_0\ \forall w_1\ \forall w_2\ \forall w_3\ \forall w_4\ [M(x, x, x, w_0) \vee$$
$$M(x, x, y, w_1) \vee M(x, y, x, w_2) \vee M(x, y, y, w_3) \vee$$
$$M(x, y, z, w_4)],$$

the point being that there are still only *two* (now sharp!) existential quantifiers.

Universal quantifiers can be "sharpened" as follows:

$$\forall w\ N(x, y, z, w)$$

is equivalent to:

$$[N(x,y,z,x) \wedge N(x,y,z,y) \wedge N(x,y,z,z) \wedge \\ \forall w \neq x,y,z\ N(x,y,z,w)].$$

Assuming a domain of at least *four* elements we can distribute this out to:

$$\forall w \neq x,y,z\ [N(x,y,z,x) \wedge N(x,y,z,y) \wedge \\ N(x,y,z,z) \wedge N(x,y,z,w)]$$

Now if N has a purely universal prefix, it can be distributed to the front of the matrix (without introducing any new variables) and the next quantifier can then be sharpened. And so on.

3 | I see I overlooked the case where there are *two* or more universal quantifiers *preceding* the pair of existential quantifiers. This is probably not very important because consideration of the initial string of universal

b quantifiers can generally ~~g~~e treated as, say, Ackerman⟦n⟧ does in his book[c] (the only reference I have handy at the moment.) Or we can argue that

$$\forall x_0\ \forall x_1\ \forall x_2\ K\ (x_0, x_1, x_2)$$

is valid if and only if *all* of the following are valid

$$\forall x_0\ K\ (x_0, x_0, x_0),$$
$$\forall x_0 \forall x_{[\![1]\!]} \neq x_0\ K\ (x_0, x_1, x_1),$$
$$\forall x_0 \forall x_2 \neq x_0\ K\ (x_0, x_0, x_2),$$
$$\forall x_0 \forall x_{[\![1]\!]} \neq x_0\ K\ (x_0, x_1, x_0),$$
$$\forall x_0 \forall x_1 \neq x_0 \forall x_2 \neq x_0, x_1\ K\ (x_0, x_1, x_2).$$

That is to say, the original problem is equivalent ⟦to⟧ the conjunction of several "sharp" problems. To be more precise: the original formula (*) had $n + m + 2$ variables. Putting together the various remarks about sharpening quantifiers we find a finite set of formulas each with a sharp prefix which in conjunction is equivalent to (*) for domains with *more than* $n + m + 2$ elements. (No, I see this bound is too small, but there is some easily calculated, probably exponential bound that will suffice.) Hence, to decide which domains (*) is valid in, we have only to check a

[c]Scott apparently refers to the theorem of *Ackermann 1954*, chapter VI §4, which is applied to the Gödel class in chapter VII §2. A more perspicuous proof of the theorem occurs in *Dreben and Goldfarb 1979*, pp. 164–168.

few finite domains and then apply some general method on sharp formulas to finish the problem for sufficiently large domains.

By now you will have seen the point of all this: the sharp formulas corresponding to (*) are still in the same Gödel-Kalmar-Schütte prefix class. Furthermore, in a formula with a sharp prefix (which makes distinct variables have distinct values) all occurrences of equality can obviously be eliminated from the matrix. Finally Gödel claims that his original method applies unchanged to sharp formulas. I have not checked this. Once several years ago I spent several days understanding Gödel's paper—though I must say that I never "really" understood the combinatorial role of the number 7. I have heard that Dreben has a general approach to the decision problem which includes all the positive results. Of course, there is no guarantee that the general method will apply to sharp formulas. I do hope you both will check this over and let us know whether it works out. If so, I wonder whether Gödel's idea of sharp formulas is useful in a broader context. Has the idea been used before?

| Let me ask another question that has been on my mind for a long time: does all the work that has been expended on positive cases of the decision problem amount to anything? I mean: can we apply any of the general results to interesting "mathematical" problems. I made one very slight application once when I noted (in an abstract in the JSL)[d] that a careful reduction to Skolem normal form shows that the decision problem for predicate calculus in *two* variables (they can be used over and over again in non-prenex formulas) reduces to the problem for the Gödel prefix. (Henkin subsequently proved a more general result by a different method involving cylindric algebras.[e]) But that is not particularly interesting. What I tried very hard to do at one time was to reduce problems about the Kripke-style semantics for modal propositional calculi to some known solvable case. This would be very helpful for finite-model-property results if only it worked out—but it didn't. The trouble is that one always wants some axioms on the alternative relation between possible worlds, and these axioms always seem to contain too many quantifiers 4

[d] *Scott 1962*. Since Scott relies on the result of *Gödel 1933i*, his proof does not, as the abstract claims, cover the case with identity. But that two-variable logic with identity has the finite model property is proved in *Mortimer 1975*.

[e] It is unclear what this result could be. Henkin's main result in the area is a decidability result for cylindric algebras of dimension 2 (*Henkin, Monk and Tarski 1971*, lemma 2.5.4, and *1985*, theorem 4.2.7), which corresponds to the decidability of a certain fragment of logic with two variables (cf. *Henkin and Tarski 1961*, pp. 106–07). Cylindric-algebraic methods do prove an equivalent of the Scott-Mortimer result; see *Henkin, Monk and Tarski 1985*, theorem 4.2.9, and, more perspicuously, *Andréka and Németi 200?*. (This note and the previous one are much indebted to Hajnal Andréka, Steven Givant, and especially Leon Henkin for helpful correspondence and comments.)

(as one has in saying that a relation is *transitive.*) So I became rather discouraged. If you have any hopeful applications I would certainly like to hear about them.

I send best regards from Professor Gödel and myself, and I hope that I might see either or both of you in the near future.

Sincerely yours,

Dana S. Scott

3. Hao Wang to Stephen C. Kleene[a]

11 May 1977

Professor S. C. Kleene
Department of Mathematics
Van Vleck Hall
University of Wisconsin
Madison, Wisconsin 53706

Dear Professor Kleene,

Thank you for the telephone conversations. I now proceed to reproduce the substance of the letter as instructed and approved by Professor Gödel.

Professor Gödel has recently spoken to me about your paper on his work. While he finds several points unsatisfactory, he suggests that I should write to you just on the most serious aspect which concerns his introduction of the model of constructible sets for set theory and his proof of the consistency of GCH. The more suggestive proof is given in his *Proceedings* paper (*The consistency-proof of the generalized continuum hypothesis*, vol. 25, 1939, pp. 220–224[b]). He feels that your discussion on p. 773 is mistaken, giving the impression that the finite axiomatization is his major contribution. This is wrong in two respects. As he states explicitly on the first page of his monograph, the finite axiom system is due essentially to Bernays. Moreover, it is not at all necessary to use a finite axiom system, as is seen clearly from his treatment in the *Proceedings* paper which uses infinitely many axioms. More generally, he

[a]This letter and letter 4 concern objections Gödel had to statements in *Kleene 1976* and express formulations of what became the addendum to the paper published in 1978. See the introductory note to the correspondence with Wang in this volume.

[b]*Gödel 1939a*. The actual title does not contain the initial 'The'.

has discussed extensively the conceptual framework of his work in letters and personal communications published in my book *From mathematics to philosophy* (1974, pp. 9–12).[c]

His work on the continuum hypothesis, in Professor Gödel's own opinion, is best summarized as follow.

"Some years before the *Proceedings* paper of 1939, he first defined the system of constructible sets in the manner described in that paper, proved that the axioms of set theory (including the axiom of choice) hold for this system, and conjectured that the continuum hypothesis also holds for it. A few years later he proved this conjecture (even for the generalized continuum hypothesis) | by an argument similar to that used in the *Proceedings* paper of 1939, however, using a submodel of the constructible sets countable in the lowest case. His lectures reported in his monograph are only a modification of this proof using no metamathematics.

2

Speaking more exactly, Gödel discovered the constructible sets before his visit to Princeton in the fall of 1935 and he told von Neumann the definition (in the form of the *Proceedings* paper) and the conjecture he had during his stay at Princeton in 1935. Shortly afterwards Gödel fell ill and did not prove the conjecture until 1937. The conjecture was proved in the strong form including the GCH."

At present Professor Gödel usually stays at home: 145 Linden Lane, Princeton, New Jersey 08540.

With best wishes,

Yours sincerely,
Hao Wang

---

[c] *Wang 1974*.

4. Hao Wang to Stephen C. Kleene

22 June 1977

Professor S. C. Kleene
Department of Mathematics
University of Wisconsin
480 Lincoln Drive
Madison, Wisconsin 53706

Dear Professor Kleene,

I have sent a copy of your reply of 24 May to Professor Gödel and have had several discussions with him. He has told me that he appreciates your paper on his work and that the paper is perfectly acceptable otherwise with the one strong correction on the continuum hypothesis.

In order to avoid unnecessary details, Professor Gödel has proposed the following formulation of an addendum to your paper.

"Gödel called my attention to the fact that his main acheivement regarding the consistency of the GCH is not quite accurately reported in my paper. His main acheivement, he says, really is that he first introduced the concept of constructible sets into set theory defining it as in his Proceedings paper of 1939,[a] proved that the axioms of set theory (including the axiom of choice) hold for it, and conjectured that the continuum hypothesis also will hold. He told these things to von Neumann during his stay at Princeton in 1935. The discovery of the proof of this conjecture on the basis of his definition is not too difficult. Gödel gave the proof (also for the GCH) not until two years later because he had fallen ill in the meantime. This proof was using a submodel of the constructible sets in the lowest case countable, similar to the one commonly given today."

In reply to the last paragraph of your letter, Professor Gödel makes explicit two not very important objections.

"The first objection concerns the fact that it is not made clear to the reader that, when Gödel wrote his completeness paper, he did not know Skolem's work⟦,⟧ which was not mentioned in Hilbert and Ackermann's book of 1928.

2 | The second objection concerns the fact that the proof in his undecidability paper is rather awkward. But the explanation lies in the fact that it is completely formalized. This objection is, however, more important than the first one. It might be a good idea to mention it."

This completes my transcript of Professor Gödel's instructions concerning your paper and your letter.

With best wishes,

Yours sincerely,

Hao Wang

---

[a] *Gödel 1939a.*

Appendix B
(*1974a*)
Alternate version of Remark 3 of *1972a*

As a companion piece to the expanded English version of *1958*, Gödel planned to publish in *Dialectica* three short remarks on the incompleteness results. These appeared in volume II of these *Works* as *1972a*. That publication had been based on galley proofs of what Gödel had sent in 1970, with some handwritten revisions. Gödel never returned the proofs. However, Gödel wrote another version of Remark 3 criticizing an argument from *Turing 1937* that Gödel interpreted as intending to show that "mental procedures cannot carry any farther than mechanical procedures." The new version appeared in *Wang 1974*, pp. 325–326, within section 7, "Gödel on minds and machines," of chapter X.

Gödel sent this section to Ted Honderich, editor of the series in which *Wang 1974* appeared, with a letter of 19 July 1972 (in this volume). From this letter and the letter to Honderich of 27 June (also in this volume), it is clear that Gödel had rewritten the section rather thoroughly from Wang's last draft. Together with what is known of the galleys of *1972a*, that confirms the assertion of Wang (*1996*, p. 195) that *1974a* is the later version. A priori, the handwritten changes to the galleys might be still later. But why would Gödel at that point have made small changes to a text of which he had already sent in a rewritten version for publication?

On the substantive issues addressed in this remark, the reader is referred to the introductory note to Remark 3 of *1972a* in these *Works*, vol. II, pp. 292–304. However, attention should be called to the argument of Wilfried Sieg (*1994*, p. 98) challenging the interpretation of the argument of *Turing 1937* on which Gödel's criticism is based. The reader should judge for himself how much the considerable differences of wording of the two versions amount to real differences of substance.[a]

Charles Parsons

[a]Corresponding to the footnote in Remark 3 of *1972a* is the following note attached to the previous sentence in *Wang 1974*:

> To be added as a footnote at the word 'mathematics' on p. 73, line 3, of *The undecidable*, op. cit. ⟦*Davis 1965*; also these *Works*, vol. I, p. 370, line 22.⟧

325 | Turing, in *Proc. Lond. Math. Soc.* 42 (1936), p. 250,[a] gives an argument which is supposed to show that mental procedures cannot carry any farther than mechanical procedures. However, this argument is inconclusive, because it depends on the supposition that a finite mind is capable of only a finite number of distinguishable states. What Turing disregards completely is the fact that *mind, in its use, is not static, but constantly developing.* This is seen, e.g., from the infinite series of ever stronger axioms of infinity in set theory, each of which expresses a new idea or insight. A similar process takes place with regard to the primitive terms. E.g., the iterative concept of set became clear only in the past few decades. Several more primitive ideas now appear on the horizon, e.g., the selfreflexive concept of proper class. Therefore, although at each stage of the mind's development the number of its possible states is finite, there is no reason why this number should not converge to infinity in the course of its development. Now there may exist systematic methods of accelerating, specializing, and uniquely determining this development, e.g. by asking the right questions on the basis of a mechanical procedure. But it must be admitted that the precise definition of a procedure of this kind would require a substantial deepening of our understanding of the basic operations of the mind. Vaguely defined 326 procedures of this kind, however, are known, e.g., the | process of defining recursive wellorderings of integers representing larger and larger ordinals or the process of forming stronger and stronger axioms of infinity in set theory.

[a] *Turing 1937*

Textual notes

The individual copy-texts and the concomitant textual issues not addressed in editorial footnotes to the letters are discussed under the individual correspondents.

In these notes, the pairs of numbers on the left indicate page and line in this volume. (Line numbers do not count titles at the top of a page.) We follow our usual editorial apparatus as described in the Information for the reader.

Arend Heyting

In letter 14 (15 November 1932), there are various numerals in red pencil on the recipient's copy in the Heyting Archief. According to Mark van Atten, they appear to be in Heyting's handwriting and lie in the left hand margin at points that Heyting may have been marking in preparation for reply. Some of these numerals are boxed (here indicated in boldface); we surmise that these are the ones with which he had dealt. These numerals are sometimes accompanied by red underlining nearby in the text. The numbers, locations and accompanying underlining are as follows:

	Numeral	Location	Underlined
58, 24	1	1 Hier bitte	endlichen Mengen zu streichen
60, 1	2	Unabhängig- 2 keit[a]	über Unabhängigkeit u. Voll- ständigkeit
60, 4	3	3 III p. 12	Ackermannsche Beweis
60, 5–6			für die Zahlentheorie nicht ausreichen
60, 10	4	4 III. Zusatz	
60, 11			Zahlentheorie
60, 17	**5**	**5** V. p. 6	
60, 19	**6**	**6** V. p. 12	
60, 23	**7**	**7** ~~III §1 p 3~~	
60, 29	**8**	**8** III §1 p 4	
61, 1	**9**	**9** V p 23	
61, 5	**10**	**10** Für sehr	
61, 7	11	11 das Verhaltnis	
61, 11	**12**	**12** Zu Abschn.	
61, 13	**13**	**13** für nötig?	
61, 15	**14**	**14** der Paradoxien	

[a]In the original, 'Unabhängigkeit' occurred at a line break.

Ernest Nagel

Letter 9 has, in addition to the transcribed shorthand note at the top, a crossed-out note that says "Über ver-" or "Über ihr".

Appendix A

565, 11–12	von Herrn Gödel genannte	commas before and after were crossed out

References

Aanderaa, Stål

See Dreben, Burton, Peter Andrews and Stål Aanderaa.

Ackermann, Wilhelm

1924 Begründung des "tertium non datur" mittels der Hilbertschen Theorie der Widerspruchsfreiheit, *Mathematische Annalen 93*, 1–36.

1940 Zur Widerspruchsfreiheit der Zahlentheorie, *Mathematische Annalen 117*, 162–194.

1952 Widerspruchsfreier Aufbau einer typenfreien Logik. (Erweitertes System), *Mathematische Zeitschrift 55*, 364–384.

1953 Widerspruchsfreier Aufbau einer typenfreien Logik, II, *Mathematische Zeitschrift 57*, 155–166.

1954 *Solvable cases of the decision problem* (Amsterdam: North-Holland).

1956 Zur Axiomatik der Mengenlehre, *Mathematische Annalen 131*, 336–345.

See also Hilbert, David, and Wilhelm Ackermann.

Addison, John W., Jr.

1958 Separation principles in the hierarchies of classical and effective descriptive set theory, *Fundamenta mathematicae 46*, 123–135.

See also Henkin et alii.

Addison, John W., Leon Henkin and Alfred Tarski

1965 (eds.) *The theory of models. Proceedings of the 1963 International Symposium at Berkeley* (Amsterdam: North-Holland).

Albers, Donald J., Gerald L. Alexanderson and Constance Reid

1990 (eds.) *More mathematical people* (Boston, San Diego and New York: Harcourt Brace Jovanovich).

Alexanderson, Gerald L.

See Albers, Donald J., Gerald L. Alexanderson and Constance Reid.

Anderson, C. Anthony

1998 Alonzo Church's contributions to philosophy and intensional logic, *The bulletin of symbolic logic 4*, 129–171.

Andrei Sakharov Archives and Human Rights Center

2000 Alexander Esenin-Volpin, in *Faces of resistance in the USSR* (photo exhibit), http://www.brandeis.edu/departments/sakharov/Exhibit (accessed 3 March 2002).

Andréka, Hajnal, and István Németi

200? Simple proof for decidability of the universal theory of cylindric set algebras of dimension 2, forthcoming; preprint, Research Group in Algebraic Logic, Institute of Mathematics, Hungarian Academy of Sciences; available online at ftp://math-inst.hu/pub/algebraic-logic/Contents.html (accessed 3 March 2002).

Andrews, Peter

See Dreben, Burton, Peter Andrews and Stål Aanderaa.

Asquith, Peter D., and Philip Kitcher

1985 (eds.) *PSA 1984: Proceedings of the 1984 biennial meeting of the Philosophy of Science Association*, vol. 2, (East Lansing, MI: Philosophy of Science Association).

Awodey, Steven, and André Carus

2001 Carnap, completeness, and categoricity: the *Gabelbarkeitssatz* of 1928, *Erkenntnis 54*, 145–172.

Ayer, Alfred Jules

1959 (ed.) *Logical positivism* (Glencoe, Ill.: Free Press).

Barbosa, Jorge Emmanuel Ferreira

1973 Sobre a consistência de *u*-sistematizaçes transfinitamente impredicativas a consistência da matemática o metatheorema da escolha e a hipótese generalizada do continuum, *Boletim de análise e lógica matemática 5, 1.*

Barendregt, Hendrik P.

1997 The impact of the lambda calculus in logic and computer science, *The bulletin of symbolic logic 3*, 181–215.

Bar-Hillel, Yehoshua

1970 (ed.) *Mathematical logic and foundations of set theory* (Amsterdam: North-Holland).

Bar-Hillel, Yehoshua, E. I. J. Poznanski, Michael O. Rabin and Abraham Robinson

1961 (eds.) *Essays on the foundations of mathematics, dedicated to A. A. Fraenkel on his seventieth anniversary* (Jerusalem: Magnes Press; Amsterdam: North-Holland).

Barwise, Jon

1975 *Admissible sets and structures: an approach to definability theory* (Berlin, Heidelberg and New York: Springer).

Bauer-Mengelberg, Stefan

1965 Review of *Gödel 1962a*, *The journal of symbolic logic 30*, 359–362.

Becker, Oskar

1930 Zur Logik der Modalitäten, *Jahrbuch für Philosophie und phänomenologische Forschung 11*, 497–548.

Behmann, Heinrich

1922 Beiträge zur Algebra der Logik, insbesondere zum Entscheidungsproblem, *Mathematische Annalen 86*, 163–229.

1927 *Mathematik und Logik*, Mathematisch-Physikalische Bibliothek, vol. 71 (Leipzig and Berlin: Teubner).

1931 Zu den Widersprüchen der Logik und der Mengenlehre, *Jahresbericht der Deutschen Mathematiker-Vereinigung 40*, 37–48.

1931a Zur Richtigstellung einer Kritik meiner Auflösung der logisch-mengentheoretischen Widersprüche, *Erkenntnis 2*, 305–306.

1959 Der Prädikatenkalkül mit limitierten Variablen. Grundlegung einer natürlichen exakten Logik, *The journal of symbolic logic 24*, 112–140.

Benacerraf, Paul, and Hilary Putnam

1964 (eds.) *Philosophy of mathematics: selected readings* (Englewood Cliffs, NJ: Prentice-Hall; Oxford: Blackwell).

1983 Second edition of *Benacerraf and Putnam 1964* (Cambridge: Cambridge University Press).

Bergamini, David, René Dubos, Henry Margenau and C. P. Snow

1963 (eds.) *Mathematics*, *Life* Science Library (New York: Time, Inc.).

Bernays, Paul

1910 Das Moralprinzip bei Sidgwick und bei Kant, *Abhandlungen der Fries'schen Schule, Neue Folge, 3. Band, 3. Heft*, 501–582 (also paginated by *Heft*, 1–82).

1913 Über den transzendentalen Idealismus, *Abhandlungen der Fries'schen Schule, Neue Folge, 4. Band, 2. Heft*, 365–394 (also paginated by *Heft*, 1–30).

1922 Über Hilberts Gedanken zur Grundlegung der Arithmetik, *Jahresbericht der Deutschen Mathematiker-Vereinigung 31*, 10–19; English translation by Paolo Mancosu in *Mancosu 1998*, 215–222.

1926 Axiomatische Untersuchung des Aussagen-Kalküls der *Principia mathematica*, *Mathematische Zeitschrift 25*, 305–320.

1928 Über Nelsons Stellungnahme in der Philosophie der Mathematik, *Die Naturwissenschaften, 16. Jahrgang, Heft 9*, 142–145.

1930 Die Philosophie der Mathematik und die Hilbertsche Beweistheorie, *Blätter für deutsche Philosophie 4*, 326–367; reprinted in *Bernays 1976*, 17–61; English translation by Paolo Mancosu in *Mancosu 1998*, 234–265.

1933 Methoden des Nachweises von Widerspruchsfreiheit und ihre Grenzen, in *Saxer 1933*, 342–343.

1935 Sur le platonisme dans les mathématiques, *L'enseignement mathématique 34*, 52–69; English translation by Charles D. Parsons in *Benacerraf and Putnam 1964*, 274–286.

1937 A system of axiomatic set theory, Part I, *The journal of symbolic logic 2*, 65–77; reprinted in *Müller 1976*, 1–13.

1940 Review of *Lautman 1938*, *The journal of symbolic logic 5*, 20–22.

1940a Review of *Gödel 1938*, *The journal of symbolic logic 5*, 117–118.

1941 A system of axiomatic set theory. Part II, *The journal of symbolic logic 6*, 1–17; reprinted in *Müller 1976*, 14–30.

1941a Sur les questions méthodologiques actuelles de la théorie hilbertienne de la démonstration, in *Gonseth 1941*, 144–152.

1946 Review of *Gödel 1944*, *The journal of symbolic logic 11*, 75–79.

1953 Über die Fries'sche Annahme einer Wiederbeobachtung der unmittelbaren Erkenntnis, in *Specht and Eichler 1953*, 113–131.

1954a Bemerkungen zu der Betrachtung von Alexander Wittenberg: Über adequäte Problemstellung in der Grundlagenforschung, *Dialectica 8*, 147–151.

1955 Zur Frage der Anknüpfung an die Kantische Erkenntnistheorie; eine kritische Erörterung, *Dialectica 9*, 23–65, 195–221.

1957 Von der Syntax der Sprache zur Philosophie der Wissenschaften, *Dialectica 11*, 233–246.

1959 Betrachtungen zu Ludwig Wittgensteins 'Bemerkungen über die Grundlagen der Mathematik', *Ratio 2*, 1–18, also in English as: Comments on Ludwig Wittgenstein's Remarks on the Foundations of Mathematics, in *Ratio 2*, English edition, 1–22; reprinted in *Bernays 1976*, 119–141.

1961 Zur Frage der Unendlichkeitsschemata in der axiomatischen Mengenlehre, in *Bar-Hillel et alii 1961*, 3–49; English translation by John L. Bell and M. Plänitz in *Müller 1976*, 121–172.

1961a Zur Rolle der Sprache in erkenntnistheoretischer Hinsicht, *Synthese 18*, 185–200; reprinted in *Bernays 1976*, 155–169.

1961b Die hohen Unendlichkeiten und die Axiomatik der Mengenlehre, *Polish Academy of Sciences 1961*, 11–20.
1962 Review of *Leblanc 1962*, *The journal of symbolic logic 27*, 248–249.
1962a Remarks about formalization and models, in *Nagel, Suppes and Tarski 1962*, 176–180.
1964 Reflections on Karl Popper's epistemology, in *Bunge 1964*, 32–44.
1966 Gedanken zu dem Buch "Bildung und Mathematik (Mathematik als exemplarisches Gymnasialfach)" von Alexander Wittenberg, *Dialectica 20*, 27–42.
1969 Bemerkungen zur Philosophie der Mathematik, in *Akten des XIV. Internationalen Kongresses für Philosophie, Wien, Sept. 1968*, vol. III, 192–198; reprinted in *Bernays 1976*, 170–175.
1970 On the original Gentzen consistency proof for number theory, in *Myhill, Kino and Vesley 1970*, 409–417.
1971 Zum Symposium über die Grundlagen der Mathematik, *Dialectica 25*, 171–195; reprinted in *Bernays 1976*, 189–213.
1974 Concerning rationality, in *Schilpp 1974*, 597–604.
1976 *Abhandlungen zur Philosophie der Mathematik* (Darmstadt: Wissenschaftliche Buchgesellschaft).
1976a See *Müller 1976*.
1976b Kurze Biographie, in *Müller 1976*, xiv–xvi; English translation in *Müller 1976*, xi–xiii.

See also Hilbert, David, and Paul Bernays.

Bernays, Paul, and Abraham Adolf Fraenkel
1958 *Axiomatic set theory* (Amsterdam: North-Holland).

Bernays, Paul, and Moses Schönfinkel
1928 Zum Entscheidungsproblem der mathematischen Logik, *Mathematische Annalen 99*, 343–372.

Betsch, Christian
1926 *Fiktionen in der Mathematik* (Stuttgart: Fr. Frommanns Verlag).

Biggers, Earl Derr
1925 *The house without a key* (New York: Avenel Books).

Boolos, George
1976 On deciding the truth of certain statements involving the notion of consistency, *The journal of symbolic logic 41*, 778–781.

Boone, William Werner

1952 *Several simple unsolvable problems of group theory related to the word problem* (doctoral dissertation, Princeton University).

1954 Certain simple, unsolvable problems of group theory I, *Koninklijke Nederlandse Akademie van Wetenschappen, Proceedings, Series A: Mathematical Sciences 57*, 231–237; also *Indagationes mathematicae 16*, 231–237.

1954a Certain simple, unsolvable problems of group theory II, *Koninklijke Nederlandse Akademie van Wetenschappen, Proceedings, Series A: Mathematical sciences 57*, 492–497; also *Indagationes mathematicae 16*, 492–497.

1955 Certain simple unsolvable problems of group theory III, *Koninklijke Nederlandse Akademie van Wetenschappen, Proceedings, Series A: Mathematical sciences 58*, 252–256; also *Indagationes mathematicae 17*, 252–256.

1955a Certain simple unsolvable problems of group theory IV, *Koninklijke Nederlandse Akademie van Wetenschappen, Proceedings, Series A: Mathematical sciences 58*, 571–577; also *Indagationes mathematicae 17*, 571–577.

1957 Certain simple unsolvable problems of group theory V, *Koninklijke Nederlandse Akademie van Wetenschappen, Proceedings Series A: Mathematical sciences 60*, 22–27; also *Indagationes mathematicae 19*, 22–27.

1957a Certain simple unsolvable problems of group theory VI, *Koninklijke Nederlandse Akademie van Wetenschappen, Proceedings, Series A: Mathematical sciences 60*, 227–232; also *Indagationes mathematicae 19*, 227–232.

1959 The word problem, *Annals of mathematics (2) 70*, 207–265.

1966 Word problems and recursively enumerable degrees of unsolvability. A first paper on Thue systems, *Annals of mathematics (2) 83*, 520–571.

1966a Word problems and recursively enumerable degrees of unsolvability. A sequel on finitely presented groups, *Annals of mathematics 84*, 49–84.

Booth, David, and Renatus Ziegler

1996 *Finsler set theory: platonism and circularity* (Basel, Boston and Berlin: Birkhäuser).

Braithwaite, Richard Bevan

1962 Introduction, in *Gödel 1962a*, 1–32.

Breger, Herbert
1992 A restoration that failed: Paul Finsler's theory of sets, in *Gillies 1992*, 249–264.

Brouwer, Luitzen Egbertus Jan
1925 Über die Bedeutung des Satzes vom ausgeschlossenen Dritten in der Mathematik, insbesondere in der Funktionentheorie, *Journal für die reine und angewandte Mathematik 154*, 1–7.

Brutian, Georg
1968 On the conception of polylogic, *Mind* (n.s.) *77*, 351–359.

Büchi, Julius Richard
1989 *Finite automata, their algebras and grammars; towards a theory of formal expressions*, edited by Dirk Siefkes (New York, Berlin and Heidelberg: Springer).
1990 *The collected works of J. Richard Büchi*, edited by Saunders Mac Lane and Dirk Siefkes (Berlin: Springer).

Büchi, Sylvia
1990 The life of J. Richard Büchi, in *Büchi 1990*, 4–6.

Buldt, Bernd
See Köhler et alii.

Bulloff, Jack J., Thomas C. Holyoke and S. W. Hahn
1969 (eds.) *Foundations of mathematics. Symposium papers commemorating the sixtieth birthday of Kurt Gödel* (New York: Springer).

Bunge, Mario
1964 *The critical approach to science and philosophy, in honor of Karl R. Popper* (Glencoe, IL: The Free Press; London: Collier-Macmillan).

Burckhardt, J. J.
1980 *Die Mathematik an der Universität Zürich 1916–1950, unter den Professoren R. Fueter, A. Speiser, P. Finsler* (Basel, Boston and Stuttgart: Birkhäuser).

Buss, Samuel R.
1986 *Bounded arithmetic*, Studies in proof theory 3 (Naples: Bibliopolis).
1995 On Gödel's theorems on lengths of proofs II: lower bounds for recognizing k symbol provability, in *Clote and Remmel 1995*, 57–90.

Butts, Robert E., and Jaakko Hintikka
1977 (eds.) *Logic, foundations of mathematics and computability theory* (Dordrecht and Boston: Reidel).

Cantor, Georg
1895 Beiträge zur Begründung der transfiniten Mengenlehre. I, *Mathematische Annalen 46*, 481–512.
1932 *Gesammelte Abhandlungen mathematischen und philosophischen Inhalts. Mit erläuternden Anmerkungen sowie mit Ergänzungen aus dem Briefwechsel Cantor–Dedekind*, edited by Ernst Zermelo (Berlin: Springer); reprinted in 1962 (Hildesheim: Olms).

Carnap, Rudolf
1930a Bericht über Untersuchungen zur allgemeinen Axiomatik, *Erkenntnis 1*, 303–307.
1931 Die logizistische Grundlegung der Mathematik, *Erkenntnis 2*, 91–105; English translation by Erna Putnam and Gerald J. Massey in *Benacerraf and Putnam 1964*, 31–41.
1932 Überwindung der Metaphysik durch logische Analyse der Sprache, *Erkenntnis 2*, 219–241; English translation by Arthur Pap in *Ayer 1959*, 60–81.
1934 Die Antinomien und die Unvollständigkeit der Mathematik, *Monatshefte für Mathematik und Physik 41*, 263–284.
1934a *Logische Syntax der Sprache* (Vienna: Springer); translated into English by Amethe Smeaton as *Carnap 1937*.
1935 Ein Gültigkeitskriterium für die Sätze der klassischen Mathematik, *Monatshefte für Mathematik und Physik 42*, 163–190.
1935c *Philosophy and logical syntax* (London: Kegan Paul, Trench, Trubner).
1937 *The logical syntax of language* (London: Kegan Paul, Trench, Trubner; New York: Harcourt, Brace); English translation of *Carnap 1934a*, with revisions.
1963 Intellectual autobiography, in *Schilpp 1963*, 3–84.
2000 *Untersuchungen zur allgemeinen Axiomatik*, edited by Thomas Bonk and Jesús Mosterín (Darmstadt: Wissenschaftliche Buchgesellschaft).
See also Hahn et alii.

Carnes, Mark C.
See Garraty, John A., and Mark C. Carnes.

Chang, Chen Chung
See Henkin et alii.

Chen, Kien-Kwong
1933 Axioms for real numbers, *Tôhoku mathematical journal 37*, 94–99.

Chevalley, Claude
1934 Sur la pensée de J. Herbrand, *L'Enseignement mathématique 34*, 97–102; reprinted in *Herbrand 1968*, 17–20; translated into English by Warren Goldfarb in *Herbrand 1971*, 25–28.

Chevalley, Claude, and Albert Lautman
1931 Notice biographique sur Jacques Herbrand, *Annuaire de l'Association amicale de secours des anciens élèves de l'École normale supérieure*, 66–68; reprinted in *Herbrand 1968*, 13–15; translated into English by Warren Goldfarb in *Herbrand 1971*, 21–23.

Church, Alonzo
1932 A set of postulates for the foundation of logic, *Annals of mathematics (2) 33*, 346–366.
1933 A set of postulates for the foundation of logic (second paper), *Annals of mathematics (2) 34*, 839–864.
1934 The Richard paradox, *American mathematical monthly 41*, 356–361.
1935 A proof of freedom from contradiction, *Proceedings of the National Academy of Sciences, U.S.A. 21*, 275–281.
1965 Review of *Braithwaite 1962*, *The journal of symbolic logic 30*, 357–359.
1968 Paul J. Cohen and the continuum problem, *Proceedings of the International Congress of Mathematicians (Moscow-1966)*, 15–20.

Chwistek, Leon
1929 Neue Grundlagen der Logik und Mathematik, *Mathematische Zeitschrift 30*, 704–724.
1930 Une méthode métamathématique d'analyse, in *Sprawozdanie z 1. Kongresu Matematyków Krajów Słowiańskich (Comptes-rendus du 1. Congrès des Mathématiciens des Pays Slaves), Warszawa, 1929* (Warsaw), 254–263.
1932 Neue Grundlagen der Logik und Mathematik. Zweite Mitteilung, *Mathematische Zeitschrift 34*, 527–534.
1933 Die nominalistische Grundlegung der Mathematik, *Erkenntnis 3*, 367–388.

Clote, Peter
1999 Computation models and function algebras, in *Griffor 1999*, 589–681.

Clote, Peter, and Jan Krajíček
1993 (eds.) *Arithmetic, proof theory, and computational complexity*, Oxford logic guides 23 (Oxford: Clarendon Press).

Clote, Peter, and Jeffrey B. Remmel
1995 (eds.) *Feasible mathematics II*, Progress in computer science and applied logic, vol. 13 (Boston: Birkhäuser).

Cohen, Morris Raphael, and Ernest Nagel
1934 *An introduction to logic and scientific method* (New York: Harcourt, Brace and Company).

Cohen, Paul J.
1960 On a conjecture of Littlewood and idempotent measures, *American journal of mathematics 82*, 191–212.
1963 The independence of the continuum hypothesis, *Proceedings of the National Academy of Sciences, U.S.A. 50*, 1143–1148.
1963a A minimal model for set theory, *Bulletin of the American Mathematical Society 69*, 537–540.
1964 The independence of the continuum hypothesis, II, *Proceedings of the National Academy of Sciences, U.S.A. 51*, 105–110.
1965 Independence results in set theory, in *Addison, Henkin and Tarski 1965*, 39–54.
1966 *Set theory and the continuum hypothesis* (New York: Benjamin).

Cooper, Necia Grant
1989 (ed.) *From cardinals to chaos: reflections on the life and legacy of Stanislaw Ulam* (Cambridge: Cambridge University Press).

Craig, William
See Henkin et alii.

Dauben, Joseph Warren
1995 *Abraham Robinson: the creation of nonstandard analysis; a personal and mathematical odyssey* (Princeton: Princeton University Press).

Davis, Martin
1965 (ed.) *The undecidable: basic papers on undecidable propositions, unsolvable problems and computable functions* (Hewlett, NY: Raven Press).
1982 Why Gödel didn't have Church's thesis, *Information and control 54*, 3–24.
1998 Review of *Dawson 1997*, *Philosophia mathematica (3) 6*, 116–128.

Davis, Martin, Hilary Putnam and Julia Robinson
1961 The decision problem for exponential diophantine equations, *Annals of mathematics (2) 74*, 425–436.

Dawson, John W., Jr.
1983 The published work of Kurt Gödel: an annotated bibliography, *Notre Dame journal of formal logic 24*, 255–284; addenda and corrigenda, *ibid. 25*, 283–287.
1984 Discussion on the foundation of mathematics, *History and philosophy of logic 5*, 111–129.
1984a Kurt Gödel in sharper focus, *The mathematical intelligencer 6 (4)*, 9–17.
1985 The reception of Gödel's incompleteness theorems, in *Asquith and Kitcher 1985*, 253–271; reprinted in *Shanker 1988a*, 74–95, and in *Drucker 1985*, 84–100.
1985a Completing the Gödel–Zermelo correspondence, *Historia mathematica 12*, 66–70.
1993 Prelude to recursion theory: the Gödel–Herbrand correspondence, in *Wolkowski 1993*, 1–13.
1997 *Logical dilemmas: the life and work of Kurt Gödel* (Wellesley, MA: A K Peters); translated into German by Jakob Kellner as *Dawson 1999*.
1999 *Kurt Gödel: Leben und Werk*, Computerkultur XI (Vienna and New York: Springer).

Dekker, Jacob C. E.
1962 (ed.) *Recursive function theory*, Proceedings of symposia in pure mathematics, vol. 5 (Providence, RI: American Mathematical Society).

Denton, John
See Dreben, Burton, and John Denton.

DePauli-Schimanovich, Werner
See Köhler et alii.

Dilworth, Robert Palmer
1961 (ed.) *Lattice theory*, Proceedings of the Second Symposium in Pure Mathematics (Providence, RI: American Mathematical Society).

Dingler, Hugo
1931 *Philosophie der Logik und Arithmetik* (Munich: Reinhardt).

Drake, Frank R., and John K. Truss
1988 (eds.) *Logic colloquium '86* (Amsterdam: North-Holland).

Dreben, Burton

1952 On the completeness of quantification theory, *Proceedings of the National Academy of Sciences, U.S.A. 38*, 1047–1052.

1963 Corrections to Herbrand, *Notices of the American Mathematical Society 10*, 285.

Dreben, Burton, Peter Andrews and Stål Aanderaa

1963 False lemmas in Herbrand, *Bulletin of the American Mathematical Society 69*, 699–706.

1963a Errors in Herbrand, *Notices of the American Mathematical Society 10*, 285.

Dreben, Burton, and John Denton

1966 A supplement to Herbrand, *The journal of symbolic logic 31*, 393–398.

Dreben, Burton, and Warren Goldfarb

1979 *The decision problem: solvable classes of quantificational formulas* (Reading: Addison-Wesley).

Dreben, Burton, and Jean van Heijenoort

1967 Introductory note to *Skolem 1928*, in *van Heijenoort 1967*, 508–512.

Drucker, Thomas L.

1985 *Perspectives on the history of mathematical logic* (Boston: Birkhäuser).

Dubislav, Walter

1926 *Über die Definition* (Berlin: H. Weiss).

1931 *Die Definition* (third, enlarged edition of *Dubislav 1926*), Beihefte der *Erkenntnis*, vol. I (Leipzig: Meiner).

1981 Fourth (unrevised) edition of *Dubislav 1931* with an introduction by Wilhelm K. Essler (Hamburg: Meiner).

Du Bois Reymond, Paul

1880 Der Beweis des Fundamentalsatzes der Integralrechnung: $\int_a^b F'(x)\,dx = F(b) - F(a)$, *Mathematische Annalen 16*, 115–128.

Dubos, René

See Bergamini et alii.

Dummett, Michael A. E.

1993 *Origins of analytical philosophy* (London: Duckworth; Cambridge: Harvard University Press, 1994).

Dunham, Bradford
See Wang, Hao and Bradford Dunham.

Easton, William B.
1970 Powers of regular cardinals, *Annals of mathematical logic 1*, 139–178.

Edwards, Paul
1967 (ed.) *The encyclopedia of philosophy* (New York: Macmillan and The Free Press; London: Collier-Macmillan).

Ehrenfeucht, Andrzej, and Georg Kreisel
1967 Review of *Esenin-Vol'pin 1961*, *The journal of symbolic logic 32*, 517.

Eichler, Willi
See Specht, Minna, and Willi Eichler.

Enderton, Herbert B.
1995 In memoriam, Alonzo Church, 1903–1995, *The bulletin of symbolic logic 1*, 486–488.
1998 Alonzo Church and the reviews, *The bulletin of symbolic logic 4*, 172–180.

Erdős, Paul, and Alfred Tarski
1961 On some problems involving inaccessible cardinals, in *Bar-Hillel et alii 1961*, 50–82.

Esenin-Vol'pin, Alexander Sergeievich (Essenin-Volpin, Yessenin-Volpin, Ésénine-Volpine; Есенин-Вольпин, Александр Сергеевич)
1961 Le programme ultra-intuitionniste des fondements des mathématiques, in *Polish Academy of Sciences 1961*, 201–223.
1970 The ultra-intuitionistic criticism and the antitraditional program for foundations of mathematics, in *Myhill, Kino and Vesley 1970*, 3–45.

Everett, C. J., and Stanisław M. Ulam
1945 Projective algebra I, *American journal of mathematics 68*, 77–88; reprinted in *Ulam 1974*, 231–242.

Ewald, William B.
1996 *From Kant to Hilbert: a source book in the foundations of mathematics*, 2 vols. (Oxford: Clarendon Press).

Feferman, Anita Burdman
1993 *Politics, logic and love: the life of Jean van Heijenoort* (Boston and London: Jones and Bartlett; Wellesley: AK Peters); reprinted as *A. B. Feferman 2001*.

1999 How the unity of science saved Alfred Tarski, in *Woleński and Köhler 1999*, 43–52.

2001 *From Trotsky to Gödel: the life of Jean van Heijenoort*, reprint of *A. B. Feferman 1993* (Natick: AK Peters).

Feferman, Solomon

1962 Transfinite recursive progressions of axiomatic theories, *The journal of symbolic logic 27*, 259–316.

1965 Some applications of the notions of forcing and generic sets, *Fundamenta mathematicae 56*, 325–345.

1969 Set-theoretical foundations of category theory (with an appendix by Georg Kreisel), in *Mac Lane 1969*, 201–247.

1982 Inductively presented systems and the formalization of metamathematics, in *van Dalen, Lascar and Smiley 1982*, 95–128.

1984 Toward useful type-free theories. I., *The journal of symbolic logic 49*, 75–111.

1984a Kurt Gödel: conviction and caution, *Philosophia naturalis 21*, 546–562; reprinted in *Shanker 1988a*, 96–114, and in *Feferman 1998*, 150–164.

1998 *In the light of logic* (New York and Oxford: Oxford University Press).

1999 Tarski and Gödel between the lines, in *Woleński and Köhler 1999*, 53–63.

Feferman, Solomon, and Azriel Lévy

1963 Independence results in set theory by Cohen's method. II (abstract), *Notices of the American Mathematical Society 10*, 593.

Feferman, Solomon, and Clifford Spector

1962 Incompleteness along paths in progressions of theories, *The journal of symbolic logic 27*, 383–390.

Feigl, Herbert

1929 *Theorie und Erfahrung in der Physik* (Karlsruhe: G. Braun); English translation of chapter III by Gisela Lincoln (revised and edited by Robert S. Cohen) in *Feigl 1981*, 116–144.

1981 *Inquiries and provocations: selected writings 1929–1974*, edited by Robert S. Cohen, Vienna Circle collection, vol. 14 (Dordrecht: Reidel).

Fine, Arthur, and Jarrett Leplin

1988 (eds.) *PSA 1988: Proceedings of the 1988 biennial meeting of the Philosophy of Science Association* (East Lansing, MI: Philosophy of Science Association).

Finsler, Paul

1926 Formale Beweise und die Entscheidbarkeit, *Mathematische Zeitschrift 25*, 676–682; reprinted in *Unger 1975*, 11–17; English translation by Stefan Bauer-Mengelberg in *van Heijenoort 1967*, 440–445; another English translation, by David Booth, Renatus Ziegler and David Renshaw, in *Booth and Ziegler 1996*, 50–55.

1944 Gibt es unentscheidbare Sätze?, *Commentarii mathematici helvetici 16*, 310–320; reprinted in *Unger 1975*, 97–107; English translation by David Booth, Renatus Ziegler and David Renshaw in *Booth and Ziegler 1996*, 63–72.

Floyd, Juliet

1995 On saying what you really want to say: Wittgenstein, Gödel, and the trisection of the angle, in *Hintikka 1995*, 373–425.

Flügge, Siegfried

1939 Kann der Energieinhalt technisch nutzbar gemacht werden?, *Die Naturwissenschaften, 27. Jahrgang 23/24*, 402–410.

Follett, Wilson

1966 *Modern American usage: a guide*, edited and completed by Jacques Barzun, in collaboration with Carlos Baker, Frederick W. Dupee, Dudley Fitts, James D. Hart, Phillis McGinley and Lionel Trilling (New York: Hill and Wang).

1998 First revised edition of *Follett 1966*, revised by Erik Wensberg (New York: Hill and Wang).

Fraenkel, Abraham Adolf

1922 Der Begriff 'definit' und die Unabhängigkeit des Auswahlaxioms, *Sitzungsberichte der Preussischen Akademie der Wissenschaften, Physikalisch-mathematische Klasse*, 253–257; English translation by Beverly Woodward in *van Heijenoort 1967*, 284–289.

See also Bernays, Paul, and Abraham Adolf Fraenkel.

Frank, Philipp

1947 *Einstein, his life and times*, English translation of *Frank 1949* by George Rosen, edited and revised by Shuichi Kusaka (New York: Alfred A. Knopf).

1949 *Einstein, sein Leben und seine Zeit* (Munich, Leipzig and Freiburg: P. List); reprinted as *Frank 1979*; English translation as *Frank 1947*.

1979 Reprint of *Frank 1949*, with foreword by Albert Einstein (Braunschweig, Wiesbaden: Vieweg).

Frege, Gottlob

1879 *Begriffsschrift, eine der arithmetischen nachgebildete Formelsprache des reinen Denkens* (Halle: Nebert); English translation by Stefan Bauer-Mengelberg in *van Heijenoort 1967*, 1–82.

1893 *Grundgesetze der Arithmetik, begriffsschriftlich abgeleitet*, vol. 1 (Jena: H. Pohle); reprinted in 1962 (Hildesheim: Olms); partial English translation by Philip E. B. Jourdain and Johann Stachelroth in *Frege 1915*, *Frege 1916* and *Frege 1917*, with an introduction by Jourdain, reprinted with alterations in *Frege 1952*, 117–138; translation of excerpts also by Michael Beaney in *Frege 1997*, 194–223.

1903 *Grundgesetze der Arithmetik, begriffsschriftlich abgeleitet*, vol. 2 (Jena: Pohle); reprinted in 1962 (Hildesheim: Olms); partial English translation by Peter T. Geach and Max Black in *Frege 1952*, 159–244, partly reprinted, with additional selections translated by Michael Beaney, in *Frege 1997*, 258–289.

1915 The fundamental laws of arithmetic, *The monist 25*, 481–494; English translation of part of *Frege 1893*.

1916 The fundamental laws of arithmetic: psychological logic, *The monist 26*, 181–199; English translation of part of *Frege 1893*.

1917 Class, function, concept, relation, *The monist 27*, 114–127; English translation of part of *Frege 1893*.

1918 Der Gedanke; eine logische Untersuchung, *Beiträge zur Philosophie des deutschen Idealismus I*, 58–77; reprinted in *Frege 1967*, 342–362; English translation by Peter T. Geach and Robert H. Stoothoff in *Frege 1984*, 351–372, reprinted in *Frege 1997*, 325–345.

1918a Die Verneinung: eine logische Untersuchung, *Beiträge zur Philosophie der deutschen Idealismus I*, 143–157; reprinted in *Frege 1967*, 362–378; English translation by Peter T. Geach in *Frege 1984*, 373–389, reprinted in *Frege 1997*, 346–361.

1952 *Translations from the philosophical writings of Gottlob Frege*, edited by Peter T. Geach and Max Black (Oxford: Basil Blackwell; New York: Philosophical Library).

1967 *Kleine Schriften*, edited by Ignacio Angelelli (Hildesheim: Georg Olms).

1976 *Wissenschaftlicher Briefwechsel*, edited by Gottfried Gabriel, Hans Hermes, Friedrich Kambartel, Christian Thiel and Albert Veraart, vol. 2 of *Nachgelassene Schriften und wissenschaftlicher Briefwechsel*, edited by Hans Hermes, Friedrich Kambartel and Friedrich Kaulbach (Hamburg: Felix Meiner, 1969–1976).

1980 *Philosophical and mathematical correspondence*, edited by Gottfried Gabriel, Hans Hermes, Friedrich Kambartel, Christian Thiel and Albert Veraart, abridged from *Frege 1976* by Brian McGuinness, and translated by Hans Kaal (Chicago: University of Chicago Press; Oxford: Basil Blackwell).

1984 *Collected papers on mathematics, logic and philosophy*, edited by Brian McGuinness, translated by Max Black, V. H. Dudman, Peter T. Geach, Hans Kaal, E.-H. W. Kluge, Brian McGuinness and Robert H. Stoothoff (Oxford and New York: Basil Blackwell).

1997 *The Frege reader*, edited by Michael Beaney (Oxford, Malden: Blackwell).

Friedberg, Richard

1957 Two recursively enumerable sets of incomparable degrees of unsolvability (solution of Post's problem, 1944), *Proceedings of the National Academy of Sciences, U.S.A. 43*, 236–238.

Friedman, Harvey

1975 One hundred two problems in mathematical logic, *The journal of symbolic logic 40*, 113–129.

Friedman, Michael

2000 *A parting of the ways. Carnap, Cassirer, and Heidegger* (Chicago and La Salle: Open Court).

Fries, Jakob Friedrich

1803 *Philosophische Rechtslehre und Kritik aller positiven Gesetzgebung, mit Beleuchtung der gewöhnlichen Fehler in der Bearbeitung des Naturrechts* (Jena: Mauke); reprinted in *Fries 1968–*, vol. 9 (Abteilung 2, Bd. 1, 1–206).

1924 *System der Metaphysik. Ein Handbuch für Lehrer und zum Selbstgebrauch.* (Heidelberg: Christian Friedrich Winter).

1968– *Sämtliche Schriften*, edited by Gert König and Lutz Geldsetzer (Aalen: Scientia Verlag).

Gabelsberger, Franz Xaver

1834 *Anleitung zur deutschen Redezeichenkunst oder Stenographie* (Munich: Georg Franz); republished in 1908 (Wölfenbüttel: Heckner).

Gabriel, Pierre

1962 Des catégories abéliennes, *Bulletin de la Société Mathématique de France 90*, 323–448.

Garey, Michael R., and David S. Johnson
1979 *Computers and intractability—a guide to the theory of NP-completeness* (San Francisco: W. H. Freeman).

Garraty, John A., and Mark C. Carnes
1999 (eds.) *American national biography* (New York and Oxford: Oxford University Press).

Geiser, James
1974 A formalization of Essenin-Volpin's proof theoretical studies by means of nonstandard analysis, *The journal of symbolic logic 39*, 81–87.

Gentzen, Gerhard
1936 Die Widerspruchsfreiheit der reinen Zahlentheorie, *Mathematische Annalen 112*, 493–565; English translation by M. E. Szabo in *Gentzen 1969*, 132–213.
1969 *The collected papers of Gerhard Gentzen*, edited and translated into English by M. E. Szabo (Amsterdam: North-Holland).

George, Alexander
1994 (ed.) *Mathematics and mind* (New York and Oxford: Oxford University Press).

Geroch, Robert
1973 Energy extraction, *Sixth Texas Symposium on Relativistic Astrophysics, Annals of the New York Academy of Sciences 224*, edited by Dennis J. Hegyi, 108–117.

Gillies, Donald
1992 (ed.) *Revolutions in mathematics* (New York: Oxford University Press).

Girard, Jean-Yves
See Nagel, Ernest, James R. Newman, Kurt Gödel and Jean-Yves Girard.

Gödel, Kurt
1929 *Über die Vollständigkeit des Logikkalküls* (doctoral dissertation).
1930 Die Vollständigkeit der Axiome des logischen Funktionenkalküls, *Monatshefte für Mathematik und Physik 37*, 349–360.
1930a Über die Vollständigkeit des Logikkalküls, *Die Naturwissenschaften 18*, 1068.
1930b Einige metamathematische Resultate über Entscheidungsdefinitheit und Widerspruchsfreiheit, *Anzeiger der Akademie der Wissenschaften in Wien 67*, 214–215.

*1930c Vortrag über Vollständigkeit des Funktionenkalküls, in *Gödel 1995*, 16–29, with English translation by Jean van Heijenoort, John W. Dawson, Jr., William Craig and Warren Goldfarb.

1931 Über formal unentscheidbare Sätze der *Principia mathematica* und verwandter Systeme I, *Monatshefte für Mathematik und Physik 38*, 173–198; translated into English by Jean van Heijenoort as *Gödel 1967*.

1931a Diskussion zur Grundlegung der Mathematik (Gödel's remarks in *Hahn et alii 1931*), *Erkenntnis 2*, 147–151; English translation by John W. Dawson, Jr. in *Dawson 1984*, 125–128.

1931b Review of *Neder 1931*, *Zentralblatt für Mathematik und ihre Grenzgebiete 1*, 5–6.

1931c Review of *Hilbert 1931*, *ibid. 1*, 260.

1931d Review of *Betsch 1926*, *Monatshefte für Mathematik und Physik (Literaturberichte) 38*, 5.

1931e Review of *Becker 1930*, *ibid. 38*, 5–6.

1931f Review of *Hasse and Scholz 1928*, *ibid. 38*, 37.

1931g Review of *von Juhos 1930*, *ibid. 38*, 39.

*1931? ⟦Über unentscheidbare Sätze⟧, in *Gödel 1995*, 30–35, with English translation by Stephen C. Kleene, John W. Dawson, Jr. and William Craig.

1932 Zum intuitionistischen Aussagenkalkül, *Anzeiger der Akademie der Wissenschaften in Wien 69*, 65–66; reprinted, with additional comment, as *1933n*.

1932a Ein Spezialfall des Entscheidungsproblems der theoretischen Logik, *Ergebnisse eines mathematischen Kolloquiums 2*, 27–28; reprinted in *Menger 1998*, 145–146.

1932b Über Vollständigkeit und Widerspruchsfreiheit, *ibid. 3*, 12–13; reprinted in *Menger 1998*, 168–169.

1932c Eine Eigenschaft der Realisierung des Aussagenkalküls, *ibid. 3*, 20–21; reprinted in *Menger 1998*, 176–177.

1932d Review of *Skolem 1931*, *Zentralblatt für Mathematik und ihre Grenzgebiete 2*, 3.

1932e Review of *Carnap 1931*, *ibid. 2*, 321.

1932f Review of *Heyting 1931*, *ibid. 2*, 321–322.

1932g Review of *von Neumann 1931*, *ibid. 2*, 322.

1932h Review of *Klein 1931*, *ibid. 2*, 323.

1932i Review of *Hoensbroech 1931*, *ibid. 3*, 289.

1932j Review of *Klein 1932*, *ibid. 3*, 291.

1932k Review of *Church 1932*, *ibid. 4*, 145–146.

1932l Review of *Kalmár 1932*, *ibid. 4*, 146.

1932m Review of *Huntington 1932*, *ibid. 4*, 146.

1932n Review of *Skolem 1932*, *ibid. 4*, 385.

1932o Review of *Dingler 1931*, *Monatshefte für Mathematik und Physik* (*Literaturberichte*) *39*, 3.

1933 Untitled remark following *Parry 1933*, *Ergebnisse eines mathematischen Kolloquiums 4*, 6; reprinted in *Menger 1998*, 188.

1933a Über Unabhängigkeitsbeweise im Aussagenkalkül, *ibid. 4*, 9–10; reprinted in *Menger 1998*, 191–192.

1933b Über die metrische Einbettbarkeit der Quadrupel des R_3 in Kugelflächen, *ibid. 4*, 16–17; reprinted in *Menger 1998*, 198–199.

1933c Über die Waldsche Axiomatik des Zwischenbegriffes, *ibid. 4*, 17–18; reprinted in *Menger 1998*, 199–200.

1933d Zur Axiomatik der elementargeometrischen Verknüpfungsrelationen, *ibid. 4*, 34; reprinted in *Menger 1998*, 216.

1933e Zur intuitionistischen Arithmetik und Zahlentheorie, *ibid. 4*, 34–38; reprinted in *Menger 1998*, 216–220.

1933f Eine Interpretation des intuitionistischen Aussagenkalküls, *ibid. 4*, 39–40; reprinted in *Menger 1998*, 221–222.

1933g Bemerkung über projektive Abbildungen, *ibid. 5*, 1; reprinted in *Menger 1998*, 229.

1933h (with K. Menger and A. Wald) Diskussion über koordinatenlose Differentialgeometrie, *ibid. 5*, 25–26; reprinted in *Menger 1998*, 253–254.

1933i Zum Entscheidungsproblem des logischen Funktionenkalküls, *Monatshefte für Mathematik und Physik 40*, 433–443.

1933j Review of *Kaczmarz 1932*, *Zentralblatt für Mathematik und ihre Grenzgebiete 5*, 146.

1933k Review of *Lewis 1932*, *ibid. 5*, 337–338.

1933l Review of *Kalmár 1933*, *ibid. 6*, 385–386.

1933m Review of *Hahn 1932*, *Monatshefte für Mathematik und Physik* (*Literaturberichte*) *40*, 20–22.

1933n Reprint of *Gödel 1932*, with additional comment, *Ergebnisse eines mathematischen Kolloquiums 4*, 40; reprinted in *Menger 1998*, 222.

*1933o The present situation in the foundations of mathematics, in *Gödel 1995*, 45–53.

*1933? Vereinfachter Beweis eines Steinitzchen Satzes, in *Gödel 1995*, with English translation by John W. Dawson, Jr., Israel Halperin and William Craig, 56–61.

1934 *On undecidable propositions of formal mathematical systems* (mimeographed lecture notes, taken by Stephen C. Kleene and J. Barkley Rosser); reprinted with revisions in *Davis 1965*, 39–74.

1934a Review of *Skolem 1933*, *Zentralblatt für Mathematik und ihre Grenzgebiete 7*, 97–98.

1934b Review of *Quine 1933*, *ibid.* *7*, 98.

1934c Review of *Skolem 1933a*, *ibid.* *7*, 193–194.

1934d Review of *Chen 1933*, *ibid.* *7*, 385.

1934e Review of *Church 1933*, *ibid.* *8*, 289.

1934f Review of *Notcutt 1934*, *ibid.* *9*, 3.

1935 Review of *Skolem 1934*, *ibid.* *10*, 49.

1935a Review of *Huntington 1934*, *ibid.* *10*, 49.

1935b Review of *Carnap 1934*, *ibid.* *11*, 1.

1935c Review of *Kalmár 1934*, *ibid.* *11*, 3–4.

1936 Untitled remark following *Wald 1936*, *Ergebnisse eines mathematischen Kolloquiums* *7*, 6; reprinted in *Menger 1998*, 324.

1936a Über die Länge von Beweisen, *ibid.* *7*, 23–24; reprinted in *Menger 1998*, 341.

1936b Review of *Church 1935*, *Zentralblatt für Mathematik und ihre Grenzgebiete* *12*, 241–242.

1938 The consistency of the axiom of choice and of the generalized continuum hypothesis, *Proceedings of the National Academy of Sciences, U.S.A.* *24*, 556–557.

*1938a Vortrag bei Zilsel, in *Gödel 1995*, with English translation by Charles Parsons and Wilfried Sieg, 86–113.

1939 The consistency of the generalized continuum hypothesis, *Bulletin of the American Mathematical Society* *45*, 93.

1939a Consistency proof for the generalized continuum hypothesis, *Proceedings of the National Academy of Sciences, U.S.A.* *25*, 220–224; errata in *1947*, footnote 23.

*1939b Vortrag Göttingen, in *Gödel 1995*, with English translation by John W. Dawson, Jr. and William Craig.

*193? ⟦Undecidable diophantine propositions⟧, in *Gödel 1995*, 164–175.

1940 *The consistency of the axiom of choice and of the generalized continuum hypothesis with the axioms of set theory* (lecture notes taken by George W. Brown), Annals of mathematics studies, vol. 3 (Princeton: Princeton University Press); reprinted with additional notes in 1951 and with further notes in 1966.

*1940a Lecture ⟦on the⟧ consistency ⟦of the⟧ continuum hypothesis, in *Gödel 1995*, 175–185.

*1941 In what sense is intuitionistic logic constructive?, in *Gödel 1995*, 189–200.

1944 Russell's mathematical logic, in *Schilpp 1944*, 123–153; reprinted, with some alterations, as *Gödel 1964a* and as *Gödel 1972b*.

1946 Remarks before the Princeton bicentennial conference on problems in mathematics; first published in *Davis 1965*, 84–88; reprinted, with some alterations, as *Gödel 1968*.

*1946/9-B2 Some observations about the relationship between theory of relativity and Kantian philosophy, version B2, in *Gödel 1995*, 230–246.

*1946/9-C1 Some observations about the relationship between theory of relativity and Kantian philosophy, version C1, in *Gödel 1995*, 247–259.

1947 What is Cantor's continuum problem?, *American mathematical monthly 54*, 515–525; errata, *55*, 151; revised and expanded as *Gödel 1964*.

1949 An example of a new type of cosmological solutions of Einstein's field equations of gravitation, *Reviews of modern physics 21*, 447–450.

1949a A remark about the relationship between relativity theory and idealistic philosophy, in *Schilpp 1949*, 555–562.

*1949b Lecture on rotating universes, in *Gödel 1995*, 269–287.

*1951 Some basic theorems on the foundations of mathematics and their philosophical implications, in *Gödel 1995*, 304–323.

1952 Rotating universes in general relativity theory, *Proceedings of the International Congress of Mathematicians; Cambridge, Massachusetts, U. S. A. August 30–September 6, 1950*, I (Providence, RI: American Mathematical Society, 1952), 175–181.

*1953/9-III Is mathematics syntax of language?, version III, in *Gödel 1995*, 334–356.

*1953/9-V Is mathematics syntax of language?, version V, in *Gödel 1995*, 356–362.

1955 Eine Bemerkung über die Beziehungen zwischen der Relativitätstheorie und der idealistischen Philosophy (German translation of *Gödel 1949a* by Hans Hartmann), in *Schilpp 1955*, 406–412.

1958 Über eine bisher noch nicht benüzte Erweiterung des finiten Standpunktes, *Dialectica 12*, 280–287; revised and expanded in English as *Gödel 1972*; translated into English by Wilfrid Hodges and Bruce Watson as *Gödel 1980*.

*1961/? The modern development of the foundations of mathematics in the light of philosophy, in *Gödel 1995*, in German, with English translation by Eckehart Köhler, Hao Wang, John W. Dawson, Jr., Charles Parsons and William Craig, 374–387.

1962 Postscript to *Spector 1962*, 27.

1962a *On formally undecidable propositions of Principia mathematica and related systems*, translation of *Gödel 1931* by B. Melzer, with introduction by R. B. Braithwaite (Edinburgh and London: Oliver and Boyd; New York: Basic Books).

1964 Revised and expanded version of *Gödel 1947*, in *Benacerraf and Putnam 1964*, 258–273.

1964a Reprint, with some alterations, of *Gödel 1944*, in *Benacerraf and Putnam 1964*, 211–232.

1965 Expanded version of *Gödel 1934*, in *Davis 1965*, 39–74.

1967 English translation of *Gödel 1931*, in *van Heijenoort 1967*, 596–616.

1968 Reprint, with some alterations, of *Gödel 1946*, in *Klibansky 1968*, 250–253.

*1970 Ontological proof, in *Gödel 1995*, 403–404.

*1970a Some considerations leading to the probable conclusion that the true power of the continuum is $\aleph_2$, in *Gödel 1995*, 420–422.

*1970b A proof of Cantor's continuum hypothesis from a highly plausible axiom about orders of growth, in *Gödel 1995*, 422–423.

*1970c ⟦Unsent letter to Alfred Tarski⟧, in *Gödel 1995*, 424–425.

1972 On an extension of finitary mathematics which has not yet been used (to have appeared in *Dialectica*; first published in *Gödel 1990*, 271–280), revised and expanded English translation of *Gödel 1958*.

1972a Some remarks on the undecidability results (to have appeared in *Dialectica*; first published in *Gödel 1990*, 305–306).

1972b Reprint, with some alterations, of *Gödel 1944*, in *Pears 1972*, 192–226.

1974 Untitled remarks, in *Robinson 1974*, x.

1974a Alternate version of remark 3 of *1972a*, in *Wang 1974*, 325–326.

1980 On a hitherto unexploited extension of the finitary standpoint, English translation of *Gödel 1958*, *Journal of philosophical logic 9*, 133–142.

1986 *Collected works*, volume I: *Publications 1929–1936*, edited by Solomon Feferman, John W. Dawson, Jr., Stephen C. Kleene, Gregory H. Moore, Robert M. Solovay and Jean van Heijenoort (New York and Oxford: Oxford University Press).

1990 *Collected works*, volume II: *Publications 1938–1974*, edited by Solomon Feferman, John W. Dawson, Jr., Stephen C. Kleene, Gregory H. Moore, Robert M. Solovay and Jean van Heijenoort (New York and Oxford: Oxford University Press).

1995 *Collected works*, volume III: *Unpublished essays and lectures*, edited by Solomon Feferman, John W. Dawson, Jr., Warren Goldfarb, Charles Parsons and Robert M. Solovay (New York and Oxford: Oxford University Press).

See also Hahn et alii.

See also Nagel, Ernest, James R. Newman, Kurt Gödel and Jean-Yves Girard.

Goldfarb, Warren

1971 Review of *Skolem 1970*, *The journal of philosophy 68*, 520–530.

1979 Logic in the twenties: the nature of the quantifier, *The journal of symbolic logic 44*, 351–368.

1984 The Gödel class with identity is unsolvable, *Bulletin of the American Mathematical Society* (n.s.) *10*, 113–115.

1984a The unsolvability of the Gödel class with identity, *The journal of symbolic logic 49*, 1237–1252.

1993 Herbrand's error and Gödel's correction, *Modern logic 3*, 103–118.

2003 In memoriam: Burton Spencer Dreben, 1927–1999, *The bulletin of symbolic logic* (to appear).

See also Dreben, Burton, and Warren Goldfarb.

Gonseth, Ferdinand

1941 (ed.) *Les entretiens de Zurich, 6–9 decémbre 1938* (Zurich: Leemann).

Grattan-Guinness, Ivor

1979 In memoriam Kurt Gödel: his 1931 correspondence with Zermelo on his incompletability theorem, *Historia mathematica 6*, 294–304.

Greenberg, Marvin Jay

1974 *Euclidean and non-Euclidean geometries: development and history* (San Francisco: Freeman).

1980 Second edition of *Greenberg 1974*.

Griffor, E. R.

1999 (ed.) *Handbook of computability theory* (Amsterdam: Elsevier).

Gross, Herbert

1971 Nachruf: Paul Finsler, *Elemente der Mathematik 26*, 19–21.

Grossi, Marie L., Montgomery Link, Katalin Makkai and Charles Parsons

1998 A bibliography of Hao Wang, *Philosophia mathematica (3) 6*, 25–38.

Günther, Gotthard

1933 *Grundzüge einer neuen Theorie des Denkens in Hegels Logik* (Leipzig: Meiner).

1937 Wahrheit, Wirklichkeit, und Zeit: die transzendentalen Bedingungen einer Metaphysik der Geschichte, *Travaux du IXe congrès international de philosophie VIII*, 105–113 (Paris: Hermann); reprinted in *Günther 1976*, 1–9.

1940 Logistik und Transzendentallogik, *Die Tatwelt 16*, 135–147; reprinted in *Günther 1976*, 11–23.

1952 Die "zweite" Maschine. Nachwort to Isaac Asimov, *Ich, der Robot* (Düsseldorf: Rauch Verlag); reprinted with some omissions in *Günther 1963*, 179–203, and in full in *Günther 1976*, 91–114.

1953 Die philosophische Idee einer nicht-aristotelischen Logik, in *Proceedings of the XIth International Congress of Philosophy, Brussels, August 20–26, 1953. Vol. V: logic, philosophical analysis, philosophy of mathematics*, 44–50 (Amsterdam: North-Holland; Louvain: Nauwelaerts); reprinted in *Günther 1976*, 24–30.

1954 Achilles and the tortoise, *Astounding science fiction 53, no. 5*, 76–88; *no. 6*, 84–97; *54, no. 1*, 80–95.

1957 Metaphysik, Logik und die Theorie der Reflexion, *Archiv für Philosophie 7/1,2*, 1–44.

1957a *Das Bewusstsein der Maschinen: eine Metaphysik der Kybernetik* (Krefeld and Baden-Baden: Agis-Verlag).

1957b Ideen zu einer Metaphysik des Todes. Grundsätzliche Bemerkungen zu Arnold Metzger's "Freiheit und Tod", *Archiv für Philosophie 7/3, 4*, 335–347; reprinted in *Günther 1980*, 1–13.

1958 Die aristotelische Logik des Seins und die nicht-aristotelische Logik der Reflexion, *Zeitschrift für philosophische Forschung 12*, 360–407.

1959 *Idee und Grundriss einer nicht-Aristotelischen Logik: Erster Band: die Idee und ihre philosophischen Voraussetzungen* (Hamburg: Felix Meiner).

1963 Second edition of *Günther 1957a*, with additional texts.

1975 Selbstdarstellung im Spiegel Amerikas, in *Pongratz 1975, II*, 1–76.

1976 *Beiträge zur Grundlegung einer operationsfähigen Dialektik. Band I. Metakritik der Logik. Nicht-Aristotelische Logik. Reflexion. Stellenwertheorie. Dialektik. Cybernetic Ontology. Morphogrammatik.* (Hamburg: Meiner).

1978 Second edition of *Günther 1933*, with a new preface (Hamburg: Meiner).

1978a Second revised and expanded edition of *Günther 1959*, with a new preface and an appendix by Rudolf Kaehr (Hamburg: Meiner).

1979 *Beiträge zur Grundlegung einer operationsfähigen Dialektik. Zweiter Band: Wirklichkeit als Poly-Kontexturalität. Reflexion—Logische Paradoxie—Mehrwertige Logik—Denken—Wollen—Promielle Relation—Kenogrammatik—Dialektik der natürlichen Zahl—Dialektiker Materialismus* (Hamburg: Meiner).

1980 *Beiträge zur Grundlegung einer operationsfähigen Dialektik. Dritter Band: Philosophie der Geschichte und der Technik. Wille—Schöpfung—Arbeit. Strukturanalyse der Vermittlung. Mehrwertigkeit—Stellen- und Kontextwertlogik—Kenogrammatik.* (Hamburg: Meiner).

1991 *Idee und Grundriß einer nicht-Aristotelischen Logik: die Idee und ihre philosophischen Voraussetzungen*, third edition of *Günther 1959*, with a new preface by Claus Baldus and Bernard Mitteraurer, with additional texts by Günther but without the appendix of *1978a* (Hamburg: Felix Meiner).

Haas, Elke, and Gerrit Haas

1982 Heinrich Behmann (1891–1970), *Allgemeine Zeitschrift für Philosophie 7/1*, 59–65.

Hahn, Hans

1932 *Reelle Funktionen* (Leipzig: Akademische Verlagsgesellschaft).

Hahn, Hans, Rudolf Carnap, Kurt Gödel, Arend Heyting, Kurt Reidemeister, Arnold Scholz and John von Neumann

1931 Diskussion zur Grundlegung der Mathematik, *Erkenntnis 2*, 135–151; English translation by John W. Dawson, Jr. in *Dawson 1984*, 116–128.

Hahn, S. W.

See Bulloff, Jack J., Thomas C. Holyoke and S. W. Hahn.

Hajnal, András

1956 On a consistency theorem connected with the generalized continuum problem, *Zeitschrift für mathematische Logik und Grundlagen der Mathematik 2*, 131–136.

1961 On a consistency theorem connected with the generalized continuum problem, *Acta mathematica Academiae Scientiarum Hungaricae 12*, 321–376.

Hanf, William Porter

1963 *Some fundamental problems concerning languages with infinitely long expressions* (doctoral dissertation, University of California, Berkeley).

1964 Incompactness in languages with infinitely long expressions, *Fundamenta mathematicae 53*, 309–324.

Hartmanis, Juris

1989 Gödel, von Neumann and the P=?NP problem, *Bulletin of the European Association for Computer Science 38*, 101–107.

Hasse, Helmut, and Heinrich Scholz

1928 *Die Grundlagenkrisis der griechischen Mathematik* (Charlottenburg: Metzner).

Hausdorff, Felix

1907 Untersuchungen über Ordnungstypen, *Berichte über die Verhandlungen der Königlich Sächsischen Gesellschaft der Wissenschaften zu Leipzig, Mathematische-Physische Klasse 59*, 84–159.

1909 Die Graduierung nach der Endverlauf, *Abhandlungen der Königlich Sächsischen Gesellschaft der Wissenschaften, Mathematisch-Physischen Klasse 31*, 295–334.

1914 *Grundzüge der Mengenlehre* (Leipzig: Veit); reprinted in 1949 (New York: Chelsea).

1927 *Mengenlehre*, second revised edition of *Hausdorff 1914* (Berlin: Gruyter).

Hechler, Stephen M.

1974 On the existence of certain cofinal subsets of ${}^{\omega}\omega$, in *Jech 1974*, 155–173.

Heidegger, Martin

1929 *Was ist Metaphysik?* (Bonn: Friedrich Cohen); reprinted in *Heidegger 1967*, 1–19.

1943 Fourth edition of *1929*, with a new postscript (Frankfurt: Vittorio Klostermann); postscript reprinted in *Heidegger 1967*, 99–108.

1949 Fifth edition of *Heidegger 1929*, with the postscript to *Heidegger 1943* and a new introduction (Frankfurt: Vittorio Klostermann); introduction reprinted in *Heidegger 1967*, 195–211.

1967 *Wegmarken* (Frankfurt: Vittorio Klostermann).

1976 Enlarged edition of *Heidegger 1967*, with marginal notes of the author and pagination of *1967* in the margins, *Gesamtausgabe I. Abteilung: Veröffentlichte Schriften 1914–1970*, Band 9 (Frankfurt: Vittorio Klostermann).

1998 *Pathmarks*, edited by William O'Neill; translation of *Heidegger 1976* with pagination of *1967* in text (Cambridge and New York: Cambridge University Press).

Heinzmann, Gerhard
1982 *Schematisierte Strukturen. Eine Untersuchung über den "Idoneïsmus" Ferdinand Gonseths auf dem Hintergrund eines konstruktivistischen Ansatzes* (Bern and Stuttgart: Verlag Paul Haupt).

Heinzmann, Gerhard and Joëlle Proust
1988 Carnap et Gödel: Échange de lettres autour de la définition de l'analyticité. Introduction, traduction et notes, *Logique et analyse 31*, 257–291.

Heisenberg, Werner
1931 Kausalgesetz und Quantenmechanik, *Erkenntnis 2*, 172–182.

Henkin, Leon
See Addison, John W., Leon Henkin and Alfred Tarski.
See also Henkin et alii.
See also Henkin, Leon, J. Donald Monk and Alfred Tarski.
See also Henkin, Leon, and Alfred Tarski.
See also Suppes et alii.

Henkin, Leon, John W. Addison, Chen Chung Chang, William Craig, Dana S. Scott and Robert L. Vaught
1974 (eds.) *Proceedings of the Tarski symposium*, Proceedings of symposia in pure mathematics, vol. 25 (Providence, RI: American Mathematical Society).

Henkin, Leon, J. Donald Monk and Alfred Tarski
1971 *Cylindric algebras*, part I (Amsterdam: North Holland).
1985 *Cylindric algebras*, part II (Amsterdam, New York and Oxford: North Holland).

Henkin, Leon, and Alfred Tarski
1961 Cylindric algebras, in *Dilworth 1961*, 83–113.

Herbrand, Jacques
1929 Sur le problème fondamental des mathématiques, *Comptes rendus hebdomadaires des séances de l'Academie des Science (Paris) 189*, 554–556, erratum, 720; reprinted in *Herbrand 1968*, 31–33; English translation by Warren Goldfarb in *Herbrand 1971*, 41–43.

1930 *Recherches sur la théorie de la démonstration* (doctoral dissertation, University of Paris); also *Prace Towarzystwa Naukowego Warszawskiego, wydział III*, no. 33; reprinted in *Herbrand 1968*, 35–153; English translation by Warren Goldfarb in *Herbrand 1971*, 44–202.

1930a Les bases de la logique hilbertienne, *Revue de métaphysique et de morale 37*, 243–255; reprinted in *Herbrand 1968*, 155–166; English translation by Warren Goldfarb in *Herbrand 1971*, 203–214.

1931 Sur la non-contradiction de l'arithmétique, *Journal für die reine und angewandte Mathematik 166*, 1–8; reprinted in *Herbrand 1968*, 221–232; English translation by Jean van Heijenoort in *van Heijenoort 1967*, 618–628, and in *Herbrand 1971*, 282–298.

1931a Sur le problème fondamental de la logique mathématique, *Sprawozdania z posiedzeń Towarzystwa Naukowego Warszawskiego wydział III, 24*, 12–56; reprinted in *Herbrand 1968*, 167–207; English translation by Warren Goldfarb in *Herbrand 1971*, 215–271.

1931b Unsigned note on *Herbrand 1930*, *Annales de l'Université de Paris 6*, 186–189; reprinted in *Herbrand 1968*, 209–214; English translation by Warren Goldfarb in *Herbrand 1971*, 272–276.

1931c Notice pour Jacques Hadamard, in *Herbrand 1968*, 215–219; English translation by Warren Goldfarb in *Herbrand 1971*, 277–281.

1968 *Écrits logiques*, edited by Jean van Heijenoort (Paris: Presses Universitaires de France).

1971 *Logical writings*, English translation of *Herbrand 1968* by Warren Goldfarb (Cambridge: Harvard University Press; Dordrecht: Reidel).

Hertz, Paul

1922 Über Axiomensysteme für beliebige Satzsysteme, *Mathematische Annalen 87*, 246–269.

1923 Über Axiomensysteme für beliebige Satzsysteme, *Mathematische Annalen 89*, 76–102.

1929 Über Axiomensysteme für beliebige Satzsysteme, *Mathematische Annalen 101*, 457–514.

Heyting, Arend

1930 Die formalen Regeln der intuitionistischen Logik, *Sitzungsberichte der Preussischen Akademie der Wissenschaften, physikalisch-mathematische Klasse*, 42–56; English translation by Paolo Mancosu in *Mancosu 1998*, 311–327.

1930a Die formalen Regeln der intuitionistischen Mathematik, *ibid.*, 57–71.

1930b Sur la logique intuitionniste, *Académie royale de Belgique, Bulletins de la classe des sciences (5) 16*, 957–963; English translation by Amy L. Rocha in *Mancosu 1998*, 306–310.

1930c Die formalen Regeln der intuitionistischen Mathematik III, *Sitzungsberichte der Preussischen Akademie der Wissenschaften, Physikalisch-mathematische Klasse*, 158–169.

1931 Die intuitionistische Grundlegung der Mathematik, *Erkenntnis 2*, 106–115; English translation by Erna Putnam and Gerald J. Massey in *Benacerraf and Putnam 1964*, 42–49.

1934 *Mathematische Grundlagenforschung. Intuitionismus. Beweistheorie*, Ergebnisse der Mathematik und ihrer Grenzgebiete 3 (Berlin: Springer).

1959 (ed.) *Constructivity in mathematics. Proceedings of the colloquium held at Amsterdam, 1957* (Amsterdam: North-Holland).

See also Hahn et alii.

Hilbert, David

1899 *Grundlagen der Geometrie. Festschrift zur Feier der Enthüllung des Gauss–Weber Denkmals in Göttingen* (Leipzig: Teubner).

1918 Axiomatisches Denken, *Mathematische Annalen 78*, 405–415; reprinted in *Hilbert 1935*, 146–156; English translation by William B. Ewald in *Ewald 1996*, vol. II, 1107–1115.

1926 Über das Unendliche, *Mathematische Annalen 95*, 161–190; English translation by Stefan Bauer-Mengelberg in *van Heijenoort 1967*, 367–392.

1928 Die Grundlagen der Mathematik, *Abhandlungen aus dem mathematischen Seminar der Hamburgischen Universität 6*, 65–85; English translation by Stefan Bauer-Mengelberg and Dagfinn Føllesdal in *van Heijenoort 1967*, 464–479.

1931 Die Grundlegung der elementaren Zahlenlehre, *Mathematische Annalen 104*, 485–494; reprinted in part in *Hilbert 1935*, 192–195; English translation by William B. Ewald in *Ewald 1996*, vol. II, 1149–1157; reprinted in *Mancosu 1998*, 266–273.

1931a Beweis des tertium non datur, *Nachrichten von der Gesellschaft der Wissenschaften zu Göttingen, mathematisch-physikalische Klasse*, 120–125.

1935 *Gesammelte Abhandlungen*, vol. 3 (Berlin: Springer).

1962 *Grundlagen der Geometrie* (Stuttgart: Teubner); ninth edition of *Hilbert 1899*, revised and expanded by Paul Bernays.

Hilbert, David, and Wilhelm Ackermann
1928 *Grundzüge der theoretischen Logik* (Berlin: Springer).
1938 Second, revised edition of *Hilbert and Ackermann 1928*; translated into English by Lewis M. Hammond, George G. Leckie and F. Steinhardt as *Hilbert and Ackermann 1950.*
1950 *Principles of mathematical logic*, English translation of *Hilbert and Ackermann 1938* (New York: Chelsea).

Hilbert, David, and Paul Bernays
1934 *Grundlagen der Mathematik*, vol. I (Berlin: Springer).
1939 *Grundlagen der Mathematik*, vol. II (Berlin: Springer).
1968 Second edition of *Hilbert and Bernays 1934.*
1970 Second edition of *Hilbert and Bernays 1939.*

Hintikka, Jaakko
1995 (ed.) *From Dedekind to Gödel, essays on the development of the foundations of mathematics* (Dordrecht and Norwell: Kluwer Academic Publishers).
See also Butts, Robert E., and Jaakko Hintikka.

Hoensbroech, Franz G.
1931 Beziehungen zwischen Inhalt und Umfang von Begriffen, *Erkenntnis 2*, 291–300.

Hofmann, Paul
1931 *Das Problem des Satzes vom ausgeschlossenen Dritten* (Berlin: Pan-Verlagsgesellschaft); also published as *1931a* without the introduction.
1931a Das Problem des Satzes vom ausgeschlossenen Dritten, *Kant-Studien 36*, 83–125.

Hölder, Otto
1924 *Die mathematische Methode. Logisch erkenntnistheoretische Untersuchungen im Gebiete der Mathematik, Mechanik und Physik* (Berlin: Julius Springer).

Holyoke, Thomas C.
See Bulloff, Jack J., Thomas C. Holyoke and S. W. Hahn.

Huntington, Edward V.
1932 A new set of independent postulates for the algebra of logic with special reference to Whitehead and Russell's *Principia mathematica*, *Proceedings of the National Academy of Sciences, U.S.A. 18*, 179–180.
1934 Independent postulates related to C. I. Lewis's theory of strict implication, *Mind (n.s.) 43*, 181–198.

Janik, Allen S., and Stephen E. Toulmin

1973 *Wittgenstein's Vienna* (New York: Touchstone/Simon and Schuster).

Jech, Thomas

1974 (ed.) *Axiomatic set theory*, Proceedings of symposia in pure mathematics, vol. 13, part 2 (Providence, RI: American Mathematical Society).

1978 *Set theory* (New York: Academic Press).

1997 Second edition of *Jech 1978* (Berlin: Springer).

Jensen, Ronald Björn, and Carol Karp

1971 Primitive recursive set functions, in *Scott 1971*, 143–176.

Johnson, David S.

See Garey, Michael R., and David S. Johnson.

Johnston, William M.

1972 *The Austrian mind* (Berkeley, Los Angeles and London: University of California Press).

Joja, Athanase

See Suppes et alii.

Jørgensen, Jørgen

1931 *A treatise of formal logic. Its evolution and main branches, with its relations to mathematics and philosophy* (Copenhagen: Levin and Munksgaard; London: Oxford); republished in 1962 (New York: Russell and Russell).

Jourdain, Philip E. B.

1912 The development of the theories of mathematical logic and the principles of mathematics, *Quarterly journal of pure and applied mathematics 43*, 219–314; reprinted in *Frege 1976*, 275–301, reprinted in part in *Frege 1980*, 179–206.

Jung, Carl Gustav

1961 *Memories, dreams, reflections*, English translation of *Jung 1962* (New York: Pantheon; Toronto: Random House).

1962 *Erinnerungen, Träume, Gedanken*, recorded and edited by Aniela Jaffé (Zürich: Rascher); English translation by Richard and Clara Winston as *Jung 1961*.

Junker, Friedrich Heinrich

1919 *Höhere Analysis: I. Differentialrechnung: II. Integralrechnung*, Sammlung Göschen, vols. 87 and 88 (Leipzig and Berlin: Göschen'sche Verlagshandlung).

Kaczmarz, Stefan
1932 Axioms for arithmetic, *The journal of the London Mathematical Society 7*, 179–182.

Kalmár, László
1932 Ein Beitrag zum Entscheidungsproblem, *Acta litterarum ac scientiarum Regiae Universitatis Hungaricae Francisco-Josephinae, sectio scientiarum mathematicarum 5*, 222–236.
1933 Über die Erfüllbarkeit derjenigen Zählausdrücke, welche in der Normalform zwei benachbarte Allzeichen enthalten, *Mathematische Annalen 108*, 466–484.
1934 Über einen Löwenheimschen Satz, *Acta litterarum ac scientiarum Regiae Universitatis Hungaricae Francisco-Josephinae, sectio scientiarum mathematicarum 7*, 112–121.
1955 Über ein Problem, betreffend die Definition des Begriffes der allgemein-rekursiven Funktion, *Zeitschrift für mathematische Logik und Grundlagen der Mathematik 1*, 93–96.

Kanamori, Akihiro
1994 *The higher infinite* (Berlin, Heidelberg and New York: Springer).
1997 *The higher infinite*, corrected second printing (Heidelberg: Springer).

Kant, Immanuel
1781 *Critik der reinen Vernunft* (Riga: Hartknoch).
1787 *Kritik der reinen Vernunft*, second revised edition of *Kant 1781*.
1788 *Kritik der praktischen Vernunft* (Riga: Hartknoch); reprinted in *Kant 1902–*, vol. V; English translation by Mary J. Gregor in *Kant 1996*.
1790 *Kritik der Urteilskraft* (Berlin and Libau: Lagarde und Friederich); reprinted in *Kant 1902–*, vol. V; English translation by Paul Guyer and Eric Matthews in *Kant 2000*.
1902– *Kants gesammelte Schriften*, edited by the Prussian Academy of Sciences, later the German Academy of Sciences (Berlin: Georg Reimer, later Walter de Gruyter).
1996 *Practical philosophy*, edited and translated by Mary J. Gregor, general introduction by Allen W. Wood, The Cambridge edition of the Works of Immanuel Kant (Cambridge and New York: Cambridge University Press).
1997 *Critique of pure reason*, translated and edited by Paul Guyer and Allen W. Wood, The Cambridge edition of the Works of Immanuel Kant (Cambridge and New York: Cambridge University Press).

2000 *Critique of the power of judgement*, edited by Paul Guyer, The Cambridge edition of the Works of Immanuel Kant (Cambridge and New York: Cambridge University Press).

Karp, Carol
See Jensen, Ronald Björn, and Carol Karp.

Kaufmann, Felix
1930 *Das Unendliche in der Mathematik und seine Ausschaltung. Eine Untersuchung über die Grundlagen der Mathematik* (Leipzig and Vienna: Franz Deuticke).

Keisler, H. Jerome, and Alfred Tarski
1964 From accessible to inaccessible cardinals: results holding for all accessible cardinal numbers and the problem of their extension to inaccessible ones, *Fundamenta mathematicae 53*, 225–308.

Kitcher, Philip
See Asquith, Peter D., and Philip Kitcher.

Kleene, Stephen C.
1935 A theory of positive integers in formal logic, *American journal of mathematics 57*, 153–173, 219–244.
1936 General recursive functions of natural numbers, *Mathematische Annalen 112*, 727–742; reprinted in *Davis 1965*, 236–252; for an erratum, a simplification and an addendum, see *Davis 1965*, 253.
1938 On notation for ordinal numbers, *The journal of symbolic logic 3*, 150–155.
1943 Recursive predicates and quantifiers, *Transactions of the American Mathematical Society 53*, 41–73; reprinted in *Davis 1965*, 254–287; for a correction and an addendum, see *Davis 1965*, 254 and 287.
1955 On the forms of predicates in the theory of constructive ordinals (second paper), *American journal of mathematics 77*, 405–428.
1955a Arithmetical predicates and function quantifiers, *Transactions of the American Mathematical Society 79*, 312–340.
1955b Hierarchies of number-theoretic predicates, *Bulletin of the American Mathematical Society 61*, 193–213.
1976 The work of Kurt Gödel, *The journal of symbolic logic 41*, 761–778; addendum, *ibid. 43* (1978), 613; reprinted in *Shanker 1988a*, 48–71.
1987b Reflections on Church's thesis, *Notre Dame journal of formal logic 28*, 490–498.

Kleene, Stephen C., and J. Barkley Rosser
1935 The inconsistency of certain formal logics, *Annals of mathematics (2) 36*, 630–636.

Klein, Carsten
See Köhler et alii.

Klein, Fritz
1931 Zur Theorie der abstrakten Verknüpfungen, *Mathematische Annalen 105*, 308–323.
1932 Über einen Zerlegungssatz in der Theorie der abstrakten Verknüpfungen, *Mathematische Annalen 106*, 114–130.

Klibansky, Raymond
1968 (ed.) *Contemporary philosophy, a survey. I, Logic and foundations of mathematics* (Florence: La Nuova Italia Editrice).

Kino, Akiko
See Myhill, John, Akiko Kino and Richard E. Vesley.

Köhler, Eckehart
See Woleński, Jan and Eckehart Köhler .

Köhler, Eckehart, Bernd Buldt, Werner DePauli-Schimanovich, Carsten Klein, Michael Stöltzner and Peter Weibel
2002 (eds.) *Wahrheit und Beweisbarkeit. Leben und Werk Kurt Gödels*, Band 1: *Dokumente und historische Analysen*, Band 2: *Kompendium zu Gödels Werk* (Vienna: Hölder–Pichler–Tempsky).

Kondô, Motokiti
1938a Sur l'uniformisation des complémentaires analytiques et les ensembles projectifs de la seconde classe, *Japanese journal of mathematics 15*, 197–230.
1958 Sur les ensembles nommables et le fondement de analyse mathématique, I, *Japanese journal of mathematics 28*, 1–116.

König, Julius (König, Gyula)
1905 Zum Kontinuum-Problem, *Mathematische Annalen 60*, 177–180, 462.
1914 *Neue Grundlagen der Logik, Arithmetik und Mengenlehre* (Leipzig: Veit).

Krajíček, Jan
See Clote, Peter, and Jan Krajíček.

Kreisel, Georg
1958a Wittgenstein's remarks on the foundations of mathematics, *The British journal for the philosophy of science 9*, 135–158.

1959 Interpretation of analysis by means of constructive functionals of finite types, in *Heyting 1959*, 101–128.

1960 Ordinal logics and the characterization of informal concepts of proof, *Proceedings of the International Congress of Mathematicians, 14–21 August 1958* (Cambridge: Cambridge University Press), 289–299.

1962b Review of *Davis, Putnam and Robinson 1961*, *Mathematical reviews 24A*, 573.

1965 Mathematical logic, in *Saaty 1965*, vol. 3, 95–195.

1968 A survey of proof theory, *The journal of symbolic logic 33*, 321–388.

1968a Functions, ordinals, species, in *van Rootselaar and Staal 1968*, 143–158.

1970a Principles of proof and ordinals implicit in given concepts, in *Myhill, Kino and Vesley 1970*, 489–516.

1971a Observations on popular discussions of foundations, in *Scott 1971*, 189–198.

1980 Kurt Gödel, 28 April 1906–14 January 1978, *Biographical memoirs of Fellows of the Royal Society 26*, 148–224; corrections, *ibid. 27*, 697, and *28*, 718.

See also Ehrenfeucht, Andrzej, and Georg Kreisel.

Krikorian, Yervant Hovhaness

1944 (ed.) *Naturalism and the human spirit* (New York: Columbia University Press).

Kunen, Kenneth

1980 *Set theory: an introduction to independence proofs* (Amsterdam: North-Holland).

Kuratowski, Kazimierz (Kuratowski, Casimir)

1936 Les ensembles projectifs et l'induction transfinie, *Fundamenta mathematicae 27*, 269–276.

1937 Sur la géométrisation des types d'ordre dénombrable, *Fundamenta mathematicae 28*, 167–185.

1937a Les suites transfinies d'ensembles et les ensembles projectifs, *Fundamenta mathematicae 28*, 186–196.

1937b Sur les suites analytiques d'ensembles, *Fundamenta mathematicae 29*, 54–59.

1937c Les types définissables et les ensembles boreliens, *Fundamenta mathematicae 29*, 97–100.

Lascar, Daniel
See van Dalen, Dirk, Daniel Lascar and Timothy J. Smiley.

Lautman, Albert
1938 *Essai sur les notions de structure et d'existence en mathématiques*, Actualités scientifiques et industrielles (Paris: Hermann et Cie), 590–591.
See also Chevalley, Claude, and Albert Lautman.

Lawvere, F. William
1964 An elementary theory of the category of sets, *Proceedings of the National Academy of Sciences, U.S.A. 52*, 1506–1511.

Leblanc, Hugues
1962 Études sur les régles d'inférence dites régles de Gentzen, *Dialogue 1*, 56–66.

Leplin, Jarrett
See Fine, Arthur, and Jarrett Leplin.

Lettvin, Jerome Y., H. R. Maturana, Warren S. McCulloch and Walter H. Pitts
1959 What the frog's eye tells the frog's brain, *Proceedings of the IRE 47*, 1940–1951; reprinted in *McCulloch 1965*, 230–255.

Lévy, Azriel
1957 Indépendance conditionnelle de $V = L$ et d'axioms qui se rattachent au système de M. Gödel, *Comptes rendus hebdomadaires des séances de l'Académie des Sciences, Paris 245*, 1582–1583.
1959 On Ackermann's set theory, *The journal of symbolic logic 24*, 154–166.
1960 Axiom schemata of strong infinity in axiomatic set theory, *Pacific journal of mathematics 10*, 223–238.
1960a Principles of reflection in axiomatic set theory, *Fundamenta mathematicae 49*, 1–10.
1960b A generalization of Gödel's notion of constructibility, *The journal of symbolic logic 25*, 147–155.
1963 Independence results in set theory by Cohen's method. I, III and IV (abstracts), *Notices of the American Mathematical Society 10*, 592–593.

1970 Definability in axiomatic set theory II, in *Bar-Hillel 1970*, 129–145.

See also Feferman, Solomon, and Azriel Lévy.

Lévy, Azriel, and Robert M. Solovay

1967 Measurable cardinals and the continuum hypothesis, *Israel journal of mathematics 5*, 234–248.

Lewis, Clarence I.

1918 *A survey of symbolic logic* (Berkeley: University of California Press); reprinted by Dover (New York).

1932 Alternative systems of logic, *The monist 42*, 481–507.

Link, Montgomery

See Grossi et alii.

Lipshitz, Leonard, Dirk Siefkes and P. Young

1984 J. Richard Büchi (1924–1984), *Newsletter of the Association for Symbolic Logic*, November 1984; reprinted in *Büchi 1990*, 2–3.

Lorenzen, Paul

1951b Maß und Integral in der konstruktiven Analysis, *Mathematische Zeitschrift 54*, 275–290.

1955 *Einführung in die operative Logik und Mathematik* (Berlin: Springer).

Luzin, Nikolai (Lusin, Nicolas; Лузин, Николай Николаевич)

1930 *Leçons sur les ensembles analytiques et leurs applications* (Paris: Gauthier-Villars); reprinted with corrections, 1972 (New York: Chelsea).

1930a Analogies entre les ensembles mesurables $\mathcal{B}$ et les ensembles analytiques, *Fundamenta mathematicae 16*, 48–76.

Mac Lane, Saunders

1961 Locally small categories and the foundations of set theory, in *Polish Academy of Sciences 1961*, 25–43.

1969 (ed.) *Reports of the Midwest Category Seminar III*, Lecture notes in mathematics 106 (Berlin: Springer).

Macrae, Norman

1992 *John von Neumann* (New York: Pantheon Books).

Magidor, Menachem

1976 How large is the first strongly compact cardinal? Or: a study on identity crises, *Annals of mathematical logic 10*, 33–57.

Makkai, Katalin

See Grossi et alii.

Mancosu, Paolo

1998 (ed.) *From Brouwer to Hilbert: the debate on the foundations of mathematics in the 1920s* (New York: Oxford University Press).

1999 Between Russell and Hilbert: Behmann on the foundations of mathematics, *The bulletin of symbolic logic 5*, 303–330.

1999a Between Vienna and Berlin: the immediate reception of Gödel's incompleteness theorems, *History and philosophy of logic 20*, 33–45.

2002 On the constructivity of proofs. A debate among Behmann, Bernays, Gödel, and Kaufmann, in *Sieg, Sommer and Talcott 2002*, 349–371.

Margenau, Henry

See Bergamini et alii.

Markov, A. A. (Марков, А. А.)

1947 On the impossibility of certain algorithms in the theory of associative systems (Russian), *Doklady Akademii Nauk S.S.S.R.* (n.s.) *55*, 587–590; translated into English in *Comptes rendus (Doklady) de l'académie de l'URSS* (n.s.), *55*, 583–586.

Matiyasevich, Yuri Vladimirovich (Matijacevič; Матиясевич, Юрий Владимирович)

1970 Enumerable sets are diophantine (Russian), *Doklady Akademii Nauk S.S.S.R. 191*, 279–282; English translation, with revisions, in *Soviet mathematics Doklady 11* (1970), 354–358.

Maturana, H. R.

See Lettvin et alii.

Mauldin, R. Daniel

1981 (ed.) *The Scottish book, mathematics from the Scottish Cafe* (Boston: Birkhauser).

Mazurkiewicz, Stefan

1927 Sur une propriété des ensembles $C(A)$, *Fundamenta mathematicae 10*, 172–174; reprinted in *Mazurkiewicz 1969*, 152–154.

1969 *Travaux de topologie et ses applications*, edited by K. Borsuk, R. Engelking, B. Knaster, K. Kuratowski, J. Łoś and R. Sikorski (Warsaw: PWN).

McAloon, Kenneth

1966 *Some applications of Cohen's method* (doctoral dissertation, University of California at Berkeley).

McCulloch, Warren S.
1965 *Embodiments of mind* (Cambridge: MIT Press).
See also Lettvin et alii.

McCulloch, Warren S., and Walter Pitts
1943 A logical calculus of the ideas immanent in nervous activity, *Bulletin of mathematical biophysics 5*, 115–133; reprinted in *McCulloch 1965*, 19–39.

McKinsey, John C. C., and Alfred Tarski
1944 The algebra of topology, *Annals of mathematics (2) 45*, 141–191.
1946 On closed elements in closure algebras, *Annals of mathematics (2) 47*, 122–162.

Menger, Karl
1933 *Krise und Neuaufbau in den exakten Wissenschaften. Die neue Logik* (Leipzig and Vienna: Deuticke).
1953 *Calculus, a modern approach* (mimeographed notes, enlarged second edition) (Chicago: IIT Bookstore).
1953a The ideas of variable and function, *Proceedings of the National Academy of Sciences, U.S.A. 39*, 956–961.
1955 *Calculus, a modern approach* (Boston: Ginn).
1970 Projective and related structures, part 2 of *Menger and Blumenthal 1970*, 135–223.
1994 *Reminiscences of the Vienna Circle and the mathematical colloquium*, edited by Louise Golland, Brian McGuinness and Abe Sklar (Dordrecht: Kluwer Academic).
1998 *Ergebnisse eines mathematischen Kolloquiums*, edited by E. Dierker and K. Sigmund (Vienna and New York: Springer).
See also *Gödel 1933h*.

Menger, Karl, and Leonard M. Blumenthal
1970 *Studies in geometry* (San Francisco: Freeman).

Menger, Karl, and Georg Nöbeling
1932 (eds.) *Kurventheorie*, Mengentheoretische Geometrie in Einzeldarstellungen II (Leipzig and Berlin: B. G. Teubner).

Menzler-Trott, Eckart
2001 *Gentzens Problem. Mathematische Logik im nationalsozialistischen Deutschland* (Basel, Boston and Berlin: Birkhäuser).

Meyerson, Émile
1931 *Du chéminement de la pensée*, 3 vols. (Paris: F. Alcan).

Mitchell, Janet
1980 (ed.) *A community of scholars* (Princeton, NJ: The Institute for Advanced Study).

Mittelstraß, Jürgen
1980 (ed. in collaboration with Gereon Wolters) *Enzyklopädie Philosophie und Wissenschaftstheorie* (Mannheim, Vienna and Zürich: Bibliographisches Institut, B. I. Wissenschaftsverlag).

Moisil, Gr. C.
See Suppes et alii.

Monk, J. Donald
See Henkin, Leon, J. Donald Monk and Alfred Tarski.

Moore, Gregory H.
1980 Beyond first-order logic: the historical interplay between mathematical logic and axiomatic set theory, *History and philosophy of logic 1*, 95–137.
1988 The origins of forcing, in *Drake and Truss 1988*, 143–173.

Mortimer, Michael
1975 On languages with two variables, *Zeitschrift für mathematische Logik und Grundlagen der Mathematik 21*, 135–140.

Mostowski, Andrzej
1939 Über die Unabhängigkeit des Wohlordnungssatzes vom Ordnungsprinzip, *Fundamenta mathematicae 32*, 201–252.
1964 Widerspruchsfreiheit und Unabhängigkeit der Kontinuumhypothese, *Elemente der Mathematik 19*, 121–125.
1965 *Thirty years of foundational studies: lectures on the development of mathematical logic and the study of the foundations of mathematics in 1930–1964* (= no. 17 of *Acta philosophica fennica*); reprinted in 1966 (New York: Barnes and Noble; Oxford: Blackwell).

Müller, Gert H.
1961 Über die unendliche Induktion, in *Polish Academy of Sciences 1961*, 75–95.
1961a Nicht-Standard Modelle der Zahlentheorie, *Mathematische Zeitschrift 77*, 414–438.
1962 Über Rekursionsformen (Habilitationsschrift, University of Heidelberg); published in *Automatentheorie und formale Sprache: Bericht einer Tagung des mathematischen Forschungsinstituts Oberwohlfach, October 1969.*
1976 (ed.) *Sets and classes: on the work of Paul Bernays* (Amsterdam: North-Holland).

Murray, F. J.
See von Neumann, John, and F. J. Murray.

Mycielski, Jan, and Hugo Steinhaus
1962 A mathematical axiom contradicting the axiom of choice, *Bulletin de l'Académie polonaise des sciences, série des sciences mathématiques, astronomiques et physiques 10*, 1–3.

Myhill, John
1973 Embedding classical logic in intuitionistic logic, *Zeitschrift für mathematische Logik und Grundlagen der Mathematik 19*, 93–96.

Myhill, John, Akiko Kino and Richard E. Vesley
1970 (eds.) *Intuitionism and proof theory* (Amsterdam: North-Holland).

Myhill, John, and Dana S. Scott
1971 Ordinal definability, in *Scott 1971*, 271–278.

Nagel, Ernest
1944 Logic without ontology, in *Krikorian 1944*, 210–241; reprinted in *Nagel 1956*, 55–92.
1956 *Logic without metaphysics and other essays in the philosophy of science* (Glencoe, IL: The Free Press).
1979 *Teleology revisited and other essays in the philosophy and history of science* (New York: Columbia University Press).
See also Cohen, Morris Raphael, and Ernest Nagel.

Nagel, Ernest, and James R. Newman
1956 Goedel's proof, *Scientific American 194, 6*, 71–86; reprinted in *Scientific American 1968*, 221–230.
1956a Goedel's proof, in *Newman 1956*, vol. 3, 1668–1695.
1958 *Gödel's proof* (New York: New York University Press).
1959 British edition of *Nagel and Newman 1958* (London: Routledge & Kegan Paul).
2001 Revised edition of *Nagel and Newman 1958*, edited and with a new foreword by Douglas R. Hofstadter (New York: New York University Press).

Nagel, Ernest, James R. Newman, Kurt Gödel and Jean-Yves Girard
1989 *Le théorème de Gödel*, French translations of *Nagel and Newman 1958* and *Gödel 1931*, by Jean-Baptiste Scherrer, together with an essay by Girard (Paris: Éditions du Seuil).

Nagel, Ernest, Patrick Suppes and Alfred Tarski

1962 (eds.) *Logic, methodology, and philosophy of science. Proceedings of the 1960 International Congress* (Stanford, CA: Stanford University Press).

Neder, Ludwig

1931 Über den Aufbau der Arithmetik, *Jahresbericht der Deutschen Mathematiker-Vereinigung 40*, 22–37.

Nelson, Leonard

1908 Über das sogenannte Erkenntnisproblem, *Abhandlungen aus der Fries'schen Schule, Neue Folge, 3. Band, 4. Heft*, 413–818 (also paginated by *Heft*, 1–406) (Göttingen: Vandenhoeck and Ruprecht); reprinted in *Nelson 1970–74*, vol. II, 59–393.

1914 Die kritische Ethik bei Kant, Schiller und Fries, eine Revision ihrer Prinzipen, *Abhandlungen aus der Fries'schen Schule, Neue Folge, 4. Band, 3. Heft*, 483–691; reprinted in *Nelson 1970–74*, 27–192.

1917 *Vorlesungen über die Grundlagen der Ethik*, Bd. I: *Kritik der praktischen Vernunft* (Leipzig: Veit); reprinted as volume IV of *Nelson 1970–74*.

1924 *Vorlesungen über die Grundlagen der Ethik*, Bd. III: *Rechtslehre und Politik* (Leipzig: Der neue Geist); reprinted as volume VI of *Nelson 1970–74*.

1932 *Vorlesungen über die Grundlagen der Ethik*, Bd. II: *System der philosophischen Ethik und Pädogogik* (Göttingen: Verlag Öffentliches Leben); reprinted as volume V of *Nelson 1970–74*, edited by Grete Hermann and Minna Specht; English translation of part 1 by L. H. Grunebaum as *Nelson 1956*.

1956 *System of ethics*, English translation of part 1 of *Nelson 1932* (New Haven: Yale University Press).

1962 *Fortschritte und Rückschritte der Philosophie von Hume und Kant bis Hegel und Fries*, edited by Julius Kraft (Frankfurt am Main: Verlag Öffentliches Leben); reprinted as volume VII of *Nelson 1970–74*; English translation by N. Humphrey Palmer as *Nelson 1970* and *1971*.

1964 *System der philosophischen Rechtslehre und Politik* (Frankfurt am Main: Verlag Öffentliches Leben); reprint of *Nelson 1924*.

1970 *Progress and regress in philosophy, from Hume and Kant to Hegel and Fries*, vol. I, translation of part of *Nelson 1962*, edited posthumously by Julius Kraft (Oxford: Basil Blackwell).

1970–74 *Gesammelte Schriften, in neun Bänden*, edited by Paul Bernays, Willi Eichler, Arnold Gysin, Gustav Heckmann, Grete Henry-Hermann, Fritz von Hippel, Stephan Körner, Werner Kroebel and Gerhard Weisser, 9 vols. (Hamburg: Felix Meiner).

1971 *Progress and regress in philosophy, from Hume and Kant to Hegel and Fries*, vol. II, translation of part of *Nelson 1962*, edited posthumously by Julius Kraft (Oxford: Basil Blackwell).

Németi, István

See Andréka, Hajnal, and Istvan Németi.

Neurath, Otto

1979 *Wissenschaftliche Weltauffassung, Sozialismus und logischer Empirismus*, edited by Rainer Hegselmann (Frankfurt am Main: Suhrkamp Verlag).

Newman, James R.

1956 (ed.) *The world of mathematics* (New York: Simon and Schuster).

See also Nagel, Ernest, and James R. Newman.

See also Nagel, Ernest, James R. Newman, Kurt Gödel and Jean-Yves Girard.

Niekus, N. H., H. van Riemsdijk and Anne S. Troelstra

1981 Bibliography of A. Heyting, *Nieuw archief voor wiskunde (3) 29*, 24–35.

Nöbeling, Georg

See Menger, Karl, and Georg Nöbeling.

Notcutt, Bernard

1934 A set of axioms for the theory of deduction, *Mind* (n.s.) *43*, 63–77.

Novikov, Petr S. (Новиков, Петр С.)

1955 On the algorithmic unsolvability of the word problem in group theory (Russian), *Trudy Matematicheskogo Instituta imeni V. A. Steklova 44*, 1–143.

Ostrowski, Alexander

1920 Über Dirichletsche Reihen und algebraische Differentialgleichungen, *Mathematische Zeitschrift 8*, 241–298.

Parikh, Rohit

1971 Existence and feasibility in arithmetic, *The journal of symbolic logic 36*, 494–508.

Parry, William T.

1933 Ein Axiomensystem für eine neue Art von Implikation (analytische Implikation), *Ergebnisse eines mathematischen Kolloquiums 4*, 5–6; reprinted in *Menger 1998*, 187–188.

Parsons, Charles

1977 What is the iterative concept of set?, in *Butts and Hintikka 1977*, 335–367; reprinted in *Parsons 1983*, 268–297.

1979 Mathematical intuition, *Proceedings of the Aristotelian Society* (n.s.), *80*, 145–168.

1983 *Mathematics in philosophy: selected essays* (Ithaca: Cornell University Press).

1995a Platonism and mathematical intuition in Kurt Gödel's thought, *The bulletin of symbolic logic 1*, 44–74.

1996 In memoriam: Hao Wang, 1921–1995, *The bulletin of symbolic logic 2*, 108–111.

1998 Hao Wang as philosopher and interpreter of Gödel, *Philosophia mathematica (3) 6*, 3–24.

1998a Finitism and intuitive knowledge, in *Schirn 1998*, 249–270.

See also Grossi et alii.

Pears, David F.

1972 (ed.) *Bertrand Russell: a collection of critical essays* (Garden City, NY: Anchor-Doubleday).

Peckhaus, Volker

1990 *Hilbertprogramm und kritische Philosophie. Das Göttinger Modell interdisziplinärer Zusammenarbeit zwischen Mathematik und Philosophie* (Göttingen: Vandenhoeck & Ruprecht).

Pilet, Paul-Emile

1977 Ferdinand Gonseth—sa vie, son oeuvre, *Dialectica 31*, 23–33.

Pitcher, Everett

1988 *A history of the second fifty years, American Mathematical Society, 1939–1988* (Providence, RI: American Mathematical Society).

Pitts, Walter

See Lettvin et alii.

See also McCulloch, Warren S., and Walter Pitts.

Polish Academy of Sciences

1961 *Infinitistic methods. Proceedings of the symposium on foundations of mathematics, Warsaw, 2–9 September 1959*, International Mathematical Union and the Mathematics Institute of the Polish Academy of Sciences (Oxford and New York: Pergamon; Warsaw: PWN).

Pongratz, Ludwig J.

1975 *Philosophie in Selbstdarstellungen* II (Hamburg: Meiner).

Popper, Karl Raimund

1947 Functional logic without axioms or primitive rules of inference, *Koninklijke Nederlandse Akademie van Wetenschappen, Proceedings, Series A: Mathematical Sciences 50*, 1214–1224; also *Indagationes mathematicae 9*, Fasc. 5, 561–571.

1955 Two autonomous axiom systems for the calculus of probabilities, *The British journal for the philosophy of science 6*, 51–57.

1959 The propensity interpretation of probability, *The British journal for the philosophy of science 10*, 25–42.

1974 Intellectual autobiography, in *Schilpp 1974*, 3–181.

Post, Emil L.

1921 Introduction to a general theory of elementary propositions, *American journal of mathematics 43*, 163–185; reprinted in *van Heijenoort 1967*, 264–283 and in *Post 1994*, 21–43.

1921a On a simple class of deductive systems, *Bulletin of the American Mathematical Society 27*, 396–397; reprinted in *Post 1994*, 545.

1936 Finite combinatory processes—formulation 1, *The journal of symbolic logic 1*, 102–105; reprinted in *Davis 1965*, 288–291.

1941 Absolutely unsolvable problems and relatively undecidable propositions: account of an anticipation, in *Davis 1965*, 338–433; reprinted in *Post 1994*, 375–441.

1943 Formal reductions of the general combinatorial decision problem, *American journal of mathematics 65*, 197–215; reprinted in *Post 1994*, 442–460.

1944 Recursively enumerable sets of positive integers and their decision problems, *Bulletin of the American Mathematical Society 50*, 284–316; reprinted in *Post 1994*, 461–494.

1947 Recursive unsolvability of a problem of Thue, *The journal of symbolic logic 12*, 1–11; reprinted in *Post 1994*, 503–513.

1965 Absolutely unsolvable problems and relatively undecidable propositions—account of an anticipation, in *Davis 1965*, 340–433; reprinted in *Post 1994*, 375–441.

1994 *Solvability, provability, definability: the collected works of Emil L. Post*, edited by Martin Davis (Boston: Birkhäuser).

Poutsma, Hendrik

1929 *A grammar of late modern English, for the use of continental, especially Dutch, students*: Part I, The sentence, second half; the composite sentence, second edition (Groningen: P. Noordhoff).

Powell, William Chambers

1972 *Set theory with predication* (doctoral thesis, State University of New York at Buffalo).

Poznanski, E. I. J.

See Bar-Hillel et alii.

Presburger, Mojżesz

1930 Über die Vollständigkeit eines gewissen Systems der Arithmetik ganzer Zahlen, in welchem die Addition als einzige Operation hervortritt, *Sprawozdanie z I Kongresu matematyków krajów słowiańskich, Warszawa 1929* (Warsaw, 1930), 92–101, 395.

Proust, Joëlle

See Heinzman, Gerhard, and Joëlle Proust.

Pudlák, Pavel

1987 Improved bounds to the length of proofs of finitistic consistency statements, in *Simpson 1987*, 309–332.

1996 On the lengths of proofs of consistency, *Collegium logicum, Annals of the Kurt Gödel Society 2*, 65–86.

Putnam, Hilary

1960a Review of *Nagel and Newman 1958*, *Philosophy of science 27*, 205–207.

See also Benacerraf, Paul, and Hilary Putnam.

See also Davis, Martin, Hilary Putnam and Julia Robinson.

Pyenson, Lewis

1999 Neugebauer, Otto Eduard, in *Garraty and Carnes 1999*, vol. 16, pp. 302–303.

Quine, Willard van Orman

1933 A theorem in the calculus of classes, *The journal of the London Mathematical Society 8*, 89–95.

1937 New foundations for mathematical logic, *American mathematical monthly 44*, 70–80; reprinted in *Quine 1953*, 80–101.

1953 *From a logical point of view. 9 logico-philosophical essays* (Cambridge: Harvard University Press).
1955a A proof procedure for quantification theory, *The journal of symbolic logic 20*, 141–149; reprinted in *Quine 1966a*, 196–204.
1963a *Set theory and its logic* (Cambridge: Belknap Press of Harvard University Press).
1966a *Selected logic papers* (New York: Random House).
1969 Revised edition of *Quine 1963a* (Cambridge: Belknap Press of Harvard University Press).
1980 Third printing of *Quine 1953* (with some revisions) (Cambridge: Harvard University Press).
1995 Enlarged edition of *Quine 1966a.*

Rabin, Michael O.
See Bar-Hillel et alii.

Rautenberg, Wolfgang
1968 Die Unabhängigkeit der Kontinuumhypothese—Problematik und Diskussion, *Mathematik in der Schule 6*, 18–37.

Ravaglia, Mark
2002 *Explicating finitist reasoning* (doctoral thesis, Carnegie Mellon University).

Reid, Constance
1970 *Hilbert* (New York: Springer).
See also Albers, Donald J., Gerald Alexanderson and Constance Reid.

Reidemeister, Kurt
See Hahn et alii.

Reinhardt, William N.
1974 Remarks on reflection principles, large cardinals, and elementary embeddings, in *Jech 1974*, 189–205.
1974a Set existence principles of Shoenfield, Ackermann, and Powell, *Fundamenta mathematicae 84*, 5–34.

Remmel, Jeffrey B.
See Clote, Peter, and Jeffrey B. Remmel.

Robinson, Abraham
1961 Non-standard analysis, *Köninklijke Nederlandse Akademie van Wetenschappen, Proceedings, Series A: Mathematical Sciences 64*, 432–440; reprinted in *Robinson 1979*, vol. 2, 3–11.
1966 *Non-standard analysis* (Amsterdam: North-Holland).
1973 Nonstandard arithmetic and generic arithmetic, in *Suppes et alii 1973*, 137–154; reprinted in *Robinson 1979*, vol. 1, 280–297.

1974 Second edition of *Robinson 1966*.
1975 Concerning progress in the philosophy of mathematics, in *Rose and Shepherdson 1975*, 41–52; reprinted in *Robinson 1979*, vol. 2, 556–567.
1979 *Selected papers*, edited by H. Jerome Keisler, Stephan Körner, W. A. J. Luxemburg and A. D. Young; vol. 1, *Model theory and algebra*; vol. 2, *Non-standard analysis and philosophy*, edited and with introduction by W. A. J. Luxemburg and Stephan Körner; vol. 3, *Aeronautics* (New Haven, CT: Yale University Press).
See also Bar-Hillel et alii.

Robinson, Julia
See Davis, Martin, Hilary Putnam and Julia Robinson.

Rodríguez-Consuegra, Francisco A.
1995 (ed.) *Kurt Gödel: Unpublished philosophical essays* (Basel, Boston and Berlin: Birkhäuser).

Rose, Harvey E.
1984 *Subrecursion: functions and hierarchies* (Oxford: Clarendon Press).

Rose, Harvey E., and John C. Shepherdson
1975 (eds.) *Logic colloquium '73* (Amsterdam: North-Holland).

Rosser, J. Barkley
1936 Extensions of some theorems of Gödel and Church, *The journal of symbolic logic 1*, 87–91; reprinted in *Davis 1965*, 230–235.
1937 Gödel theorems for non-constructive logics, *The journal of symbolic logic 2*, 129–137.
See also Kleene, Stephen C., and J. Barkley Rosser.

Rosser, J. Barkley and Atwell R. Turquette
1952 *Many-valued logics* (Amsterdam: North-Holland).

Russell, Bertrand
1903 *The principles of mathematics* (London: Allen and Unwin).
1918 The philosophy of logical atomism, *The monist 28, 4*, 495–527; reprinted in *Russell 1986*, 160–180.
1919 *Introduction to mathematical philosophy* (London: Allen and Unwin; New York: Macmillan).
1920 Second edition of *Russell 1919*.
1924 Reprint of *Russell 1920*.
1937 Second edition of *Russell 1903*, with new introduction (New York: Norton).

1940 *An inquiry into meaning and truth* (London: Allen and Unwin).

1945 *A history of western philosophy* (New York: Simon and Schuster).

1958 Philosophical analysis (in German), *Zeitschrift für philosophische Forschung 12, 1,* 3–16.

1968 *The autobiography of Bertrand Russell, 1914–1944* (London: Allen and Unwin; Boston: Little, Brown and Co.)

1986 *The collected papers of Bertrand Russell*, vol. 8, *The philosophy of logical atomism and other essays, 1914–1919*, edited by John G. Slater (London, Boston and Sydney: George Allen and Unwin).

See also Whitehead, Alfred North, and Bertrand Russell.

Saaty, Thomas L.

1965 (ed.) *Lectures on modern mathematics* (New York: Wiley).

Savage, C. Wade

1988 Herbert Feigl: 1902–1988, in *Fine and Leplin 1988*, vol. 2, 15–22; available online at http://www.mcps.umn.edu/fbiofr.htm (accessed 15 November 2001).

Saxer, Walter

1933 (ed.) *Verhandlungen des Internationalen Mathematiker-Kongresses Zürich 1932,* II. Band: *Sektionsvorträge* (Zürich and Leipzig: Orell Füssli Verlag).

Schelling, Friedrich Wilhelm von

1809 *Philosophische Untersuchungen über das Wesen der Menschlichen Freiheit und die damit zusammenhängenden Gegenstände* (Reutlingen: J. N. Enßlin); reprinted in *Schelling 1927*, 223–308.

1856–61 *Friedrich Wilhelm Joseph von Schellings Sämmtliche Werke*, edited by Karl Friedrich August Schelling (Stuttgart and Augsburg: Cotta).

1927 *Schellings Werke, Nach der Originalausgabe in neuer Unordnung herausgegeben*, edited by Manfred Schröter from *Schelling 1856–61*, vol. 4, *Schriften zur Philosophie der Freiheit, 1804–1815* (München: Beck).

1964 *Philosophische Untersuchungen über das Wesen der menschlichen Freiheit und die damit zusammenhängenden Gegenstände* (Stuttgart: Reclam).

Schilpp, Paul A.

1941 (ed.) *The philosophy of Alfred North Whitehead*, Library of living philosophers, vol. 3 (Evanston: Northwestern University); second edition (New York: Tudor, 1951).

1944 (ed.) *The philosophy of Bertrand Russell*, Library of living philosophers, vol. 5 (Evanston: Northwestern University); third edition (New York: Tudor, 1951).

1949 (ed.) *Albert Einstein, philosopher-scientist*, Library of living philosophers, vol. 7 (Evanston: Library of living philosophers); third edition (New York: Tudor, 1951).

1955 (ed.) *Albert Einstein als Philosoph und Naturforscher*, German translation by Hans Hartmann (with additions) of *Schilpp 1949* (Stuttgart: Kohlhammer).

1959 The abdication of philosophy, *Proceedings and addresses of the American Philosophical Association 32*, 19–39; reprinted with biographical notes in *The Texas Quarterly 3, no. 2*, 1–20.

1963 (ed.) *The philosophy of Rudolf Carnap*, Library of living philosophers, vol. 11 (La Salle: Open Court; London: Cambridge University Press).

1974 (ed.) *The philosophy of Karl Popper*, Library of living philosophers, vol. 14 (La Salle: Open Court).

Schirn, Matthias

1998 (ed.) *The philosophy of mathematics today* (Oxford: Clarendon Press).

Schoenflies, Arthur

1927 Die Krisis in Cantors mathematischem Schaffen, *Acta mathematica 50*, 1–23.

Scholz, Arnold

See Hahn et alii.

Scholz, Heinrich

See Hasse, Helmut, and Heinrich Scholz.

Schönfinkel, Moses

See Bernays, Paul, and Moses Schönfinkel.

Schopenhauer, Arthur

1850 "Ueber die anscheinende Absichtlicheit im Schicksale des Einzelnen", Transcendente Spekulation über die anscheinende Absichtlichkeit im Schicksale des Einzelnen, in *Schopenhauer 1972*, vol. 5, 211–237.

1972 *Parerga und Paralipomena: kleine philosophische Schriften.* Volumes 5 and 6 of *Sämtliche Werke*, edited by Arthur Hübscher, third edition (Wiesbaden: Brockhaus).

Schütte, Kurt

1951 Beweistheoretische Erfassung der unendlichen Induktion in der Zahlentheorie, *Mathematische Annalen 122*, 369–389.

1960 *Beweistheorie*, Die Grundlehren der mathematischen Wissenschaften in Einzeldarstellungen mit besonderen Berücksichtigung der Anwendungsgebiete, Bd. 103 (Berlin: Springer); revised, expanded and translated into English by J. N. Crossley as *Schütte 1977*.

1977 *Proof theory* (Berlin: Springer).

Scientific American

1968 *Mathematics in the modern world* (San Francisco and London: W. H. Freeman).

Scott, Dana S.

1961 Measurable cardinals and constructible sets, *Bulletin de l'Académie polonaise des sciences, série des sciences mathématiques, astronomiques, et physiques 9*, 521–524.

1962 A decision method for validity of sentences in two variables, *The journal of symbolic logic 27*, 477.

1967 A proof of the independence of the continuum hypothesis, *Mathematical systems theory 1*, 89–111.

1971 (ed.) *Axiomatic set theory*, Proceedings of symposia in pure mathematics, vol. 13, part 1 (Providence, RI: American Mathematical Society).

See also Myhill, John, and Dana S. Scott.

Seelig, Carl

1952 *Albert Einstein und die Schweiz* (Zürich, Stuttgart and Vienna: Europa Verlag).

1960 *Albert Einstein: Leben und Werk eines Genies unserer Zeit* (Zürich: Bertelsmann Lesering–Europa Verlag).

Shanker, Stuart G.

1987 *Wittgenstein and the turning-point in the philosophy of mathematics* (Albany: State University of New York Press).

1988 Wittgenstein's remarks on the significance of Gödel's theorem, in *Shanker 1988a*, 155–256.

1988a (ed.) *Gödel's theorem in focus* (London: Croom Helm).

Shelah, Saharon
1984 Can you take Solovay's inaccessible away?, *Israel journal of mathematics 48*, 1–47.

Shepherdson, John C.
1953 Inner models for set theory, part III, *The journal of symbolic logic 18*, 145–167.
1959 Review of *Lévy 1957*, *The journal of symbolic logic 24*, 226.
See also Rose, Harvey E., and John C. Shepherdson.

Shoenfield, Joseph R.
1971 Unramified forcing, in *Scott 1971*, 357–381.

Siefkes, Dirk
1985 The work of J. Richard Büchi, in *Drucker 1985*, 176–189; reprinted in *Büchi 1990*, 7–17.
See also Lipshitz, Leonard, Dirk Siefkes and P. Young.

Sieg, Wilfried
1988 Hilbert's program sixty years later, *The journal of symbolic logic 53*, 338–348.
1994 Mechanical procedures and mathematical experience, in *George 1994*, 71–117.
1997 Step by recursive step: Church's analysis of effective calculability, *The bulletin of symbolic logic 3*, 154–180.
1999 Hilbert's programs: 1917–1922, *The bulletin of symbolic logic 5*, 1–44.
2002 Calculations by man and machine: conceptual analysis, in *Sieg, Sommer and Talcott 2002*, 390–409.

Sieg, Wilfried, Richard Sommer and Carolyn Talcott
2002 (eds.) *Reflections on the foundations of mathematics: Essays in honor of Solomon Feferman*, Lecture notes in logic 15 (Urbana: Association for Symbolic Logic; Natick: A K Peters).

Silver, Jack H.
1975 On the singular cardinals problem, in *Proceedings of the International Congress of Mathematicians, Vancouver 1974*, vol. I, 265–268.

Simpson, Stephen G.
1987 (ed.) *Logic and combinatorics: proceedings of the AMS-IMS-SIAM joint summer research conference held August 4–10, 1985*, Contemporary mathematics 65 (Providence: American Mathematical Society).

Sinaceur, Hourya

2000 Address at the Princeton University bicentennial conference on problems of mathematics (December 17–19, 1946), by Alfred Tarski, edited with additional material and an introduction by Hourya Sinaceur, *The bulletin of symbolic logic 6*, 1–44.

Sinaceur, Mohammed Allal

1991 (ed.) *Penser avec Aristote* (Toulouse: Erès).

Skolem, Thoralf

1920 Logisch-kombinatorische Untersuchungen über die Erfüllbarkeit oder Beweisbarkeit mathematischer Sätze nebst einem Theoreme über dichte Mengen, *Skrifter utgit av Videnskapsselskapet i Kristiania*, I. *Matematisk-naturvidenskapelig klasse*, no. 4, 1–36; reprinted in *Skolem 1970*, 103–136; partial English translation by Stefan Bauer-Mengelberg in *van Heijenoort 1967*, 252–263.

1923a Einige Bemerkungen zur axiomatischen Begründung der Mengenlehre, *Matematikerkongressen i Helsingfors den 4–7 Juli 1922, Den femte skandinaviska matematikerkongressen, Redogörelse* (Helsinki: Akademiska Bokhandeln), 217–232; reprinted in *Skolem 1970*, 137–152; English translation by Stefan Bauer-Mengelberg in *van Heijenoort 1967*, 290–301.

1928 Über die mathematische Logik, *Norsk matematisk tidsskrift 10*, 125–142; reprinted in *Skolem 1970*, 189–206; English translation by Stefan Bauer-Mengelberg and Dagfinn Føllesdal in *van Heijenoort 1967*, 508–524.

1929 Über einige Grundlagenfragen der Mathematik, *Skrifter utgitt av Det Norske Videnskaps-Akademi i Oslo, I. Matematisk-naturvidenskapelig klasse*, no. 4, 1–49; reprinted in *Skolem 1970*, 227–273.

1931 Über einige Satzfunktionen in der Arithmetik, *Skrifter utgitt av Det Norske Videnskaps-Akademi i Oslo, I. Matematisk-naturvidenskapelig klasse*, no. 7, 1–28; reprinted in *Skolem 1970*, 281–306.

1932 Über die symmetrisch allgemeinen Lösungen im identischen Kalkül, *Skrifter utgitt av Det Norske Videnskaps-Akademi i Oslo, I. Matematisk-naturvidenskapelig klasse*, no. 6, 1–32; also appeared in *Fundamenta mathematicae 18*, 61–76; reprinted in *Skolem 1970*, 307–336.

1933 Ein kombinatorischer Satz mit Anwendung auf ein logisches Entscheidungsproblem, *Fundamenta mathematicae 20*, 254–261; reprinted in *Skolem 1970*, 337–344.

1933a Über die Unmöglichkeit einer vollständigen Charakterisierung der Zahlenreihe mittels eines endlichen Axiomensystems, *Norsk matematisk forenings skrifter, series 2*, no. 10, 73–82; reprinted in *Skolem 1970*, 345–354.

1934 Über die Nicht-charakterisierbarkeit der Zahlenreihe mittels endlich oder abzählbar unendlich vieler Aussagen mit ausschließlich Zahlenvariablen, *Fundamenta mathematicae 23*, 150–161; reprinted in *Skolem 1970*, 355–366.

1936 *Utvalgte kapitler av den matematiske logikk.* (*Efterforleesninger ved Universitetet i Oslo i månedene januar, februar og mail 1936*). (*Selected chapters of mathematical logic. Based on lectures at the University of Oslo in the months of January, February and May 1936*). Christian Michelsens Institutt for Videnskap og Åndsfrihet, Beretninger VI, 6. (Bergen: A. S. John Griegs).

1937 Über die Zurückführbarkeit einiger durch Rekursionen definierter Relationen auf "arithmetische", *Acta litterarum ac scientiarum Regiae Universitatis Hungaricae Francisco-Josephinae. Sectio scientiarum mathematicarum 8*, 73–88; reprinted in *Skolem 1970*, 425–440.

1940 Einfacher Beweis der Unmöglichkeit eines allgemeinen Lösungsverfahrens für arithmetische Probleme, *Det Kongelige Norske Videnskabers Selskabs Forhandlinger BD XIII*, 1–4; reprinted in *Skolem 1970*, 451–454.

1962 *Abstract set theory* (Notre Dame, IN: University of Notre Dame).

1970 *Selected works in logic*, edited by Jens Erik Fenstad (Oslo: Universitetsforlaget).

Smalheiser, Neil H.

2000 Walter Pitts, *Perspectives in biology and medicine 43:2*, 217–226.

Smiley, Timothy J.

See van Dalen, Dirk, Daniel Lascar and Timothy J. Smiley.

Snow, C. P.

See Bergamini et alii.

Solovay, Robert M.

1965 $2^{\aleph_0}$ can be anything it ought to be, in *Addison, Henkin and Tarski 1965*, 435.

1965b The measure problem (abstract), *Notices of the American Mathematical Society 12*, 217.

1970 A model of set theory in which every set of reals is Lebesgue measurable, *Annals of mathematics (2) 92*, 1–56.
1971 Real-valued measurable cardinals, in *Scott 1971*, 397–428.
1974 Strongly compact cardinals and the GCH, in *Henkin et alii 1974*, 365–372.
See also Lévy, Azriel, and Robert M. Solovay.

Sommer, Richard
See Sieg, Wilfried, Richard Sommer and Carolyn Talcott.

Specht, Minna, and Willi Eichler
1953 (eds.) *Leonard Nelson zum Gedächtnis* (Frankfurt: Verlag Öffentliches Leben).

Spector, Clifford
1962 Provably recursive functionals of analysis: a consistency proof of analysis by an extension of principles formulated in current intuitionistic mathematics, in *Dekker 1962*, 1–27.
See also Feferman, Solomon, and Clifford Spector.

Speiser, Andreas
1952 *Elemente der Philosophie und der Mathematik* (Basel: Birkhäuser).

Staal, J. Frits
See van Rootselaar, Bob, and J. Frits Staal.

Steinhaus, Hugo
See Mycielski, Jan, and Hugo Steinhaus.

Stöltzner, Michael
See Köhler et alii.

Suppes, Patrick
See Nagel, Ernest, Patrick Suppes and Alfred Tarski.

Suppes, Patrick, Leon Henkin, Athanase Joja and Gr. C. Moisil
1973 (eds.) *Logic, methodology and philosophy of science IV. Proceedings of the Fourth International Congress for Logic, Methodology and Philosophy of Science, Bucharest, 1971*, Studies in logic and the foundations of mathematics 74 (Amsterdam and London: North-Holland; New York: American Elsevier).

Surányi, János
1959 *Reduktionstheorie des Entscheidungsproblems im Prädikatenkalkül der ersten Stufe* (Budapest: Verlag der ungarischen Akademie der Wissenschaften).

Tait, William W.

1961 Nested recursion, *Mathematische Annalen 143*, 236–250.

1965b The substitution method, *The journal of symbolic logic 30*, 175–192.

1981 Finitism, *The journal of philosophy 78*, 524–546.

2001 Gödel's unpublished papers on foundations of mathematics, *Philosophia mathematica (3) 9*, 87–126.

2002 Remarks on finitism, in *Sieg, Sommer and Talcott 2002*, 410–419.

Takeuti, Gaisi

1961 Remarks on Cantor's absolute, *Journal of the Mathematical Society of Japan 13*, 197–206.

1961a Remarks on Cantor's absolute II, *Proceedings of the Japan Academy 37*, 437–439.

1961b Axioms of infinity of set theory, *Journal of the Mathematical Society of Japan 13*, 220–233.

1967 Consistency proofs of subsystems of classical analysis, *Annals of mathematics (2) 86*, 299-348.

1975 *Proof theory* (Amsterdam: North-Holland).

1987 Second edition of *Takeuti 1975*.

Talcott, Carolyn

See Sieg, Wilfried, Richard Sommer and Carolyn Talcott.

Tarski, Alfred

1924 Sur les principes de l'arithmétique des nombres ordinaux (transfinis), *Polskie Towarzystwo Matematyczne (Cracow), Rocznik (=Annales de la Société Polonaise de Mathématique) 3*, 148–149.

1930 Über einige fundamentale Begriffe der Metamathematik, *Sprawozdania a posiedzeń Towarzystwa Naukowego Warszawskiego, wydział III, 23*, 22–29; reprinted in *Tarski 1986*, vol. 1, 311–320; English translation by Joseph H. Woodger, with revisions, in *Tarski 1956*, 30–37.

1930a Fundamentale Begriffe der Methodologie der deduktiven Wissenschaften I, *Monatshefte für Mathematik und Physik 37*, 361–404; reprinted in *Tarski 1986*, vol. 1, 345–390; English translation by Joseph H. Woodger in *Tarski 1956*, 60–109.

1932 Der Wahrheitsbegriff in den Sprachen der deduktiven Disziplinen, *Anzeiger der Akademie der Wissenschaften in Wien 69*, 23–25; reprinted in *Tarski 1986*, vol. 1, 613–617.

1933 Einige Betrachtungen über die Begriffe der ω-Widerspruchsfreiheit und der ω-Vollständigkeit, *Monatshefte für Mathematik und Physik 40*, 97–112; reprinted in *Tarski 1986*, vol. 1, 619–636; English translation by Joseph H. Woodger in *Tarski 1956*, 279–295.

1933a Pojecie prawdy w jezykach nauk dedukcyjnych (The concept of truth in the languages of deductive sciences), *Prace Towarzystwa Naukowego Warszawskiego, wydział III*, no. 34; translated into German by L. Blaustein as *Tarski 1935*; English translation by Joseph H. Woodger in *Tarski 1956*, 152–278.

1935 Der Wahrheitsbegriff in den formalisierten Sprachen, *Studia philosophica* (Lemberg) *1*, 261–405; German translation of *Tarski 1933a*; reprinted in *Tarski 1986*, vol. 2, 51–198.

1938a Der Aussagenkalkül und die Topologie, *Fundamenta mathematicae 31*, 103–134; reprinted in *Tarski 1986*, vol. 2, 473–506.

1949a Arithmetical classes and types of Boolean algebras. Preliminary report, *Bulletin of the American Mathematical Society 55*, 64.

1956 *Logic, semantics, metamathematics: papers from 1923 to 1938*, translated into English and edited by Joseph H. Woodger (Oxford: Clarendon Press).

1962 Some problems and results relevant to the foundations of set theory, in *Nagel, Suppes and Tarski 1962*, 125–135; reprinted in *Tarski 1986*, vol. 4, 113–125.

1983 Revised second edition of *Tarski 1956*, edited by John Corcoran (Indianapolis: Hackett).

1986 *Collected papers*, edited by Steven R. Givant and Ralph N. McKenzie (Basel, Boston and Stuttgart: Birkhäuser).

1999 Letters to Kurt Gödel, 1942–1947 (translated and edited by Jan Tarski), in *Woleński and Köhler 1999*, 261–273.

See also Henkin, Leon, and Alfred Tarski.

See also Henkin, Leon, J. Donald Monk and Alfred Tarski.

See also Keisler, H. Jerome, and Alfred Tarski.

See also Nagel, Ernest, Patrick Suppes and Alfred Tarski.

Tarski, Alfred, Andrzej Mostowski and Raphael M. Robinson

1953 *Undecidable theories* (Amsterdam: North-Holland).

Thiel, Christian

1980 Behmann, Heinrich, in *Mittelstraß 1980*, vol. 1, 274–275.

2002 Gödels Anteil am Streit über Behmanns Behandlung der Antinomien, in *Köhler et alii 2002*, vol. 2, 387–394.

Toulmin, Stephen E.
See Janik, Allen S., and Stephen E. Toulmin.

Trautman, Andrzej
1962 Conservation laws in general relativity, in *Witten 1962*, 169–198.

Troelstra, Anne S.
1981 Arend Heyting and his contribution to intuitionism, *Nieuw archief voor wiskunde (3) 29*, 1–23.
See also Niekus, N. H., H. van Riemsdijk and Anne S. Troelstra.

Troelstra, Anne S., and Dirk van Dalen
1988 *Constructivism in mathematics*, vols. I and II (Amsterdam: North-Holland).

Tseng Ting-Ho
1938 *La philosophie mathématique et la théorie des ensembles* (doctoral dissertation, University of Paris).

Tucker, J. V.
1963 Constructivity and grammar, *Proceedings of the Aristotelian Society 63*, 45–66.

Turing, Alan Mathison
1937 On computable numbers, with an application to the Entscheidungsproblem, *Proceedings of the London Mathematical Society (2) 42*, 230–265; correction, *ibid. 43*, 544–546; reprinted as *Turing 1965*.
1939 Systems of logic based on ordinals, *Proceedings of the London Mathematical Society (2) 45*, 161–228; reprinted in *Davis 1965*, 155–222.
1950 The word problem in semi-groups with cancellation, *Annals of mathematics (2) 52*, 491–505.
1954 Solvable and unsolvable problems, *Science news 31* (London: Penguin), 7–23; reprinted in *Turing 1992*, 99–115 and *Turing 1992a*, 187–203.
1965 Reprint of *Turing 1937*, in *Davis 1965*, 116–154.
1992 *Pure mathematics*, edited by J. L. Britton, *Collected works of A. M. Turing*, vol. 1 (Amsterdam, London, New York and Tokyo: North-Holland).
1992a *Mechanical intelligence*, edited by D. C. Ince, *Collected works of A. M. Turing*, vol. 3 (Amsterdam, London, New York and Tokyo: North-Holland).

Turquette, Atwell R.
See Rosser, J. Barkley, and Atwell R. Turquette.

Ulam, Stanisław M.
1930 Zur Masstheorie in der allgemeinen Mengenlehre, *Fundamenta mathematicae 16*, 140–150; reprinted in *Ulam 1974*, 9–19.
1958 John von Neumann, 1903–1957, *Bulletin of the American Mathematical Society 3 (2)* (May supplement), 1–49.
1960 *A collection of mathematical problems* (New York: Interscience Publishers); reprinted in *Ulam 1974*, 505–670.
1964 Combinatorial analysis in infinite sets and some physical theories, *SIAM review 6 (4), 343–355.*
1974 *Sets, numbers, and universes, selected works*, edited by William A. Beyer, Jan Mycielski and Gian-Carlo Rota (Cambridge and London: The MIT Press).
1976 *Adventures of a mathematician* (New York: Charles Scribner's).
See also Everett, C. J., and Stanisław Ulam.

Unger, Georg
1975 (ed.) *Aufsätze zur Mengenlehre* (Wissenschaftliche Buchgesellschaft: Darmstadt).

Urmson, J. O.
1956 *Philosophical analysis; its development between the two world wars* (Oxford: Clarendon Press).

van Dalen, Dirk
See Troelstra, Anne S., and Dirk van Dalen.

van Dalen, Dirk, Daniel Lascar and Timothy J. Smiley
1982 (eds.) *Logic colloquium '80* (Amsterdam: North-Holland).

van Heijenoort, Jean
1967 (ed.) *From Frege to Gödel: a source book in mathematical logic, 1879–1931* (Cambridge: Harvard University Press).
1978 *With Trotsky in exile: from Prinkipo to Coyoacán* (Cambridge and London: Harvard University Press).
1985 *Selected essays* (Naples: Bibliopolis).
See also Dreben, Burton, and Jean van Heijenoort.

van Riemsdijk, H.
See Niekus, N. H., H. van Riemsdijk and Anne S. Troelstra.

van Rootselaar, Bob, and J. Frits Staal
1968 (eds.) *Logic, methodology and philosophy of science III. Proceedings of the Third International Congress for Logic, Methodology and Philosophy of Science, Amsterdam 1967* (Amsterdam: North-Holland).

Vaught, Robert L.
1956 On models of some strong set theories, *Bulletin of the American Mathematical Society 62*, 601–602.

Verein Ernst Mach
1929 (eds.) *Wissenschaftliche Weltauffassung: der Wiener Kreis* (Vienna: A. Wolf); reprinted in *Neurath 1979*, 81–101.

Vesley, Richard E.
See also Myhill, John, Akiko Kino and Richard E. Vesley.

Vihan, Premysl
1995 The last months of Gerhard Gentzen in Prague, *Collegium logicum, Annals of the Kurt Gödel Society 1*, 1–7.

von Hayek, Friedrich August
1960 *The constitution of liberty* (Chicago: University of Chicago Press).

von Juhos, Béla
1930 *Das Problem der mathematischen Wahrscheinlichkeit* (Munich: Reinhardt).

von Neumann, John
1923 Zur Einführung der transfiniten Zahlen, *Acta litterarum ac scientiarum Regiae Universitatis Hungaricae Francisco-Josephinae. Sectio scientiarum mathematicarum 1*, 199–208; reprinted in *von Neumann 1961*, 24–33; English translation by Jean van Heijenoort in *van Heijenoort 1967*, 347–354.
1925 Eine Axiomatisierung der Mengenlehre, *Journal für die reine und angewandte Mathematik 154*, 219–240; correction, *ibid. 155*, 128; reprinted in *von Neumann 1961*, 34–56; English translation by Stefan Bauer-Mengelberg and Dagfinn Føllesdal in *van Heijenoort 1967*, 393–413.
1926 Zur Prüferschen Theorie der idealen Zahlen, *Acta litterarum ac scientiarum Regiae Universitatis Hungaricae Francisco-Josephinae. Sectio scientarum mathematicarum 2*, 193–227; reprinted in *von Neumann 1961*, 69–103.
1927 Zur Hilbertschen Beweistheorie, *Mathematische Zeitschrift 26*, 1–46; reprinted in *von Neumann 1961*, 256–300.

1928 Über die Definition durch transfinite Induktion und verwandte Fragen der allgemeinen Mengenlehre, *Mathematische Annalen 99*, 373–391; reprinted in *von Neumann 1961*, 320–338.

1928a Die Axiomatisierung der Mengenlehre, *Mathematische Zeitschrift 27*, 669–752; reprinted in *von Neumann 1961*, 339–422.

1929 Über eine Widerspruchfreiheitsfrage in der axiomatischen Mengenlehre, *Journal für die reine und angewandte Mathematik 160*, 227–241; reprinted in *von Neumann 1961*, 494–508.

1931 Die formalistische Grundlegung der Mathematik, *Erkenntnis 2*, 116–121; reprinted in *von Neumann 1961a*, 234–239; English translation by Erna Putnam and Gerald J. Massey in *Benacerraf and Putnam 1964*, 50–54.

1936 Continuous geometry, part I, mimeographed notes by L. Roy Wilcox, The Institute for Advanced Study, 1936; reproduced in *von Neumann 1960.*

1937 Continuous geometry, parts II and III, notes by L. Roy Wilcox, planographed by Edwards Brothers, Inc., Ann Arbor, Michigan; reproduced in *von Neumann 1960.*

1958 *The computer and the brain, the 1956 Silliman Lecture*, with preface by Klara von Neumann (New Haven: Yale University Press).

1960 *Continuous geometry* (Princeton: Princeton University Press).

1961 *Collected works*, vol. I: *logic, theory of sets and quantum mechanics*, edited by A. H. Taub (New York and Oxford: Pergamon).

1961a *Collected works*, vol. II: *operators, ergodic theory and almost periodic functions in a group*, edited by A. H. Taub (New York and Oxford: Pergamon).

2000 Second edition of *von Neumann 1958*, with a foreword by Paul M. Churchland and Patricia S. Churchland (New Haven: Yale Nota Bene).

See also Hahn et alii.

von Neumann, John, and F. J. Murray

1936 On rings of operators, *Annals of mathematics (2) 37*, 116–229.

Wajsberg, Mordchaj

1938 Untersuchungen über den Aussagenkalkül von A. Heyting, *Wiadomości matematyczne 46*, 45–101.

Wald, Abraham

1936 Über die Produktionsgleichungen der ökonomischen Wertlehre (II. Mitteilung), *Ergebnisse eines mathematischen Kolloquiums 7*, 1–6; reprinted in *Menger 1998*, 319–324.

See also *Gödel 1933h.*

Wang, Hao
1970 A survey of Skolem's work in logic, in *Skolem 1970*, 17–52.
1974 *From mathematics to philosophy* (London: Routledge and Kegan Paul; New York: Humanities Press).
1974a Metalogic, in *Encyclopedia Britannica*, fifteenth edition, vol. 11, 1078–1086 (Chicago: Encyclopedia Britannica, Inc.); partly incorporated into chapter V of *Wang 1974*, remainder reprinted in *Wang 1990*; 325–330.
1977 Large sets, in *Butts and Hintikka 1977*, 309–333.
1978 Kurt Gödel's intellectual development, *The mathematical intelligencer 1*, 182–184.
1981 Some facts about Kurt Gödel, *The journal of symbolic logic 46*, 653–659.
1987 *Reflections on Kurt Gödel* (Cambridge: MIT Press).
1990 *Computation, logic, philosophy, a collection of essays* (Beijing: Science Press; Dordrecht: Kluwer Academic Publishers).
1991 Kurt Gödel et certaines de ses conceptions philosophiques: l'esprit, la matière, et les mathèmatiques, in *Sinaceur 1991*, 441–451.
1996 *A logical journey. From Gödel to philosophy* (Cambridge: MIT Press).

Wang, Hao, and Bradford Dunham
1973 A recipe for Chinese typewriters, IBM technical report RC4521, 5 September 1973.
1976 Chinese version of *Wang and Dunham 1973*, *Dousou bimonthly 14* (March 1976), 56–62.

Wegel, Heinrich
1956 Axiomatische Mengenlehre ohne Elemente von Mengen, *Mathematische Annalen 131*, 435–462.

Weibel, Peter
See Köhler et alii.

Weinzierl, Ulrich
1982 *Carl Seelig, Schriftsteller* (Vienna: Löcker).

Weizmann, Karl Ludwig
1915 *Lehr- und Übungsbuch der Gabelsbergerschen Stenographie*, twelfth edition (Vienna: Manzsche k.u.k. Hof-, Verlags- und Universitäts-Buchhandlung).

Wette, Eduard
1974 Contradiction within pure number theory because of a system-internal 'consistency'-deduction, *International logic review. Rassegna internazionale di logica 9*, 51–62.

Weyl, Hermann
1918 *Das Kontinuum. Kritische Untersuchungen über die Grundlagen der Analysis* (Leipzig: Veit); translated into English by Stephen Pollard and Thomas Bole as *Weyl 1987.*
1927 *Philosophie der Mathematik und Naturwissenschaft, Handbuch der Philosophie* (Munich: Oldenbourg).
1932 Second edition of *Weyl 1918.*
1949 *Philosophy of mathematics and natural science*, revised and augmented English edition of *Weyl 1927* based on a translation by Olaf Helmer (Princeton: Princeton University Press).
1987 *The continuum. A critical examination of the foundation of analysis*, English translation of *Weyl 1918* (Kirksville, MO: Thomas Jefferson University Press).

Whitehead, Alfred North, and Bertrand Russell
1910 *Principia mathematica*, vol. 1 (Cambridge: Cambridge University Press).
1925 Second edition of *Whitehead and Russell 1910.*

Wiedemann, Hans-Rudolf
1989 *Briefe großer Naturforscher und Ärzte in Handschriften* (Lübeck: Verlag Graphische Werkstätten).

Wiener, Norbert
1948 *Cybernetics, or control and communication in the animal and the machine* (Cambridge, MA: Technology Press; New York: John Wiley and Sons; Paris: Hermann et cie).
1961 Second, enlarged edition of *Wiener 1948* (Cambridge, MA: The MIT Press).

Witten, Louis
1962 (ed.) *Gravitation: an introduction to current research* (New York and London: John Wiley & Sons).

Wittenberg, Alexander Israel
1953 Über adäquate Problemstellung in der mathematischen Grundlagenforschung, *Dialectica 7*, 232–254.
1954 Über adäquate Problemstellung in der mathematischen Grundlagenforschung. Eine Antwort, *Dialectica 8*, 152–157.
1957 *Vom Denken in Begriffen, Mathematik als Experiment des reinen Denkens* (Basel and Stuttgart: Birkhäuser).
1963 *Bildung und Mathematik, Mathematik als exemplarisches Gymnasialfach* (Stuttgart: Ernst Klett).

Wittgenstein, Ludwig

1921 Logische-philosophische Abhandlung, *Annalen der Naturphilosophie 14*, 185–262; reprinted as *Logische-philosophische Abhandlung. Tractatus logico-philosophicus* in *Wittgenstein 1984*, vol. 1, 7–85; English translation by C. K. Ogden in *Wittengenstein 1922*.

1922 *Tractatus logico-philosophicus*, English translation of *Wittgenstein 1921*, with corrected German text (New York: Harcourt, Brace; London: Kegan Paul).

1956 *Bemerkungen über die Grundlagen der Mathematik: Remarks on the foundations of mathematics*, edited by G. H. von Wright, R. Rhees and G. E. M. Anscombe, with translation by G. E. M. Anscombe (Oxford: Basil Blackwell).

1984 *Werkausgabe*, Suhrkamp Taschenbuch Wissenschaft 501–508 (Frankfurt: Suhrkamp).

Woleński, Jan and Eckehart Köhler

1999 *Alfred Tarski and the Vienna circle. Austro-Polish connections in logical empiricism* (Dordrecht: Kluwer Academic Publishers).

Wolkowski, Zbigniew W.

1993 (ed.) *First international symposium on Gödel's theorems* (Singapore: World Scientific Publishing Co.).

Young, P.

See Lipshitz, Leonard, Dirk Siefkes and P. Young.

Zach, Richard

1999 Completeness before Post: Bernays, Hilbert, and the development of propositional logic, *Bulletin of symbolic logic 5*, 331–366.

2001 *Hilbert's finitism: historical, philosophical and metamathematical perspectives* (doctoral dissertation, University of California, Berkeley); available online at http://www.ucalgary.ca-/~rzach/publications/html (accessed 3 March 2002).

Zermelo, Ernst

1929 Über den Begriff der Definitheit in der Axiomatik, *Fundamenta mathematicae 14*, 339–344.

1930 Über Grenzzahlen und Mengenbereiche: Neue Untersuchungen über die Grundlagen der Mengenlehre, *Fundamenta mathematicae 16*, 29–47; English translation by Michael Hallett in *Ewald 1996*, vol. II, 1219–1233.

1932 Über Stufen der Quantifikation und die Logik des Unendlichen, *Jahresbericht der Deutschen Mathematiker-Vereinigung 41*, part 2, 85–88.

Ziegler, Renatus
See Booth, David, and Renatus Ziegler.

Index

Pages in boldface type indicate a section devoted to a particular correspondent. The index does not include the detailed collection outline of the finding aid.

Printed in the United States
By Bookmasters